Heinz Lüneburg

Translation Planes

Springer-Verlag
Berlin Heidelberg New York

Heinz Lüneburg

FB Mathematik
Universität Kaiserslautern
Postfach 3049
6750 Kaiserslautern
Federal Republic of Germany

AMS Subject Classification (1980): 50A40

ISBN 3-540-09614-0 Springer-Verlag Berlin Heidelberg

ISBN 0-387-09614-0 Springer-Verlag New York

Library of Congress Cataloging in Publication Data

Lüneburg, Heinz.
  Translation planes.

  Bibliography: p.
  Includes indexes.
  1. Translation planes.   I. Title.
QA477.L83   516′.4   79-16647

9 8 7 6 5 4 3 2 1

Foreword

Wir unterhielten uns einmal darüber, daß man sich in einer fremden Sprache nur unfrei ausdrücken kann und im Zweifelsfall lieber das sagt, was man richtig und einwandfrei zu sagen hofft, als das, was man eigentlich sagen will. Molnár nickte bestätigend: "Es ist sehr traurig", resümierte er. "Ich habe oft mitten im Satz meine Weltanschauung ändern müssen ... "

Friedrich Torberg, Die Tante Jolesch

The last two decades have witnessed great progress in the theory of translation planes. Being interested in, and having worked a little on this subject, I felt the need to clarify for myself what had been happening in this area of mathematics. Thus I lectured about it for several semesters and, at the same time, I wrote what is now this book. It is my very personal view of the story, which means that I selected mainly those topics I had touched upon in my own investigations. Thus finite translation planes are the main theme of the book. Infinite translation planes, however, are not completely disregarded.

As all theory aims at the mastering of the examples, these play a central role in this book. I believe that this fact will be welcomed by many people. However, it is not a beginner's book of geometry. It presupposes considerable knowledge of projective planes and algebra, especially group theory. The books by Gorenstein, Hughes and Piper, Huppert, Passman, and Pickert mentioned in the bibliography will help to fill any gaps the reader may have.

Finally, I would like to thank all those people who have helped me in the course of writing the book. Special thanks are due to Mike Kallaher, Brian Mortimer and Chris Norman for reading and commenting on parts of the manuscript, and to Joachim Assion and Arno Cronheim for allowing me to incorporate unpublished material of theirs into this book.

Han-sur-Lesse, June 1979   Heinz Lüneburg

Contents

Standard Symbols

A_n: alternating group of degree n

$\mathfrak{C}_G(g)$: centralizer of g in G

$GF(q)$: Galois field with q elements

$GL(V)$, $GL(n, q)$, $GL(V, K)$: group of all bijective linear mappings of V onto itself emphasizing the dimension n and the underlying field K, $GF(q)$ if necessary

$\mathfrak{N}_G(X)$: normalizer of X in G

$PG(3, q)$: projective geometry of dimension 3 over $GF(q)$

$P \text{ I } l$: the point P is incident with the line (circle, etc.) l

$\text{rank}(V)$, $\text{rk}_K(V)$: rank of the K-vector space V, also called the dimension of V by other authors

$SL(n, q)$: group of all $n \times n$ matrices of determinant 1 over $GF(q)$

S_n: symmetric group of degree n

$\text{Syl}_p(G)$: set of Sylow p-subgroups of G

$\mathfrak{Z}(G)$: centre of G

CHAPTER I

Introduction

In this chapter we shall collect the basic results about translation planes which will be used throughout the book. An exception is section 4 where we apply the general theory for the first time giving a characterization of pappian planes.

1. André's Description of Translation Planes

An affine plane $\mathfrak{A}$ is called a *translation plane*, if its translation group T operates transitively on the set of points of $\mathfrak{A}$. Let U be a point on l_∞ and let $T(U)$ denote the group of all translations with centre U. As every non-identity translation has exactly one centre, $\pi = \{T(U) \mid U \mathrel{I} l_\infty\}$ is a partition of T. If κ is a collineation of $\mathfrak{A}$, then $\kappa^{-1}T(U)\kappa = T(U^\kappa)$. This yields in particular that $T(U)$ is a normal subgroup of T for all $U \mathrel{I} l_\infty$.

For a group admitting a non-trivial partition (i.e. a partition with at least two components) all components of which are normal subgroups of the given group we have:

1.1 Theorem (Kontorowitsch). *Let π be a non-trivial partition of the group G. If all components of π are normal subgroups of G, then G is abelian.*

PROOF. Let $g \in G$. There exists $X \in \pi$ such that $g \in X$. Let $h \in G\backslash X$. Then there exists $Y \in \pi$ with $h \in Y$. Since $h \notin X$, we have $X \cap Y = \{1\}$. The normality of Y in G implies $g^{-1}hg \in Y$ and thus $h^{-1}g^{-1}hg \in Y$. Likewise $h^{-1}g^{-1}hg \in X$. Therefore $h^{-1}g^{-1}hg \in X \cap Y = \{1\}$. As a consequence $G\backslash X \subseteq \mathfrak{C}_G(g)$. The partition π being non-trivial implies that

1

$G \setminus X \neq \emptyset$. Hence $G = \langle G \setminus X \rangle$. Therefore $G = \mathfrak{C}_G(g)$. In other words $g \in \mathfrak{Z}(G)$. As this is true for all $g \in G$, the group G is abelian. $\qquad \square$

The remarks before Theorem 1.1 together with this theorem yield:

1.2 Corollary. *If $\mathfrak{A}$ is a translation plane, then* T *is abelian.*

We note a further important property of the translation group of a translation plane:

1.3 Theorem. *Let $\mathfrak{A}$ be a translation plane and* T *its translation group. If U and V are distinct points on l_∞, then* $T = T(U)T(V)$.

PROOF. Let O be an affine point of $\mathfrak{A}$ and let $\tau \in T$. Put $X = O^\tau V \cap OU$. Then there exist $\rho \in T(U)$ and $\sigma \in T(V)$ such that $O^\rho = X$ and $X^\sigma = O^\tau$. Hence $O^{\rho\sigma} = O^\tau$. Since T operates sharply transitively on the set of affine points of $\mathfrak{A}$, we have $\rho\sigma = \tau$. $\qquad \square$

Assume that G is a group and that π is a partition of G. Denote by $\mathfrak{L}$ the set of all right cosets of all the components of π and put $\pi(G) = (G, \mathfrak{L}, \in)$. Then $\pi(G)$ is an incidence structure with points the elements of G and lines the elements of $\mathfrak{L}$.

A non-trivial partition π of the group G is called a *spread*, if $G = AB$ for all $A, B \in \pi$ with $A \neq B$.

1.4 Theorem (André 1954). *Let π be a spread of the group G. Then the following is true:*

a) $\pi(G)$ *is a translation plane.*
b) *If $g \in G$, then the mapping $\tau(g)$ defined by $x^{\tau(g)} = xg$ is a translation of* $\pi(G)$.
c) τ *is an isomorphism from G onto the translation group* T *of $\pi(G)$.*
d) G *is abelian.*

PROOF. Let g and h be distinct points of $\pi(G)$. Then $gh^{-1} \neq 1$. Hence there is a unique $U \in \pi$ such that $gh^{-1} \in U$. Thus Uh is a line joining g and h. Let $V \in \pi$ and $x \in G$ and assume $g, h \in Vx$. Then $gh^{-1} \in V$ whence $gh^{-1} \in V \cap U$. Since $1 \neq gh^{-1}$, we obtain $V = U$. This proves that Uh is the only line joining g and h.

Let $g, h \in G$ and $U \in \pi$ and assume $g \notin Uh$. Then $Ug \cap Uh = \emptyset$. Let Vx be a second line through g. Obviously $Vx = Vg$. If $V \neq U$, then $G = UV$ as π is a spread. Therefore, there exist $u \in U$ and $v \in V$ such that $u^{-1}v = hg^{-1}$. This implies $vg = uh$ showing that $Vg \cap Uh \neq \emptyset$. Thus Ug is the only line through g which does not meet Uh.

As π is non-trivial, there are two different components U and V in π. Let $1 \neq u \in U$ and $1 \neq v \in V$. Then $1, u$ and v are non-collinear as is easily seen. This establishes that $\pi(G)$ is an affine plane.

Obviously, $\tau(g)$ is a collineation of $\pi(G)$. Furthermore, $\tau(g)$ maps every line of $\pi(G)$ onto a parallel line, since either $Ux = Uxg$ or $Ux \cap Uxg = \emptyset$. Therefore, $\tau(g)$ is a dilatation. If $g \neq 1$, then $x \neq x^{\tau(g)}$ for all $x \in G$, whence it follows that $\tau(g)$ is a translation. Since the group of right translations of G operates transitively on G, we obtain that $\pi(G)$ is a translation plane. Thus a) and b) are proved.

τ is obviously a monomorphism from G into T. As $\tau(G)$ operates transitively and T operates regularly on the set of points of $\pi(G)$, we have that τ is surjective proving c).

c) and 1.2 imply d). $\qquad\qquad\qquad\qquad\qquad\qquad\qquad\qquad\square$

1.5 Theorem (André 1954). *If $\mathfrak{A}$ is a translation plane and* T *its translation group, then $\mathfrak{A}$ is isomorphic to $\pi(\mathrm{T})$ where $\pi = \{\mathrm{T}(U) \mid U\, \mathrm{I}\, l_\infty\}$.*

PROOF. Let O be an affine point of $\mathfrak{A}$. For each point P of $\mathfrak{A}$ there is a unique $\tau_P \in \mathrm{T}$ with $O^{\tau_P} = P$. Hence τ is a bijection from the set of points of $\mathfrak{A}$ onto T. Let $U\, \mathrm{I}\, l_\infty$. Then $P\, \mathrm{I}\, OU$ if and only if $\tau_P \in \mathrm{T}(U)$. Thus τ maps the set of lines through O onto π. Assume that l is a line of $\mathfrak{A}$ such that $U = l \cap l_\infty$. If $X\, \mathrm{I}\, l$, then $X^{\tau_X^{-1}} = O$. Hence $l^{\tau_X^{-1}} = OU$. If P is any point of $\mathfrak{A}$, then $P\, \mathrm{I}\, l$, if and only if $\tau_P \tau_X^{-1} \in \mathrm{T}(U)$. $\qquad\qquad\square$

Because of 1.4 and 1.5, we may assume in the sequel that a given translation plane has the form $\pi(V)$, where V is an additively written abelian group and π is a spread of V.

Let V be an abelian group and π a partition of V. Denote by $\mathrm{K}(V, \pi)$ the set of all endomorphisms κ of V satisfying $X^\kappa \subseteq X$ for all $X \in \pi$. We call $\mathrm{K}(V, \pi)$ the *kernel* of $\pi(V)$.

1.6 Theorem (André 1954). *Let V be an abelian group and let π be a non-trivial partition of V. Then*:

a) *If $0 \neq \kappa \in \mathrm{K}(V, \pi)$, then κ is injective. In particular, $\mathrm{K}(V, \pi)$ is a ring without zero-divisors.*

b) *If π is a spread, then $\mathrm{K}(V, \pi)$ is a division ring.*

PROOF. a) Let $\kappa \in \mathrm{K}(V, \pi)$ and assume $\mathrm{kern}(\kappa) \neq \{0\}$. Furthermore, let $0 \neq u \in \mathrm{kern}(\kappa)$. Then there is exactly one $X \in \pi$ such that $u \in X$. Let $v \in V \backslash X$. Then $u + v \notin X$. Hence there exist $Y, Z \in \pi \backslash \{X\}$ with $v \in Y$ and $u + v \in Z$. If $Y = Z$, then $u \in Y$ and hence $u \in Y \cap X$. This yields $Y = X$, as $u \neq 0$. Thus $v \in X$: a contradiction. Therefore, $Y \neq Z$ and hence $Y \cap Z = \{0\}$. Now

$$v^\kappa = u^\kappa + v^\kappa = (u + v)^\kappa \in Y \cap Z = \{0\}.$$

Therefore $V \backslash X \subseteq \mathrm{kern}(\kappa)$. As π is non-trivial, $V \backslash X$ generates V. Therefore $V = \mathrm{kern}(\kappa)$. This proves that κ is injective provided $\kappa \neq 0$.

Let $\kappa, \lambda \in \mathrm{K}(V, \pi)$ and assume $\kappa\lambda = 0$. If $\kappa = 0$ there is nothing to prove. Assume $\kappa \neq 0$. Then there exists $v \in V$ with $v^\kappa \neq 0$. But $v^{\kappa\lambda} = 0$. Hence $\lambda = 0$ by what we have already proved.

b) Let $0 \neq \kappa \in \mathrm{K}(V, \pi)$. In order to show that κ is surjective we pick

$v \in V \setminus \{0\}$. Then there is exactly one $X \in \pi$ such that $v \in X$. Since π is non-trivial, there exists $Y \in \pi \setminus \{X\}$. Let $0 \neq u \in Y$. We have $u^\kappa \neq 0$ by a). Furthermore, $u^\kappa - v \neq 0$, since otherwise $v \in Y \cap X = \{0\}$. Hence there is exactly one $Z \in \pi$ such that $u^\kappa - v \in Z$. Since $u^\kappa \notin X$, we get $Z \neq X$. We infer from 1.4 that $\pi(V)$ is an affine plane. This implies that there exists $w \in (Z + u) \cap X$ because $Z \neq X$. This yields in particular that $w - u \in Z$ and hence $w^\kappa - u^\kappa \in Z$. Since $u^\kappa - v \in Z$, we have $w^\kappa - v \in Z$. Both v and w belong to X. Therefore, $w^\kappa - v \in X \cap Z = \{0\}$. This proves that κ is surjective. This together with a) says that κ is an automorphism of V. Let $X \in \pi$. Then $X = X^\kappa$ as it is easily seen. Therefore $X^{\kappa^{-1}} = X^{\kappa \kappa^{-1}} = X$ and thus $\kappa^{-1} \in K(V, \pi)$.   □

Let π be a spread of V. By 1.6 b), we see that V is a $K(V, \pi)$-vector space and the components of π are subspaces of this vector space. For these subspaces we prove:

1.7 Theorem (André 1954). *Let π be a spread of V. If $X, Y \in \pi$, then X and Y are isomorphic subspaces of the $K(V, \pi)$-vector space V.*

PROOF. $\pi(V)$ is an affine plane by 1.4. Hence there are at least three lines through 0, i.e. the spread π contains at least three components. Hence there exists $Z \in \pi \setminus \{X, Y\}$. As π is a spread, $V = X \oplus Z = Y \oplus Z$ from which it follows that $X \cong Y$.   □

1.8 Corollary. *Let π be a spread of V. If V is finitely generated as a $K(V, \pi)$-vector space, then there is a positive integer n such that $\mathrm{rank}(X) = n$ for all $X \in \pi$ and $\mathrm{rank}(V) = 2n$.*

This follows immediately from 1.7. Likewise we get from 1.7:

1.9 Corollary. *If $\mathfrak{A}$ is a finite translation plane, then the order of $\mathfrak{A}$ is a power of a prime.*

1.10 Theorem. *Let π be a spread of V and π' be a spread of V'. If σ is an isomorphism of $\pi(V)$ onto $\pi'(V')$ with $0^\sigma = 0'$, then σ is a bijective semilinear mapping of the $K(V, \pi)$-vector space V onto the $K(V', \pi')$-vector space V'.*

PROOF. The mapping τ defined by $x^{\tau(v)} = x + v$ is, by 1.4, an isomorphism of V onto the translation group T of $\pi(V)$. The corresponding mapping of V' onto T' will be denoted by τ'. Obviously, $\sigma^{-1} \tau(v) \sigma \in T'$ for all $v \in V$. Furthermore,

$$0'^{\sigma^{-1} \tau(v) \sigma} = 0^{\tau(v) \sigma} = v^\sigma = 0'^{\tau'(v^\sigma)}.$$

Thus $\sigma^{-1} \tau(v) \sigma = \tau'(v^\sigma)$ for all $v \in V$. This implies

$$\tau'((u + v)^\sigma) = \sigma^{-1} \tau(u + v) \sigma = \sigma^{-1} \tau(u) \tau(v) \sigma = \sigma^{-1} \tau(u) \sigma \sigma^{-1} \tau(v) \sigma$$
$$= \tau'(u^\sigma) \tau'(v^\sigma) = \tau'(u^\sigma + v^\sigma),$$

whence we infer $(u + v)^\sigma = u^\sigma + v^\sigma$ for all $u, v \in V$. This proves that σ is additive.

Put $\kappa^\alpha = \sigma^{-1}\kappa\sigma$ for all $\kappa \in K(V, \pi)$. Then $\kappa^\alpha \in K(V', \pi')$, as it is easily seen. Furthermore

$$(v^\kappa)^\sigma = v^{\sigma\sigma^{-1}\kappa\sigma} = v^{\sigma\kappa^\alpha}$$

and hence

$$v^{\sigma(\kappa+\lambda)^\alpha} = (v^{\kappa+\lambda})^\sigma = (v^\kappa + v^\lambda)^\sigma = v^{\kappa\sigma} + v^{\lambda\sigma}$$
$$= v^{\sigma\kappa^\alpha} + v^{\sigma\lambda^\alpha} = v^{\sigma\kappa^\alpha + \sigma\lambda^\alpha} = v^{\sigma(\kappa^\alpha+\lambda^\alpha)}$$

and

$$v^{\sigma(\kappa\lambda)^\alpha} = v^{\kappa\lambda\sigma} = v^{\kappa\sigma\lambda^\alpha} = v^{\sigma\kappa^\alpha\lambda^\alpha}.$$

Since σ is bijective, we have $(\kappa + \lambda)^\alpha = \kappa^\alpha + \lambda^\alpha$ and $(\kappa\lambda)^\alpha = \kappa^\alpha\lambda^\alpha$.

Finally, replacing σ by σ^{-1}, one sees that α is bijective. Thus, σ is a bijective semilinear mapping.   □

We shall denote by $\Gamma L(V, K)$ the group of all bijective semilinear mappings of the K-vector space V onto itself. If $\pi(V)$ is a translation plane and G its collineation group, then $G = TG_0$, where G_0 is the stabilizer of the point 0. Theorem 1.10 says that G_0 is contained in $\Gamma L(V, K(V, \pi))$. Sometimes it suffices to know that G_0 is a subgroup of the group of units of $\mathrm{End}_Z(V)$, a fact which is established by the first part of the proof of 1.10.

1.11 Theorem. *Let V and V' be groups and π and π' be spreads of V and V' resp. Then the following statements are equivalent:*

a) *$\pi(V)$ and $\pi'(V')$ are isomorphic.*
b) *There is a bijective semilinear mapping of V onto V' which maps π onto π'.*
c) *There is an isomorphism from $V(+)$ onto $V'(+)$ mapping π onto π'.*

PROOF. a) implies b): Let ρ be an isomorphism of $\pi(V)$ onto $\pi'(V')$. Then there is a translation τ of $\pi'(V')$ such that $0^{\rho\tau} = 0'$. By 1.10, the mapping $\rho\tau$ is a bijective semilinear mapping from V onto V'. Since $\rho\tau$ is also an isomorphism from $\pi(V)$ onto $\pi'(V')$ which maps 0 onto 0', we obtain that π is mapped by $\rho\tau$ onto π'.

b) implies c) trivially.

c) implies a): This follows immediately from the definition of $\pi(V)$ resp. $\pi'(V')$.   □

Let $\mathfrak{A}$ be an affine plane and P be a point of $\mathfrak{A}$. Then $\Delta(P)$ will denote the group of all homologies with centre P and axis l_∞.

1.12 Theorem (André 1954). *If π is a spread of V, then $\Delta(0)$ is the multiplicative group of $K(V, \pi)$.*

PROOF. $\Delta(0)$ is a subgroup of the group of units of $\operatorname{End}_{\mathbb{Z}}(V)$ by 1.10. Since $\Delta(0)$ fixes each individual line through 0, we have $\Delta(0) \subseteq K(V, \pi)^*$. Conversely, if $\kappa \in K(V, \pi)^*$, then κ induces a collineation in $\pi(V)$ that fixes all the lines through 0. Hence $\Delta(0) = K(V, \pi)^*$. $\square$

The simplest examples of translation planes are obtained as follows. Let V be a vector space of rank 2 over K and let π be the set of subspaces of rank 1 of V. Then π is a spread of V. Hence $\pi(V)$ is a translation plane. $K(V, \pi)$ contains all the mappings $v \to vk$, where $k \in K$. This implies that $K(V, \pi)$ is isomorphic to K and that V is of rank 2 over $K(V, \pi)$. We shall call $\pi(V)$ *desarguesian* if π is a spread of V and if $\pi(V)$ is isomorphic to one of the examples just constructed, i.e., if V is of rank 2 over $K(V, \pi)$.

The next theorem follows immediately from this definition.

1.13 Theorem. *If $\mathfrak{A}$ is a finite translation plane of prime order, then $\mathfrak{A}$ is desarguesian.*

Let V be a K-vector space and denote by $\mathfrak{U}_i(V)$ the set of its subspaces of rank i. If V has rank 3 over K, then $\mathfrak{E} = (\mathfrak{U}_1(V), \mathfrak{U}_2(V), \subseteq)$ is a projective plane. If one line G of $\mathfrak{E}$ is taken as the line at infinity, then one obtains an affine plane $\mathfrak{E}_G$ which is, in fact, a desarguesian affine plane, as we shall now show. To this end, choose $P \in \mathfrak{U}_1(V)$ with $V = P \oplus G$. Since P is of rank 1, there exists $p \in P$ with $P = pK$. Define $\varphi \in \operatorname{Hom}_K(V, K)$ by $\varphi(pk + g) = k$ for all $k \in K$ and all $g \in G$. If $g \in G$, we define, using φ, a transvection $\tau(g)$ by $v^{\tau(g)} = v + g\varphi(v)$. Trivial computations show that τ is a monomorphism of $G(+)$ into $\operatorname{GL}(V)$. The image of $G(+)$ under τ will be denoted by $\operatorname{Trans}(V, G)$. The group $\operatorname{Trans}(V, G)$ induces in $\mathfrak{E}_G$ a group D of dilatations, since G is fixed elementwise by all the transvections in $\operatorname{Trans}(V, G)$. We show that D is actually the group T of translations of $\mathfrak{E}_G$. Let Q be a point of $\mathfrak{E}_G$ and let $C = G \cap (P + Q)$. By the modularity of the lattice of subspaces of V, we have $P + Q = Q + C$. Thus there exist $q \in Q$ and $c \in C$ such that $p = q + c$. Furthermore, $q \neq 0$, since $P \cap G = \{0\}$. Hence $Q = qK$. Finally, $p^{\tau(c)} = p + c\varphi(p) = p + c = q$ implies that $P^{\tau(c)} = Q$. Therefore D operates transitively on the set of points of $\mathfrak{E}_G$.

Assume $P^{\tau(g)} = P$. Then $p + g = p + g\varphi(p) \in P$, whence we obtain $g \in P \cap G = \{0\}$. Thus $\tau(g) = 1$. This proves firstly $\operatorname{Trans}(V, G) \cong D$ and secondly that D operates sharply transitively on the set of points of $\mathfrak{E}_G$. Since D is a group of dilatations, $D = $ T. In particular, $\mathfrak{E}_G$ is a translation plane.

Let $cK = C \in \mathfrak{U}_1(G)$. Then a simple computation shows $\operatorname{T}(C) = \{\tau(ck) \mid k \in K\}$ (here we have identified $\operatorname{Trans}(V, G)$ with T). If $\pi_0 = \{\operatorname{T}(C) \mid C \in \mathfrak{U}_1(G)\}$ and $\pi = \mathfrak{U}_1(G)$, then it follows that τ is an isomorphism of $\pi(G)$ onto $\pi_0(\operatorname{T})$. From this and 1.5, we infer:

1.14 Theorem. *Let V be a vector space of rank 3 over the division ring K and* *let* $\mathfrak{E} = (\mathfrak{U}_1(V), \mathfrak{U}_2(V), \subseteq)$. *If* $G \in \mathfrak{U}_2(V)$ *and* $\pi = \mathfrak{U}_1(G)$, *then* $\mathfrak{E}_G$ *and* $\pi(G)$ *are isomorphic. In particular,* $\mathfrak{E}_G$ *is desarguesian.*

We obtain furthermore:

1.15 Theorem. *Let* $\mathfrak{A}$ *be a desarguesian affine plane. Then there is up to isomorphism exactly one division ring* K *with the property*: *If* V *is a* K-*vector space of rank 3 and if* G *is a line of* $\mathfrak{E} = (\mathfrak{U}_1(V), \mathfrak{U}_2(V), \subseteq)$, *then* $\mathfrak{A}$ *and* $\mathfrak{E}_G$ *are isomorphic.*

2. An Alternative Description of Translation Planes

The description of translation planes which follows is well-known, but it seems not to have been used as a tool in their investigation. Nevertheless it is useful, as we shall see.

2.1 Lemma. *Let* V *be a vector space and* X, Y, Z, X', Y', Z' *subspaces of* V *such that* $V = X \oplus Y = Y \oplus Z = Z \oplus X = X' \oplus Y' = Y' \oplus Z' = Z' \oplus X'$. *Then there is a* $\sigma \in \mathrm{GL}(V)$ *such that* $X^\sigma = X'$, $Y^\sigma = Y'$ *and* $Z^\sigma = Z'$.

PROOF. We infer from the assumptions that there is a linear isomorphism μ from X onto Y such $Z = \{x + x^\mu \mid x \in X\}$ and likewise a linear isomorphism μ' from X' onto Y' with $Z' = \{x' + x'^{\mu'} \mid x' \in X'\}$. If V is finitely generated, we deduce furthermore $\mathrm{rank}(X) = \frac{1}{2}\mathrm{rank}(V)$ $= \mathrm{rank}(X')$. If V is not finitely generated, then $\mathrm{rank}(X) = \mathrm{rank}(V)$ $= \mathrm{rank}(X')$. In either case, there is a linear isomorphism ρ from X onto X'. The mapping $\mu^{-1}\rho\mu'$ is then an isomorphism from Y onto Y'. Define σ by $(x + y)^\sigma = x^\rho + y^{\mu^{-1}\rho\mu'}$. Then $\sigma \in \mathrm{GL}(V)$ and satisfies $X^\sigma = X'$ and $Y^\sigma = Y'$. Furthermore, $(x + x^\mu)^\sigma = x^\rho + x^{\rho\mu'} \in Z'$. Thus $Z^\sigma = Z'$.   □

Let V be a K-vector space and let π be a spread of V which consists of K-subspaces of V, i.e. a spread for which $K \subseteq \mathrm{K}(V, \pi)$ holds. Then $V = X \oplus Y$ with $X, Y \in \pi$. Since $X \cong Y$ by 1.7, we may assume that $V = X \oplus X$ is the outer direct sum of two copies of X. Put $Z = \{(x, x) \mid x \in X\}$. We may assume by 2.1 that $Z \in \pi$. Put $V(0) = \{(x, 0) \mid x \in X\}$, $V(\infty) = \{(0, x) \mid x \in X\}$ and $V(1) = Z$. If $U \in \pi$ and if $U \neq V(0), V(\infty)$, then there is exactly one $\sigma \in \mathrm{GL}(X)$ such that $U = \{(x, x^\sigma) \mid x \in X\}$. Set $V(\sigma) = \{(x, x^\sigma) \mid x \in X\}$ for $\sigma \in \mathrm{GL}(X)$ and $\Sigma(\pi) = \{\sigma \mid \sigma \in \mathrm{GL}(X), V(\sigma) \in \pi\}$. Then we have:

2.2 Theorem. *Let* X *be a* K-*vector space and assume that* π *is a spread of* $V = X \oplus X$ *such that* $K \subseteq \mathrm{K}(V, \pi)$ *and* $V(0), V(\infty) \in \pi$. *Then the following holds*:

a) *If $x, y \in X \setminus \{0\}$, then there is exactly one $\sigma \in \Sigma(\pi)$ such that $x^\sigma = y$.*
b) *If $\rho, \sigma \in \Sigma(\pi)$ and if $\rho \neq \sigma$, then $\rho - \sigma \in \mathrm{GL}(X)$.*

PROOF. a) Since $x \neq 0 \neq y$, there is exactly one $U \in \pi \setminus \{V(0), V(\infty)\}$ with $(x, y) \in U$. Furthermore, there is exactly one $\sigma \in \Sigma(\pi)$ such that $U = V(\sigma)$. Hence $y = x^\sigma$. Assume $y = x^\sigma = x^\tau$ and $\sigma, \tau \in \Sigma(\pi)$. Then $(x, y) \in V(\sigma) \cap V(\tau)$. Hence $V(\sigma) = V(\tau)$ by the uniqueness of U, whence $\sigma = \tau$.

 b) Let $0 = x^{\rho - \sigma}$. Then $x^\rho = x^\sigma$. Therefore $(x, x^\rho) = (x, x^\sigma) \in V(\rho) \cap V(\sigma)$. Since $\rho \neq \sigma$, we have $V(\rho) \cap V(\sigma) = \{(0, 0)\}$. Thus $x = 0$. Hence $\rho - \sigma$ is injective.

 Since π is a spread and since $\rho \neq \sigma$, the vector space V is the direct sum of $V(\rho)$ and $V(\sigma)$. Let $y \in X$. Then there exist $x, z \in X$ such that $(0, y) = (x, x^\rho) - (z, z^\sigma)$. As a result $x = z$ and $y = x^\rho - x^\sigma = x^{\rho - \sigma}$. Thus $\rho - \sigma$ is also surjective. Therefore $\rho - \sigma \in \mathrm{GL}(X)$. $\square$

2.3 Theorem. *Let X be a K-vector space and let Σ be a subset of $\mathrm{GL}(X)$ that enjoys the following properties:*

a) *If $x, y \in X \setminus \{0\}$, then there is a $\sigma \in \Sigma$ such that $x^\sigma = y$.*
b) *If $\rho, \sigma \in \Sigma$ and if $\rho \neq \sigma$, then $\rho - \sigma \in \mathrm{GL}(X)$.*
 Then $\pi = \{V(0), V(\infty)\} \cup \{V(\sigma) \mid \sigma \in \Sigma\}$ is a spread of $V = X \oplus X$ such that $K \subseteq \mathrm{K}(V, \pi)$.

PROOF. Let $(x, y) \in V$. If $x = 0$ or $y = 0$, then $(x, y) \in V(0) \cup V(\infty)$. Assume $x \neq 0 \neq y$. Then there exists $\sigma \in \Sigma$ such that $y = x^\sigma$. Thus $(x, y) \in V(\sigma)$ proving $V = \bigcup_{U \in \pi} U$.

 Obviously, $V(0) \cap V(\sigma) = V(\infty) \cap V(\sigma) = \{(0, 0)\}$ for all $\sigma \in \Sigma$. Let $(x, y) \in V(\rho) \cap V(\sigma)$ and assume $(0, 0) \neq (x, y)$. Then $y = x^\rho = x^\sigma$ and hence $x \neq 0 \neq y$. From this together with $0 = x^{\rho - \sigma}$ and b) we infer $\rho = \sigma$. This shows that π is a partition of V.

 Let $\rho, \sigma \in \Sigma$ and let $\rho \neq \sigma$. Assume furthermore that $(x, y) \in V$. By b), there are $a, b \in X$ such that $a^{\rho - \sigma} = y$ and $b^{\sigma - \rho} = x^\rho$. Then it follows that

$$\left(a + b + x, (a + b + x)^\rho\right) - \left(a + b, (a + b)^\sigma\right)$$
$$= (x, a^\rho + b^\rho + x^\rho - a^\sigma - b^\sigma) = (x, a^{\rho - \sigma}) = (x, y).$$

Thus $V(\rho) + V(\sigma) = V$.

 Similarly, $V = V(0) + V(\infty) = V(0) + V(\rho) = V(\infty) + V(\rho)$ for all $\rho \in \Sigma$.

 The last statement is trivial, since all the components of π are K-subspaces of V. $\square$

2.4 Lemma. *Let X be a K-vector space and $\rho, \sigma \in \mathrm{GL}(X)$. Then $\rho \sigma^{-1}$ operates fixed point free on $X \setminus \{0\}$, if and only if $\rho - \sigma$ is injective.*

PROOF. Let $u \in X$. Then $0 = u^{\rho - \sigma}$, if and only if $u^{\rho \sigma^{-1}} = u$. $\square$

2.5 Corollary. *Let X be a finitely generated K-vector space and let $\Sigma \subseteq \mathrm{GL}(X)$. Assume that Σ has the properties:*

a) *If $x, y \in X \setminus \{0\}$ then there is a $\sigma \in \Sigma$ such that $x^\sigma = y$.*
b) *If ρ and σ are distinct elements of Σ then $\rho\sigma^{-1}$ operates fixed point free on $X \setminus \{0\}$.*

Then $\pi = \{V(0), V(\infty)\} \cup \{V(\sigma) \mid \sigma \in \Sigma\}$ is a spread of $V = X \oplus X$ such that $K \subseteq K(V, \pi)$.

PROOF. This follows immediately from 2.4 and 2.3, since injective endomorphisms of finitely generated vector spaces are bijective. □

REMARK. As we remarked earlier, we may replace π by a spread π' such that $V(1) \in \pi'$. That $V(1)$ belongs to π is equivalent to saying that $1 \in \Sigma(\pi)$. Thus we may always assume that $1 \in \Sigma(\pi)$.

2.6 Theorem. *Let X be a K-vector space and let π_1 and π_2 be spreads of $V = X \oplus X$ consisting of K-subspaces of V. For $j = 1$ or 2 let X_{ij} be three distinct components of π_j. Furthermore, let λ and μ be elements of $\mathrm{GL}(V)$ such that $X_{11}^\lambda = X_{12}^\mu = V(0)$, $X_{21}^\lambda = X_{22}^\mu = V(1)$ and $X_{31}^\lambda = X_{32}^\mu = V(\infty)$. (Such λ and μ exist by 2.1.) Then there is an isomorphism κ from $\pi_1(V)$ onto $\pi_2(V)$ such that $X_{i1}^\kappa = X_{i2}$ for $i = 1, 2, 3$ if and only if $\Sigma(\pi_1^\lambda)$ and $\Sigma(\pi_2^\mu)$ are conjugate in $\Gamma L(X)$.*

PROOF. Let κ be such an isomorphism. Then $\lambda^{-1}\kappa\mu$ is an isomorphism from $\pi_1^\lambda(V)$ onto $\pi_2^\mu(V)$ such that $V(0)^{\lambda^{-1}\kappa\mu} = V(0)$, $V(1)^{\lambda^{-1}\kappa\mu} = V(1)$ and $V(\infty)^{\lambda^{-1}\kappa\mu} = V(\infty)$. This together with 1.10 implies that there is an $\alpha \in \Gamma L(X)$ such that $(x, y)^{\lambda^{-1}\kappa\mu} = (x^\alpha, y^\alpha)$ for all $x, y \in X$.

Let $\sigma \in \Sigma(\pi_1^\lambda)$. Since $V(\sigma)^{\lambda^{-1}\kappa\mu} \in \pi_2^\mu$, there exists $\tau \in \Sigma(\pi_2^\mu)$ with $(x^\alpha, x^{\sigma\alpha}) \in V(\tau)$ for all $x \in X$. Hence $\alpha^{-1}\sigma\alpha = \tau$. This proves $\alpha^{-1}\Sigma(\pi_1^\lambda)\alpha \subseteq \Sigma(\pi_2^\mu)$. Conversely, given $\tau \in \Sigma(\pi_2^\mu)$, there is a σ in $\Sigma(\pi_1^\lambda)$ such that $V(\sigma)^{\lambda^{-1}\kappa\mu} = V(\tau)$. By our previous argument $\alpha^{-1}\sigma\alpha = \tau$. Thus $\alpha^{-1}\Sigma(\pi_1^\lambda)\alpha = \Sigma(\pi_2^\mu)$.

Let $\alpha \in \Gamma L(X)$ be such that $\alpha^{-1}\Sigma(\pi_1^\lambda)\alpha = \Sigma(\pi_2^\mu)$. Define ρ by $(x, y)^\rho = (x^\alpha, y^\alpha)$. Then ρ is an isomorphism from $\pi_1^\lambda(V)$ onto $\pi_2^\mu(V)$ with $V(i)^\rho = V(i)$ for $i = 0, 1, \infty$. Put $\kappa = \lambda\rho\mu^{-1}$. Then κ is an isomorphism from $\pi_1(V)$ onto $\pi_2(V)$ such that $X_{i1}^\kappa = X_{i2}$ for $i = 1, 2, 3$. □

2.7 Theorem. *Let X be a K-vector space and let π be a spread of $V = X \oplus X$ such that $K \subseteq K(V, \pi)$ and $V(0), V(1), V(\infty) \in \pi$. For all $\kappa \in K(V, \pi)$ define $\alpha(\kappa)$ by $(x, 0)^\kappa = (x^{\alpha(\kappa)}, 0)$. Then $(x, y)^\kappa = (x^{\alpha(\kappa)}, y^{\alpha(\kappa)})$ for all $x, y \in X$. Moreover, α is an isomorphism of $K(V, \pi)$ onto $\mathfrak{C}_{\mathrm{End}_Z(X)}(\Sigma(\pi))$.*

PROOF. Let $(0, y)^\kappa = (0, y^{\beta(\kappa)})$. Since κ is additive, we obtain $(x, y)^\kappa = (x^{\alpha(\kappa)}, y^{\beta(\kappa)})$ for all $x, y \in X$. We deduce from $V(1)^\kappa \subseteq V(1)$ that $\alpha = \beta$. Trivial computations show that α is additive and multiplicative.

Since $(x^{\alpha(\kappa)}, x^{\sigma\alpha(\kappa)}) \in V(\sigma)^\kappa \subseteq V(\sigma)$, we have $x^{\sigma\alpha(\kappa)} = x^{\alpha(\kappa)\sigma}$ for all $x \in X$. Hence $\alpha(\kappa) \in \mathfrak{C}_{\mathrm{End}_Z(X)}(\Sigma(\pi))$. Conversely, let $\eta \in \mathfrak{C}_{\mathrm{End}_Z(X)}(\Sigma(\pi))$

and define κ by $(x, y)^\kappa = (x^\eta, y^\eta)$. Obviously, $\kappa \in K(V, \pi)$ and $\alpha(\kappa) = \eta$. Hence α is surjective and so, as $K(V, \pi)$ is a division ring, α is also injective.   □

2.8 Corollary. *If $\Sigma(\pi)$ is commutative, then $\Sigma(\pi)$ is an abelian group and $\pi(V)$ is the desarguesian plane over $K(V, \pi)$. Furthermore, $K(V, \pi)$ is a field.*

This follows immediately from 2.7.
Desarguesian planes over commutative fields are called *pappian*.

3. Homologies and Shears of Translation Planes

Let $\pi(V)$ be a translation plane and let $v \in V$. Then the group $\Delta(v)$ of all homologies with centre v and axis l_∞ is conjugate to $\Delta(0)$. Thus $\Delta(v)$ is isomorphic to the multiplicative group of $K(V, \pi)$. Since the multiplicative group of a finite division ring is cyclic, we have:

3.1 Theorem. *Let $\mathfrak{A}$ be a finite translation plane. If P is a point of $\mathfrak{A}$, then the group of all homologies of $\mathfrak{A}$ with centre P and axis l_∞ is cyclic.*

Let X be a K-vector space and let π be a spread of $V = X \oplus X$ such that $K \subseteq K(V, \pi)$ and $V(0), V(1), V(\infty) \in \pi$. Denote by (0) the point at infinity on $V(0)$ and by (∞) the point at infinity on $V(\infty)$. Furthermore, let $\Delta((0), V(\infty))$ be the group of all homologies with centre (0) and axis $V(\infty)$ and define $\Delta((\infty), V(0))$ similarly. If $\delta \in \Delta((0), V(\infty))$, then δ is semilinear by 1.10. We infer from $(0, x)^\delta = (0, x)$ for all $x \in X$ that δ is even linear. From this and $V(0)^\delta = V(0)$ we deduce the existence of $\alpha(\delta) \in GL(X)$ such that $(x, 0)^\delta = (x^{\alpha(\delta)}, 0)$ for all $x \in X$. Thus $(x, y)^\delta = (x^{\alpha(\delta)}, y)$ for all $x, y \in X$. Let $\sigma \in \Sigma(\pi) = \Sigma$. Then $(x, x^\sigma)^\delta = (x^{\alpha(\delta)}, x^\sigma)$. As δ is a collineation fixing $V(0)$ and $V(\infty)$, there exists $\tau \in \Sigma$ with $V(\sigma)^\delta = V(\tau)$. Hence $(x^{\alpha(\delta)}, x^\sigma) = (x^{\alpha(\delta)}, x^{\alpha(\delta)\tau})$ for all $x \in X$. Therefore $\sigma = \alpha(\delta)\tau$ proving $\Sigma \subseteq \alpha(\delta)\Sigma$. Obviously, α is a monomorphism from $\Delta((0), V(\infty))$ in $GL(X)$. Thus $\Sigma \subseteq \alpha(\delta^{-1})\Sigma = \alpha(\delta)^{-1}\Sigma$ and hence $\Sigma = \alpha(\delta)\Sigma$. As $V(1) \in \pi$, we have $1 \in \Sigma$, whence $\alpha(\delta) \in \Sigma$.

Conversely, let $\rho \in \Sigma$ be such that $\Sigma = \rho\Sigma$. Define δ by $(x, y)^\delta = (x^\rho, y)$. One verifies easily that $\delta \in \Delta((0), V(\infty))$ and that $\alpha(\delta) = \rho$. Put $\Sigma_l = \{\rho | \rho \in \Sigma, \Sigma = \rho\Sigma\}$. Then we have:

3.2 Theorem. *Define for all $\delta \in \Delta((0), V(\infty))$ the mapping $\alpha(\delta)$ by $(x, 0)^\delta = (x^{\alpha(\delta)}, 0)$. Then α is an isomorphism from $\Delta((0), V(\infty))$ onto Σ_l. Furthermore, $(x, y)^\delta = (x^{\alpha(\delta)}, y)$ for all $x, y \in X$.*

Let $\delta \in \Delta((\infty), V(0))$. As above, δ is linear. As $V(\infty)^\delta = V(\infty)$, there exists $\beta(\delta) \in GL(X)$ such that $(0, x)^\delta = (0, x^{\beta(\delta)})$ for all $x \in X$. This implies $(x, y)^\delta = (x, y^{\beta(\delta)})$ for all $x, y \in X$. Consequently, $(x, x^\sigma)^\delta = (x,$

$x^{\sigma\beta(\delta)}$) for all $x \in X$ and all $\sigma \in \Sigma$. Since δ is a collineation fixing $V(0)$ and $V(\infty)$, we have $\sigma\beta(\delta) \in \Sigma$ for all $\sigma \in \Sigma$. Hence $\Sigma\beta(\delta) \subseteq \Sigma$. Since β is a monomorphism from $\Delta((\infty), V(0))$ into $\mathrm{GL}(X)$, we obtain $\Sigma\beta(\delta)^{-1} = \Sigma\beta(\delta^{-1}) \subseteq \Sigma$. Therefore, $\Sigma\beta(\delta) = \Sigma$. Putting $\Sigma_r = \{\rho \mid \rho \in \Sigma, \Sigma\rho = \Sigma\}$ we obtain as above:

3.3 Theorem. *For all* $\delta \in \Delta((\infty), V(0))$, *define the mapping* $\beta(\delta)$ *by* $(0, x)^{\delta} = (0, x^{\beta(\delta)})$. *Then* β *is an isomorphism from* $\Delta((\infty), V(0))$ *onto* Σ_r. *Furthermore*, $(x, y)^{\delta} = (x, y^{\beta(\delta)})$ *for all* $x, y \in X$.

Let Δ be a group of homologies of a translation plane and assume that every homology in Δ has centre $P \mathrm{I} l_{\infty}$ and axis l. We have $l \neq l_{\infty}$, as P is not on l. Using 2.1 and 3.2 we obtain that Δ is isomorphic to a group of fixed point free linear mappings of a vector space. If Δ contains an involution, then it is multiplication by -1, since all other involutory linear mappings have 1 as an eigenvalue. This implies in particular that Δ contains at most one involution. This proves that the characteristic of $\mathrm{K}(V, \pi)$ is different from 2, if $\pi(V)$ admits an involutory homology with affine axis.

3.4 Corollary. *Let* $\mathfrak{A}$ *be a translation plane. If* $P \mathrm{I} l_{\infty}$ *and if* l *is a line of* $\mathfrak{A}$ *with* $P \nmid l$, *then there exists at most one involutory* (P, l)-*homology. If there exists an involutory* (P, l)-*homology, then the characteristic of the kernel of* $\mathfrak{A}$ *is different from 2.*

A finite group is called a Z-group, if and only if all its Sylow subgroups are cyclic. Using results of Burnside and Zassenhaus (see e.g. Passman [1968, Theorem 18.1, p. 193, Theorem 18.2, p. 196, Theorem 18.6, p. 204]), we obtain:

3.5 Corollary. *Let* $\mathfrak{A}$ *be a translation plane and* Δ *a finite group of homologies all of which have the same centre and the same axis. Then we have:*

a) *If* p *is a prime* $\geqslant 3$ *and if* Π *is a Sylow* p-*subgroup of* Δ, *then* Π *is cyclic.*
b) *The Sylow 2-subgroups of* Δ *are either cyclic or generalized quaternion groups.*
c) *Each subgroup of order* pq *of* Δ *where* p *and* q *are primes is cyclic.*
d) *If* Δ *is soluble, then* Δ *contains a normal Z-subgroup* Δ_0 *such that* Δ/Δ_0 *is isomorphic to a subgroup of* S_4.
e) *If* Δ *is non-soluble, then* Δ *contains a normal subgroup* Δ_0 *of index 1 or 2 and* Δ_0 *is the direct product of a group which is isomorphic to* $\mathrm{SL}(2, 5)$ *and a Z-group the order of which is relatively prime to 30.*

3.6 Corollary. *Let* $\mathfrak{A}$ *be a translation plane and* Δ *a finite group of homologies all of which have the same centre and the same axis.*

a) *If Δ contains a normal subgroup which is isomorphic to* SL(2,3), *then Δ contains precisely one subgroup which is isomorphic to* SL(2,3).
b) *If Δ is non-soluble, then Δ contains exactly one subgroup which is isomorphic to* SL(2,5).

PROOF. a) Let S be a normal subgroup of Δ which is isomorphic to SL(2,3). Then Δ is soluble by 3.5. Hence there is a normal Z-subgroup Δ_0 such that Δ/Δ_0 is isomorphic to a subgroup of S_4. As the Sylow 2-subgroup of Δ_0 is cyclic, Δ_0 contains a normal 2-complement Γ (see e.g. Huppert [1967, IV.2.8, p. 420]). Since Γ is a normal Hall subgroup of Δ_0, it is normal in Δ. Therefore, $S\Gamma$ is a normal subgroup of Δ. Furthermore, $\Delta/S\Gamma$ is a 2-group. This implies that all subgroups of Δ which are isomorphic to SL(2,3) are contained in $S\Gamma$, since SL(2,3) is generated by its Sylow 3-subgroups. Let S_1 be one such group. As the Sylow 3-subgroups of Δ are cyclic, we have that 3 does not divide $|\Gamma|$. Therefore, $|\Gamma|$ being odd, $S_1 \cap \Gamma = \{1\}$. Thus $S\Gamma = S_1\Gamma$. Finally, SS_1 is a subgroup of $S\Gamma$ the order of which is relatively prime to $|\Gamma|$. Hence $|SS_1|$ divides $|S|$ showing that $S = S_1$.

b) By 3.5 b), Δ contains a normal subgroup Δ_0 of index 1 or 2 and Δ_0 is the direct product of a group S which is isomorphic to SL(2,5) and a Z-group T the order of which is relatively prime to the order of S. As above, S is the only subgroup of Δ_0 which is isomorphic to SL(2,5). Since SL(2,5) does not contain a subgroup of index 2, all subgroups of Δ which are isomorphic to SL(2,5) are contained in Δ_0. $\square$

3.7 Corollary. *Let $\mathfrak{A}$ be a finite translation plane and let P be a point on l_∞ and l a line of $\mathfrak{A}$ which does not pass through P. If $\Delta(P,l)$ is not soluble, then the order of $\mathfrak{A}$ is a square.*

PROOF. If SL(2,5) $\subseteq$ GL(n,q), then SL(2,5) $\subseteq$ SL(n,q), since SL(2,5) is equal to its commutator subgroup and since GL(n,q)/SL(n,q) is cyclic. Furthermore, if SL(2,5) operates fixed point free on the vector space of rank n over $GF(q)$ and if $1 \neq \zeta \in \mathfrak{Z}(\mathrm{SL}(2,5))$, then $v^\zeta = -v$ for all v. This implies $1 = \det \zeta = (-1)^n$. Therefore n is even. Using this remark and 3.5 e), we obtain the corollary. $\square$

3.8 Lemma. *Let $\mathfrak{P}$ be a projective plane. If P and Q are points and l and m are lines of $\mathfrak{P}$ such that P I m and Q I l, then $\Delta(P,l)$ and $\Delta(Q,m)$ centralize each other, unless $P = Q$ and $l = m$.*

PROOF. Assume $P \neq Q$ or $l \neq m$. Then $\Delta(P,l) \cap \Delta(Q,m) = \{1\}$. Let $\delta \in \Delta(P,l)$ and $\epsilon \in \Delta(Q,m)$. Then $\delta^{-1}\epsilon\delta \in \Delta(Q^\delta, m^\delta)$. As P I m and Q I l, we have $m^\delta = m$ and $Q^\delta = Q$. Thus $\delta^{-1}\epsilon\delta \in \Delta(Q,m)$ and hence $\epsilon^{-1}\delta^{-1}\epsilon\delta \in \Delta(Q,m)$. As the assumptions are symmetric in δ and ϵ, we also have $\delta^{-1}\epsilon^{-1}\delta\epsilon \in \Delta(P,l)$ and, therefore, $\epsilon^{-1}\delta^{-1}\epsilon\delta = (\delta^{-1}\epsilon^{-1}\delta\epsilon)^{-1} \in \Delta(P,l)$. Thus $\epsilon^{-1}\delta^{-1}\epsilon\delta \in \Delta(P,l) \cap \Delta(Q,m) = \{1\}$. $\square$

Let $\Gamma_0 \subseteq \Delta((0), V(\infty))$ and $\Gamma_\infty \subseteq \Delta((\infty), V(0))$ be subgroups of $\Delta((0), V(\infty))$ and $\Delta((\infty), V(0))$ respectively. By 3.8, $\Gamma_0\Gamma_\infty$ is a group and Γ_0 and

Γ_∞ are normal subgroups of $\Gamma_0\Gamma_\infty$. Assume that the set of orbits of Γ_0 on l_∞ is the same as the set of orbits of Γ_∞ on l_∞. It follows that the orbit of $V(\sigma)$ for $\sigma \in \Sigma(\pi)$ under Γ_0 is the same as the orbit of $V(\sigma)$ under Γ_∞. Let Σ_0 be the image of Γ_0 under the mapping defined in 3.2 and define Σ_∞ similarly using 3.3. Then we have $\Sigma_0\sigma = \sigma\Sigma_\infty$ for all $\sigma \in \Sigma(\pi)$. In particular $\Sigma_0 = \Sigma_\infty$, as $1 \in \Sigma(\pi)$, and hence $\Sigma \subseteq \mathfrak{N}_{GL(X)}(\Sigma_0)$.

Assume conversely that Σ_0 is a subgroup of $\Sigma_l \cap \Sigma_r$ and that Σ normalizes Σ_0. Using α and β as defined in 3.2 and 3.3 and putting $\Gamma_0 = \alpha^{-1}(\Sigma_0)$ and $\Gamma_\infty = \beta^{-1}(\Sigma_0)$, we obtain that the set of orbits of Γ_0 on l_∞ is the same as the set of orbits of Γ_∞ on l_∞. This proves:

3.9 Theorem. *Let $\Gamma_0 \subseteq \Delta((0), V(\infty))$ and $\Gamma_\infty \subseteq \Delta((\infty), V(0))$ be subgroups of $\Delta((0), V(\infty))$ and $\Delta((\infty), V(0))$ respectively and assume that the set of orbits of Γ_0 on l_∞ is equal to the set of orbits of Γ_∞ on l_∞. Then $\alpha(\Gamma_0) = \beta(\Gamma_\infty)$ where α and β are the mappings defined in 3.2 and 3.3 respectively. Furthermore Σ normalizes $\alpha(\Gamma_0)$. Conversely, let Σ_0 be a subgroup of $\Sigma_r \cap \Sigma_l$ such that Σ normalizes Σ_0, then the set of orbits of $\alpha^{-1}(\Sigma_0)$ on l_∞ is the same as the set of orbits of $\beta^{-1}(\Sigma_0)$ on l_∞.*

Let $\sigma \in \Sigma$ and let ρ be a perspectivity with axis $V(\sigma)$ which interchanges $V(0)$ and $V(\infty)$. As $(x, x^\sigma)^\rho = (x, x^\sigma)$ for all $x \in X$, we have that ρ is linear. Therefore, there exist $\alpha, \beta \in GL(X)$ with $(x, 0)^\rho = (0, x^\alpha)$ and $(0, x)^\rho = (x^\beta, 0)$. This implies $(x, y)^\rho = (y^\beta, x^\alpha)$. We have in particular $(x, x^\sigma) = (x, x^\sigma)^\rho = (x^{\sigma\beta}, x^\alpha)$. We infer from these equations that $\beta = \sigma^{-1}$ and $\alpha = \sigma$. Thus $(x, y)^\rho = (y^{\sigma^{-1}}, x^\sigma)$. Let $\tau \in \Sigma$. Then $(x, x^\tau)^\rho = (x^{\tau\sigma^{-1}}, x^\sigma) = (x^{\tau\sigma^{-1}}, x^{\tau\sigma^{-1}\sigma\tau^{-1}\sigma})$. As ρ is a collineation, $\sigma\tau^{-1}\sigma \in \Sigma$ for all $\tau \in \Sigma$. Putting $\Sigma^{-1} = \{\tau^{-1} | \tau \in \Sigma\}$, it follows that $\sigma\Sigma^{-1}\sigma \subseteq \Sigma$. Let $\xi \in \Sigma$. Then there exists $\eta \in \Sigma$ with $V(\eta)^\rho = V(\xi)$. This yields $\sigma\eta^{-1}\sigma = \xi$ and hence $\sigma\Sigma^{-1}\sigma = \Sigma$.

Assume conversely that $\sigma \in \Sigma$ is such that $\sigma\Sigma^{-1}\sigma \subseteq \Sigma$. Define ρ by $(x, y)^\rho = (y^{\sigma^{-1}}, x^\sigma)$. Then we have $\rho \in GL(V)$. As $\sigma\Sigma^{-1}\sigma \subseteq \Sigma$, we have $\pi^\rho \subseteq \pi$. Furthermore, $\rho^2 = 1$ which yields $\pi = \pi^{\rho^2} \subseteq \pi^\rho$. Therefore, $\pi^\rho = \pi$. This proves that ρ is a collineation of $\pi(V)$. Finally, it is easily seen that ρ is a perspectivity with axis $V(\sigma)$ which interchanges $V(0)$ and $V(\infty)$. Thus we have shown:

3.10 Theorem. *Let X be a K-vector space and let π be a spread of $V = X \oplus X$ with $K \subseteq \mathrm{K}(V, \pi)$ and $V(0), V(1), V(\infty) \in \pi$. If $\sigma \in \Sigma(\pi)$, then the following conditions are equivalent:*

a) $\sigma\Sigma(\pi)^{-1}\sigma = \Sigma(\pi)$.
b) $\sigma\Sigma(\pi)^{-1}\sigma \subseteq \Sigma(\pi)$.
c) *The mapping ρ defined by $(x, y)^\rho = (y^{\sigma^{-1}}, x^\sigma)$ is a perspectivity of $\pi(V)$ with axis $V(\sigma)$ which interchanges $V(0)$ and $V(\infty)$.*

Conversely, if ρ is a perspectivity with axis $V(\sigma)$ which interchanges $V(0)$ and $V(\infty)$, then $(x, y)^\rho = (y^{\sigma^{-1}}, x^\sigma)$ for all $x, y \in X$.

Assume that every $V(\sigma)$ is the axis of a perspectivity which interchanges $V(0)$ and $V(\infty)$. Then we have:

(1) $\sigma\Sigma^{-1}\sigma = \Sigma$ *for all $\sigma \in \Sigma$.*
    As $1 \in \Sigma$, it follows that
(2) $\Sigma^{-1} = \Sigma$.
    Let $\sigma \in \Sigma$. Then $\sigma^0 = 1 \in \Sigma$. Assume $\sigma^j \in \Sigma$ for all $j \leqslant i$. Then we have $\sigma^{i+1} = \sigma\sigma^{i-1}\sigma \in \sigma\Sigma\sigma = \Sigma$. Using this and (2) yields
(3) $\sigma^z \in \Sigma$ *for all $\sigma \in \Sigma$ and all $z \in \mathbb{Z}$.*
    Finally, we have:
(4) $\Sigma_r = \Sigma_l$ *and Σ normalizes Σ_r.*

PROOF. Let $\sigma \in \Sigma$ and $(x, y)^\rho = (y^{\sigma^{-1}}, x^\sigma)$. As ρ interchanges $V(0)$ and $V(\infty)$, we have $\rho^{-1}\Delta((0), V(\infty))\rho = \Delta((\infty), V(0))$. Let $\delta \in \Delta((0), V(\infty))$. Then there exists $\mu \in \Sigma_l$ such that $(x, y)^\delta = (x^\mu, y)$. Therefore, $(x, y)^{\rho^{-1}\delta\rho} = (x, y^{\sigma^{-1}\mu\sigma})$. This yields $\sigma^{-1}\Sigma_l\sigma = \Sigma_r$ for all $\sigma \in \Sigma$. As $1 \in \Sigma$, the assertions follow. $\qquad\square$

If Σ is a group, then we do have $\sigma\Sigma^{-1}\sigma = \Sigma$ for all $\sigma \in \Sigma$, but Σ need not be a group in order to satisfy condition a) of 3.10 for all $\sigma \in \Sigma$, as R. Burn [1968] has shown. His examples are all infinite. The finite case will be studied in more detail later on.

Let $\mathfrak{P}$ be a projective plane and let P be a point and l a line of $\mathfrak{P}$. The plane $\mathfrak{P}$ is called (P,l)-*transitive*, if for any two points X and Y with $P \neq X, Y$ and $X, Y \nmid l$ and $PX = PY$ there exists $\delta \in \Delta(P,l)$ such that $X^\delta = Y$.

A translation plane $\mathfrak{A}$ is called a *nearfield plane*, if $\mathfrak{A}$ possesses two distinct points $P, Q \mathrel{I} l_\infty$ and an affine point O such that $\mathfrak{A}$ is (P, OQ)- as well as (Q, OP)-transitive.

Using 2.1, 3.2 and 3.10, we obtain:

3.11 Theorem. *Let $\mathfrak{A}$ be a nearfield plane and let P and Q have the same meaning as in the above definition. If l is a line of $\mathfrak{A}$ which does not pass through either of the points P and Q, then there exists a perspectivity with axis l of $\mathfrak{A}$ which interchanges P and Q.*

Using 2.1, 3.2 and 3.3, we get:

3.12 Theorem (Gingerich). *Let $\mathfrak{A}$ be a translation plane. If P and Q are distinct points on l_∞ and if O is an affine point of $\mathfrak{A}$, then $\mathfrak{A}$ is (P, OQ)-transitive, if and only if $\mathfrak{A}$ is (Q, OP)-transitive.*

3.13 Theorem. *Let X be a K-vector space and let π be a spread of $V = X \oplus X$ with $K \subseteq K(V, \pi)$ and $V(0), V(1), V(\infty) \in \pi$. If $\tau \in \Delta((\infty), V(\infty))$, then there exists $\sigma \in \Sigma \cup \{0\}$ such that $\sigma + (\Sigma \cup \{0\}) = \Sigma \cup \{0\}$ and $(x, y)^\tau = (x, x^\sigma + y)$ for all $x, y \in X$. Conversely, if $\sigma \in \Sigma \cup \{0\}$ and*

$\sigma + (\Sigma \cup \{0\}) = \Sigma \cup \{0\}$, *then the mapping τ defined by $(x, y)^\tau = (x, x^\sigma + y)$ is an element of $\Delta((\infty), V(\infty))$. The set of all $\sigma \in \Sigma \cup \{0\}$ which satisfy $\sigma + (\Sigma \cup \{0\}) = \Sigma \cup \{0\}$ is a subgroup of the additive group of $\mathrm{End}_K(X)$ which is isomorphic to $\Delta((\infty), V(\infty))$.*

PROOF. We have $(x, y)^\tau = (x, 0)^\tau + (0, y)^\tau = (x, 0)^\tau + (0, y)$. Furthermore, $(x, 0)^\tau \in (V(\infty) + (x, 0))^\tau = V(\infty) + (x, 0)$, since $V(\infty) + (x, 0)$ is a line through the centre of τ. Thus, $(x, 0)^\tau = (x, x')$ for some $x' \in X$. As τ is a collineation, there exists $\sigma \in \Sigma \cup \{0\}$ with $V(0)^\tau = V(\sigma)$. As $(x, x') = (x, 0)^\tau \in V(0)^\tau = V(\sigma)$, we have $x' = x^\sigma$. Hence $(x, y)^\tau = (x, x^\sigma + y)$ for all $x, y \in X$.

Let $\rho \in \Sigma \cup \{0\}$. Then there is a $\delta \in \Sigma \cup \{0\}$ with $V(\rho)^\tau = V(\delta)$. From this we infer $(x, x^\delta) = (x, x^\rho)^\tau = (x, x^\sigma + x^\rho) = (x, x^{\sigma + \rho})$. Therefore, $\sigma + \rho = \delta \in \Sigma \cup \{0\}$. This proves $\sigma + (\Sigma \cup \{0\}) \subseteq \Sigma \cup \{0\}$. If $\delta \in \Sigma \cup \{0\}$, then there is a $\rho \in \Sigma \cup \{0\}$ with $V(\rho)^\tau = V(\delta)$. As above, $\sigma + \rho = \delta$. Hence $\sigma + (\Sigma \cup \{0\}) = \Sigma \cup \{0\}$.

The remaining parts of 3.13 are as easily proved. $\qquad\square$

3.14 Theorem. *Let $\mathfrak{A}$ be a translation plane. $\mathfrak{A}$ is desarguesian, if and only if $\mathfrak{A}$ has two distinct points P and Q on l_∞ and an affine point O such that $\mathfrak{A}$ is (P, OQ)- and (Q, OQ)-transitive.*

PROOF. The necessity of these conditions follows from 1.12 and 1.14. Hence we may assume that $\mathfrak{A}$ satisfies the conditions of the theorem. Furthermore, we may assume that $\mathfrak{A}$ has standard form and that $\mathfrak{A}$ is $((0), V(\infty))$- and $((\infty), V(\infty))$-transitive. 3.13 yields that $L = \Sigma \cup \{0\}$ is a subgroup of the additive group of $\mathrm{End}_K(X)$ and, by 3.2, we have that Σ is a group. We infer that L is a division ring.

Let $0 \neq e \in X$. Then $X = \{e^\sigma \mid \sigma \in L\}$ by 2.2. For $\delta \in L$ define δ' by $(e^\sigma)^{\delta'} = e^{\delta\sigma}$. Let $x, y \in X$. Then $x = e^\sigma$ and $y = e^\tau$ with $\sigma, \tau \in L$. Thus we have

$$(x + y)^{\delta'} = e^{(\sigma + \tau)\delta'} = e^{\delta(\sigma + \tau)} = e^{\delta\sigma + \delta\tau} = e^{\delta\sigma} + e^{\delta\tau} = x^{\delta'} + y^{\delta'}.$$

Furthermore,

$$x^{\rho\delta'} = e^{\sigma\rho\delta'} = e^{\delta\sigma\rho} = e^{\sigma\delta'\rho} = x^{\delta'\rho}.$$

Therefore, $\delta' \in \mathfrak{C}_{\mathrm{End}_Z(X)}(\Sigma)$. The theorem now follows from 2.7. $\qquad\square$

A projective plane $\mathfrak{P}$ is called a *nearfield plane with respect to its line l* if the affine plane $\mathfrak{P}_l$ is a nearfield plane.

3.15 Corollary. *Let $\mathfrak{P}$ be a projective plane. If $\mathfrak{P}$ is a nearfield plane with respect to two distinct lines l and m, then $\mathfrak{P}$ is desarguesian.*

PROOF. The collineation group of $\mathfrak{P}_l$ has at most two orbits on l by 3.2 and 3.11. If it had two orbits, then one orbit would have length 2 in contradiction to the $(l \cap m, m)$-transitivity of $\mathfrak{P}$. Therefore, $\mathfrak{P}_l$ satisfies the hypotheses of 3.14, whence $\mathfrak{P}_l$ and hence $\mathfrak{P}$ is desarguesian. $\qquad\square$

3.16 Corollary. *Let $\mathfrak{P}$ be a non-desarguesian projective plane. If $\mathfrak{P}$ is a nearfield plane, then $\mathfrak{P}$ is not selfdual.*

PROOF. Let Γ be the collineation group of $\mathfrak{P}$. Then Γ fixes a line l of $\mathfrak{P}$ by 3.15. As Γ has no fixed point and as Γ is also the collineation group of $\mathfrak{P}^d$, the dual of $\mathfrak{P}$, the planes $\mathfrak{P}$ and $\mathfrak{P}^d$ cannot be isomorphic. $\qquad\square$

3.17 Lemma. *Let $\mathfrak{A}$ be an affine plane and let P and Q be distinct points of $\mathfrak{A}$. If ρ is an involutory (P, l_∞)- and σ an involutory (Q, l_∞)-homology, then $\rho\sigma$ is a translation with centre $PQ \cap l_\infty$.*

PROOF. $\rho\sigma$ is a perspectivity with axis l_∞. Let C be the centre of $\rho\sigma$. Then $C\ I\ PQ$, as $(PQ)^{\rho\sigma} = PQ$. Furthermore, $C^\rho = C^\sigma$, as $C^{\rho\sigma} = C$ and σ is involutory. This yields $(C^\rho)^{\rho\sigma} = C^{\rho^2\sigma} = C^\sigma = C^\rho$. If C^ρ is not on l_∞, then C^ρ is the centre of $\rho\sigma$, since $\rho\sigma \neq 1$, as $P \neq Q$. Therefore $C^\rho = C$. This implies $P = C = C^\rho = C^\sigma$. Hence $P = Q$, a contradiction. Thus $C^\rho\ I\ l_\infty$ and therefore $C = C^{\rho^2}\ I\ l_\infty$. $\qquad\square$

3.18 Lemma. *Let $\mathfrak{P}$ be a projective plane, l a line and P and Q two distinct points of $\mathfrak{P}$ which are not on l. If $\mathfrak{P}$ is (P, l)- and (Q, l)-transitive, then $\mathfrak{P}$ is also $(PQ \cap l, l)$-transitive.*

PROOF. Put $\Gamma = \langle \Delta(P, l), \Delta(Q, l)\rangle$. As $P \neq Q$, the group Γ acts doubly transitively on $PQ \setminus \{R\}$ where $R = PQ \cap l$. Let X and Y be distinct points on PQ which are different from R. Then there exists $\gamma \in \Gamma$ such that $X^\gamma = Y$ and $Y^\gamma = X$. As γ^2 fixes X and Y, we have $\gamma^2 = 1$, since γ^2 is a perspectivity with axis l. If $\gamma \notin \Delta(R, l)$, we infer from 3.17 that $\Delta(R, l) \neq \{1\}$. Thus $\Delta(R, l) \neq \{1\}$ in either case. As $\Delta(R, l)$ is normal in Γ, the double transitivity of Γ on $PQ \setminus \{R\}$ implies that $\Delta(R, l)$ is transitive on $PQ \setminus \{R\}$. $\qquad\square$

The reader will certainly have realized that the above proof works only if every line of $\mathfrak{P}$ carries at least 4 points. But the theorem is trivially true for the plane of order 2.

3.19 Theorem (André 1955). *Let $\mathfrak{A}$ be a non-desarguesian translation plane and let P and Q be two distinct points on l_∞ and O an affine point of $\mathfrak{A}$ and assume that $\mathfrak{A}$ is (P, OQ)-transitive. If (R, l) is a non-incident point-line pair of $\mathfrak{A}$ with $R \neq P, Q$ and if $\mathfrak{A}$ is (R, l)-transitive, then $\mathfrak{A}$ is finite and the order of $\mathfrak{A}$ is 9.*

PROOF. We may assume that $\mathfrak{A}$ has standard form $\pi(V)$ with $P = (0)$ and $Q = (\infty)$. As $\mathfrak{A}$ is non-desarguesian, $l \neq l_\infty$. We infer from this and 3.14 that R is on l_∞. Furthermore, 3.18 and 3.14 yield $P, Q\ \nmid\ l$. By 3.11, there exists a perspectivity λ with axis l interchanging P and Q. Again by 3.18 and 3.14, $R^\lambda = R$. As $R\ \nmid\ l$, the characteristic of $K(V, \pi)$ is different from 2.

Let $\sigma \in \Sigma$ and $(x, y)^\rho = (y^{\sigma^{-1}}, x^\sigma)$. Then ρ is a perspectivity with axis $V(\sigma)$ which interchanges $V(0)$ and $V(\infty)$. Let $\sigma \neq \tau \in \Sigma$ and $V(\tau)^\rho = V(\tau)$. Then

$$(x, x^\tau)^\rho = (x^{\tau\sigma^{-1}}, x^\sigma) = (x^{\tau\sigma^{-1}}, x^{\tau\sigma^{-1}\tau})$$

for all $x \in X$. Thus $(\tau\sigma^{-1})^2 = 1$. As $\tau \neq \sigma$, we have by 2.2 and 2.4 that $\tau\sigma^{-1}$ operates fixed point free on X. Therefore, $\tau\sigma^{-1} = -1$ and hence $\tau = -\sigma$. This implies that $V(-\sigma) \cap l_\infty$ is the centre of ρ.

As $\mathfrak{A}$ is $((0), V(\infty))$-transitive, we infer from the (R, l)-transitivity with the aid of 3.18, 3.14 and the above argument that $\mathfrak{A}$ is $(V(-\sigma) \cap l_\infty, V(\sigma))$-transitive for all $\sigma \in \Sigma$. Let $\sigma \neq 1, -1$. Then there is a $\mu \in \Delta(V(-\sigma) \cap l_\infty, V(\sigma))$ with $V(1)^\mu = V(-1)$. By 3.18 and 3.14, we have $V(-1)^\mu = V(1)$. Since $\Delta(V(-\sigma) \cap l_\infty, V(\sigma))$ contains only one involution by 3.4, we have $(x, y)^\mu = (y^{\sigma^{-1}}, x^\sigma)$ by 3.10. Therefore, $V(-1) = V(1)^\mu = \{(x^{\sigma^{-1}}, x^\sigma) \mid x \in X\}$. This yields $x^\sigma = -x^{\sigma^{-1}}$. Hence $\sigma^2 = -1$.

As $\mathfrak{A}$ is non-desarguesian, Σ is not abelian by 2.8. Therefore, Σ contains at least 8 elements, since Σ is a 2-group. Furthermore, $\Sigma/\{1, -1\}$ is an elementary abelian 2-group. If Σ contains more than 8 elements, then $\Sigma/\{1, -1\}$ contains a subgroup of order 8. Thus Σ contains a subgroup Σ_0 of order 16. As $\sigma^4 = 1$ for all $\sigma \in \Sigma$, the group Σ_0 is not cyclic. Moreover, Σ_0 contains only one involution. Thus Σ_0 is the quaternion group of order 16. This implies that Σ_0 and hence Σ contains an element of order 8, a contradiction. $\qquad \square$

We shall see later on that the nearfield plane of order 9 is indeed an exception.

4. A Characterization of Pappian Planes

The following characterization of pappian planes is a restatement of a theorem of Pickert (Pickert [1959]. See also Burn [1968] and Schröder [1968]).

4.1 Theorem. *Let $\mathfrak{A}$ be an affine plane. $\mathfrak{A}$ is pappian, if and only if there exist two distinct points P and Q on l_∞ and an affine point O such that the following holds:*

a) *For all lines l of $\mathfrak{A}$ with $P, Q \nmid l$ there is a perspectivity σ_l with axis l switching P and Q.*

b) *If l, m, n are three lines through O with $P, Q \nmid l, m, n$, then there exists a line p through O such that $P, Q \nmid p$ and $\sigma_l \sigma_m \sigma_n = \sigma_p$.*

PROOF. Assume that $\mathfrak{A}$ is pappian. We may assume that $\mathfrak{A} = \pi(V)$ with $V = X \oplus X$ and that $V(0) = OP$ and $V(\infty) = OQ$ where P and Q are any two given points on l_∞. The set $\Sigma = \Sigma(\pi)$ is an abelian group, as $\mathfrak{A}$ is pappian. Since Σ is a group, $\mathfrak{A}$ has property a) by 3.10.

Let $V(\alpha)$, $V(\beta)$, $V(\gamma)$ be three lines through O which are distinct from $V(0)$ and $V(\infty)$. Let σ_ξ be the perspectivity with axis $V(\xi)$ which interchanges $V(0)$ and $V(\infty)$. By 3.10, we have

$$(x, y)^{\sigma_\alpha \sigma_\beta \sigma_\gamma} = \left(y^{\alpha^{-1} \beta \gamma^{-1}}, x^{\alpha \beta^{-1} \gamma} \right).$$

As Σ is a group, $\delta = \alpha \beta^{-1} \gamma \in \Sigma$, and Σ being abelian yields $\delta^{-1} = \alpha^{-1} \beta \gamma^{-1}$. Thus $\sigma_\alpha \sigma_\beta \sigma_\gamma = \sigma_\delta$. This proves one part of 4.1.

To prove the converse, we need some preparation.

4.2 Lemma. *Let $\mathfrak{A}$ be an affine plane and let l be an affine line of $\mathfrak{A}$. If there is an involutory (D, l_∞)-homology σ_D for all points D on l, then $\mathfrak{A}$ is $(l \cap l_\infty, l_\infty)$-transitive. Moreover, if C is any point on l, then $\mathrm{T}(l \cap l_\infty) = \{\sigma_C \sigma_D \mid D \mathrel{I} l\}$.*

PROOF. Let C be any point on l. By 3.17, we have $\mathrm{T}(l \cap l_\infty) \supseteq \{\sigma_C \sigma_D \mid D \mathrel{I} l\}$. Let m be a line parallel to and distinct from l and let X and Y be two points on m. Put $D = l \cap YX^{\sigma_C}$. As $\sigma_C \sigma_D \in \mathrm{T}(l \cap l_\infty)$, we have $X^{\sigma_C \sigma_D} \mathrel{I} m$. On the other hand $X^{\sigma_C \sigma_D} \mathrel{I} X^{\sigma_C} D = YX^{\sigma_C}$. Hence $X^{\sigma_C \sigma_D} = YX^{\sigma_C} \cap m = Y$. $\qquad\square$

4.3 Corollary (Baer 1944). *Let $\mathfrak{A}$ be an affine plane. If every point P of $\mathfrak{A}$ is the centre of an involutory (P, l_∞)-homology, then $\mathfrak{A}$ is a translation plane and the group generated by all those involutory homologies contains the translation group of $\mathfrak{A}$.*

PROOF. This follows immediately from 4.2. $\qquad\square$

4.4 Lemma. *Let $\mathfrak{P}$ be a finite projective plane of order n and let $\mathfrak{Q}$ be a proper subplane of order m of $\mathfrak{P}$.*

a) *If every point of $\mathfrak{P}$ is incident with a line of $\mathfrak{Q}$, then $m^2 = n$.*
b) *If there exists a point of $\mathfrak{P}$ which does not lie on a line of $\mathfrak{Q}$, then $m^2 + m \leqslant n$.*

PROOF. a) Let P be a point of $\mathfrak{Q}$. Since $\mathfrak{Q}$ is a proper subplane of $\mathfrak{P}$, there exists a line l through P which does not belong to $\mathfrak{Q}$. There are $m + 1$ lines of $\mathfrak{Q}$ passing through P. The remaining m^2 lines $l_1, \ldots, l_{m^2}$ of $\mathfrak{Q}$ meet l in points which are different from P. As l is not a line of $\mathfrak{Q}$, there is only one point of $\mathfrak{Q}$ which is on l, namely P. Therefore, the mapping $i \to l \cap l_i$ is injective. This mapping is also surjective, since every point on l is on a line of $\mathfrak{Q}$. Thus $i \to l \cap l_i$ is a bijection from $\{1, 2, \ldots, m^2\}$ onto $\{Q \mid Q \mathrel{I} l, Q \neq P\}$. Hence $n = m^2$.

b) Let P be a point of $\mathfrak{P}$ which is not on a line of $\mathfrak{Q}$. Then $X \to XP$ is an injective mapping of the set of points of $\mathfrak{Q}$ into the set of lines through P. Therefore $m^2 + m + 1 \leqslant n + 1$. ☐

A proper subplane $\mathfrak{Q}$ of a projective plane $\mathfrak{P}$ is called a *Baer subplane* of $\mathfrak{P}$, if every point of $\mathfrak{P}$ is on a line of $\mathfrak{Q}$ and if every line of $\mathfrak{P}$ carries a point of $\mathfrak{Q}$. Lemma 4.4 implies that a proper subplane $\mathfrak{Q}$ of a finite projective plane $\mathfrak{P}$ is a Baer subplane, if every point of $\mathfrak{P}$ is on a line of $\mathfrak{Q}$. This is not true for infinite planes.

A non-identity collineation fixing a Baer subplane pointwise is called a *Baer collineation*.

4.5 Lemma (Baer 1946a). *If $\mathfrak{P}$ is a projective plane and $\kappa \neq 1$ is a collineation of $\mathfrak{P}$, then the following statements are equivalent:*

a) *For all points P of $\mathfrak{P}$ the points $P, P^\kappa, P^{\kappa^2}$ are collinear.*
b) *For all lines l of $\mathfrak{P}$ the lines $l, l^\kappa, l^{\kappa^2}$ are confluent.*
c) *κ is either a Baer collineation or a perspectivity.*

PROOF. a) is equivalent to b): Assume that a) holds. Let l be a line of $\mathfrak{P}$ and let P be a point on l and l^κ. If $P = P^\kappa$, then $P = P^{\kappa^2}$ and hence $P \mathrel{I} l, l^\kappa, l^{\kappa^2}$. We may thus assume that $P \neq P^\kappa$. By assumption, $P, P^\kappa, P^{\kappa^2}$ are collinear. Hence $PP^\kappa = P^\kappa P^{\kappa^2}$. Obviously, $l^\kappa = PP^\kappa$. Thus $l^{\kappa^2} = P^\kappa P^{\kappa^2} = PP^\kappa = l^\kappa$. Therefore, $l = l^\kappa = l^{\kappa^2}$. This proves b). By duality, b) implies a).

Next we show that a) and b) together imply c). First we prove: κ has at least two fixed points and at least two fixed lines. As $\kappa \neq 1$, there exists a line l with $l \neq l^\kappa$. Put $P = l \cap l^\kappa$. Then $P^\kappa = P$, since l, l^κ and l^{κ^2} are confluent. Assume that P is the only fixed point of κ. Then XX^κ is a line for all points $X \neq P$. Furthermore, $XX^\kappa = X^\kappa X^{\kappa^2} = (XX^\kappa)^\kappa$ by a). Hence XX^κ is a fixed line. As two distinct fixed lines meet in a fixed point, all the lines through P are fixed. Therefore, κ is a central collineation, but central collineations have more than one fixed point. This contradiction shows that κ has at least two fixed points. That κ has two fixed lines is proved dually.

Every point X of $\mathfrak{P}$ lies on a fixed line of κ: If $X \neq X^\kappa$, then XX^κ is a fixed line through X. If $X = X^\kappa$, then there exists a fixed point Y of κ with $Y \neq X$. Then XY is the desired fixed line through X. Dually we have that every line of $\mathfrak{P}$ carries a fixed point.

If there are four fixed points of κ no three of which are collinear, then the sets of fixed points and fixed lines of κ form a proper subplane $\mathfrak{Q}$ of $\mathfrak{P}$. (Remember $\kappa \neq 1$.) As every line of $\mathfrak{P}$ carries a fixed point and every point is on a fixed line, $\mathfrak{Q}$ is a Baer subplane of $\mathfrak{P}$.

We may thus assume that κ does not fix a quadrangle pointwise. Let P, Q, R be three non-collinear fixed points of κ. Let X be a point with $X \mathrel{\nmid} PQ, QR, RP$. Then $X \neq X^\kappa$. Thus XX^κ is a fixed line of κ. Assume $P, Q, R \mathrel{\nmid} XX^\kappa$. Then $PQ \cap XX^\kappa$, $QR \cap XX^\kappa$, P and R are four fixed points in general position: a contradiction. Therefore, we may assume wlog that $P \text{ I } XX^\kappa$. Let l be a line through P and assume $l \neq l^\kappa$. Then there exists a point Y on l with $Y \nmid PQ, QR, RP$. In particular, $Y \neq Y^\kappa$. The line YY^κ is fixed by κ. Furthermore, $l = PY$. As $l \neq l^\kappa$, we have $YY^\kappa \neq l = PY$. This implies $P \nmid YY^\kappa$. By what we have proved already, $Q \text{ I } YY^\kappa$ or $R \text{ I } YY^\kappa$. Furthermore, $YY^\kappa \neq QR, RP, PQ$, as $Y \nmid QR, RP, PQ$. This implies that $P, Q, R, XX^\kappa \cap YY^\kappa$ are four fixed points in general position. This contradiction shows that all lines through P are fixed by κ. Therefore, κ is a perspectivity.

It remains to consider the case that all fixed points of κ are collinear. As κ has more than one fixed point, there is a unique line l carrying all the fixed points of κ. Furthermore, $l^\kappa = l$. Thus κ induces a collineation in the affine plane $\mathfrak{P}_l$. Since every point of $\mathfrak{P}_l$ is on a fixed line and since distinct fixed lines are parallel, we infer that κ is a translation of $\mathfrak{P}_l$.

It is easily seen that c) implies a).   □

4.6 Corollary (Baer 1946a). *Let κ be an involutory collineation of a projective plane. Then κ is either a Baer collineation or a perspectivity.*

4.7 Corollary. *Let $\mathfrak{P}$ be a projective plane and $\mathfrak{Q}$ a proper subplane of $\mathfrak{P}$. If κ is a perspectivity of $\mathfrak{P}$, then κ fixes $\mathfrak{Q}$, if and only if the centre and the axis of κ belong to $\mathfrak{Q}$ and there is a point X of $\mathfrak{Q}$ which is distinct from the centre and not on the axis of κ such that X^κ also belongs to $\mathfrak{Q}$.*

The proof is an easy exercise.

4.8 Lemma (Ostrom 1956 & Lüneburg 1965c). *Let $\mathfrak{P}$ be a projective plane. Let P and Q be points and l and m lines of $\mathfrak{P}$ such that $P \nmid 1$, $P \text{ I } m$, $Q \text{ I } 1$, $Q \nmid m$. If ρ is an involution in $\Delta(P, l)$ and σ an involution in $\Delta(Q, m)$, then $\rho\sigma$ is an involution in $\Delta(l \cap m, PQ)$. Moreover, ρ is the only involution in $\Delta(P, l)$.*

PROOF. $\rho\sigma = \sigma\rho$ by 3.8. Thus $\tau = \rho\sigma$ is an involution.

Assume that X is a point with $X = X^\tau$ and $X \nmid l, m, PQ$. It follows that $X^\rho = X^\sigma$. Furthermore, $X^\rho \text{ I } PX$ and $X^\sigma \text{ I } QX$, as ρ and σ are perspectivities with centre P resp. Q. Therefore, $X^\rho = X^\sigma = PX \cap QX = X$: a contradiction. Hence τ is not a Baer involution. Consequently, by 4.6, we have that τ is a perspectivity. Since P and Q are distinct, neither of them can be the centre of τ. This implies $\tau \in \Delta(l \cap m, PQ)$.

Let $\rho' \in \Delta(P, l)$ be an involution. Then $\rho'\sigma \in \Delta(l \cap m, PQ)$. Therefore, $X^\rho = X^\sigma = X^{\rho'}$ for all $X \text{ I } PQ$. This yields $\rho = \rho'$.   □

4.9 Theorem (Burn 1968). *Let $\mathfrak{A}$ be an affine plane and let P and Q be two distinct points on l_∞. If there is to every line l of $\mathfrak{A}$ with $P, Q \nmid l$ a perspectivity σ_l with axis l switching P and Q, then $\mathfrak{A}$ is a translation plane. Moreover, if $\mathfrak{A}$ is not the affine plane of order 2, then the group generated by all the σ_l's contains the translation group of $\mathfrak{A}$.*

PROOF. Let l be a line of $\mathfrak{A}$ with $P, Q \, \mathbb{I} \, l$. Let ρ be a perspectivity with axis l interchanging P and Q. Then $\rho\sigma_l$ is a perspectivity with axis l fixing P and Q. This implies $\rho\sigma_l = 1$ and hence $\rho = \sigma_l$. Therefore σ_l is the only perspectivity with axis l interchanging P and Q. In particular, $\sigma_l^2 = 1$.

Assume that there exists a line l such that σ_l is a homology. Let C be the centre of σ_l. Then $C \, \mathbb{I} \, l_\infty$. Set $m = OC$ where O is an affine point on l. Furthermore, put $\rho = \sigma_l \sigma_m \sigma_l$. As $C \, \mathbb{I} \, m$, the perspectivity ρ has axis m. Moreover, ρ switches P and Q. Thus $\rho = \sigma_m$. This shows that σ_l and σ_m commute. Let D be the centre of σ_m. Obviously, $D = l \cap l_\infty$. We infer from 4.8 that $\sigma_l \sigma_m$ is an involutory (O, l_∞)-homology. As O is an arbitrary point on l, the plane $\mathfrak{A}$ is (D, l_∞)-transitive by 4.2, and $T(D)$ is contained in the group generated by all the σ_x's. Using σ_m for one particular m through C instead of σ_l the above argument yields that $\mathfrak{A}$ is also (C, l_∞)-transitive and $T(C)$ is contained in the group generated by all the σ_x's. This proves 4.9 in this case.

We may assume now that all the σ_l's are shears. Let $C \, \mathbb{I} \, l_\infty$ and $C \neq P, Q$. Furthermore, let l be an affine line through C, and X and Y two distinct affine points on l. Set $m = (PX \cap QY)(PY \cap QX)$ and $\sigma = \sigma_l$. Then we have

$$m^\sigma = (P^\sigma X^\sigma \cap Q^\sigma Y^\sigma)(P^\sigma Y^\sigma \cap Q^\sigma X^\sigma) = (QX \cap PY)(QY \cap PX) = m.$$

This yields $\sigma\tau = \tau\sigma$ where $\tau = \sigma_m$. As $l \neq m$, we have $C \, \mathbb{I} \, m$ and the centre of τ is on l. Thus C is the centre of τ. Put $\rho = \sigma\tau$. Since P and Q are fixed points of ρ, we obtain that $\rho \in T(C)$.

The definition of m yields $PX \cap QY \, \mathbb{I} \, m$. Furthermore, $PX = P(PX \cap PY)$ and $QY = Q(PX \cap QY)$. Therefore,

$$(PX)^\tau = (P(PX \cap QY))^\tau = P^\tau(PX \cap QY)^\tau = Q(PX \cap QY) = QY.$$

Analogously, $(QX)^\tau = PY$. Hence

$$X^\rho = X^{\sigma\tau} = X^\tau = (PX \cap QX)^\tau = QY \cap PY = Y.$$

This proves that $\mathfrak{A}$ is (C, l_∞)-transitive and that $T(C)$ is contained in the group generated by all the σ_l's. The rest is now trivial. $\qquad\square$

We are now able to finish the proof of 4.1. As $\mathfrak{A}$ satisfies a), $\mathfrak{A}$ is a translation plane by 4.9. We may thus assume that $\mathfrak{A} = \pi(V)$ has normal form such that $P = V(0) \cap l_\infty$, $Q = V(\infty) \cap l_\infty$. As $\mathfrak{A}$ is a translation plane, the point $V(0) \cap V(\infty)$ enjoys property b).

Let $\alpha, \beta \in \Sigma(\pi)$. As $1 \in \Sigma(\pi)$, we obtain by 3.10 that the mappings ρ, σ and τ defined by $(x, y)^\rho = (y, x)$, $(x, y)^\sigma = (y^\alpha, x^{\alpha^{-1}})$ and $(x, y)^\tau = (y^{\beta^{-1}}, x^\beta)$ are perspectivities of the required kind. Thus, by b), there is a $\gamma \in \Sigma(\pi)$ such that $(x, y)^{\rho\sigma\tau} = (y^{\gamma^{-1}}, x^\gamma)$. On the other hand $(x, y)^{\rho\sigma\tau} = (y^{\alpha^{-1}\beta^{-1}}, x^{\alpha\beta})$. This yields $\gamma^{-1} = \alpha^{-1}\beta^{-1}$ and $\gamma = \alpha\beta$ and hence $\alpha\beta = \beta\alpha$. Thus $\Sigma(\pi)$ is abelian. 2.8 now gives the desired result.

5. Quasifields

An important tool for the investigation of translation planes are the quasifields which we shall define now and whose connection with translation planes we shall study in this section.

Let Q be a set with two binary operations $+$ and $\circ$. We call $Q(+, \circ)$ a *quasifield*, if the following conditions are satisfied:

1) $Q(+)$ is an abelian group.
2) If $a, b, c \in Q$, then $(a + b) \circ c = a \circ c + b \circ c$.
3) $a \circ 0 = 0$ for all $a \in Q$.
4) For $a, c \in Q$ with $a \neq 0$, there exists exactly one $x \in Q$ such that $a \circ x = c$.
5) For $a, b, c \in Q$ with $a \neq b$ there exists exactly one $x \in Q$ such that $x \circ a - x \circ b = c$.
6) There exists an element $1 \in Q \backslash \{0\}$ such that $1 \circ a = a \circ 1 = a$ for all $a \in Q$.

Using 2) we obtain $0 \circ a = (0 + 0) \circ a = 0 \circ a + 0 \circ a$. This yields $0 \circ a = 0$ by 1). From this we infer $0 = (a - a) \circ b = a \circ b + (-a) \circ b$. Hence $(-a) \circ b = -a \circ b$, where we put $-a \circ b = -(a \circ b)$.

Consider the set Σ of all mappings of the form $x \to x \circ m$ with $m \neq 0$. Using 2) we see that all these mappings are endomorphisms of $Q(+)$. Furthermore, Σ operates transitively on $Q \backslash \{0\}$ by 4). Using 5) with $a = m$ and $b = 0$, we obtain that Σ consists of automorphisms only. Finally, Schur's lemma implies that $K = \mathbb{C}_{\text{End}_Z(Q(+))}(\Sigma)$ is a division ring. Therefore, $\Sigma \subseteq \text{GL}(Q(+), K)$. If $\rho, \sigma \in \Sigma$ and if $x^\rho = x \circ a$ and $x^\sigma = x \circ b$, then $x^{\rho - \sigma} = x \circ a - x \circ b$. If $\rho \neq \sigma$, then $a \neq b$ and $\rho - \sigma \in \text{GL}(Q(+), K)$ by 5). This yields by 2.3:

5.1 Theorem (M. Hall, Jr. 1943). *Let $Q(+, \circ)$ be a quasifield and let $V = Q \oplus Q$, the direct sum of the abelian group $Q(+)$ with itself. If $V(m) = \{(x, x \circ m) \,|\, x \in Q\}$ and $V(\infty) = \{(0, x) \,|\, x \in Q\}$, then $\pi = \{V(m) \,|\, m \in Q \cup \{\infty\}\}$ is a spread of V. Furthermore, $\Sigma(\pi) = \{(x \mapsto x \circ m) \,|\, m \in Q \backslash \{0\}\}$.*

The translation plane defined by a quasifield Q via 5.1 will be denoted by $\mathfrak{A}(Q)$.

Next we show that every translation plane is isomorphic to a plane $\mathfrak{A}(Q)$ where Q is a suitable quasifield. Let X be a K-vector space and π a spread of $V = X \oplus X$ with $V(0), V(1), V(\infty) \in \pi$. Put $\Sigma = \Sigma(\pi)$ and let 1 be an element of $X \backslash \{0\}$. For all $x \in X$ we put $x \circ 0 = 0$. Let $y \in X \backslash \{0\}$. Then there is exactly one $\sigma \in \Sigma$ such that $1^\sigma = y$. We put $x \circ y = x^\sigma$. We have that $X(+)$ is an abelian group, i.e. 1) is valid. Furthermore, 3) is true trivially. Obviously, 2) is true for $c = 0$. Let $c \neq 0$ and $1^\sigma = c$ with $\sigma \in \Sigma$. Then $(a + b) \circ c = (a + b)^\sigma = a^\sigma + b^\sigma = a \circ c + b \circ c$. Thus 2) is true universally. Let $a, c \in X$ and assume $a \neq 0$. If $c = 0$, then $x = 0$ is the only solution of $a \circ x = c$. Hence we may assume that $c \neq 0$. In this case there is exactly one $\sigma \in \Sigma$ such that $a^\sigma = c$. Put $x = 1^\sigma$. Then $a \circ x = c$ and x is the only solution. Therefore 4), is valid. 5) follows similarly, since $\rho - \sigma \in GL(X, K)$ by 2.2 b), provided $\rho, \sigma \in \Sigma$ and $\rho \neq \sigma$. Finally, 6) is easily checked. As $V(\sigma) = \{(x, x \circ 1^\sigma) \mid x \in X\}$, we have:

5.2 Theorem (M. Hall, Jr. 1943). *Let $\mathfrak{X}$ be a translation plane, let P and R be distinct points on l_∞ and let O be an affine point of $\mathfrak{X}$. Then there exists a quasifield Q and an isomorphism σ from $\mathfrak{X}$ onto $\mathfrak{A}(Q)$ with $(OP)^\sigma = V(0)$ and $(OR)^\sigma = V(\infty)$.*

Let Q be a set with two binary operations $+$ and $\circ$. We shall call $Q(+, \circ)$ a *weak quasifield*, if it satisfies 1), 2), 3), 4) and 6).

Let $Q(+, \circ)$ be a weak quasifield and put $\Sigma = \{(x \to x \circ a) \mid a \in Q \setminus \{0\}\}$ and $K = \mathfrak{C}_{\mathrm{End}_\mathbb{Z}(Q(+))}(\Sigma)$. As Σ operates irreducibly on $Q(+)$ by 4), we obtain that K is a division ring by Schur's lemma. We call K the *outer kernel* of $Q(+, \circ)$. By the definition of K, we see that $Q(+)$ is a vector space over K.

5.3 Theorem. *Let Q be a weak quasifield and K its outer kernel. If Q is a vector space of finite rank over K, then Q is a quasifield. In particular, if Q is a finite weak quasifield, then Q is a quasifield.*

PROOF. We have to prove 5). Let a, b be distinct elements of Q and consider the mapping α defined by $x^\alpha = x \circ a - x \circ b$. Then

$$(x + y)^\alpha = (x + y) \circ a - (x + y) \circ b$$
$$= x \circ a - x \circ b + y \circ a - y \circ b = x^\alpha + y^\alpha.$$

(Here we have used $(-a) \circ b = -(a \circ b)$ which is a consequence of 1) and 2).) Thus α is additive.

Let $\kappa \in K$. Then $x^{\kappa\alpha} = x^\kappa \circ a - x^\kappa \circ b = (x \circ a)^\kappa - (x \circ b)^\kappa = x^{\alpha\kappa}$. Hence α is linear. Furthermore, α is injective by 4). Thus α is bijective, as the K-rank of Q is finite. $\qquad\square$

Let Q be a quasifield, $\Sigma = \{(x \to x \circ m) \mid m \in Q \setminus \{0\}\}$ and K the outer kernel of Q. The kernel of $\mathfrak{A}(Q)$ is the set of all mappings $(x, y) \to (x^\kappa, y^\kappa)$ with $\kappa \in K$ by 2.7. We want to give a description of the kernel of $\mathfrak{A}(Q)$ within Q. For this purpose we define a mapping φ from K into Q by $\varphi(\kappa) = 1^\kappa$. For $\kappa, \lambda \in K$ we obtain $\varphi(\kappa) + \varphi(\lambda) = 1^\kappa + 1^\lambda = 1^{\kappa+\lambda} = \varphi(\kappa + \lambda)$. Therefore, φ is a homomorphism from $K(+)$ into $Q(+)$. The equation $1^\kappa = 0$ yields $\kappa = 0$. Thus φ is a monomorphism.

Let $\kappa \in K$ and $x, y \in Q$. If $x = 0$, then $\varphi(\kappa) \circ x = 0 = 0^\kappa = x^\kappa$. If $x = 1^\sigma$ with $\sigma \in \Sigma$, then $\varphi(\kappa) \circ x = 1^\kappa \circ x = 1^{\kappa\sigma} = 1^{\sigma\kappa} = x^\kappa$. Therefore, $\varphi(\kappa) \circ x = x^\kappa$ for all $x \in Q$. If $y = 0$, then $\varphi(\kappa) \circ (x \circ y) = (\varphi(\kappa) \circ x) \circ y$. Assume

$y = 1^\tau$ with $\tau \in \Sigma$. Then $\varphi(\kappa) \circ (x \circ y) = (x \circ y)^\kappa = x^{\tau\kappa} = x^{\kappa\tau} = x^\kappa \circ y$ $= (\varphi(\kappa) \circ x) \circ y$. Thus we have $\varphi(\kappa) \circ (x \circ y) = (\varphi(\kappa) \circ x) \circ y$ for all $x, y \in Q$. Furthermore, $\varphi(\kappa) \circ (x + y) = (x + y)^\kappa = x^\kappa + y^\kappa = \varphi(\kappa) \circ x + \varphi(\kappa) \circ y$. Finally, $\varphi(\kappa\lambda) = 1^{\kappa\lambda} = \varphi(\kappa)^\lambda = \varphi(\lambda) \circ \varphi(\kappa)$.

Let

$$k(Q) = \{c \mid c \in Q, c \circ (x + y) = c \circ x + c \circ y$$

$$\text{and } c \circ (x \circ y) = (c \circ x) \circ y \text{ for all } x, y \in Q\}.$$

For $c \in k(Q)$ we define κ by $x^\kappa = c \circ x$. It is easily seen that $\kappa \in K$ and that $\varphi(\kappa) = c$. Calling $k(Q)$ the *kernel* of Q we thus have:

5.4 Theorem. *Let Q be a quasifield, $k(Q)$ its kernel and K its outer kernel. Then the mapping φ defined by $\varphi(\kappa) = 1^\kappa$ for all $\kappa \in K$ is an antiisomorphism from K onto $k(Q)$. Furthermore, $\Delta(0) = \{((x, y) \to (c \circ x, c \circ y)) \mid c \in k(Q) \backslash \{0\}\}$ is the group of all homologies of $\mathfrak{A}(Q)$ with centre 0 and axis l_∞.*

This theorem shows that $\mathfrak{A}(Q)$ is desarguesian, if and only if $k(Q) = Q$. We define two further substructures of Q, namely

$$n_r(Q) = \{a \mid 0 \neq a \in Q, (x \circ y) \circ a = x \circ (y \circ a) \text{ for all } x, y \in Q\}$$

and

$$n_m(Q) = \{a \mid 0 \neq a \in Q, (x \circ a) \circ y = x \circ (a \circ y) \text{ for all } x, y \in Q\}.$$

$n_r(Q)$ is called the *right* and $n_m(Q)$ the *middle nucleus* of Q. The left nucleus which can be defined similarly has not yet found a geometric interpretation except of course for those cases where it coincides with the multiplicative group of the kernel of Q.

It is a trivial exercise to prove

5.5 Theorem.

a) $\Sigma_r = \{(x \to x \circ a) \mid a \in n_r(Q)\}$.
b) $\Sigma_l = \{(x \to x \circ a) \mid a \in n_m(Q)\}$.

If Q is a quasifield, then we put

$$d(Q) = \{s \mid s \in Q, x \circ (s + y) = x \circ s + x \circ y \text{ for all } x, y \in Q\}.$$

5.6 Theorem. *Let Q be a quasifield and define the mapping τ by $(x, y)^{\tau(s)} = (x, x \circ s + y)$ for all $s \in d(Q)$. Then τ is an isomorphism from $d(Q)$ onto $\Delta((\infty), V(\infty))$.*

This follows easily from 3.13.

CHAPTER II

Generalized André Planes

We now turn to the investigation of a large class of translation planes which has been fairly well studied. Planes belonging to this class occur in several interesting instances.

6. Some Number Theoretic Tools

Let Φ_n be the n-th cyclotomic polynomial. For $a, b \in \mathbb{C}$ we put $\Phi_n(a, b) = b^{\varphi(n)} \Phi_n(ab^{-1})$, where φ is the Euler φ-function. Furthermore, we define

$$L(n) = \inf\{|\Phi_n(a, b)| \,\big|\, a, b \in \mathbb{C}, |a| \geqslant |b| + 1 \geqslant 2\}.$$

In this section, p will always denote a prime number.

6.1 Lemma. *If $n \neq 1, 2, 3, 6$, then $L(n) > \prod_{p/n} p$.*

PROOF. We have $\Phi_n(a, b) = \prod_\epsilon (a - \epsilon b)$, where the product is taken over all primitive n-th roots of unity. Therefore,

$$|\Phi_n(a, b)| \geqslant (|a| - |b|)^{\varphi(n)}.$$

Furthermore,

$$|a^p| - |b^p| = (|a| - |b| + |b|)^p - |b|^p = \sum_{i=0}^{p} \binom{p}{i}(|a| - |b|)^i |b|^{p-i} - |b|^p$$

$$= \sum_{i=1}^{p} \binom{p}{i}(|a| - |b|)^i |b|^{p-i}$$

$$\geqslant (|a| - |b|)^p + p(|a| - |b|)^{p-1}|b| \geqslant 1 + p.$$

Hence

$$|\Phi_n(a^p, b^p)| \geqslant (1 + p)^{\varphi(n)}.$$

The roots of Φ_m are the primitive m-th roots of unity. Therefore, if η is a root of Φ_{np}, we have that η is a root of $\Phi_n(x^p)$. Thus Φ_{np} is a divisor of $\Phi_n(x^p)$. The degree of $\Phi_n(x^p)$ is $p\varphi(n)$. If p is a divisor of n, we have the equation $\varphi(pn) = p\varphi(n)$. In this case $\Phi_{np} = \Phi_n(x^p)$. Hence we have:

(1)   If p divides n, then $L(np) \geqslant L(n)$ and $L(np) \geqslant (1 + p)^{\varphi(n)}$.

Specializing yields

(2) $L(2^r) \geqslant 3^{2^{r-2}}$ for $r \geqslant 2$, $L(9) \geqslant 16$, $L(12) \geqslant 9$ and $L(18) \geqslant 16$.

Assume that p does not divide n. Then each primitive (pn)-th root of unity is the product of a uniquely determined primitive p-th root of unity and a uniquely determined primitive n-th root of unity. Therefore, we have in this case

$$\Phi_{np}(a, b) = \prod_{\substack{\epsilon^p = 1 \\ \epsilon \neq 1}} \Phi_n(a, \epsilon b).$$

Thus $L(np) \geqslant L(n)^{p-1}$, if p does not divide n. This yields

(3) If p does not divide n, if $p \geqslant 3$ and if $L(n) \geqslant 3$, then $L(np) \geqslant pL(n)$.

Next we prove:

(4)   If $p \geqslant 5$, then $L(2p) \geqslant L(p) > 2p$.

As 2 does not divide p, we have $\Phi_{2p}(a, b) = \Phi_p(a, -b)$. This implies $L(2p) \geqslant L(p)$.

Since we are considering only those a and b for which $|a| \geqslant |b| + 1 \geqslant 2$ holds, $a \neq b$. Thus

$$|\Phi_p(a, b)| = |a^p - b^p| |a - b|^{-1} \geqslant (|a|^p - |b|^p)(|a| + |b|)^{-1}.$$

Put $x = |a|$ and $y = |b|$. Then $x \geqslant y + 1 \geqslant 2$. Assume

(*) $(x^p - y^p)(x + y)^{-1} \leqslant ((y + 1)^p - y^p)(2y + 1)^{-1}.$

Then

$$(x^p - y^p)(2y + 1) \leqslant ((y + 1)^p - y^p)(x + y).$$

This yields

$$x^p(2y + 1) - y^p(y + 1) - y^{p+1} \leqslant (y + 1)^p(x + y) - y^p x - y^{p+1}.$$

Therefore,

$$x^p(2y + 1) \leqslant (y + 1)^p(x + y) + y^p(y + 1 - x).$$

As $x \geqslant y + 1$, we obtain

$$x^p(2y + 1) \leqslant (y + 1)^p(x + y)$$

and thus

$$(y + 1)^p y \geq x^p(2y + 1) - (y + 1)^p x.$$

It follows from $y + 1 \leq x$ that $-(y + 1)^{p-1} \geq -x^{p-1}$ and, therefore, $-(y + 1)^p x \geq -(y + 1)x^p$. This yields

$$(y + 1)^p y \geq x^p(2y + 1) - (y + 1)x^p = x^p y.$$

This implies $(y + 1)^p \geq x^p$ and hence $y + 1 \geq x$. Thus $x = y + 1$. This shows that equality holds in (*). Therefore, we have

$$(x^p - y^p)(x + y)^{-1} \geq ((y + 1)^p - y^p)(2y + 1)^{-1}$$

for all x, y such that $x \geq y + 1 \geq 2$.

As $y \geq 1$ and $p \geq 5$, we have $py \geq p$, $py^{p-1} \geq p$, $\binom{p}{2}y^{p-2} \geq 2py$ and $\binom{p}{p-2}y^2 \geq 2py$. Hence

$$(y + 1)^p - y^p = \sum_{i=0}^{p-1} \binom{p}{i} y^i > p + 2py + 2py + p = 2p(2y + 1).$$

This proves $((y + 1)^p - y^p)(2y + 1)^{-1} > 2p$ and hence (4).

We assume now that 6.1 is false. Let n be the smallest integer distinct from 1, 2, 3 and 6 such that $L(n) \leq \prod_{p/n} p$. Let q be a prime divisor of n and put $n = qm$. If q is also a divisor of m, we have $L(n) \geq L(m)$ by (1). Furthermore, $\prod_{p/m} p = \prod_{p/n} p$. Therefore, $m = 1, 2, 3$ or 6 by the minimality of n. This yields $q = 2$ or 3. If $q = 2$, then $m = 2$ or 6 and $n = 4$ or 12 which is impossible by (2). If $q = 3$, then $m = 3$ or 6 and $n = 9$ or 18 which is likewise impossible by (2). Thus q does not divide m. This proves that n is square free. From this we infer that we may assume $q \geq 5$, since otherwise $n = 1, 2, 3$ or 6. Moreover, $m \neq 1$ and 2 by (4). Thus $m \geq 3$. If $m \neq 3, 6$, then $L(m) > \prod_{p/m} p = m \geq 3$ and hence $L(n) \geq qL(m) > \prod_{p/n} p$ by (3). Therefore, $m = 3$ or 6. By (3) and (4), $L(3q) \geq 3L(q) > 3q$. Therefore, $m = 6$. Again by (3) and (4), we reach the final contradiction $L(n) = L(6q) \geq 3L(2q) > 6q$. Hence 6.1 is proved. □

6.2 Theorem (Zsigmondy 1892). *Let a and n be integers greater than 1. Then there exists a prime p which divides $a^n - 1$ but not $a^i - 1$ for any $i \in \{1, 2, \ldots, n - 1\}$ except in the cases where $n = 2$ and $a + 1$ is a power of 2 or $n = 6$ and $a = 2$.*

PROOF. If $n = 2$, then the theorem follows from $(a - 1, a + 1) \leq 2$. Thus we may assume $n \geq 3$. Let Φ_n be the n-th cyclotomic polynomial. Then $\Phi_n(a)$ divides $a^n - 1$. As a consequence $a^n \equiv 1 \bmod p$ for every prime divisor p of $\Phi_n(a)$. Let f be the order of $a \bmod p$, where p is a prime divisor of $\Phi_n(a)$. Then f divides n. Put $n = fp^i m$, where p does not divide m. Consider $r = fp^i$. As $a^r \equiv 1 \bmod p$, we have

$$(a^n - 1)(a^r - 1)^{-1} = ((a^r - 1 + 1)^m - 1)(a^r - 1)^{-1}$$

$$= \sum_{i=1}^{m} \binom{m}{i}(a^r - 1)^{i-1} \equiv m \bmod p.$$

If $m > 1$, then $\Phi_n(a)$ divides $(a^n - 1)(a^r - 1)^{-1}$. This yields that p divides m, a contradiction. Thus $n = fp^i$.

Let $t < n$ and assume that p also divides $\Phi_t(a)$. Then f divides t and hence $i \geqslant 1$. Consider the case $p \geqslant 3$ and put $s = fp^{i-1}$. Then $a^s \equiv 1 \bmod p$. Therefore,

$$(a^n - 1)(a^s - 1)^{-1} = p + \tfrac{1}{2} p(p - 1)(a^s - 1) + \sum_{i=3}^{p} \binom{p}{i}(a^s - 1)^{i-1}$$

$$\equiv p \bmod p^2.$$

This implies that $\Phi_n(a)$ is not divisible by p^2.

If $p = 2$, then $f = 1$ and $n = 2^i \geqslant 3$. In this case $i \geqslant 2$. As $\Phi_{2^i}(a) = a^{2^{i-1}} + 1$, we have that 4 does not divide $\Phi_{2^i}(a)$.

Assume now that every prime divisor of $\Phi_n(a)$ divides some $\Phi_m(a)$ with $m < n$. If p is a prime divisor of $\Phi_n(a)$, then p divides n and p^2 does not divide $\Phi_n(a)$ by what we have proved. Hence $|\Phi_n(a)| \leqslant \prod_{p/n} p$. As $\Phi_n(a) = \Phi_n(a, 1)$, we infer from 6.1 that $n = 3$ or 6. If $n = 3$, then $\Phi_3(a) = a^2 + a + 1 \geqslant 7$, a contradiction. Thus $n = 6$. Now $\Phi_6(a) = a^2 - a + 1 = a(a - 1) + 1$. Therefore, $\Phi_6(a) \leqslant 6$ yields $a = 2$. □

The integer t is called an *a-primitive divisor* of $a^n - 1$, if t divides $a^n - 1$ and if t is relatively prime to $a^i - 1$ for all $i \in \{1, 2, \ldots, n - 1\}$. Theorem 6.2 tells us that there is almost always an *a-primitive prime divisor* of $a^n - 1$.

6.3 Theorem. *Let q and n be positive integers with $q > 1$ and let p be a prime dividing $q - 1$. If p^a is the highest power of p dividing $q - 1$ and if p^b is the highest power of p dividing n, then the following is true:*

a) $q^n - 1 \equiv 0 \bmod p^{a+b}$.
b) *If $p^a \neq 2$ or $b = 0$, then* $q^n - 1 \not\equiv 0 \bmod p^{a+b+1}$.
c) *If $p^a = 2$ and $b \geqslant 1$, then* $q^n - 1 \equiv 0 \bmod 2^{b+2}$.

PROOF. Put $n = p^b m$. Then

$$q^n - 1 = (q^{p^b} - 1) \sum_{i=0}^{m-1} q^{ip^b}.$$

As $q \equiv 1 \bmod p$, we have

$$\sum_{i=0}^{m-1} q^{ip^b} \equiv m \not\equiv 0 \bmod p.$$

Therefore, we may assume that $n = p^b$.

a) The proof is by induction on b. If $b = 0$, then a) is true. Assume that

$$q^{p^b} - 1 \equiv 0 \bmod p^{a+b}.$$

Then

$$q^{p^{b+1}} - 1 = (q^{p^b} - 1) \sum_{i=0}^{p-1} q^{ip^b}.$$

Now

$$\sum_{i=0}^{p-1} q^{ip^b} \equiv p \equiv 0 \bmod p,$$

as $q \equiv 1 \bmod p$. Thus

$$q^{p^{b+1}} - 1 \equiv 0 \bmod p^{a+b+1}.$$

This proves a).

b) is true for $b = 0$. Let $b \geqslant 0$ and assume

$$q^{p^b} - 1 \not\equiv 0 \bmod p^{a+b+1}$$

and

$$q^{p^{b+1}} - 1 \equiv 0 \bmod p^{a+b+2}.$$

We infer from

$$q^{p^{b+1}} - 1 = (q^{p^b} - 1) \sum_{i=0}^{p-1} q^{ip^b}$$

that

$$\sum_{i=0}^{p-1} q^{ip^b} \equiv 0 \bmod p^2.$$

Using a) we obtain

$$q^{ip^b} \equiv 1 \bmod p^{a+b}.$$

Therefore

$$\sum_{i=0}^{p-1} q^{ip^b} \equiv p \bmod p^{a+b}.$$

This together with $\sum_{i=0}^{p-1} q^{ip^b} \equiv 0 \bmod p^2$ yields $a + b \leqslant 1$, whence it follows that $a = 1$ and $b = 0$.

From $\sum_{j=0}^{i-1} q^j \equiv i \bmod p$ and $q - 1 \equiv 0 \bmod p$ we infer

$$q^i - 1 = (q - 1)\sum_{j=0}^{i-1} q^j \equiv (q-1)i \bmod p^2.$$

Therefore

$$0 \equiv \sum_{i=0}^{p-1} q^i = p + \sum_{i=0}^{p-1}(q^i - 1) \equiv p + (q-1)\tfrac{1}{2}p(p-1) \bmod p^2.$$

This implies $p^a = p = 2$: a contradiction.

c) As $q^2 - 1 = (q-1)(q+1) \equiv 0 \bmod 8$, we have that c) is true for $b = 1$. Let $b \geqslant 1$ and assume

$$q^{2^b} - 1 \equiv 0 \bmod 2^{b+2}.$$

Then

$$q^{2^{b+1}} - 1 = (q^{2^b} - 1)(q^{2^b} + 1) \equiv 0 \bmod 2^{b+3}. \quad \square$$

6.4 Theorem. *Let q and n be positive integers with $q \geqslant 2$ and assume that every prime divisor of n divides $q - 1$. Furthermore, assume that $n \not\equiv 0$ mod 4, if $q \equiv 3$ mod 4. Then*

$$1, (q^2 - 1)(q - 1)^{-1}, (q^3 - 1)(q - 1)^{-1}, \ldots, (q^n - 1)(q - 1)^{-1}$$

is a complete residue system mod n. *In particular,*

$$(q^n - 1)(q - 1)^{-1} \equiv 0 \text{ mod } n.$$

PROOF. Let $1 \leqslant i < n$ and assume $q^i \equiv 1$ mod $(q - 1)n$. As $1 \leqslant i < n$, there exists a prime p such that p^{b+1} divides n, where p^b is the highest power of p dividing i. Let p^a be the highest power of p dividing $q - 1$. By assumption, $a \geqslant 1$. From $q^i \equiv 1$ mod $(q - 1)n$ we infer $q^i \equiv 1$ mod q^{a+b+1}. This yields $p^a = 2$ and $b \geqslant 1$ by 6.3. Thus $q \equiv 3$ mod 4 and $n \equiv 0$ mod 4, a contradiction. This shows that $q^i \not\equiv 1$ mod $(q - 1)n$, provided $1 \leqslant i < n$.

If $(q^i - 1)(q - 1)^{-1} \equiv (q^j - 1)(q - 1)^{-1}$ mod n, then $q^i \equiv q^j$ mod $(q - 1)n$. We may assume that $i \leqslant j$. Then $q^i(q^{j-i} - 1) \equiv 0$ mod $(q - 1)n$. As every prime divisor of n divides $q - 1$, we obtain that q^i and $(q - 1)n$ are relatively prime. Hence $q^{j-i} - 1 \equiv 0$ mod $(q - 1)n$. This implies $i = j$ by what we have proved already.   □

6.5 Lemma. *Let q, s, t and n be positive integers with $q = s^t \geqslant 2$ and assume that every prime divisor of n divides $q - 1$. If $1 \leqslant i \leqslant tn$ and if $(q^n - 1)n^{-1}$ divides $s^i - 1$, then $i = tn$.*

PROOF. If $n = 1$, there is nothing to prove. The case $n = 2$ and $t = 1$ is also trivial. Thus we may assume $tn \geqslant 3$. By 6.2, there exists an s-primitive prime divisor p of $q^n - 1 = s^{tn} - 1$ unless $s = 2$ and $tn = 6$. The latter case yields $(q, n) \in \{(2, 6), (4, 3), (8, 2)\}$, as $n \geqslant 2$. Since every prime divisor of n divides $q - 1$, we have $q = 4$ and $n = 3$. Hence $(q^n - 1)n^{-1} = 21$. As 21 divides $2^i - 1$ only for $i = 6$, the lemma is proved in this case. In the former case, as p is s-primitive and $n \geqslant 2$, we have that p does not divide $q - 1$ and so it also does not divide n. Hence p divides $(q^n - 1)n^{-1}$ and thus $s^i - 1$. This yields $i = tn$.   □

7. Finite Nearfield Planes

Let F be a set with two binary operations $+$ and $\circ$. We call $F(+, \circ)$ a *nearfield*, if the following conditions are satisfied:

1) $F(+)$ is an abelian group.
2) If $a, b, c \in F$, then $(a + b) \circ c = a \circ c + b \circ c$.
3) $(F \backslash \{0\})(\circ)$ is a group.
4) $a \circ 0 = 0$ for all $a \in F$.

Every nearfield is obviously a weak quasifield. Therefore, by 5.3, the finite nearfields are exactly those finite quasifields for which $(Q \setminus \{0\})(\circ)$ is a group.

Let F be finite nearfield. Then $\mathfrak{A}(F)$ is a nearfield plane by 5.1 and 5.5, and one obtains every finite nearfield plane in this manner. An analogous result is true for infinite nearfield planes. But in this case the nearfields which are admissible are those which at the same time are quasifields. A nearfield which is also a quasifield is said to be *planar*. Thus every finite nearfield is planar.

7.1 Theorem (André 1955). *Let F and F' be planar nearfields. Then $\mathfrak{A}(F)$ and $\mathfrak{A}(F')$ are isomorphic, if and only if F and F' are isomorphic.*

PROOF. $F \cong F'$ implies $\mathfrak{A}(F) \cong \mathfrak{A}(F')$. In order to prove the converse we assume $\mathfrak{A}(F) \cong \mathfrak{A}(F')$. If $\mathfrak{A}(F)$ is desarguesian, then $\mathfrak{A}(F')$ is desarguesian. It follows from 1.11, 1.15 and 5.4 that F and F' are isomorphic in this case. We may assume henceforth that $\mathfrak{A}(F)$ is non-desarguesian. It is easily seen that there is up to isomorphism only one nearfield of order 9 which is not a field (see also section 8). Therefore, we may assume $|F| > 9$. Let σ be an isomorphism from $\mathfrak{A}(F)$ onto $\mathfrak{A}(F')$. By 3.19, we have $\{P, Q\}^\sigma = \{P', Q'\}$, where $P = V(0) \cap l_\infty$, $Q = V(\infty) \cap l_\infty$, $P' = V(0') \cap l'_\infty$, $Q' = V(\infty') \cap l'_\infty$. Using 3.11 and the fact that $\mathfrak{A}(F)$ is a translation plane we see that we may assume $V(0)^\sigma = V(0')$ and $V(\infty)^\sigma = V(\infty')$. As the stabilizer of $V(0')$ and $V(\infty')$ in the collineation group of $\mathfrak{A}(F')$ operates transitively on $\{(x, y) \mid x, y \in F' \setminus \{0\}\}$, we may also assume that $(1, 1)^\sigma = (1', 1')$. We infer from $V(0)^\sigma = V(0')$ and $V(\infty)^\sigma = V(\infty')$ that there are bijections β and γ from F onto F' such that $(x, 0)^\sigma = (x^\beta, 0')$ and $(0, x)^\sigma = (0', x^\gamma)$ for all $x \in F$. As $(1, 1)^\sigma = (1', 1')$, we have $V(1)^\sigma = V(1')$. Hence $\beta = \gamma$. Furthermore, β is additive, since σ is. Let $V(m)^\sigma = V(m')$. Then $x^\beta \circ m' = (x \circ m)^\beta$ for all $x \in F$. Putting $x = 1$ we obtain $m' = m^\beta$. Therefore, $x^\beta \circ m^\beta = (x \circ m)^\beta$. This establishes that β is an isomorphism from F onto F'. $\qquad\square$

The proof of 7.1 also establishes the following result.

7.2 Theorem. *Let F be a planar nearfield and let σ be a collineation of $\mathfrak{A}(F)$ with $(0, 0)^\sigma = (0, 0)$, $(1, 0)^\sigma = (1, 0)$ and $(0, 1)^\sigma = (0, 1)$. Then there exists an automorphism β of F such that $(x, y)^\sigma = (x^\beta, y^\beta)$ for all $x, y \in F$. Conversely, if $\beta \in \operatorname{Aut}(F)$, then σ defined by $(x, y)^\sigma = (x^\beta, y^\beta)$ is a collineation of $\mathfrak{A}(F)$ fixing $(0, 0)$, $(1, 0)$ and $(0, 1)$.*

A nearfield $F(+, \circ)$ is called a *Dickson nearfield*, if there is a third binary operation $\cdot$ defined on F such that $F(+, \cdot)$ is a division ring and such that the mapping $x \to (x \circ a)a^{-1}$ is an automorphism of $F(+, \cdot)$ for all $a \in F \setminus \{0\}$. We shall determine now all finite Dickson nearfields.

Let $F(+, \circ)$ be a finite Dickson nearfield. For $x \in F$ and $m \in \mathbb{Z}$ we denote by $x^{)m}$ the m-th power of x in $F(\circ)$ and by x^m the m-th power of x in $F(\cdot)$. Furthermore, we define $\rho(a)$ by $x^{\rho(a)} = (x \circ a)a^{-1}$. The mapping ρ is a homomorphism from $F^*(\circ)$ into $\operatorname{Aut}(F(+, \cdot))$. This will be established by the following computation.

$$x^{\rho(a)\rho(b)}(a \circ b) = x^{\rho(a)\rho(b)}a^{\rho(b)}b = (x^{\rho(a)}a)^{\rho(b)}b = (x \circ a) \circ b$$
$$= x \circ (a \circ b) = x^{\rho(a \circ b)}(a \circ b).$$

Denote by Γ the image of F^* under ρ and let K be the fixed field of Γ in $F(+, \cdot)$. Let $|\Gamma| = n$ and $|K| = q$. Then $|F| = q^n$. Let U be the kernel of ρ. Then $|U| = (q^n - 1)n^{-1}$. The definition of U yields

(1) $x \circ u = xu$ for all $x \in F$ and all $u \in U$. In particular $U(\circ) = U(\cdot)$.

$F^*(\cdot)$ is cyclic. Therefore F^*/U is cyclic. Let wU be a generator of F^*/U. Then we have

(2) $$F^*(\cdot) = U \cup wU \cup \cdots \cup w^{n-1}U.$$

(1) implies that (2) is also a decomposition of $F^*(\circ)$ into left cosets modulo U.

Let $x^\gamma = x^q$. Then $\Gamma = \langle \gamma \rangle$. Therefore, there exists an integer s with $1 \leqslant s < n$ and $\rho(w^s) = \gamma$, since $F^*(\circ)/U \cong \Gamma$. This implies that $w^s, (w^s)^{)2}, \ldots, (w^s)^{)n}$ is a system of coset representatives of $F^*(\circ)/U$.

(3) $$(w^s)^{)a} = w^{s(q^a-1)(q-1)^{-1}}.$$

This is true for $a = 1$. Assume that (3) is true for some $a \geqslant 1$. Then

$$(w^s)^{)a+1} = (w^s)^{)a} \circ w^s = \left(w^{s(q^a-1)(q-1)^{-1}}\right)^q w^s = w^{s(q^{a+1}-1)(q-1)^{-1}}.$$

As a consequence, (3) is true for all a.

We infer from (3) that $s, s(q^2 - 1)(q - 1)^{-1}, \ldots, s(q^n - 1)(q - 1)^{-1}$ is a complete residue system mod n. This implies in particular $(s,n) = 1$. Therefore w^sU generates F^*/U. Hence we may assume $s = 1$. Furthermore we have

(4) $(q^n - 1)(q - 1)^{-1} \equiv 0 \bmod n$ and $(q^a - 1)(q - 1)^{-1} \not\equiv 0 \bmod n$
   for $0 < a < n$.

Next we prove:

(5)   If p is a prime divisor of n, then p divides $q - 1$.

Assume that p does not divide $q - 1$. As n divides $q^n - 1$, the prime p also divides $q^n - 1$. Therefore $(p,q) = 1$. Let p^d be the highest power of p dividing n. Then

$$q^{p^{d-1}(p-1)} = q^{\varphi(p^d)} \equiv 1 \bmod p^d.$$

Let r be the biggest prime divisor of n which does not divide $q - 1$ and let f be the order of q mod p^d. Then r does not divide $p - 1$, as $r \geqslant p$. On the other hand f divides $p^{d-1}(p - 1)$ and n, as $q^n \equiv 1 \bmod p^d$. This yields that

f divides nr^{-1}, i.e.

(a)
$$q^{nr^{-1}} - 1 \equiv 0 \bmod p^d.$$

Let x be a prime divisor of $q - 1$. Then $\sum_{i=0}^{r-1} q^{nr^{-1}i} \equiv r \bmod x$. Therefore, x does not divide $\sum_{i=0}^{r-1} q^{nr^{-1}i}$. Now

$$q^n - 1 = \left(q^{nr^{-1}} - 1\right) \sum_{i=0}^{r-1} q^{nr^{-1}i} \equiv 0 \bmod n(q - 1).$$

This establishes

(b) $q^{nr^{-1}} - 1 \equiv 0 \bmod x^e$, if x^e is the highest power of x

dividing $n(q - 1)$.

Using (a), (b) and (4) we obtain the contradiction

$$q^{nr^{-1}} - 1 \equiv 0 \bmod n(q - 1).$$

(6)   If $q \equiv 3 \bmod 4$, then $n \not\equiv 0 \bmod 4$.

Let 2^c be the highest power of 2 dividing n. Then 2^{c+1} is the highest power of 2 dividing $n(q - 1)$, as $q \equiv 3 \bmod 4$. Hence $n(q - 1)2^{-c-1}$ is odd. Using 6.3 a) we obtain that $n(q - 1)2^{-c-1}$ is a divisor of $q^{n2^{-1}} - 1$. If $c \geqslant 2$, then $2^{c-1+2} = 2^{c+1}$ divides $q^{n2^{-1}} - 1$ by 6.3 c). This together with (4) yields the contradiction

$$q^{n2^{-1}} - 1 \equiv 0 \bmod n(q - 1).$$

Hence $c \leqslant 1$.

Let q be a power of a prime and let n be a positive integer. If every prime divisor of n divides $q - 1$ and if $n \not\equiv 0 \bmod 4$ in the case of $q \equiv 3 \bmod 4$, then $\{q,n\}$ is called a *Dickson pair*. If F is a Dickson nearfield and if q and n have the above meaning, then we shall say that F is of type $\{q,n\}$. We have proved:

7.3 Theorem (Ellers & Karzel 1964). *Let F be a finite Dickson nearfield of type $\{q,n\}$. Then $\{q,n\}$ is a Dickson pair. Furthermore, there is a generator wU of $F^*(\cdot)/U$ such that*

$$F^*(\cdot) = wU \cup w^{(q^2-1)(q-1)^{-1}}U \cup \cdots \cup w^{(q^n-1)(q-1)^{-1}}U,$$

where U is the subgroup of order $(q^n - 1)n^{-1}$ of $F^*(\cdot)$, and, if $a \in w^{(q^i-1)(q-1)^{-1}}U$, then $x^{\rho(a)} = x^{q^i}$.

The next theorem gives some information about the number of Dickson nearfields of a given order.

7.4 Theorem (Lüneburg 1971). *Let p be a prime and q a power of p. If $\{q,n\}$ is a Dickson pair, then there are up to isomorphism exactly $\varphi(n)f^{-1}$ Dickson nearfields of type $\{q,n\}$, where φ is the Euler function and f is the order of p mod n.*

PROOF. Put $GF(q^n) = F(+, \cdot)$ and let U be the subgroup of order $(q^n - 1)n^{-1}$ of F^*. If wU is a generator of F^*/U, then, by 6.4,

$$F^* = wU \cup w^{(q^2-1)(q-1)^{-1}}U \cup \cdots \cup w^{(q^n-1)(q-1)^{-1}}U.$$

Define a multiplication $\circ$ in F by $a \circ 0 = 0$ and $a \circ b = a^{q^i}b$ for $b \in w^{(q^i-1)(q-1)^{-1}}U$. Trivial computations show that $F(+, \circ)$ is a Dickson nearfield of type $\{q,n\}$. Theorem 7.3 tells us that we obtain, up to isomorphism, all finite nearfields in this way. Different generators of F^*/U yield different nearfields. The question is, how many isomorphism types there are among these $\varphi(n)$ nearfields.

Let σ be the automorphism of $GF(q^n)$ defined by $x^\sigma = x^p$. Then $\mathrm{Aut}(GF(q^n)) = \langle \sigma \rangle$. Let i be chosen in such a way that $(wU)^{\sigma^i} = wU$ and $(wU)^{\sigma^j} \neq wU$ for $1 \leqslant j < i$. Thus i is the smallest positive integer such that $w^{p^i-1} \in U$. Now $w^{p^k-1} \in U$, if and only if n divides $p^k - 1$. Hence $i = f$ is the order of p mod n. In particular, i is independent of wU. Hence the set of $\varphi(n)$ nearfields constructed above splits under $\mathrm{Aut}(F)$ into $\varphi(n)f^{-1}$ orbits, each orbit consisting of f nearfields. Thus the number of isomorphism types is at most $\varphi(n)f^{-1}$. In order to show that $\varphi(n)f^{-1}$ is the exact number of isomorphism types we need the following lemma.

7.5 Lemma. *Let $\{q,n\}$ be a Dickson pair with $\{q,n\} \neq \{3,2\}$. If F is a Dickson nearfield of type $\{q,n\}$, then U is the only abelian subgroup of order $(q^n - 1)n^{-1}$ of $F^*(\circ)$.*

PROOF. Let A be an abelian subgroup of $F^*(\circ)$ with $|A| = |U|$. Furthermore, let t be a q-primitive prime divisor of $q^n - 1$. If t^s is the highest power of t dividing $q^n - 1$, then t^s divides $(q^n - 1)n^{-1}$, as $\{q,n\}$ is a Dickson pair. Therefore, if T is a Sylow t-subgroup of U, then T is a Sylow t-subgroup of $F^*(\circ)$. As U is a normal abelian subgroup of $F^*(\circ)$, we deduce that T is the only Sylow t-subgroup of $F^*(\circ)$. Hence $T \subseteq A$. Since T operates irreducibly on $F(+)$, Schur's lemma implies that the centralizer C of T in $F^*(\circ)$ is cyclic. From $A, U \subseteq C$ we thus infer $A = U$.

If there is no q-primitive prime divisor, then $\{q,n\} = \{2,6\}$ or $q + 1 = 2^r$ and $n = 2$ by 6.2. Since $\{2,6\}$ is not a Dickson pair, we have $q + 1 = 2^r$ and $n = 2$. It follows from $\{q,n\} \neq \{3,2\}$ that $r \geqslant 3$. Hence U contains a cyclic normal subgroup Z of order $2^r \geqslant 8$ of $F^*(\circ)$. The Sylow 2-subgroups of $F^*(\circ)$ are either cyclic or generalized quaternion groups. This implies that Z is the only cyclic subgroup of order 2^r in $F^*(\circ)$. As the Sylow 2-subgroups of A are cyclic, $Z \subseteq A$. Using Schur's lemma again, we obtain $A = U$. $\qquad\square$

To finish the proof of 7.4, we assume that σ is an isomorphism from $F(+, \circ)$ onto $F(+, *)$. As $\varphi(2) = 1$, we may assume that $\{q,n\} \neq \{3,2\}$.

Let $y \in U$. Then $\rho(y) = 1 = \rho^*(y)$, if ρ^* has for $F(+, *)$ the same meaning as ρ has for $F(+, \circ)$. Hence $x \circ y = xy = x * y$ for all $x \in F$. In particular, $U(\circ) = U(\cdot) = U(*)$. We infer from 7.5 that $U^\sigma = U$. This yields

$$(uv)^\sigma = (u \circ v)^\sigma = u^\sigma * v^\sigma = u^\sigma v^\sigma$$

for all $u, v \in U$. This implies that σ induces an automorphism in $GF(p)(U)$. By 6.5, we have $F = GF(p)(U)$, whence $\sigma \in \mathrm{Aut}(GF(q^n))$. □

7.6 Corollary. *Let p be a prime and q a power of p. If $\{q,n\}$ is a Dickson pair and F a nearfield of type $\{q,n\}$, then the following holds:*

a) *If $\{q,n\} \neq \{3,2\}$, then $\mathrm{Aut}(F)$ is cyclic of order nf^{-1}, where f is the order of p modulo n.*

b) *If $\{q,n\} = \{3,2\}$, then $\mathrm{Aut}(F) \cong S_3$.*

PROOF. That a) holds follows immediately from the proof of 7.4. Statement b) is an easy exercise. □

The kernel of a Dickson nearfield of type $\{q,n\}$ is isomorphic to $GF(q)$. This is easily proved, but this will also follow from the more general theorem 10.7.

Using 7.6 and 7.2 as well as the earlier development of nearfield planes, it is easy to determine the full collineation group of a nearfield plane over a finite Dickson nearfield of type $\{q,n\} \neq \{3,2\}$.

Besides the Dickson nearfields there are only seven other finite nearfields which were also found by Dickson, as was shown by Zassenhaus [1936] (see also Passman [1968]). We shall give a brief description of these nearfields.

Let p be one of the numbers 11, 29, 59. Using Dickson's list of subgroups of the $\mathrm{PSL}(2,p)$ one sees that $\mathrm{SL}(2,p)$ contains a subgroup $S \cong \mathrm{SL}(2,5)$. Let $\sigma \in S$ have eigenvalue 1. As $\det \sigma = 1$, the multiplicity of the eigenvalue 1 is 2. Since p does not divide $|S| = 120$, we have $\sigma = 1$ by Maschke's theorem. This shows that S operates regularly on $V \setminus \{0\}$, where V is the vector space of rank 2 over $GF(p)$. Now $|V \setminus \{0\}| = p^2 - 1$. Furthermore, $11^2 - 1 = 120$, $29^2 - 1 = 120 \cdot 7$ and $59^2 - 1 = 29 \cdot 120$. Let Z be the cyclic subgroup of order 1, 7 or 29 according as $p = 11$, 29 or 59 of the centre of $GL(2,p)$. Then SZ operates transitively and regularly on $V \setminus \{0\}$. In this way we obtain three of the seven exceptions.

For $p = 5$ or 11, the group $\mathrm{SL}(2,p)$ contains a subgroup $S \cong \mathrm{SL}(2,3)$. Again S operates regularly on $V \setminus \{0\}$. Let Z be the subgroup of order 1 or 5, if $p = 5$ or 11 resp., of the centre of $GL(2,p)$. As above SZ operates transitively and regularly on $V \setminus \{0\}$. Thus we obtain two more nearfields.

Finally, for $p = 7$ or 23, the group $\mathrm{SL}(2,p)$ contains subgroups T and S with $|T:S| = 2$ and $S \cong \mathrm{SL}(2,3)$. Let Z be the subgroup of order 1 or 11 according as $p = 7$ or 23 of the centre of $GL(2,p)$. Then TZ operates transitively and regularly on $V \setminus \{0\}$. This gives us the last two exceptions.

8. The Nearfield Plane of Order 9

The nearfield plane of order 9 plays a special rôle in many ways. Therefore, we shall investigate it closely in this section.

Let $F(+, \cdot)$ be the Dickson nearfield of type $\{3, 2\}$. Then $F^*(\cdot)$ is the quaternion group of order 8. As a consequence $a^2 = -1$ for all $a \in F^* \backslash \{1, -1\}$. Consider the permutation η of $F \times F$ defined by $(x, y)^\eta = (x - y, x + y)$. Then $V(0)^\eta = V(1)$, $V(1)^\eta = V(\infty)$, $V(\infty)^\eta = V(-1)$ and $V(-1)^\eta = V(0)$. Furthermore, for $a \in F^* \backslash \{1, -1\}$ we obtain

$$V(a)^\eta = \{(x - xa, x + xa) \mid x \in F\}$$
$$= \{(x - xa, x(-1)a^2 + xa) \mid x \in F\}$$
$$= \{(x - xa, (x(-1)a + x)a) \mid x \in F\}.$$

This proves $V(a)^\eta = V(a)$, as $-1 \in \mathfrak{Z}(F)$. Hence η is a collineation of $\mathfrak{A}(F)$.

Let G denote the stabilizer of $(0, 0)$ in the full collineation group of $\mathfrak{A}(F)$. Then we have by the remark made above:

8.1 Theorem (M. Hall, Jr. 1943). *G operates transitively on l_∞.*

As a consequence:

8.2 Theorem (M. Hall, Jr. 1943). *GT operates 2-transitively on the set of points of $\mathfrak{A}(F)$.*

As $\mathfrak{A}(F)$ is non-desarguesian by the remark made after theorem 5.4, we have $H = G_{(0)} = G_{(0), (\infty)}$ by 3.18 and 3.14. Hence $|G| = 10|H|$. Since H operates transitively on the set of points off the lines $V(0)$ and $V(\infty)$, we have $|G| = 10 \cdot 8^2 |H_{(1, 1)}|$. Finally, from $G_{(0), (\infty), (1, 1)} = G_{(0, 0), (1, 0), (0, 1)}$ and 7.2 as well as 7.6 b), we deduce $|H_{(1, 1)}| = 3!$. Therefore $|G| = 10 \cdot 8^2 \cdot 6 = 2^8 \cdot 3 \cdot 5$.

Let $Z = \{(0), (\infty)\}^G$. Then $|Z| = 5$ by 8.1, 3.18 and 3.14. Furthermore, G induces a permutation group $\overline{G}$ on Z which is 2-transitive as is easily seen and η induces a transposition on Z. Hence $\overline{G} \cong S_5$. Thus we have proved:

8.3 Theorem (André 1955). *G induces on Z the full symmetric group. The kernel of $^-$ has order 2^5. The order of G is $2^5 \cdot 5!$*

The next two theorems which are due to M. Hall, Jr. will be proved together.

8.4 Theorem (M. Hall, Jr. 1943). *There exists up to isomorphism only one non-desarguesian translation plane of order 9.*

8.5 Theorem (M. Hall, Jr. 1943). *There exist up to conjugacy exactly five subsets* Σ *of* GL(2,3) *with* $1 \in \Sigma$ *such that* Σ *satisfies the conditions* a) *and* b) *of 2.3.*

PROOF. First of all we show that there are at most five such Σ's. As the characteristic is 3, the elements of order 3 in GL(2,3) have fixed vectors $\neq 0$. The number of elements of order 3 in GL(2,3) is 8. Furthermore, there are $4 \cdot 3 = 12$ involutions in GL(2,3) each of which fixes a subspace of rank 1 vectorwise. Since $|GL(2,3)| = 48$, there exist thus at most $48 - 8 - 12 - 1 = 27$ elements in GL(2,3) which operate fixed point free. These elements are:

1) One element of order 2:
$$\begin{pmatrix} -1 & 0 \\ 0 & -1 \end{pmatrix}.$$

2) The elements of order 4:
$$\begin{pmatrix} 0 & 1 \\ -1 & 0 \end{pmatrix}, \begin{pmatrix} 0 & -1 \\ 1 & 0 \end{pmatrix}; \begin{pmatrix} 1 & 1 \\ 1 & -1 \end{pmatrix}, \begin{pmatrix} -1 & -1 \\ -1 & 1 \end{pmatrix}; \begin{pmatrix} 1 & -1 \\ -1 & -1 \end{pmatrix}, \begin{pmatrix} -1 & 1 \\ 1 & 1 \end{pmatrix}.$$

All these elements are conjugate.

3) The elements of order 8:
$$\begin{pmatrix} 0 & 1 \\ 1 & 1 \end{pmatrix}, \begin{pmatrix} -1 & 1 \\ 1 & 0 \end{pmatrix}; \begin{pmatrix} 0 & 1 \\ 1 & -1 \end{pmatrix}, \begin{pmatrix} 1 & 1 \\ 1 & 0 \end{pmatrix}; \begin{pmatrix} 0 & -1 \\ -1 & 1 \end{pmatrix}, \begin{pmatrix} -1 & -1 \\ -1 & 0 \end{pmatrix};$$

$$\begin{pmatrix} 0 & -1 \\ -1 & -1 \end{pmatrix}, \begin{pmatrix} 1 & -1 \\ -1 & 0 \end{pmatrix}; \begin{pmatrix} 1 & -1 \\ 1 & 1 \end{pmatrix}, \begin{pmatrix} -1 & -1 \\ 1 & -1 \end{pmatrix}; \begin{pmatrix} -1 & 1 \\ -1 & -1 \end{pmatrix},$$

$$\begin{pmatrix} 1 & 1 \\ -1 & 1 \end{pmatrix}.$$

These elements split into two conjugacy classes. One such class is:
$$\begin{pmatrix} 1 & -1 \\ 1 & 1 \end{pmatrix}, \begin{pmatrix} -1 & 1 \\ 1 & 0 \end{pmatrix}, \begin{pmatrix} 0 & 1 \\ 1 & -1 \end{pmatrix}, \begin{pmatrix} 1 & 1 \\ -1 & 1 \end{pmatrix}, \begin{pmatrix} 0 & -1 \\ -1 & -1 \end{pmatrix}, \begin{pmatrix} -1 & -1 \\ -1 & 0 \end{pmatrix}$$

4) The elements of order 6:
$$\begin{pmatrix} -1 & -1 \\ 0 & -1 \end{pmatrix}, \begin{pmatrix} -1 & 1 \\ 0 & -1 \end{pmatrix}; \begin{pmatrix} 0 & -1 \\ 1 & 1 \end{pmatrix}, \begin{pmatrix} 1 & 1 \\ -1 & 0 \end{pmatrix}; \begin{pmatrix} -1 & 0 \\ -1 & -1 \end{pmatrix},$$

$$\begin{pmatrix} -1 & 0 \\ 1 & -1 \end{pmatrix}; \begin{pmatrix} 0 & 1 \\ -1 & 1 \end{pmatrix}, \begin{pmatrix} 1 & -1 \\ 1 & 0 \end{pmatrix}.$$

All these elements are conjugate.

The pairs in this list separated by semicolons consist of matrices which are inverse to each other.

In fact, all matrices in this list operate fixed point free, but this need not be proved here, since we are only concerned with finding an upper bound for the number of Σ's.

Case 1: All elements in Σ are 2-elements.

1.1: $o(\sigma) \leqslant 4$ for all $\sigma \in \Sigma$. In this case, Σ consists of the identity matrix, the matrix $\begin{pmatrix} -1 & 0 \\ 0 & -1 \end{pmatrix}$ and all the elements of order 4.

1.2: $o(\sigma) \neq 4$ for all $\sigma \in \Sigma$. As we have two conjugacy classes of elements of order 8, we have to distinguish two cases. Assume first that $\sigma = \begin{pmatrix} 1 & -1 \\ 1 & 1 \end{pmatrix} \in \Sigma$. We investigate the elements τ with $o(\tau) = 8$ and such that $\sigma - \tau$ is in $GL(2,3)$.

$$\begin{pmatrix} 1 & -1 \\ 1 & 1 \end{pmatrix} - \begin{pmatrix} 0 & 1 \\ 1 & 1 \end{pmatrix} = \begin{pmatrix} 1 & -2 \\ 0 & 0 \end{pmatrix} \notin GL(2,3).$$

$$\begin{pmatrix} 1 & -1 \\ 1 & 1 \end{pmatrix} - \begin{pmatrix} -1 & 1 \\ 1 & 0 \end{pmatrix} = \begin{pmatrix} 2 & -2 \\ 0 & 1 \end{pmatrix} \in GL(2,3).$$

$$\begin{pmatrix} 1 & -1 \\ 1 & 1 \end{pmatrix} - \begin{pmatrix} 0 & 1 \\ 1 & -1 \end{pmatrix} = \begin{pmatrix} 1 & -2 \\ 0 & 2 \end{pmatrix} \in GL(2,3).$$

$$\begin{pmatrix} 1 & -1 \\ 1 & 1 \end{pmatrix} - \begin{pmatrix} 1 & 1 \\ 1 & 0 \end{pmatrix} = \begin{pmatrix} 0 & -2 \\ 0 & 1 \end{pmatrix} \notin GL(2,3).$$

$$\begin{pmatrix} 1 & -1 \\ 1 & 1 \end{pmatrix} - \begin{pmatrix} 0 & -1 \\ -1 & 1 \end{pmatrix} = \begin{pmatrix} 1 & 0 \\ 2 & 0 \end{pmatrix} \notin GL(2,3).$$

$$\begin{pmatrix} 1 & -1 \\ 1 & 1 \end{pmatrix} - \begin{pmatrix} -1 & -1 \\ -1 & 0 \end{pmatrix} = \begin{pmatrix} 2 & 0 \\ 2 & 1 \end{pmatrix} \in GL(2,3).$$

$$\begin{pmatrix} 1 & -1 \\ 1 & 1 \end{pmatrix} - \begin{pmatrix} 0 & -1 \\ -1 & -1 \end{pmatrix} = \begin{pmatrix} 1 & 0 \\ 2 & 2 \end{pmatrix} \in GL(2,3).$$

$$\begin{pmatrix} 1 & -1 \\ 1 & 1 \end{pmatrix} - \begin{pmatrix} 1 & -1 \\ -1 & 0 \end{pmatrix} = \begin{pmatrix} 0 & 0 \\ 2 & 1 \end{pmatrix} \notin GL(2,3).$$

We have tested so far only those matrices of order 8 which do not belong to the group generated by $\begin{pmatrix} 1 & -1 \\ 1 & 1 \end{pmatrix}$. As $\alpha - \beta \in GL(2,3)$ implies $\beta - \alpha \in GL(2,3)$: we thus have:

(A) If $\sigma, \tau \in GL(2,3)$ with $o(\sigma) = o(\tau) = 8$ and if σ and τ are not conjugate, then $\sigma - \tau \in GL(2,3)$, if and only if $\langle \sigma \rangle = \langle \tau \rangle$.

As Σ contains at least 6 elements of order 8, at least one of the elements $\begin{pmatrix} -1 & -1 \\ 1 & -1 \end{pmatrix}$, $\begin{pmatrix} -1 & 1 \\ -1 & -1 \end{pmatrix}$, $\begin{pmatrix} 1 & 1 \\ -1 & 1 \end{pmatrix}$ belongs to Σ. Furthermore, at least two of the elements $\begin{pmatrix} -1 & 1 \\ 1 & 0 \end{pmatrix}$, $\begin{pmatrix} 0 & 1 \\ 1 & -1 \end{pmatrix}$, $\begin{pmatrix} -1 & -1 \\ 1 & 0 \end{pmatrix}$, $\begin{pmatrix} 0 & -1 \\ -1 & -1 \end{pmatrix}$ also belong to Σ. As

$$\begin{pmatrix} -1 & -1 \\ 1 & -1 \end{pmatrix} - \begin{pmatrix} -1 & 1 \\ 1 & 0 \end{pmatrix} = \begin{pmatrix} 0 & -2 \\ 0 & -1 \end{pmatrix} \notin GL(2,3),$$

$$\begin{pmatrix} -1 & -1 \\ 1 & -1 \end{pmatrix} - \begin{pmatrix} 0 & 1 \\ 1 & -1 \end{pmatrix} = \begin{pmatrix} -1 & -2 \\ 0 & 0 \end{pmatrix} \notin GL(2,3),$$

$$\begin{pmatrix} -1 & -1 \\ 1 & -1 \end{pmatrix} - \begin{pmatrix} -1 & -1 \\ -1 & 0 \end{pmatrix} = \begin{pmatrix} 0 & 0 \\ 2 & -1 \end{pmatrix} \notin GL(2,3),$$

we have $\begin{pmatrix} -1 & -1 \\ 1 & -1 \end{pmatrix} \notin \Sigma$. Furthermore,

$$\begin{pmatrix} -1 & 1 \\ -1 & -1 \end{pmatrix} - \begin{pmatrix} -1 & 1 \\ 1 & 0 \end{pmatrix} = \begin{pmatrix} 0 & 0 \\ -2 & -1 \end{pmatrix} \notin GL(2,3),$$

$$\begin{pmatrix} -1 & 1 \\ -1 & -1 \end{pmatrix} - \begin{pmatrix} -1 & -1 \\ -1 & 0 \end{pmatrix} = \begin{pmatrix} 0 & 2 \\ 0 & -1 \end{pmatrix} \notin GL(2,3),$$

$$\begin{pmatrix} -1 & 1 \\ -1 & -1 \end{pmatrix} - \begin{pmatrix} 0 & -1 \\ -1 & -1 \end{pmatrix} = \begin{pmatrix} -1 & 2 \\ 0 & 0 \end{pmatrix} \notin GL(2,3)$$

Hence $\left(\begin{smallmatrix} -1 & 1 \\ -1 & -1 \end{smallmatrix}\right) \notin \Sigma$. Therefore, Σ consists of $\left(\begin{smallmatrix} 1 & 0 \\ 0 & 1 \end{smallmatrix}\right)$, $\left(\begin{smallmatrix} -1 & 0 \\ 0 & -1 \end{smallmatrix}\right)$ and all the conjugates of $\left(\begin{smallmatrix} 1 & -1 \\ 1 & 1 \end{smallmatrix}\right)$. The only other possibility is that Σ consists of the matrices $\left(\begin{smallmatrix} 1 & 0 \\ 0 & 1 \end{smallmatrix}\right)$, $\left(\begin{smallmatrix} -1 & 0 \\ 0 & -1 \end{smallmatrix}\right)$ and all the conjugates of $\left(\begin{smallmatrix} -1 & -1 \\ 1 & -1 \end{smallmatrix}\right)$.

1.3: Σ contains elements of order 4 and 8. We may assume without loss of generality that $\left(\begin{smallmatrix} 0 & 1 \\ -1 & 0 \end{smallmatrix}\right) \in \Sigma$. Testing with elements of order 8 yields:

$$\begin{pmatrix} 0 & 1 \\ -1 & 0 \end{pmatrix} - \begin{pmatrix} 0 & 1 \\ 1 & 1 \end{pmatrix} = \begin{pmatrix} 0 & 0 \\ -2 & -1 \end{pmatrix}$$

$$\begin{pmatrix} 0 & 1 \\ -1 & 0 \end{pmatrix} - \begin{pmatrix} -1 & 1 \\ 1 & 0 \end{pmatrix} = \begin{pmatrix} 1 & 0 \\ -2 & 0 \end{pmatrix}$$

$$\begin{pmatrix} 0 & 1 \\ -1 & 0 \end{pmatrix} - \begin{pmatrix} 0 & 1 \\ 1 & -1 \end{pmatrix} = \begin{pmatrix} 0 & 0 \\ -2 & 1 \end{pmatrix}$$

$$\begin{pmatrix} 0 & 1 \\ -1 & 0 \end{pmatrix} - \begin{pmatrix} 1 & 1 \\ 1 & 0 \end{pmatrix} = \begin{pmatrix} -1 & 0 \\ -2 & 0 \end{pmatrix}$$

$$\begin{pmatrix} 0 & 1 \\ -1 & 0 \end{pmatrix} - \begin{pmatrix} 0 & -1 \\ -1 & 1 \end{pmatrix} = \begin{pmatrix} 0 & 2 \\ 0 & -1 \end{pmatrix}$$

$$\begin{pmatrix} 0 & 1 \\ -1 & 0 \end{pmatrix} - \begin{pmatrix} -1 & -1 \\ -1 & 0 \end{pmatrix} = \begin{pmatrix} 1 & 2 \\ 0 & 0 \end{pmatrix}$$

$$\begin{pmatrix} 0 & 1 \\ -1 & 0 \end{pmatrix} - \begin{pmatrix} 0 & -1 \\ -1 & -1 \end{pmatrix} = \begin{pmatrix} 0 & 2 \\ 0 & 1 \end{pmatrix}$$

$$\begin{pmatrix} 0 & 1 \\ -1 & 0 \end{pmatrix} - \begin{pmatrix} 1 & -1 \\ -1 & 0 \end{pmatrix} = \begin{pmatrix} -1 & 2 \\ 0 & 0 \end{pmatrix}.$$

This proves

(B) If $\sigma, \tau \in GL(2,3)$ with $o(\sigma) = 4$ and $o(\tau) = 8$, then $\sigma - \tau \in GL(2,3)$, if and only if $\sigma \in \langle \tau \rangle$.

Thus we have $\Sigma = \langle \left(\begin{smallmatrix} 1 & -1 \\ 1 & 1 \end{smallmatrix}\right) \rangle$ in this case.

Case 2: Σ contains elements of order 6. Let τ be an element of order 6. Then τ^2 is of order 3. Therefore, $\tau^2 - 1 \notin GL(2,3)$, i.e. $(\tau - \tau^{-1})\tau \notin GL(2,3)$. Thus $\tau \in \Sigma$ implies $\tau^{-1} \notin \Sigma$. Hence Σ contains at most 4 elements of order 6. Furthermore, $-\tau$ has order 3 and hence $-(\tau + 1) = -\tau - 1 \notin GL(2,3)$. As a consequence $\left(\begin{smallmatrix} -1 & 0 \\ 0 & -1 \end{smallmatrix}\right) \notin \Sigma$. As all elements of order 6 are conjugate, we may assume $\left(\begin{smallmatrix} -1 & -1 \\ 0 & -1 \end{smallmatrix}\right) \in \Sigma$. Testing the elements of order 8, we find:

(C) If σ is an element of order 8 in Σ, then σ is one of the matrices:

$$\begin{pmatrix} -1 & 1 \\ 1 & 0 \end{pmatrix}, \begin{pmatrix} 0 & 1 \\ 1 & -1 \end{pmatrix}, \begin{pmatrix} 1 & -1 \\ -1 & 0 \end{pmatrix}, \begin{pmatrix} 1 & -1 \\ 1 & 1 \end{pmatrix}, \begin{pmatrix} 0 & -1 \\ -1 & 1 \end{pmatrix},$$

$$\begin{pmatrix} -1 & 1 \\ -1 & -1 \end{pmatrix}.$$

The first four of these are in one conjugacy class, the last two belong to the other conjugacy class.

Assume that Σ does not contain an element of order 4. Then Σ contains at least 3 elements of order 8, as Σ contains at most 4 elements of order 6 and $\left(\begin{smallmatrix} -1 & 0 \\ 0 & -1 \end{smallmatrix}\right) \notin \Sigma$. Assume $\left(\begin{smallmatrix} -1 & 1 \\ -1 & -1 \end{smallmatrix}\right) \in \Sigma$. We infer from (C) and (A) that $\left(\begin{smallmatrix} 1 & -1 \\ 1 & 1 \end{smallmatrix}\right)$, $\left(\begin{smallmatrix} 0 & -1 \\ -1 & 1 \end{smallmatrix}\right) \in \Sigma$. This is impossible by (A), since $\left(\begin{smallmatrix} 1 & -1 \\ 1 & 1 \end{smallmatrix}\right)$ and

$(\begin{smallmatrix} 0 & -1 \\ -1 & 1 \end{smallmatrix})$ are in different conjugacy classes. Hence $(\begin{smallmatrix} -1 & 1 \\ -1 & -1 \end{smallmatrix}) \notin \Sigma$. Similarly $(\begin{smallmatrix} 0 & -1 \\ -1 & 1 \end{smallmatrix}) \notin \Sigma$. Therefore, only

(D)
$$\begin{pmatrix} -1 & 1 \\ 1 & 0 \end{pmatrix}, \begin{pmatrix} 0 & 1 \\ 1 & -1 \end{pmatrix}, \begin{pmatrix} 1 & -1 \\ -1 & 0 \end{pmatrix}, \begin{pmatrix} 1 & -1 \\ 1 & 1 \end{pmatrix}$$

can occur, if we assume that Σ contains no element of order 4.

The elements of order 6 which can occur in Σ together with $(\begin{smallmatrix} -1 & -1 \\ 0 & -1 \end{smallmatrix})$ are $(\begin{smallmatrix} 1 & 1 \\ -1 & 0 \end{smallmatrix})$, $(\begin{smallmatrix} -1 & 0 \\ 1 & -1 \end{smallmatrix})$, $(\begin{smallmatrix} 0 & 1 \\ -1 & 1 \end{smallmatrix})$. At least 3 of the elements (D) occur in Σ, at least 2 of the elements $(\begin{smallmatrix} -1 & 1 \\ 1 & 0 \end{smallmatrix})$, $(\begin{smallmatrix} 0 & 1 \\ 1 & -1 \end{smallmatrix})$, $(\begin{smallmatrix} 1 & -1 \\ -1 & 0 \end{smallmatrix})$ occur. Now

$$\begin{pmatrix} -1 & 1 \\ 1 & 0 \end{pmatrix} - \begin{pmatrix} 1 & 1 \\ -1 & 0 \end{pmatrix} = \begin{pmatrix} -2 & 0 \\ 2 & 0 \end{pmatrix} \quad \notin GL(2,3)$$

$$\begin{pmatrix} -1 & 1 \\ 1 & 0 \end{pmatrix} - \begin{pmatrix} -1 & 0 \\ 1 & -1 \end{pmatrix} = \begin{pmatrix} 0 & 1 \\ 0 & 1 \end{pmatrix} \quad \notin GL(2,3)$$

$$\begin{pmatrix} -1 & 1 \\ 1 & 0 \end{pmatrix} - \begin{pmatrix} 0 & 1 \\ -1 & 1 \end{pmatrix} = \begin{pmatrix} -1 & 0 \\ 2 & -1 \end{pmatrix} \in GL(2,3)$$

$$\begin{pmatrix} 0 & 1 \\ 1 & -1 \end{pmatrix} - \begin{pmatrix} 1 & 1 \\ -1 & 0 \end{pmatrix} = \begin{pmatrix} -1 & 0 \\ 2 & -1 \end{pmatrix} \in GL(2,3)$$

$$\begin{pmatrix} 0 & 1 \\ 1 & -1 \end{pmatrix} - \begin{pmatrix} -1 & 0 \\ 1 & -1 \end{pmatrix} = \begin{pmatrix} 1 & 1 \\ 0 & 0 \end{pmatrix} \quad \notin GL(2,3)$$

$$\begin{pmatrix} 0 & 1 \\ 1 & -1 \end{pmatrix} - \begin{pmatrix} 0 & 1 \\ -1 & 1 \end{pmatrix} = \begin{pmatrix} 0 & 0 \\ 2 & -2 \end{pmatrix} \quad \notin GL(2,3)$$

$$\begin{pmatrix} 1 & -1 \\ -1 & 0 \end{pmatrix} - \begin{pmatrix} 1 & 1 \\ -1 & 0 \end{pmatrix} = \begin{pmatrix} 0 & -2 \\ 0 & 0 \end{pmatrix} \quad \notin GL(2,3)$$

$$\begin{pmatrix} 1 & -1 \\ -1 & 0 \end{pmatrix} - \begin{pmatrix} -1 & 0 \\ 1 & -1 \end{pmatrix} = \begin{pmatrix} 2 & -1 \\ -2 & 1 \end{pmatrix} \notin GL(2,3)$$

$$\begin{pmatrix} 1 & -1 \\ -1 & 0 \end{pmatrix} - \begin{pmatrix} 0 & 1 \\ -1 & 1 \end{pmatrix} = \begin{pmatrix} 1 & -2 \\ 0 & -1 \end{pmatrix} \in GL(2,3).$$

This shows that at most 2 elements of order 6 occur in Σ and hence at least 5 elements of order 8, which is impossible. Thus Σ contains an element of order 4.

Replacing Σ by a conjugate if necessary we may assume $(\begin{smallmatrix} 0 & 1 \\ -1 & 0 \end{smallmatrix}) \in \Sigma$. Playing off this matrix against the elements of order 6 yields that we have as possible candidates for elements of order 6 in Σ the matrices $(\begin{smallmatrix} -1 & 1 \\ 0 & -1 \end{smallmatrix})$, $(\begin{smallmatrix} 0 & -1 \\ 1 & 1 \end{smallmatrix})$, $(\begin{smallmatrix} -1 & 0 \\ -1 & -1 \end{smallmatrix})$, $(\begin{smallmatrix} 1 & -1 \\ 1 & 0 \end{smallmatrix})$. Testing these with $(\begin{smallmatrix} 0 & -1 \\ 1 & 0 \end{smallmatrix})$ shows that $(\begin{smallmatrix} 0 & -1 \\ 1 & 0 \end{smallmatrix}) \notin \Sigma$. This proves: If $\sigma \in \Sigma$ with $o(\sigma) = 4$, then $\sigma^{-1} \notin \Sigma$. Therefore, Σ contains at most 3 elements of order 4. Furthermore, using (B), if Σ contains two elements of order 4, then Σ does not contain an element of order 8.

Assume that Σ does contain two elements of order 4. Then Σ contains exactly 4 elements of order 6 and 3 elements of order 4. Testing yields

$$\begin{pmatrix} 1 & 1 \\ 1 & -1 \end{pmatrix} - \begin{pmatrix} -1 & 1 \\ 0 & -1 \end{pmatrix} = \begin{pmatrix} 2 & 0 \\ 1 & 0 \end{pmatrix} \quad \notin GL(2,3)$$

$$\begin{pmatrix} -1 & -1 \\ -1 & 1 \end{pmatrix} - \begin{pmatrix} 0 & -1 \\ 1 & 1 \end{pmatrix} = \begin{pmatrix} -1 & 0 \\ -2 & 0 \end{pmatrix} \notin GL(2,3).$$

Hence $\begin{pmatrix} 1 & -1 \\ 1 & -1 \end{pmatrix}$, $\begin{pmatrix} -1 & -1 \\ -1 & 1 \end{pmatrix} \notin \Sigma$. Therefore, $\begin{pmatrix} -1 & -1 \\ -1 & 1 \end{pmatrix}, \begin{pmatrix} -1 & 1 \\ 1 & 1 \end{pmatrix} \in \Sigma$, which is impossible, as these two elements are inverse to each other. Therefore, Σ contains exactly one element of order 4 and thus at least two elements of order 8. Using **(B)** we find that the only candidates are $\begin{pmatrix} 1 & -1 \\ 1 & 1 \end{pmatrix}$, $\begin{pmatrix} -1 & -1 \\ 1 & -1 \end{pmatrix}$, $\begin{pmatrix} -1 & -1 \\ -1 & 1 \end{pmatrix}$, $\begin{pmatrix} -1 & 1 \\ -1 & 1 \end{pmatrix}$. Testing the first and third of these with the candidates of order 6 one sees that they do not belong to Σ. Hence Σ consists of the matrices

$$\begin{pmatrix} 1 & 0 \\ 0 & 1 \end{pmatrix}, \begin{pmatrix} 0 & 1 \\ -1 & 0 \end{pmatrix}, \begin{pmatrix} -1 & -1 \\ 1 & -1 \end{pmatrix}, \begin{pmatrix} 1 & 1 \\ -1 & 1 \end{pmatrix}, \begin{pmatrix} -1 & 1 \\ 0 & -1 \end{pmatrix}, \begin{pmatrix} 0 & -1 \\ 1 & 1 \end{pmatrix},$$
$$\begin{pmatrix} -1 & 0 \\ -1 & -1 \end{pmatrix}, \begin{pmatrix} 1 & -1 \\ 1 & 0 \end{pmatrix}.$$

This proves that there are up to conjugacy at most five possibilities for Σ.

Next we prove that all five possibilities really occur. Σ is the cyclic group of order 8, if and only if the plane under consideration is desarguesian. Therefore, it suffices to show that the four other possibilities are realized in the plane $\mathfrak{A}(F)$ over the nearfield F of type $\{3, 2\}$. In order to establish this let $\mathfrak{X}$ be the set of triples (P, Q, R) with $P, Q, R \text{ I } l_\infty$ and $P \neq Q \neq R \neq P$. If $P \text{ I } l_\infty$, then there is a unique point $P' \neq P$ on l_∞ such that $\mathfrak{A}(F)$ is (P, OP')- as well as (P', OP)-transitive. Let $\mathfrak{X}_1$ be the set of triples $(P, Q, R) \in \mathfrak{X}$ such that $\{P, Q, R\} \cap \{P', Q', R'\} = \emptyset$. Then $\mathfrak{X}_1$ is invariant under the collineation group G of $\mathfrak{A}(F)$. Let $\mathfrak{X}_2$ be the set of $(P, Q, R) \in \mathfrak{X}$ such that $Q = P'$ and define $\mathfrak{X}_3$ and $\mathfrak{X}_4$ similarly. Then $\mathfrak{X}_2$, $\mathfrak{X}_3$, $\mathfrak{X}_4$ are also invariant under G. Therefore, G splits $\mathfrak{X}$ into at least 4 orbits. According to 2.6 and what we have proved so far, G splits $\mathfrak{X}$ into at most 4 orbits. Hence G splits $\mathfrak{X}$ into exactly 4 orbits thus proving 8.4 and 8.5.   □

9. Generalized André Planes

In this and the following sections, we shall study more closely the phenomenon that occurs in connection with the Dickson nearfield, namely that a quasifield is intimately connected with a division ring.

Let $Q(+, \circ)$ be a quasifield. $Q(+, \circ)$ will be called a *generalized André system*, if Q admits a binary operation $\cdot$ such that $Q(+, \cdot)$ is a division ring and such that the mapping $x \to (x \circ a)a^{-1}$ is an automorphism of $Q(+, \cdot)$ for all $a \in Q^*$. The translation plane $\mathfrak{A}$ will be called a *generalized André plane*, if $\mathfrak{A} \cong \mathfrak{A}(Q)$ for some generalized André system Q. According to this definition, every desarguesian plane is a generalized André plane. Furthermore, the planes over Dickson nearfields are also generalized André planes.

In this section we shall give a sufficient condition for a plane to be a generalized André plane.

9.1 Lemma. *Let V be a vector space over the field K and let A be an abelian subgroup of $\mathrm{GL}(V,K)$. If $V = \bigoplus_{i \in I} V_i$, where the V_i are isomorphic irreducible $K[A]$-modules, then $L = K[A]$ is a field and V is an L-vector space. Furthermore, $\mathfrak{C}_{\mathrm{GL}(V,K)}(A) = \mathrm{GL}(V,L)$ and $\mathfrak{N}_{\mathrm{GL}(V,K)}(A) \subseteq \Gamma\mathrm{L}(V,L:K)$, where $\Gamma\mathrm{L}(V,L:K)$ denotes the group of all semilinear automorphisms of the L-vector space V having automorphisms of L which fix K elementwise as companion automorphisms. If L is finite, then $\mathfrak{N}_{\mathrm{GL}(V,K)}(A) = \Gamma\mathrm{L}(V,L:K)$. Finally, $\mathrm{rk}_K(V) = \mathrm{rk}_L(V)[L:K]$, where $[L:K]$ denotes the rank of L over K.*

PROOF. As A is an abelian subgroup of $\mathrm{GL}(V,K)$, the ring $L = K[A]$ is commutative. Let $0 \neq v \in V_i$ and let N be the annihilator of v in L. Then N is a maximal right ideal of L, since V_i is irreducible. Furthermore N annihilates all of V_i. Let $z \in V_j$. Then there exists an L-isomorphism σ from V_j onto V_i. Therefore, for all $x \in N$,

$$zx = zx\sigma^{-1}\sigma = (z\sigma^{-1})x\sigma = 0\sigma = 0.$$

Hence $x = 0$. Thus $\{0\}$ is a maximal right ideal. As L is a commutative ring with 1, this implies that L is a field.

Because every K-linear mapping which centralizes A also centralizes $K[A] = L$, we have $\mathfrak{C}_{\mathrm{GL}(V,K)}(A) \subseteq \mathrm{GL}(V,L)$. On the other hand $\mathrm{GL}(V,L) \subseteq \mathrm{GL}(V,K)$, as $K \subseteq L$. Furthermore, $A \subseteq L^* = \mathfrak{Z}(\mathrm{GL}(V,L))$. Therefore $\mathfrak{C}_{\mathrm{GL}(V,K)}(A) = \mathrm{GL}(V,L)$.

Let $\alpha_i \in A$, $k_i \in K$ and $\nu \in \mathfrak{N}_{\mathrm{GL}(V,K)}(A)$. Then $\nu^{-1}\sum_{i=1}^n k_i\alpha_i\nu = \sum_{i=1}^n k_i\nu^{-1}\alpha_i\nu$. Therefore, $\nu^{-1}\lambda\nu \in L$ for all $\lambda \in L$. Hence $\lambda \to \nu^{-1}\lambda\nu$ is an automorphism of L which fixes K pointwise. This establishes $\nu \in \Gamma\mathrm{L}(V, L:K)$, i.e. $\mathfrak{N}_{\mathrm{GL}(V,K)}(A) \subseteq \Gamma\mathrm{L}(V,L:K)$.

If L is finite, A is characteristic in L^*, as L^* is cyclic. This yields $\mathfrak{N}_{\mathrm{GL}(V,K)}(A) = \Gamma\mathrm{L}(V,L:K)$.

The statement about the ranks is trivial. $\square$

9.2 Theorem (Lüneburg 1976a). *Let X be a vector space over the field K and let π be a spread of $V = X \oplus X$ with $K \subseteq \mathrm{K}(V,\pi)$ and $V(0), V(1), V(\infty) \in \pi$. Furthermore, let $X(+, \circ)$ be the quasifield constructed with the aid of $\Sigma(\pi)$ as in section 5. If A is an abelian collineation group of $\pi(V)$ with $A \subseteq \mathrm{GL}(V,K)$ and $V(0)^A = V(0)$, $V(\infty)^A = V(\infty)$ and if for all $\sigma \in \Sigma(\pi)$, the subspace $V(\sigma)$ is an irreducible $K[A_\sigma]$-module, where A_σ denotes the stabilizer of $V(\sigma)$ in A, then $X(+, \circ)$ is a generalized André system and $\pi(V)$ is a generalized André plane. The division ring $X(+, \cdot)$ belonging to $X(+, \circ)$ is a field.*

Before we prove 9.2 we state and prove the following:

9.3 Corollary. *Let $\mathfrak{A}$ be a translation plane and let A be an abelian collineation group of $\mathfrak{A}$ which fixes two points P and Q on l_∞ and an affine point O. If A has the property that the stabilizer A_W of any point W I l_∞*

which is distinct from P and Q induces on $\mathrm{T}(W)$ an irreducible group of automorphisms, then $\mathfrak{A}$ is a generalized André plane.

PROOF. Using 1.5, 1.10 and 2.1 one obtains the situation of 9.2 by taking for K the prime field of the kernel of $\mathfrak{A}$. $\qquad\square$

We shall now state a sequence of lemmas which will finally prove 9.2. We shall use the notation $V(m)$ instead of $V(\sigma)$ where $m = 1^\sigma \in X$ and A_m instead of A_σ.

9.4 Lemma. *Let* $m \in X\setminus\{0\}$. *Then* $K[A_m]$ *operates irreducibly on* $V(0)$ *and* $V(\infty)$.

PROOF. Let $u, x \in X$ with $u \neq 0$. Then $(u, u \circ m), (x, x \circ m) \in V(m)$. As $K[A_m]$ operates irreducibly on $V(m)$, there exist $\alpha_1, \ldots, \alpha_n \in A_m$ and $k_1, \ldots, k_n \in K$ such that

$$(x, x \circ m) = \sum_{i=1}^n (u, u \circ m)^{\alpha_i} k_i .$$

Therefore

$$(x, 0) + (0, x \circ m) = \sum_{i=1}^n (u, 0)^{\alpha_i} k_i + \sum_{i=1}^n (0, u \circ m)^{\alpha_i} k_i .$$

As A fixes $V(0)$ and $V(\infty)$, we obtain $(x, 0) = \sum_{i=1}^n (u,0)^{\alpha_i} k_i$. Thus $V(0)$ is generated as a $K[A_m]$-module by $(u, 0)$. Since this is true for all $u \in X\setminus\{0\}$, the $K[A_m]$-module $V(0)$ is irreducible. That $V(\infty)$ is an irreducible $K[A_m]$-module is proved similarly. $\qquad\square$

9.5 Lemma. $V(0)$ *and* $V(\infty)$ *are isomorphic* $K[A_m]$-*modules.*

PROOF. Define the mapping μ from $V(0)$ onto $V(\infty)$ by $(x,0)^\mu = (0, x \circ m)$. Then μ is K-linear. Furthermore, $(x,0) + (x,0)^\mu \in V(m)$. Let $\alpha \in A_m$. We infer from $(x,0) + (x,0)^\mu \in V(m)$ that $(x,0)^\alpha + (x,0)^{\mu\alpha} \in V(m)$. On the other hand, $(x,0)^\alpha \in V(0)$. Hence $(x,0)^\alpha + (x,0)^{\alpha\mu} \in V(m)$. Therefore $(x,0)^{\mu\alpha} - (x,0)^{\alpha\mu} \in V(m) \cap V(\infty)$. Thus $(x,0)^{\mu\alpha} = (x,0)^{\alpha\mu}$. $\qquad\square$

9.6 Lemma. $K_m = K[A_m]$ *is a field and* V *is a vector space of rank* 2 *over* K_m. *Furthermore,* $\mathfrak{C}_{\mathrm{GL}(V,K)}(A_m) = \mathrm{GL}(V, K_m) \cong \mathrm{GL}(2, K_m)$.

PROOF. This follows from 9.1, 9.4 and 9.5. $\qquad\square$

For $\chi \in K_1$ we define $\varphi(\chi)$ by $(\varphi(\chi), 0) = (1,0)^\chi$. Then φ is a bijection of K_1 onto X. As

$$(\varphi(\chi + \lambda), 0) = (1,0)^{\chi+\lambda} = (1,0)^\chi + (1,0)^\lambda = (\varphi(\chi) + \varphi(\lambda), 0),$$

the mapping φ is additive.

Define a multiplication in X by $xy = \varphi(\varphi^{-1}(x)\varphi^{-1}(y))$. Then φ is an isomorphism from K_1 onto $X(+, \cdot)$.

Furthermore

$$(x,0)^\chi = (1,0)^{\varphi^{-1}(x)\varphi^{-1}\varphi(x)} = (1,0)^{\varphi^{-1}(x\varphi(\chi))} = (x\varphi(\chi),0).$$

Define ψ by $\psi(x,0) = (0,x)$. Then ψ is an K_1-isomorphism from $V(0)$ onto $V(\infty)$. Hence

$$(0,x)^\chi = (\psi(x,0))^\chi = \psi((x,0)^\chi) = \psi(x\varphi(\chi),0) = (0,x\varphi(\chi)).$$

Thus $(x,y)^\chi = (x\varphi(\chi), y\varphi(\chi))$.

9.7 Lemma. *A is a subgroup of the group B of all mappings $(x,y) \to (xa, yb)$ with $a,b \in X \setminus \{0\}$.*

PROOF. $A \subseteq \mathbb{C}_{GL(V,K)}(A_1) = GL(V,K_1)$ by 9.6. Furthermore, $V(0)^A = V(0)$ and $V(\infty)^A = V(\infty)$. This yields that to every $\alpha \in A$ there are elements $\chi, \lambda \in K_1$ such that $(x,0)^\alpha = (x,0)^\chi$ and $(0,y)^\alpha = (0,y)^\lambda$ for all $x,y \in X$. Hence $(x,y)^\alpha = (x\varphi(\chi), y\varphi(\lambda))$.

9.8 Lemma. $K_m \setminus \{0\} \subseteq B$ *for all $m \in X \setminus \{0\}$.*

This follows from 9.7 and $A_m \subseteq A$.

9.9 Lemma. *For $m \in X \setminus \{0\}$ define $\alpha(m)$ by $x^{\alpha(m)} = (x \circ m)m^{-1}$. Then α is a mapping from $X \setminus \{0\}$ into $\mathrm{Aut}(X(+, \cdot))$. Moreover, K_m consists of all the mappings $(x,y) \to (xa, ya^{\alpha(m)})$ with $a \in X \setminus \{0\}$. Finally, F_m is isomorphic to $X(+, \cdot)$.*

PROOF. As $m \neq 0$, the mapping $\alpha(m)$ is a permutation of X. Obviously $(1,m) \in V(m)$. Let $a \in Q$. Then $(a, a \circ m) \in V(m)$. Since $V(m)$ is a K_m-subspace of rank 1, there exists $\chi \in K_m$ with $(a, a \circ m) = (1,m)^\chi$. By 9.8, there exist $u, v \in X$ with $(x,y)^\chi = (xu, yv)$ for all $x,y \in X$. Thus $(a, a \circ m) = (u, mv)$ and hence $a = u$ and $v = a^{\alpha(m)}$ (remember that $X(+, \cdot)$ is commutative) proving that the mapping $(x,y) \to (xa, ya^{\alpha(m)})$ belongs to K_m. Conversely, if $\chi \in K_m$, then $(1,m)^\chi = (a, a \circ m)$ for some $a \in X$ and hence $(x,y)^\chi = (xa, ya^{\alpha(m)})$. This proves that K_m consists of all the mappings $(x,y) \to (xa, ya^{\alpha(m)})$.

Now

$$(x+y)^{\alpha(m)} = ((x+y) \circ m)m^{-1} = (x \circ m + y \circ m)m^{-1}$$

$$= (x \circ m)m^{-1} + (y \circ m)m^{-1} = x^{\alpha(m)} + y^{\alpha(m)}.$$

Moreover, if $(x,y)^\chi = (xa, ya^{\alpha(m)})$, $(x,y)^\lambda = (xb, yb^{\alpha(m)})$ and $(x,y)^{\chi\lambda} = (xc, yc^{\alpha(m)})$, then

$$(xab, ya^{\alpha(m)}b^{\alpha(m)}) = (x,y)^{\chi\lambda} = (xc, yc^{\alpha(m)})$$

which yields $c = ab$ and $a^{\alpha(m)}b^{\alpha(m)} = (ab)^{\alpha(m)}$. Hence $\alpha(m) \in \mathrm{Aut}(X(+, \cdot))$.

Finally, the mapping $(a, a^{\alpha(m)}) \to a$ provides us with an isomorphism from K_m onto $X(+, \cdot)$. $\square$

PROOF OF 9.2. 9.9 shows that $X(+, \circ)$ is a generalized André system. Furthermore, $X(+, \cdot)$ is a field. Thus 9.2 is proved. $\qquad\square$

The next lemma is essentially André's.

9.10 Lemma. *Let $Q(+, \circ)$ be a generalized André system and assume that the division ring $Q(+, \cdot)$ belonging to it is a field. If $a, b \in Q \setminus \{0\}$ and if $c \circ (a + b) = c \circ a + c \circ b$ for all $c \in Q$, then $\alpha(a) = \alpha(b)$.*

PROOF. If $a + b = 0$, then

$$0 = c \circ (a + b) = c \circ a + c \circ b = c^{\alpha(a)}a + c^{\alpha(b)}b = (c^{\alpha(a)} - c^{\alpha(b)})a$$

for all $c \in A$. Thus $\alpha(a) = \alpha(b)$, as $a \neq 0$.

If $a + b \neq 0$, then

$$c^{\alpha(a+b)}(a + b) = c \circ (a + b) = c \circ a + c \circ b = c^{\alpha(a)}a + c^{\alpha(b)}b.$$

Replace c by cd. Then

$$(cd)^{\alpha(a+b)}(a + b)^2 = (c^{\alpha(a)}d^{\alpha(a)}a + c^{\alpha(b)}d^{\alpha(b)}b)(a + b)$$

$$= c^{\alpha(a)}d^{\alpha(a)}a^2 + c^{\alpha(a)}d^{\alpha(a)}ab + c^{\alpha(b)}d^{\alpha(b)}ba + c^{\alpha(b)}d^{\alpha(b)}b^2.$$

As $Q(+, \cdot)$ is commutative

$$(cd)^{\alpha(a+b)}(a + b)^2 = c^{\alpha(a+b)}(a + b)d^{\alpha(a+b)}(a + b)$$

$$= (c^{\alpha(a)}a + c^{\alpha(b)}b)(d^{\alpha(a)}a + d^{\alpha(b)}b)$$

$$= c^{\alpha(a)}d^{\alpha(a)}a^2 + c^{\alpha(a)}d^{\alpha(b)}ab + c^{\alpha(b)}d^{\alpha(a)}ba + c^{\alpha(b)}d^{\alpha(b)}b^2.$$

As $ab \neq 0$, we infer

$$0 = (c^{\alpha(a)} - c^{\alpha(b)})(d^{\alpha(a)} - d^{\alpha(b)}).$$

Since this is true for all $c, d \in Q$, it follows that $\alpha(a) = \alpha(b)$. $\qquad\square$

9.11 Theorem (Foulser 1967a). *Let $Q(+, \circ)$ be a generalized André system. If the division ring $Q(+, \cdot)$ belonging to $Q(+, \circ)$ is a field, then the following statements are equivalent:*

a) $Q(+, \circ) = Q(+, \cdot)$.
b) $Q(+, \circ)$ *is a division ring.*
c) *There exists $b \in Q \setminus \{0\}$ such that $c \circ (a + b) = c \circ a + c \circ b$ for all $a, c \in Q$.*

PROOF. It suffices to show that c) implies a). As a consequence of 9.10 we have $\alpha(a) = \alpha(b)$ for all $a \in A \setminus \{0\}$. In particular $\alpha(b) = \alpha(1) = 1$. Thus $\alpha(a) = 1$ for all $a \in Q \setminus \{0\}$. Hence $x \circ a = xa$ for all $x, a \in Q$. $\qquad\square$

Remark. Let $\mathfrak{A}$ be a desarguesian plane over a non-commutative division ring. Then 9.2 and 9.11 imply that $\mathfrak{A}$ does not satisfy the assumptions of 9.2. Hence 9.2 does not characterize the generalized André planes. I do not know whether a plane over a generalized André system for which $Q(+, \cdot)$

is a field always admits such a large abelian collineation group. However, if $Q(+, \circ)$ is finite, then the existence of such an abelian collineation group can be proved, as we shall see in section 11. In this context see also R. Rink [1977].

The following corollary is essentially due to Foulser [1967a, Lemma 4.1].

9.12 Corollary. *Let $\mathfrak{A}$ be a translation plane, let P, Q be distinct points on l_∞ and O a point not on l_∞. If A is an abelian collineation group of $\mathfrak{A}$ with $P^A = P$, $Q^A = Q$, $O^A = O$ and if A_W induces an irreducible group of automorphisms on $T(W)$ for all $W \, \mathrm{I} \, l_\infty$ which are different from P and Q, then the following statements are equivalent:*

a) *$\mathfrak{A}$ is pappian.*
b) *$\Delta(P, OP) \neq \{1\}$.*
c) *$\Delta(Q, OQ) \neq \{1\}$.*

PROOF. It follows from 1.15 that a) implies b) and c). As the conditions are symmetric in P and Q, it suffices to prove that c) implies a).

$\mathfrak{A}$ is a generalized André plane by 9.3. Therefore, we may assume that $\mathfrak{A} = \pi(V)$ where $V = X \oplus X$ is a direct sum of a K-vector space X with itself. Furthermore, we may assume $K \subseteq K(V, \pi)$ and $V(0), V(1), V(\infty) \in \pi$. Using 2.1, we may finally assume that $Q = V(\infty) \cap l_\infty$ and $P = V(0) \cap l_\infty$. Let $X(+, \circ)$ be the quasifield constructed with the aid of $\Sigma(\pi)$. Then $X(+, \circ)$ is a generalized André system and $X(+, \cdot)$ is a field by 9.2. As $\Delta(Q, OQ) = \Delta((\infty), V(\infty)) \neq \{1\}$, we infer from 5.6 and 9.11 that $X(+, \circ) = X(+, \cdot)$. $\qquad\square$

9.13 Lemma (Foulser 1967a). *If $Q(+, \circ)$ is a generalized André system, then*

a) $\mathrm{n_r}(Q) = \{a \,|\, a \in Q \setminus \{0\}, \alpha(x \circ a) = \alpha(x)\alpha(a)$ *for all* $x \in Q \setminus \{0\}\}$.
b) $\mathrm{n_m}(Q) = \{a \,|\, a \in Q \setminus \{0\}, \alpha(a \circ x) = \alpha(a)\alpha(x)$ *for all* $x \in Q \setminus \{0\}\}$.

PROOF. a) We have $(x \circ y) \circ a = (x^{\alpha(y)}y) \circ a = x^{\alpha(y)\alpha(a)}y^{\alpha(a)}a$ and $x \circ (y \circ a) = x^{\alpha(y \circ a)}y^{\alpha(a)}a$. Hence $a \in \mathrm{n_r}(Q)$, if and only if $\alpha(y)\alpha(a) = \alpha(y \circ a)$ for all $y \in Q \setminus \{0\}$.

b) is proved similarly. $\qquad\square$

9.14 Lemma. *If $Q(+, \circ)$ is a generalized André system, then $\{a \,|\, a \in Q, a^{\alpha(m)} = a$ for all $m \in Q \setminus \{0\}\} \subseteq \mathrm{k}(Q)$.*

PROOF. Assume that $a^{\alpha(m)} = a$ holds for all $m \in Q \setminus \{0\}$. Then $a \circ x = ax$ for all $x \in Q$. Hence

$$a \circ (x + y) = a(x + y) = ax + ay = a \circ x + a \circ y$$

and

$$a \circ (x \circ y) = a(x^{\alpha(y)}y) = a^{\alpha(y)}x^{\alpha(y)}y = (ax) \circ y = (a \circ x) \circ y. \qquad \square$$

10. Finite Generalized André Planes

In this section we follow mainly D. Foulser [1967a, section 2].

Let p be a prime and $q = p^s$. Furthermore, put $K = GF(q)$ and $F = GF(q^d)$ and let ρ be the automorphism of F which is defined by $x^\rho = x^q$. Then ρ generates the Galois group of F over K. For $k \in \mathbb{N}$ we put $I_k = \{0, 1, \ldots, k-1\}$. Let λ be a mapping from I_{q^d-1} into I_d with $\lambda(0) = 0$ and let w be a generator of F. We define an operation $\circ$ on F by $a \circ 0 = 0$ and $a \circ w^i = a^{\rho^{\lambda(i)}} w^i$. Then we have: $F(+)$ is an abelian group, $a \circ 1 = 1 \circ a = a$ and $(a+b) \circ c = a \circ c + b \circ c$ for all $a, b, c \in F$. We denote $F(+, \circ)$ by F_λ. Obviously, F_λ is a generalized André system, if and only if F_λ is a quasifield.

To avoid tedious repetitions, we shall also write $\lambda(a)$ instead of $\lambda(i)$, if $a = w^i$.

10.1 Lemma. F_λ is a quasifield, if and only if λ satisfies the condition: If $i, j \in I_{q^d-1}$ and $i \equiv j \bmod q^t - 1$, where $t = (d, \lambda(i) - \lambda(j))$, then $i = j$.

PROOF. Assume that F_λ is a quasifield. Let $i, j \in I_{q^d-1}$ be such that $i \equiv j \bmod q^t - 1$, where $t = (d, \lambda(i) - \lambda(j))$. We may assume $\lambda(i) \geqslant \lambda(j)$. As $t = (d, \lambda(i) - \lambda(j))$, we have $q^t - 1 = (q^d - 1, q^{\lambda(i)-\lambda(j)} - 1)$. Since $q^{\lambda(j)}$ and $q^d - 1$ are relatively prime, it follows that

$$q^t - 1 = (q^d - 1, q^{\lambda(j)}(q^{\lambda(i)-\lambda(j)} - 1)).$$

As $q^t - 1$ divides $j - i$, there is therefore an integer k such that

$$j - i \equiv k(q^{\lambda(i)} - q^{\lambda(j)}) \bmod q^d - 1.$$

Thus

$$w^k \circ w^i = w^{kq^{\lambda(i)}+i} = w^{kq^{\lambda(i)}+j} = w^k \circ w^j.$$

As F_λ is a quasifield, $w^i = w^j$ and hence $i = j$.

Conversely, let the condition on λ be satisfied. Because of 5.3, we have only to prove that F_λ is a weak quasifield. In order to do this, consider the mapping $x \to a \circ x$ and let $a \circ w^i = a \circ w^j$. As $a = w^k$ for some k, we have

$$w^{kq^{\lambda(i)}+i} = w^{kq^{\lambda(j)}+j}.$$

Hence

$$i + kq^{\lambda(i)} \equiv j + kq^{\lambda(j)} \bmod q^d - 1.$$

We may assume $\lambda(i) \geqslant \lambda(j)$. Then

$$j - i \equiv kq^{\lambda(j)}(q^{\lambda(i)-\lambda(j)} - 1) \bmod q^d - 1.$$

If $t = (d, \lambda(i) - \lambda(j))$, then we obtain

$$j - i \equiv 0 \bmod q^t - 1,$$

whence $i = j$ by assumption. Therefore, the mapping $x \to a \circ x$ is injective and hence surjective, as F is finite. $\qquad\square$

10.2 Corollary. *Let $K = \mathrm{GF}(2)$ and $F = \mathrm{GF}(2^d)$. If F_λ is a quasifield with $\mathrm{k}(F_\lambda) = K$, then $d = 1$ or d is divisible by at least three distinct primes.*

PROOF. Let $t = (\lambda(i) - \lambda(j), d) = 1$. Then $2^t - 1 = 1$ and hence $i \equiv j$ mod $2^t - 1$. By 10.1, we have $i = j$ and thus $\lambda(i) = \lambda(j)$. Therefore $1 = (0, d) = d$.

Hence $(\lambda(i) - \lambda(j), d) \neq 1$ for all i and j, if $d \neq 1$. In particular, $(\lambda(i), d) = (\lambda(i) - \lambda(0), d) \neq 1$ for all i. Let p be a prime dividing d and assume that p divides $\lambda(i)$ for all i. Then

$$\langle \rho^{\lambda(i)} \,|\, i \in I_{2^d - 1} \rangle \neq \langle \rho \rangle.$$

This implies by 9.14 that $GF(2)$ is not the kernel of F_λ. Therefore, there exists $i \in I_{2^d - 1}$ such that p does not divide $\lambda(i)$. As $(\lambda(i), d) \neq 1$, there exists a prime $q \neq p$ dividing $(\lambda(i), d)$. Hence d is divisible by at least two primes. Moreover, there exists $j \in I_{2^d - 1}$ such that q does not divide $(\lambda(j), d)$. Assume that p and q are the only prime divisors of d. Then p divides $(\lambda(j), d)$ as $(\lambda(j), d) \neq 1$ and as q does not divide $(\lambda(j), d)$. Similarly, q divides $(\lambda(i), d)$. Furthermore, either p or q divides $(\lambda(i) - \lambda(j), d)$. As everything is symmetric in p and q, we may assume that p divides $(\lambda(i) - \lambda(j), d)$. Then p divides $\lambda(i) - \lambda(j)$. As p divides $\lambda(j)$, it divides $\lambda(i)$, a contradiction. Hence there exists a third prime dividing d. $\qquad\square$

10.3 Corollary. *There is no F_λ with $|F_\lambda| = 64$ and $|\mathrm{k}(F_\lambda)| = 2$.*

Examples of F_λ's with $\mathrm{k}(F_\lambda) = GF(2)$ are given in M. L. N. Rao & J. L. Zemmer [1969].

Assume that F_λ is a quasifield. We extend the mapping λ from I_{q^d-1} into I_d to all of $\mathbb{Z}$ by the rule: If $i \in I_{q^d-1}$ and $x \in \mathbb{Z}$ and if $x \equiv i$ mod $q^d - 1$, then $\lambda(x) = \lambda(i)$. Furthermore, we put

$$J = \{k \,|\, k \in \mathbb{Z}, \lambda(i + k) = \lambda(i) \text{ for all } i \in \mathbb{Z}\}.$$

Then J is an ideal of $\mathbb{Z}$.

Put $u = \mathrm{lcm}\{q^m - 1 \,|\, m$ divides d and $m \neq d\}$. Let $\lambda(j) \neq \lambda(i)$. We may assume $\lambda(j) < \lambda(i)$. Then $0 < \lambda(i) - \lambda(j) < d$, as $I_d = \{0, \ldots, d-1\}$. Therefore, $t = (d, \lambda(i) - \lambda(j)) < d$. Hence $q^t - 1$ divides u. This implies $i \not\equiv j$ mod u, since $i \equiv j$ mod u implies $i \equiv j$ mod $q^t - 1$ and hence $i = j$. This proves: If $i \equiv j$ mod u, then $\lambda(i) = \lambda(j)$. Therefore, $\lambda(i + u) = \lambda(i)$ for all $i \in \mathbb{Z}$ and hence $u \in J$. As all ideals of $\mathbb{Z}$ are principal $J = v\mathbb{Z}$. As $u \in J$, we see that v divides u. This yields in particular $v \leqslant u \leqslant q^d - 1$.

Put $N_v = \{w^i \mid i \in J\}$ and $N_u = \{w^i \mid i \in u\mathbb{Z}\}$. As $u\mathbb{Z} \subseteq J = v\mathbb{Z}$, we have $N_u \subseteq N_v$. Furthermore, N_u and N_v are cyclic subgroups of $(F \setminus \{0\})(\cdot)$.

If $i \in J$, then $\lambda(i) = \lambda(0) = 0$. Hence $x \circ a = xa$ for all $x \in F$ and all $a \in N_v$. Thus we have proved the following:

10.4 Lemma. N_v and N_u are cyclic subgroups of $F^*(\cdot)$ of order $v^{-1}(q^d - 1)$, resp. $u^{-1}(q^d - 1)$. Furthermore, $x \circ a = xa$ for all $x \in F$ and all $a \in N_v$. In particular $N_v(\circ) = N_v(\cdot)$. Moreover $N_u \subseteq N_v$.

Next we prove:

10.5 Lemma. Let G be an elementary abelian p-group of order p^n and let t be a p-primitive prime divisor of $p^n - 1$. If A is a non-trivial t-group of automorphisms of G, then A operates irreducibly on G and is cyclic. Furthermore, A operates regularly on $G \setminus \{0\}$.

PROOF. Let $\alpha \in A$ with $o(\alpha) = t$. Put $G_\alpha = \{x \mid x \in G, x^\alpha = x\}$. Then G_α is a subgroup of G and hence $|G_\alpha| = p^m$. As $\alpha \neq 1$, we have $m < n$. The prime t divides $|G \setminus G_\alpha| = p^m(p^{n-m} - 1)$ and hence $p^{n-m} - 1$. As t is p-primitive, this yields $n - m = n$, i.e. $G_\alpha = \{0\}$. Hence A operates regularly on $G \setminus \{0\}$. Let $H \neq \{0\}$ be an A-invariant subgroup. Then $|A|$ divides $|H \setminus \{0\}|$, as A operates regularly on $G \setminus \{0\}$. Using again that t is p-primitive, we deduce $H = G$. Hence A operates irreducibly on G. As A is a t-group, $\mathfrak{Z}(A) \neq \{1\}$. Furthermore, $\mathfrak{Z}(A)$ also operates irreducibly on G. By Schur's lemma, the centralizer of $\mathfrak{Z}(A)$ is cyclic and so, therefore, is A. $\qquad\square$

10.6 Lemma. Let G be an elementary abelian p-group of order q^d with $d > 1$. Put $u = \mathrm{lcm}\{q^r - 1 \mid r$ divides d and $r \neq d\}$. If A is a group of automorphisms of G with $|A| = u^{-1}(q^d - 1)$, then A operates irreducibly on G unless we have one of the following cases:

a) $q = 2$, $d = 6$.
b) $q = 4$, $d = 3$ and A does not operate regularly on $G \setminus \{0\}$.
c) $q = 8$, $d = 2$ and A is elementary abelian of order 9.
d) $q = 3$, $d = 2$ and A is elementary abelian of order 4.

PROOF. Put $q = p^s$. If t is a p-primitive prime divisor of $p^{sd} - 1$, then t divides $|A|$. Let S be a Sylow t-subgroup of A. Then S, and hence A, operates irreducibly on G by 10.5. Thus, if A operates reducibly, then there is no p-primitive prime divisor of $p^{sd} - 1$. By 6.2, either $p = 2$ and $sd = 6$ or $sd = 2$ and $p + 1 = 2^a$.

If $p = 2$ and $d = 6$, then $|A| = 3$ and A operates reducibly. If $p = 2$ and $d < 6$, then $s = 2$ or 3, as $d \neq 1$. Assume $s = 2$. In this case $|A| = 3 \cdot 7$. Let $H \neq \{0\}$ be an A-invariant subgroup of G. If A operates regularly on $G \setminus \{0\}$, then 21 divides $|H| - 1$. As $32 - 1$ is not divisible by 21, we obtain the contradiction $H = G$. If $s = 3$, then $|A| = 9$. Again, A cannot operate regularly on G, as $16 - 1$ and $32 - 1$ are not divisible by 9. As G is a completely reducible A-module by Maschke's theorem, $G = G_1 \oplus G_2 \oplus G_3$ with $|G_i| = 4$. This yields that A is not cyclic, since otherwise A would not be faithful.

Finally, we have to consider the case $s = 1$, $d = 2$ and $p + 1 = 2^a$. Then $|A| = 2^a$ and $G = G_1 \oplus G_2$ with $G_i^A = G_i$ and $|G_i| = p$. Let A_i be the kernel of the restriction of A to G_i. As $p \equiv 3 \bmod 4$, we have $|A_i| \geqslant 2^{a-1} \geqslant 2$. Furthermore $A_1 \cap A_2 = \{1\}$, as A operates faithfully on G. Hence $A = A_1 A_2$ and $|A_i| = 2^{a-1}$. This yields

$$2^a = |A_1| |A_2| |A_1 \cap A_2|^{-1} = 2^{2a-2},$$

whence $a = 2$. Furthermore, A is elementary abelian as $|A_i| = 2$. $\square$

10.7 Theorem (Foulser 1967a). *If F_λ is a finite generalized André system, then*

$$k(F_\lambda) = \{ a \mid a \in F, a^{\rho^{\lambda(i)}} = a \text{ for all } i \in \mathbb{Z} \}$$

$$= \{ a \mid a \in F, a \circ (x + y) = a \circ x + a \circ y \text{ for all } x, y \in F \}.$$

PROOF. Put $k_1 = \{ a \mid a \in F, a^{\rho^{\lambda(i)}} = a$ for all $i \in \mathbb{Z} \}$ and $k_2 = \{ a \mid a \in F,$ $a \circ (x + y) = a \circ x + a \circ y$ for all $x, y \in F \}$. By 9.14 and the definition of $k(F_\lambda)$, we have $k_1 \subseteq k(F_\lambda) \subseteq k_2$. We show $k_2 \subseteq k_1$. Let $x \in k_2$ and $y \in F \backslash \{0\}$. By 10.6, 10.4 and 10.3, $F = \mathrm{GF}(p)(N_v)$. Hence there are $a_i \in N_v$ such that $y = \sum a_i$. This implies

$$x^{\rho^{\lambda(y)}} y = x \circ y = x \circ \sum a_i = \sum x \circ a_i = \sum x a_i = xy$$

and hence $x^{\rho^{\lambda(y)}} = x$ for all $y \in F \backslash \{0\}$. $\square$

11. Homologies of Finite Generalized André Planes

11.1 Lemma (Foulser 1967a). *Let F_λ be a generalized André system of order q^d, where λ, q and d have the usual meaning. Then:*

a) *The groups $n_r(F_\lambda)$ and $n_m(F_\lambda)$ are isomorphic to subgroups of $\Gamma L(1, F)$.*
b) $n_r(F_\lambda) = \{ w^i \mid \lambda(kq^{\lambda(i)} + i) \equiv \lambda(k) + \lambda(i) \bmod d \text{ for all } k \in \mathbb{Z} \}$.
c) $n_m(F_\lambda) = \{ w^i \mid \lambda(iq^{\lambda(k)} + k) \equiv \lambda(i) + \lambda(k) \bmod d \text{ for all } k \in \mathbb{Z} \}$.
d) $N_v \subseteq n_r(F_\lambda) \cap n_m(F_\lambda)$.
e) $N_v = \{ w^i \mid w^i \in n_r(F_\lambda), \lambda(i) = 0 \}$.

PROOF. a) For $a \in n_r(F_\lambda)$ we define $\sigma(a)$ by $x^{\sigma(a)} = x \circ a$. Then σ is an isomorphism from $n_r(F_\lambda)$ onto $\Sigma(\pi)_r$. Furthermore, $x \circ w^i = x^{\rho^{\lambda(i)}} w^i$, whence $\Sigma(\pi)_r \subseteq \Gamma L(1, F)$. Similarly, $n_m(F_\lambda) \cong \Sigma(\pi)_l \subseteq \Gamma L(1, F)$.

b) Put $\alpha(w^k) = \rho^{\lambda(k)}$. By 9.13, $w^i \in n_r(F_\lambda)$, if and only if $\alpha(x \circ w^i)$

$= \alpha(x)\alpha(w^i)$ for all $x \in F_\lambda\setminus\{0\}$. Thus $w^i \in n_r(F_\lambda)$, if and only if

$$\alpha(w^{kq^{\lambda(i)}+i}) = \alpha(w^k)\alpha(w^i),$$

i.e., if and only if

$$\lambda(kq^{\lambda(i)} + i) \equiv \lambda(k) + \lambda(i) \bmod d$$

for all $k \in \mathbb{Z}$.

c) is proved similarly.

d) follows immediately from the definition of N_v and b) and c).

e) It follows from d) and the definition of N_v that

$$N_v \subseteq \{w^i \,|\, w^i \in n_r(F_\lambda), \lambda(i) = 0\}.$$

Let $w^i \in n_r(F_\lambda)$ with $\lambda(i) = 0$. By b) we obtain

$$\lambda(k + i) = \lambda(kq^{\lambda(i)} + i) \equiv \lambda(k) + \lambda(i) = \lambda(k) \bmod d$$

for all $k \in \mathbb{Z}$. As $\lambda(k + i), \lambda(k) < d$, we infer $\lambda(k + i) = \lambda(k)$ for all $k \in \mathbb{Z}$. Therefore $w^i \in N_v$. $\square$

We denote by Σ_0 the set of all mappings $x \to x \circ a$ with $a \in N_v$ and by Σ, as usual, the set of all mappings $x \to x \circ m$ with $m \in F_\lambda\setminus\{0\}$. Then we have:

11.2 Lemma. $\Sigma_0 \subseteq \Sigma_r \cap \Sigma_l$ and $\Sigma \subseteq \mathfrak{N}_{GL(F(+), K)}(\Sigma_0)$, where $F(+)$ is the additive group of F_λ considered as a vector space over $K = GF(q)$.

PROOF. The first assertion follows from 11.1 d). Furthermore, $\Sigma \subseteq GL(F(+), K)$. Let $\sigma \in \Sigma$ and $x^\sigma = x \circ m$ and let $\mu \in \Sigma_0$ and $x^\mu = x \circ a$. Then $m = w^k$ and $a = w^i$ for suitable k and i. As $w^i \in N_v$, we have $\lambda(i) = 0$. Furthermore,

$$x^{\sigma\mu} = (x \circ w^k) \circ w^i = x^{q^{\lambda(k)}} w^{k+i}.$$

The multiplicative group F^* of F is cyclic. Hence N_v is a characteristic subgroup of F^*. Hence there exists $w^j \in N_v$ with $w^{jq^{\lambda(k)}} = w^i$. Put $x^\nu = x \circ w^j$. Then $\nu \in \Sigma_0$ and

$$x^{\nu\sigma} = (x \circ w^j) \circ w^k = (xw^j)^{q^{\lambda(k)}} w^k = x^{q^{\lambda(k)}} w^{jq^{\lambda(k)}+k} = x^{q^{\lambda(k)}} w^{i+k} = x^{\sigma\mu}.$$

Therefore, $\sigma\mu\sigma^{-1} = \nu \in \Sigma_0$. $\square$

Σ_0 is cyclic and the set Σ_l of all mappings of the form $x \to x \circ a$ with $a \in N_u$ is a subgroup, and hence a characteristic subgroup of Σ_0. Therefore, we have:

11.3 Corollary. $\Sigma_1 \subseteq \Sigma_r \cap \Sigma_l$ and $\Sigma \subseteq \mathfrak{N}_{GL(F(+), K)}(\Sigma_1)$.

Denote by Γ_0 the set of all mappings $(x, y) \to (x^\sigma, y)$ with $\sigma \in \Sigma_1$ and by Γ_∞ the set of all mappings $(x, y) \to (x, y^\sigma)$ with $\sigma \in \Sigma_1$. Using 3.2, 3.3, 3.9 and 11.3 we obtain:

11.4 Theorem. Γ_0 is a subgroup of $\Delta((0), V(\infty))$ and Γ_∞ is a subgroup of $\Delta((\infty), V(0))$. Moreover, the set of orbits of Γ_0 on l_∞ is equal to the set of orbits of Γ_∞ on l_∞.

By 3.8, $A = \Gamma_0\Gamma_\infty = \Gamma_0 \times \Gamma_\infty$ is abelian. If A_m denotes the stabilizer of $V(m)$ in A, Theorem 11.4 yields $|A_m| = |\Gamma_0| = |\Sigma_1| = u^{-1}(q^d - 1)$ for all $m \neq 0$. Let $\alpha \in A_m$ be in the centralizer of $T(m)$, the group of all translations the centre of which is $V(m) \cap l_\infty$. Then $x^\alpha = x$ for all $x \in V(m)$, as $0^\alpha = 0$ and $T(m)$ operates transitively on $V(m)$. Thus α is a collineation with axis $V(m)$. As α fixes (0) and (∞), we deduce $\alpha = 1$. This proves that A_m operates faithfully on $T(m)$. The group A_m is cyclic, as A_m is a diagonal of A. Therefore, by 10.3 and 10.6, A_m operates irreducibly on $T(m)$. This together with 9.3 yields:

11.5 Theorem (Lüneburg 1976b). *Let $\mathfrak{A}$ be a finite translation plane. Then $\mathfrak{A}$ is a generalized André plane, if and only if $\mathfrak{A}$ admits an abelian collineation group A which fixes two distinct points P and Q on l_∞ and an affine point O such that for all $W \, \mathrm{I} \, l_\infty$ which are distinct from P and Q the stabilizer A_W of W in A induces a group of automorphisms on $T(W)$ which acts irreducibly.*

Next we turn to:

11.6 Lemma. *Let q be a power of a prime and $d \in \mathbb{N}$. Put $u = \mathrm{lcm}\{q^r - 1 \mid 0 < r < d \text{ and } r \text{ divides } d\}$. Furthermore, let $(q, d) \neq (2, 6)$. If V is a vector space of rank 1 over $\mathrm{GF}(q^d)$, if G is a subgroup of $\Gamma L(V)$ which operates regularly on $V \setminus \{0\}$ and if $u^{-1}(q^d - 1)$ divides $|G \cap \mathrm{GL}(V)|$, then G contains exactly one abelian subgroup of order $u^{-1}(q^d - 1)$, unless $(q, d) = (3, 2)$ and G is the quaternion group of order 8.*

PROOF. If $d = 1$, then $u = 1$ and hence $G = \mathrm{GL}(V)$, as $|G|$ divides $|V| - 1$. Therefore, we may assume $d > 1$. As $\mathrm{GL}(V)$ is cyclic and $u^{-1}(q^d - 1)$ divides $|G \cap \mathrm{GL}(V)|$, there exists a cyclic subgroup Z of $G \cap \mathrm{GL}(V)$ of order $u^{-1}(q^d - 1)$. Since Z is a characteristic in $G \cap \mathrm{GL}(V)$, it is normal in G.

Let $q = p^r$ where p is a prime and let t be a p-primitive prime divisor of $p^{rd} - 1$. Furthermore, let t^a be the highest power of t dividing $q^d - 1$. Then t^a divides $u^{-1}(q^d - 1)$. Let T be a Sylow t-subgroup of Z and T_0 a Sylow t-subgroup of G which contains T. By 10.5, we obtain $T = T_0$. As T is characteristic in Z and Z is normal in G, we have that T is normal in G. Hence T is the only Sylow t-subgroup of G. As T operates irreducibly on V by 10.5, Schur's lemma yields that $C = \mathfrak{C}_G(T)$ is cyclic. Furthermore, $Z \subseteq C$. Let Z_1 be an abelian subgroup of order $u^{-1}(q^d - 1)$ of G. As T is unique, $T \subseteq Z_1$. Hence $Z_1 \subseteq C$. This yields $Z = Z_1$, as C, being cyclic, contains exactly one subgroup of order $u^{-1}(q^d - 1)$.

Assume that Z is not unique. Then there is no p-primitive prime divisor of $p^{rd} - 1$. Therefore $q^d = 64$ or $q = p$, $d = 2$ and $p + 1 = 2^b$ by 6.2. Let

$q^d = 64$. As $(q,d) \neq (2,6)$, we have $(q,d) = (4,3)$ or $(q,d) = (8,2)$. Consider the case $(q,d) = (8,2)$. Then $|Z| = 3^2$. As $|G|$ divides $|V| - 1 = 3^2 \cdot 7$, we have that Z is a Sylow 3-subgroup of G. Therefore, Z is unique, since Z is normal in G. Next we consider the case $(q,d) = (4,3)$. In this case $|Z| = 3 \cdot 7$. Let Z_0 be a Sylow 7-subgroup of Z. Then Z_0 is normal in G and hence the only Sylow 7-subgroup of G, as $|G|$ divides $63 = 3^2 \cdot 7$. Using Schur's lemma again, we see that Z is unique. (Remember that we are working over GF(4).) Thus $q = p$, $d = 2$ and $p + 1 = 2^b$. Let Z_1 be an abelian subgroup of order $u^{-1}(q^d - 1) = p + 1 = 2^b$ which is distinct from Z. As Z is normal, ZZ_1 is a 2-subgroup of G. Since $|G|$ divides $p^2 - 1$, the order of ZZ_1 is 2^{b+1}. Furthermore, ZZ_1 is either cyclic or a generalized quaternion group, since ZZ_1 operates regularly on $V \setminus \{0\}$. The first case cannot occur as $Z \neq Z_1$. As Z_1 operates regularly, Z_1 is cyclic. But a generalized quaternion group of order greater than 8 contains only one cyclic subgroup of index 2. Thus $b = 2$ and $q = 3$. $\square$

11.7 Theorem. Γ_0 *is the only subgroup of order* $u^{-1}(q^d - 1)$ *in* $\Delta((0), V(\infty))$ *and* Γ_∞ *is the only subgroup of order* $u^{-1}(q^d - 1)$ *in* $\Delta((\infty), V(0))$, *unless* F_λ *is the nearfield of type* $\{3,2\}$.

This follows from 11.6 and 10.3.

12. The André Planes

Let L be a field and Γ a group of automorphisms of L. Let M be a subgroup of L^* which is invariant under Γ and assume that Γ operates trivially on L^*/M. Finally, let β be a mapping from L^*/M into Γ with $\beta(M) = 1$ and define the mapping α from L^* into Γ by $\alpha(a) = \beta(aM)$. Then we have in particular $\alpha(1) = 1$. For $a,b \in L$ define $a \circ b$ by $a \circ b = 0$, if $b = 0$, and by $a \circ b = a^{\alpha(b)}b$, if $b \neq 0$. Obviously, $L(+)$ is an abelian group, $a \circ 0 = 0$ and $(a + b) \circ c = a \circ c + b \circ c$ for all $a,b,c \in L$.

Let $a,b,m \in L^*$. Then

$$a^{\alpha(m)}bM = a^{\alpha(m)}MbM = aMbM = abM.$$

Therefore, $\alpha(a^{\alpha(m)}b) = \alpha(ab)$ for all $a,b,m \in L^*$. In particular, $\alpha(a \circ b) = \alpha(ab)$ for all $a,b \in L^*$.

Let $a,x,c \in L$ with $a \neq 0$ and assume $a \circ x = c$. If $x \neq 0$, then $a \circ x = a^{\alpha(x)}x \neq 0$. Therefore, $c = 0$ implies $x = 0$. If $c \neq 0$, then $x \neq 0$ and $a^{\alpha(x)}x = c$. Hence $x = a^{-\alpha(x)}c$. The remark made above yields $\alpha(x) = \alpha(a^{-1}c)$. Thus $x = a^{-\alpha(a^{-1}c)}c$ is uniquely determined. Conversely, if $a,c \in L$, then we put $x = a^{-\alpha(a^{-1}c)}c$. A trivial computation shows $a \circ x = c$.

Finally, $x \circ 1 = x = 1 \circ x$, as is easily seen. Hence $L(+, \circ)$ is a weak quasifield. In general, $L(+, \circ)$ will not be a quasifield. However, if Γ is finite, then $L(+, \circ)$ is a quasifield: Let K be the fixed field of Γ. Then

$[L:K]$ is finite. Furthermore, K is contained in the outer kernel of $L(+, \circ)$ as is easily seen. Applying 5.3 yields that $L(+, \circ)$ is a quasifield in this case.

If Γ is finite, then we define $\mathfrak{n}_\Gamma$ by $\mathfrak{n}_\Gamma(a) = \prod_{\gamma \in \Gamma} a^\gamma$ for all $a \in L^*$. Then $\mathfrak{n}_\Gamma(a^\gamma) = \mathfrak{n}_\Gamma(a) = \mathfrak{n}_\Gamma(a)^\gamma$ for all $a \in L^*$ and all $\gamma \in \Gamma$. If $\alpha(a) = \alpha(b)$ for all $a, b \in L^*$ with $ab^{-1} \in \ker(\mathfrak{n}_\Gamma)$, then we shall call $L(+, \circ)$ an *André system* and every translation plane which is coordinatized by an André system will be called an *André plane*.

We have $a^{\gamma-1} \in M \cap \ker(\mathfrak{n}_\Gamma)$ for all $a \in L^*$ and all $\gamma \in \Gamma$. Applying Hilbert's Satz 90 (see e.g., S. Lang [1971, p. 213]), we obtain that $\ker(\mathfrak{n}_\Gamma) \subseteq M$, if Γ is cyclic. Therefore, $L(+, \circ)$ is an André system, if L is finite.

Not all the generalized André systems described above are André systems, as the following examples show which are due to P. Roquette. Let K be the field of rationals and let p be a prime with $p \equiv 3 \bmod 8$ or $p \equiv -3 \bmod 8$. As p is odd, $\sqrt{p} \notin K(\sqrt{2})$. Therefore, $L = K(\sqrt{2}, \sqrt{p})$ is an extension of K of degree 4. Furthermore, L is a Galois extension of K and the Galois group Γ is elementary abelian of order 4. The group Γ is generated by the elements σ and τ which are defined by $(\sqrt{2})^\sigma = -\sqrt{2}$, $(\sqrt{p})^\sigma = \sqrt{p}$ and $(\sqrt{2})^\tau = \sqrt{2}$, $(\sqrt{p})^\tau = -\sqrt{p}$. Put $a = 1 + \sqrt{2}$. Then $a^{\sigma+1} = -1$ and hence

$$\mathfrak{n}_\Gamma(a) = a^{1+\sigma+\tau+\sigma\tau} = a^{(1+\sigma)(1+\tau)} = (-1)^{1+\tau} = 1.$$

Let $b, c, d \in L$ and assume $a = b^{\sigma-1} c^{\tau-1} d^{\sigma\tau-1}$. As $d^{\sigma\tau-1} = d^{\sigma(\tau-1)} d^{\sigma-1}$, we may assume $d = 1$. Then we have

$$-1 = a^{\sigma+1} = b^{\sigma^2-1} c^{(\tau-1)(\sigma+1)} = c^{(\tau-1)(\sigma+1)}.$$

As $c \in L^*$, there exist $\alpha, \beta, \gamma, \delta \in K$ with $c = \alpha + \beta\sqrt{2} + \gamma\sqrt{p} + \delta\sqrt{2}\sqrt{p}$. This yields after an easy computation

$$c^{\sigma+1} = \alpha^2 + 2\alpha\gamma\sqrt{p} - 2\beta^2 - 4\beta\delta\sqrt{p} + \gamma^2 p - 2p\delta^2.$$

Furthermore,

$$c^{\tau(\sigma+1)} = c^{(\sigma+1)\tau} = \alpha^2 - 2\alpha\gamma\sqrt{p} - 2\beta^2 + 4\beta\delta\sqrt{p} + \gamma^2 p - 2p\delta^2.$$

As $-c^{\sigma+1} = c^{\tau(\sigma+1)}$, we obtain from the two above equations

$$0 = \alpha^2 - 2\beta^2 + p(\gamma^2 - 2\delta^2).$$

Since $\alpha, \beta, \gamma, \delta$ are rational, we may assume that they are integers. We may further assume that they are relatively prime, as $c \neq 0$. Computing modulo p yields $\alpha^2 \equiv 2\beta^2 \bmod p$. From $p \equiv 3$ or $-3 \bmod 8$, we infer that 2 is not a quadratic residue modulo p. Therefore $\alpha \equiv \beta \equiv 0 \bmod p$. Put $\alpha = \epsilon p$ and $\beta = \zeta p$. Then $0 = p^2(\epsilon^2 - 2\zeta^2) + p(\gamma^2 - 2\delta^2)$ and hence $\gamma^2 \equiv 2\delta^2 \bmod p$ which yields as above $\gamma \equiv \delta \equiv 0 \bmod p$, a contradiction. This contradiction proves that $M = \langle l^{\gamma-1} \mid l \in L^*, \gamma \in \Gamma \rangle$ is a proper subgroup of $\ker(\mathfrak{n}_\Gamma)$. Using this M, it is now easy to construct an $L(+, \circ)$ which is not an André system.

Let $L(+, \circ)$ be a generalized André system constructed as at the beginning of this section. The mapping φ defined by $(x, y)^\varphi = (xa, ya^{\alpha(m)})$ is a collineation of $\mathfrak{A}(L)$ for all $a, m \in L^*$, as we shall now see. As a consequence of $(a^{-1} \circ b)a^{\alpha(m)} = a^{-\alpha(\beta)}ba^{\alpha(m)}$ we obtain $\alpha((a^{-1} \circ b)a^{\alpha(m)})$ $= \alpha(b)$. Obviously, φ is a permutation of $V = L \oplus L$ which leaves invariant $V(0)$ and $V(\infty)$. Furthermore, using $\alpha((a^{-1} \circ b)a^{\alpha(m)}) = \alpha(b)$, we find

$$(xa) \circ \left((a^{-1} \circ b)a^{\alpha(m)}\right) = x^{\alpha(b)}a^{\alpha(b)}a^{-\alpha(b)}ba^{\alpha(m)} = (x \circ b)a^{\alpha(m)}.$$

Hence $V(b)^\varphi = V((a^{-1} \circ b)a^{\alpha(m)})$. This proves that φ is a collineation.

As $(a^{-1} \circ m)a^{\alpha(m)} = m$, we have $V(m)^\varphi = V(m)$. Let A_m be the set of all mappings of the form $(x, y) \to (xa, ya^{\alpha(m)})$ and let A be the group generated by all the A_m. Then A is an abelian collineation group of $\mathfrak{A}(L)$ and A_m stabilizes $V(m)$. Moreover A_m operates transitively on $V(m)\setminus\{0\}$ which yields that A_m is the stabilizer of $V(m)$ in A. Therefore, we have proved the first part of the next theorem:

12.1 Theorem (Lüneburg). *Let $\mathfrak{A}$ be a translation plane. Then $\mathfrak{A}$ is isomorphic to one of the planes over a quasifield described above, if and only if $\mathfrak{A}$ admits an abelian collineation group A which fixes two non-parallel lines l and h of $\mathfrak{A}$ such that the stabilizer A_m of any line m through $l \cap h$ which is distinct from l and h is transitive on $m\setminus\{l \cap h\}$.*

PROOF. In order to prove the second half of the theorem, we may assume that $\mathfrak{A}$ has standard form $\pi(V)$ and that A fixes the lines $V(0)$ and $V(\infty)$. According to 9.2, V is a vector space of rank 2 over a field L and the components of π which are distinct from $V(0)$ and $V(\infty)$ have the form $V(m) = \{(x, x^{\alpha(m)}m)\,|\,x \in L\}$ with $m \in L^*$ and $\alpha(m) \in \mathrm{Aut}(L)$. By 9.9 we get that A_m consists of all the mappings $(x, y) \to (xa, ya^{\alpha(m)})$ with $a \in L^*$. In particular, A_1 consists of all the mappings $(x, y) \to (xa, ya)$. Therefore $(x, y) \to (x, ya^{\alpha(m)-1})$ is a collineation of $\pi(V)$ for all $a, m \in L^*$. Furthermore, this collineation belongs to $\Delta((\infty), V(0))$. This yields: If $b \in L^*$, then there exists $c \in L^*$ such that $(x \circ b)a^{\alpha(m)-1} = x \circ c$ for all $x \in L$. Putting $x = 1$ gives $ba^{\alpha(m)-1} = c$ and hence $x^{\alpha(b)} = x^{\alpha(c)}$ for all $x \in L$. Thus $\alpha(ba^{\alpha(m)-1}) = \alpha(b)$ for all $a, b, m \in L^*$.

Let $\Gamma = \langle \alpha(m)\,|\,m \in L^*\rangle$ and $M = \langle a^{\gamma-1}\,|\,a \in L^*,\ \gamma \in \Gamma\rangle$. We now show $M = \langle a^{\alpha(m)-1}\,|\,m, a \in L^*\rangle$. Put $N = \langle a^{\alpha(m)-1}\,|\,m, a \in L^*\rangle$. Then $N \subseteq M$. If $\gamma \in \Gamma$, then $\gamma = \gamma_1 \cdots \gamma_n$ where $\gamma_i = \alpha(m_i)$ for some $m_i \in L^*$. If $n = 1$, then $a^{\gamma-1} \in N$ for all $a \in L^*$. Let $n > 1$ and assume $a^{\gamma_2 \cdots \gamma_n - 1} \in N$ for all a. As

$$a^{\gamma_1 \cdots \gamma_n - 1} = a^{\gamma_1(\gamma_2 \cdots \gamma_n - 1)}a^{\gamma_1 - 1},$$

we obtain $a^{\gamma-1} \in N$. Therefore $N = M$.

M is invariant under Γ, as $(a^{\gamma-1})^\delta = a^{\gamma\delta-1}(a^{-1})^{\delta-1}$, and Γ operates trivially on L^*/M by the very definition of M. Finally, if $a, b, \in L^*$ and

$ba^{-1} \in M$, then

$$b = aa_1^{\alpha(m_1)-1}a_2^{\alpha(m_2)-1} \cdots a_r^{\alpha(m_r)-1}$$

and hence $\alpha(b) = \alpha(a)$, as $\alpha(xy^{\alpha(m)-1}) = \alpha(x)$.   □

12.2 Corollary (Ostrom 1969). *Let $\mathfrak{A}$ be a finite translation plane. Then $\mathfrak{A}$ is an André plane, if and only if $\mathfrak{A}$ admits an abelian collineation group A which fixes two distinct points P and Q on l_∞ and an affine point O such that A_W operates transitively on the set of affine points on OW which are distinct from O for all $W \text{ I } l_\infty$ distinct from P and Q.*

This follows immediately from 12.1 and the remark made above that the finite quasifields among the ones described above are André systems.

12.3 Theorem. *If L is a finite André system, then $n_r(L) = n_m(L)$.*

PROOF. We have $\alpha(a \circ b) = \alpha(ab) = \alpha(ba) = \alpha(b \circ a)$. The assertion now follows from 9.13 and the remark that Γ is abelian, as L is finite.   □

12.4 Theorem. *Let L be a finite generalized André system with $|k(L)| = q$ and $|L| = q^d$. Then the following statements are equivalent:*

1) *L is an André system.*
2) *v divides $q - 1$ where v is the integer defined after 10.3.*
3) *$\Delta((0), V(\infty))$ contains a cyclic subgroup Z_1 and $\Delta((\infty), V(0))$ contains a cyclic subgroup Z_2 with $|Z_1| = |Z_2| = (q^d - 1)(q - 1)^{-1}$ such that the set of orbits of Z_1 on l_∞ is equal to the set of orbits of Z_2 on l_∞.*
4) *$\Delta((\infty), V(0))$ contains a cyclic subgroup Z of order $(q^d - 1)(q - 1)^{-1}$.*

PROOF. Assume 1). Let Γ be the group generated by all the $\alpha(m)$. By 10.7, $k(L)$ is the fixed field of Γ. Hilbert's Satz 90 then implies that the kernel M of n_Γ has order $(q^d - 1)(q - 1)^{-1}$. If $a \in M$, then $\alpha(xa) = \alpha(x)$. This yields $M \subseteq N_v$ whence v divides $q - 1$.

Assume 2). Then M, as defined above, is a subgroup of N_v and hence a characteristic subgroup of N_v. Therefore, 3) follows by 11.2 and 3.9.

Assume 4). If $q^d = 9$, then either L is isomorphic to GF(9) or L is the nearfield of type $\{3, 2\}$. In either case, L is an André system. Therefore, we may assume $q^d \neq 9$. By 11.7, $\Delta((\infty), V(0))$ contains exactly one cyclic subgroup Z_∞ of order $u^{-1}(q^d - 1)$, where u is the least common multiple of the $q^i - 1$ with i a proper divisor of d. This yields $Z_\infty \subseteq Z$. Therefore, Z operates irreducibly on $V(\infty)$ by 10.6. Applying Schur's lemma, we obtain

$(*)$ $Z = \{((x, y) \to (x, ya)) \mid a \in M\}$

where M, as above, denotes the kernel of n_Γ. Let $a, b \in L^*$ and assume $ba^{-1} \in M$. Then $(x \circ a)ba^{-1} = x^{\alpha(a)}b$ for all $x \in L$. By $(*)$, there exists $c \in L^*$ with $(x \circ a)ba^{-1} = x \circ c$ for all $x \in L$. This yields $b = c$ and hence $\alpha(a) = \alpha(b)$.   □

12.5 Corollary (Rink). *If L is a generalized André system with $|k(L)| = q$ and $|L| = q^d$ where d is a prime, then L is an André system.*

This follows from 12.4 and the fact that v divides u and the remark that
$$u = \text{lcm}\{ q^i - 1 \,|\, i \text{ is a proper divisor of } d \} = q - 1$$
in this case.

13. The Hall Planes

We start with a construction due to Ostrom. It is convenient for this purpose to identify lines and subplanes of a given plane with the appropriate point sets.

Let $\mathfrak{A}$ be an affine plane of order q^2 and D a set of $q + 1$ points on l_∞. Put $D^* = l_\infty \setminus D$. Let $\mathfrak{B}$ be a set of Baer subplanes of $\mathfrak{A}$ such that:

1) If $b \in \mathfrak{B}$, then $b \cap l_\infty = D$.
2) If $P, Q \in \mathfrak{A}$ with $P \neq Q$ and $PQ \cap l_\infty \in D$, then there exists $b \in \mathfrak{B}$ with $P, Q \in b$.

As the plane which is generated by P, Q and D is contained in b and has order $\geq q$, it follows that there is exactly one $b \in \mathfrak{B}$ such that $P, Q \in b$. Furthermore, $\mathfrak{B}$ consists of all the Baer subplanes of $\mathfrak{A}$ which contain D.

The number of point-pairs (P, Q) with $P \neq Q$ and $PQ \cap l_\infty \in D$ is $q^4(q + 1)(q^2 - 1)$ and the number of pairs (P, Q) with $P \neq Q$ and $P, Q \in b$ where $b \in \mathfrak{B}$ is $q^2(q + 1)(q - 1)$. Therefore, $|\mathfrak{B}| = q^2(q + 1)$.

We define a new incidence structure $\mathfrak{A}'$ as follows:

a) The points of $\mathfrak{A}'$ are the points of $\mathfrak{A}$.
b) The lines of $\mathfrak{A}'$ are the lines l of $\mathfrak{A}$ with $l \cap l_\infty \in D^*$ and the elements of $\mathfrak{B}$.
c) $\in$ is the incidence relation.

Then $\mathfrak{A}'$ contains q^4 points and $q^2(q + 1) + q(q - 1)q^2 = q^4 + q^2$ lines. Furthermore, each line carries exactly q^2 points and each pair of points of $\mathfrak{A}$ is on exactly one line. Thus $\mathfrak{A}'$ is an affine plane of order q^2. We call $\mathfrak{A}'$ the *plane derived from $\mathfrak{A}$ via D*.

Let D' be the set of points on l'_∞ which are on lines belonging to $\mathfrak{B}$ and $\mathfrak{B}'$ the set of affine lines of $\mathfrak{A}$ which carry a point of D. Then $\mathfrak{B}'$ is a set of Baer subplanes of $\mathfrak{A}'$ and $\mathfrak{A}', \mathfrak{B}', D'$ satisfy 1) and 2): Let $l \in \mathfrak{B}'$ and let P and Q be two distinct points on l. Denote by $(PQ)'$ the line joining P and Q in $\mathfrak{A}'$. Then $(PQ)' \in \mathfrak{B}$ and therefore $|(PQ)' \cap l| = q$. A simple counting argument now proves that l is a Baer subplane of $\mathfrak{A}'$. Furthermore $l \cap l'_\infty = D'$. This together with $|\mathfrak{B}'| = q^2(q + 1)$ yields the desired conclusion. In particular, if $\mathfrak{A}''$ is the plane derived from $\mathfrak{A}'$ via D', then $\mathfrak{A}'' = \mathfrak{A}$.

13.1 Lemma. *Let* $\mathfrak{A}, \mathfrak{A}', D, D'$ *be as above. If* G *is the collineation group of* $\mathfrak{A}$ *and* G' *the collineation group of* $\mathfrak{A}'$, *then* $G_D = G'_{D'}$.

This follows immediately from the remark made above that $\mathfrak{B}$ and $\mathfrak{B}'$ are the sets of all Baer subplanes of $\mathfrak{A}$ and $\mathfrak{A}'$ resp. which intersect l_∞ in D and l'_∞ in D' resp.

13.2 Lemma. *If* $\mathfrak{A}$ *is a translation plane, then* $\mathfrak{A}'$ *is a translation plane and the translation group of* $\mathfrak{A}$ *is also the translation group of* $\mathfrak{A}'$.

PROOF. It suffices to prove the second assertion. Let T be the translation group of $\mathfrak{A}$ and $P \in D^*$. Then $T(P) \subseteq G_D = G'_{D'}$. Furthermore, if $\tau \in T(P)$, then τ is the central collineation in $\mathfrak{A}$ as well as in $\mathfrak{A}'$ the centre of τ being P. As τ has no fixed point on $\mathfrak{A}$ unless $\tau = 1$, we have that τ is a translation of $\mathfrak{A}$ and of $\mathfrak{A}'$. Therefore, the assertion of 13.2 follows from 1.3, as $|D^*| = q(q - 1) \geqslant 2$. □

13.3 Lemma. *If* κ *is a perspectivity of* $\mathfrak{A}$ *with axis* l_∞, *then* κ *is a collineation of* $\mathfrak{A}'$.

This follows immediately from 13.1. We remark that κ need not be a perspectivity of $\mathfrak{A}'$.

13.4 Lemma. *Let* κ *be a perspectivity of* $\mathfrak{A}$ *with* $D^\kappa = D$ *and axis* $m \neq l_\infty$. *Then*:

a) *If* $m \notin \mathfrak{B}'$, *then* κ *is a perspectivity of* $\mathfrak{A}'$.
b) *If* $m \in \mathfrak{B}'$, *then* κ *is a Baer collineation of* $\mathfrak{A}'$.

The proof is trivial.

13.5 Theorem. *Let* $L = GF(q^2)$ *and* $V = L \oplus L$. *Put* $\pi = \{V(0), V(\infty)\} \cup \{V(m) \mid m \in L^*\}$ *where* $V(m) = \{(x, xm) \mid x \in L\}$. *For* $t \in GF(q)^*$ *define* S_t *by* $S_t = \{x \mid x \in L^*, x^{q+1} = t\}$. *If* D_t *is the set of points at infinity on the lines* $V(m)$ *with* $m \in S_t$ *and if* $\mathfrak{B}_t = \{V(\circ m) + x \mid m \in S_t, x \in V\}$ *where* $V(\circ m) = \{(x, x^q m) \mid x \in L\}$, *then* $\pi(V), D_t, \mathfrak{B}_t$, *satisfy conditions* 1) *and* 2) *of page* 57.

PROOF. It suffices to show that $\mathfrak{B}'_{t,0} = \{V(m) \mid m \in S_t\}$ is a regulus in the projective space defined by the $GF(q)$-vector space V and that $\mathfrak{B}_{t,0} = \{V(\circ m) \mid m \in S_t\}$ is the regulus opposite to $\mathfrak{B}'_{t,0}$. In this context we shall write V_q instead of V if we consider V as a $GF(q)$-vector space. Let $X \in \mathfrak{B}_{t,0} \cup \mathfrak{B}'_{t,0}$. Then X is a line in the projective geometry belonging to V_q. Furthermore, set

$$\mathfrak{Q} = \bigcup_{X \in \mathfrak{B}'_{t,0}} X = \{(x, xm) \mid x \in L, m \in S_t\}.$$

Let $m \in S_t$. Then $(x, x^q m) = (x, xx^{q-1}m)$. Moreover, $(x^{q-1}m)^{q+1}$ $= x^{q^2-1}m^{q+1} = t$, provided $x \neq 0$. Thus $x^{q-1}m \in S_t$. Hence $(x, x^q m) \in \mathfrak{Q}$. This shows $V(\circ m) \subseteq \mathfrak{Q}$ for all $m \in S_t$.

Let $m, n \in S_t$. Then $(mn^{-1})^{q+1} = tt^{-1} = 1$. Applying Hilbert's Satz 90 yields an $a \in L$ with $mn^{-1} = a^{q-1}$. In particular $a \neq 0$ and hence $(0,0) \neq (a, am) = (a, a^q n) \in V(m) \cap V(\circ n)$. This establishes 13.5, as the lines of $\mathfrak{B}_{t,0}$ resp. $\mathfrak{B}'_{t,0}$ are pairwise skew.   □

13.6 Theorem. *Let $L = \mathrm{GF}(q^2)$ and $\pi(V)$ the desarguesian plane over L. If D is a set of $q + 1$ points on l_∞ and $\mathfrak{B}$ a set of Baer subplanes such that* 1) *and* 2) *on page* 57 *are satisfied, then there is a collineation κ of $\pi(V)$ with $D^\kappa = D_1$ and $\mathfrak{B}^\kappa = \mathfrak{B}_1$.*

This follows from the fact that $\mathfrak{B}' \cap \pi$ is a regulus in the projective geometry belonging to V_q, as is easily established with the aid of $\mathfrak{B} \cap \pi$.

13.7 Theorem. *Let $L = \mathrm{GF}(q^2)$ and $K = \mathrm{GF}(q)$. Let X and Y be subsets of K^* with $K^* = X \cup Y$ and $X \cap Y = \emptyset$. If $\pi(V)_X$ resp. $\pi(V)_Y$ are the planes which are obtained by deriving simultaneously via D_t for all $t \in X$ resp. all $t \in Y$, then $\pi(V)_X$ and $\pi(V)_Y$ are isomorphic André planes. Moreover, $\pi(V)_X$ is desarguesian, if and only if $X = \emptyset$ or $X = K^*$.*

PROOF. The construction shows that $\pi(V)_X$ and $\pi(V)_Y$ are generalized André planes. As K is contained in their kernels, both planes are André planes by 12.5.

The mapping $\varphi : (x, y) \to (x, y^q)$ is an involutory linear mapping of V_q onto itself with $\mathfrak{B}_{t,0} = \varphi \mathfrak{B}'_{t,0}$. As a consequence: $\pi(V)_X$ and $\pi(V)_Y$ are isomorphic. The last assertion follows from 9.11.   □

It follows from 13.5 and 13.6 that there is up to isomorphism only one plane $H(q)$ of order q^2 which is constructed by a single derivation from the desarguesian plane of order q^2. This plane will be called the *Hall plane* of order q^2. These planes were first discovered by M. Hall Jr. $H(2)$ is the desarguesian plane of order 4. This follows from 13.7, as $2 - 1 = 1$. In all other cases we have:

13.8 Theorem. *If $q \geqslant 3$, then $H(q)$ is non-desarguesian.*

PROOF. This follows from the last assertion of 13.7, as $q - 1 \geqslant 2$.   □

13.8 and 8.4 yield that $H(3)$ is the nearfield plane of order 9.

The next theorem which is very useful is a special case of a theorem on Frobenius groups. As the proof of this special case is much easier than the proof of the general case, we shall give it here.

13.9 Theorem. *Let $\mathfrak{B}$ be a finite projective plane and l a line of $\mathfrak{B}$. Let Δ be a group of perspectivities with axis l and denote by $\mathfrak{B}$ the set of points P with*

$\Delta(P, l) \neq \{1\}$. *If* T *is the group of all elations with axis* l *which are contained in* Δ, *then* $\mathfrak{B}$ *is an orbit of* T.

PROOF (André). We have

$$|\Delta| = |\text{T}| + \sum_{P \in \mathfrak{B}} (|\Delta(P, l)| - 1) = |\text{T}| - |\mathfrak{B}| + \sum_{P \in \mathfrak{B}} |\Delta(P, l)|.$$

As $|\Delta(P, l)| \geqslant 2$ for all $P \in \mathfrak{B}$, we obtain

$$|\Delta| \geqslant |\text{T}| + |\mathfrak{B}| \geqslant 1 + |\mathfrak{B}|.$$

Let $\mathfrak{B}_1, \ldots, \mathfrak{B}_r$ be the orbits of Δ with $|\mathfrak{B}_i| < |\Delta|$ and $\mathfrak{B}_i \not\subseteq l$. Since $\Delta_P = \Delta(P, l)$ for all $P \in \mathfrak{B}$, we have $\mathfrak{B} = \bigcup_{i=1}^r \mathfrak{B}_i$. Therefore,

$$|\Delta| = |\text{T}| - |\mathfrak{B}| + \sum_{i=1}^r \sum_{P \in \mathfrak{B}_i} |\Delta(P, l)|.$$

If $Q \in \mathfrak{B}_i$, then $|\Delta(P, l)| = |\Delta(Q, l)|$ for all $P \in \mathfrak{B}_i$. Hence

$$\sum_{P \in \mathfrak{B}_i} |\Delta(P, l)| = |\mathfrak{B}_i| |\Delta(Q, l)| = |\Delta|.$$

Thus

$$|\Delta| = |\text{T}| - |\mathfrak{B}| + r|\Delta|$$

Whence $(r - 1)|\Delta| = |\mathfrak{B}| - |\text{T}|$. As $r \geqslant 1$, we obtain

$$0 \leqslant (r - 1)|\Delta| < |\mathfrak{B}| < |\mathfrak{B}| + 1 \leqslant |\Delta|.$$

Therefore, $r = 1$.   □

13.10 Theorem. *Let the Hall plane* $H(q)$ *be derived from the desarguesian plane* $A(q^2)$ *by deriving via* D. *If* G' *is the collineation group of* $H(q)$ *and* G *the collineation group of* $A(q^2)$ *and if* $q > 3$, *then* $G' = G_D$.

PROOF. By 13.1, $G_D = G'_{D'} \subseteq G'$. Furthermore, G_D contains a cyclic subgroup Z of order $q + 1$ the elements of which are homologies with axis $l \neq l_\infty$ and centre $P \text{ I } l_\infty$. From $|D| = q + 1$ we infer $P \notin D$ and $l \cap l_\infty \notin D$. Thus the elements of Z are homologies of $H(q)$ by 13.4. As G_D contains a subgroup isomorphic to $SL(2, q)$ which operates transitively on D^*, there exists such a Z for all $P \in D^*$. Let $\gamma \in G'$ and assume $D'^\gamma \neq D'$. Then $|D'^\gamma \cup D'| \leqslant 2|D'| = 2(q + 1)$. Furthermore,

$$q^2 + 1 - 2(q + 1) = q^2 - 2q - 1 = q(q - 2) - 1 > 0,$$

as $q > 2$. Hence there is a $P \in D^* \backslash (D'^\gamma \cup D')$. Let Z be a cyclic group of order $q + 1$ the elements of which are homologies with centre P leaving D^* invariant. Then Z also fixes D'. Similarly, let Z' be a cyclic group of order $q + 1$ which leaves D'^γ invariant, the elements of which are also homologies with centre P. Let a be the axis of the homologies in Z and a' the axis of the homologies in Z'. We may assume $O \text{ I } a, a'$. If $a \neq a'$, then $\Delta(P, OP) \neq \{1\}$ by the dual of 13.7. Hence $H(q)$ is desarguesian by 9.12, a contradiction. Therefore $a = a'$. By 11.7, we thus obtain $Z = Z'$. As a consequence $D' \cap D'^\gamma = \emptyset$ and $Z^\gamma = Z$ which yields $P^\gamma = P$ and

$(a \cap l_\infty)^\gamma = a \cap l_\infty$. Using the notation of 13.5, we may assume $D = D_1$. We then deduce from the assertions made that $D'^\gamma = D_t$ for some $t \neq 1$. By deriving $H(q)$ via D'^γ we obtain a desarguesian plane, i.e. by deriving $A(q^2)$ simultaneously via D_1 and D_t we obtain a plane isomorphic to $A(q^2)$. Hence, by 13.7, $q = 3$.   □

By 13.9, the collineation group of $H(q)$ is known, if $q > 3$, whereas 8.1, 8.2, and 8.3 take care of the case $q = 3$. We state a few more properties of the collineation group of $H(q)$. The proofs are left as exercises for the reader.

13.11 Theorem. *Let l be a line of $A(q^2)$ with $l \cap l_\infty \in D$. Then l is a Baer subplane of $H(q)$ the pointwise stabilizer of which has order $q(q-1)$.*

This theorem gives another proof of the fact that $H(q)$ is non-desarguesian if $q > 2$.

13.12 Theorem. *If $P \in D^*$, then there is a unique $P' \in D^* \setminus \{P\}$ such that $\Delta(P, OP')$ contains a cyclic subgroup of order $q + 1$. Moreover, $P'' = P$.*

13.13 Theorem. *Let l be a line of $H(q)$ with $l \cap l'_\infty \in D'$. Then there exists exactly one involutory perspectivity of $H(q)$ the axis of which is l.*

13.14 Theorem. *If q is odd, then each affine line of $H(q)$ is the axis of exactly one involutory homology.*

14. The Collineation Group of a Generalized André Plane

The groups $\mathrm{PSL}(2, q)$ and $\mathrm{PGL}(2, q)$ play a prominent rôle in the theory of finite projective planes. Thus a good working knowledge of them is important to every student of this theory. We shall state here without proofs a few theorems about their subgroups. Proofs may be found in Dickson [1958, Chapt. XII] and Huppert [1967, Chapt. II, §8].

14.1 Theorem. *If U is a subgroup of $\mathrm{PSL}(2, p^r)$, then U is one of the following groups:*

(1) *An elementary abelian p-group of order p^m with $m \leqslant r$.*
(2) *A cyclic group of order z where z is a divisor of $2^r - 1$ or $2^r + 1$, if $p = 2$, and a divisor of $\frac{1}{2}(p^r - 1)$ or $\frac{1}{2}(p^r + 1)$, if $p > 2$.*
(3) *A dihedral group of order $2z$ where z is as in (2).*
(4) *A semidirect product of an elementary abelian p-group of order p^m and a cyclic group of order t where t is a divisor of $p^{(m,r)} - 1$.*

(5) *A group isomorphic to* A_4. *In this case, r is even, if* $p = 2$.
(6) *A group isomorphic to* S_4. *In this case* $p^{2r} - 1 \equiv 0$ mod 16.
(7) *A group isomorphic to* A_5. *In this case* $p^r(p^{2r} - 1) \equiv 0$ mod 5.
(8) *A group isomorphic to* PSL(2, p^m) *where m divides r.*
(9) *A group isomorphic to* PGL(2, p^m) *with 2m a divisor of r.*

   The converse is also true which means that PSL(2, p^r) *contains a subgroup of type* (x) *provided the side conditions mentioned in* (x) *are satisfied.*

   The next theorem is actually the key to the proof of 14.1

14.2 Theorem. PSL(2, q) *contains exactly* $\frac{1}{2}q(q + 1)$ *cyclic subgroups of order* $(q - 1)(q - 1, 2)^{-1}$, *exactly* $\frac{1}{2}q(q - 1)$ *cyclic subgroups of order* $(q + 1)(q - 1, 2)^{-1}$ *and exactly* $q + 1$ *elementary abelian p-groups of order q. The set of all these groups is a partition of* PSL(2, q).

   This theorem is a consequence of the following:

14.3 Theorem. PGL(2, q) *contains exactly* $\frac{1}{2}q(q + 1)$ *cyclic subgroups of order* $q - 1$, *exactly* $\frac{1}{2}q(q - 1)$ *cyclic subgroups of order* $q + 1$ *and exactly* $q + 1$ *elementary abelian p-groups of order q. The set of all these groups is a partition of* PGL(2, q).

14.4 Theorem. *If* $q^2 \equiv 1$ mod 16 *and if U is a subgroup of* PSL(2, q) *which is isomorphic to* S_4, *then U is its own normalizer in* PSL(2, q) *as well as in* PGL(2, q). *If V is a subgroup of* PSL(2, q) *which is isomorphic to* A_4, *then there exists* $\eta \in$ PGL(2, q) *with* $V^\eta \subseteq U$.

14.5 Theorem. *If* $q^2 \not\equiv 1$ mod 16 *and if U is a subgroup of* PSL(2, q) *which is isomorphic to* A_4, *then U is its own normalizer in* PSL(2, q) *and of index 2 in its normalizer in* PGL(2, q). *All subgroups of* PSL(2, q) *which are isomorphic to* A_4 *are in one conjugacy class.*

14.6 Theorem. *If* $q^2 \equiv 1$ mod 10 *and if U is a subgroup of* PSL(2, q) *which is isomorphic to* A_5, *then U is its own normalizer in* PSL(2, q) *as well as in* PGL(2, q). *All subgroups of* PGL(2, q) *which are isomorphic to* A_5 *are conjugate under* PGL(2, q).

   A 2-group G is called *semidihedral*, if G is generated by two elements a and b subject to the conditions $a^{2^{n+1}} = 1 = b^2$ and $bab = a^{2^n - 1}$.

14.7 Theorem. *If* $q \equiv 3$ mod 4, *then the Sylow 2-subgroups of* GL(2, q) *are semidihedral.*

PROOF. Let 2^{n+1} be the highest power of 2 dividing $q^2 - 1$ and let e be a primitive 2^{n+1}-th root of unity in GF(q^2). Then $\left(\begin{smallmatrix} e & 0 \\ 0 & e^q \end{smallmatrix}\right)$ has order 2^{n+1}. As

$e \notin GF(q)$, we have $e^q - e \neq 0$. Thus the matrix $\left(\begin{smallmatrix} 1 & 1 \\ e & e^q \end{smallmatrix}\right)$ is regular. Furthermore

$$\begin{pmatrix} 1 & 1 \\ e & e^q \end{pmatrix}^{-1} = (e^q - e)^{-1} \begin{pmatrix} e^q & -1 \\ -e & 1 \end{pmatrix}$$

and $e^{q+1} = -1$, as $q \equiv 3 \bmod 4$. Therefore,

$$\begin{pmatrix} 1 & 1 \\ e & e^q \end{pmatrix} \begin{pmatrix} e & 0 \\ 0 & e^q \end{pmatrix} \begin{pmatrix} 1 & 1 \\ e & e^q \end{pmatrix}^{-1} = \begin{pmatrix} 0 & 1 \\ 1 & e + e^q \end{pmatrix}.$$

Hence $\left(\begin{smallmatrix} 0 & 1 \\ 1 & e+e^q \end{smallmatrix}\right)$ is an element of order 2^{n+1} of $GL(2, q)$. The matrix $\left(\begin{smallmatrix} 0 & -1 \\ 1 & 0 \end{smallmatrix}\right)$ has order 4 and its inverse is $\left(\begin{smallmatrix} 0 & 1 \\ -1 & 0 \end{smallmatrix}\right)$. Furthermore,

$$\begin{pmatrix} 0 & -1 \\ 1 & 0 \end{pmatrix} \begin{pmatrix} 0 & 1 \\ 1 & e+e^q \end{pmatrix} \begin{pmatrix} 0 & 1 \\ -1 & 0 \end{pmatrix} = \begin{pmatrix} 0 & 1 \\ 1 & e+e^q \end{pmatrix}^{2^n - 1}.$$

Now it follows easily that $\left(\begin{smallmatrix} 0 & -1 \\ 1 & 0 \end{smallmatrix}\right)$ and $\left(\begin{smallmatrix} 0 & 1 \\ 1 & e+e^q \end{smallmatrix}\right)$ generate a semidihedral group S of order 2^{n+2}. As $q \equiv 3 \bmod 4$ and $|GL(2, q)| = q(q^2 - 1)(q - 1)$, we find that S is a Sylow 2-subgroup of $GL(2, q)$. □

Each semidihedral group S has exactly one cyclic subgroup of index 2 and all cyclic subgroups of S of order greater than 4 are contained in this cyclic subgroup. This easily checked remark will be useful in the sequel.

14.8 Lemma (Lüneburg 1974). *Let $\mathfrak{P}$ be a finite projective plane, let P and Q be two distinct points of $\mathfrak{P}$ and let l and m be lines with $P \not\,I\, l$ and $Q \not\,I\, m$. Put $P' = PQ \cap l$ and $Q' = PQ \cap m$. If $(|\Delta(P, l)|, |\Delta(Q, m)|) \geq 3$, then there exists $\eta \in H = \langle \Delta(P, l), \Delta(Q, m) \rangle$ with $\{P, P'\}^\eta = \{Q, Q'\}$.*

PROOF. Let q be the order of $\mathfrak{A}$.

Case 1: There exists a prime $p > 2$ which divides $(|\Delta(P, l)|, |\Delta(Q, m)|)$. Then p divides $q - 1$ but not $q(q + 1)$. Let S be a Sylow p-subgroup of H such that $S \cap \Delta(P, l)$ is a Sylow p-subgroup of $\Delta(P, l)$. As p does not divide $q(q + 1)$, the group S has two fixed points on PQ. We infer from $S \cap \Delta(P, l) \neq \{1\}$ that $S \cap \Delta(P, l)$ has exactly two fixed points on PQ, namely P and P'. Hence P and P' are the only fixed points of S on PQ. Likewise, we find a Sylow p-subgroup T of H such that Q and Q' are the only fixed points of T on PQ. As S and T are conjugate in H, there exists $\eta \in H$ with $\{P, P'\}^\eta = \{Q, Q'\}$.

Case 2: $(|\Delta(P, l)|, |\Delta(Q, m)|) = 2^r$. By assumption $r \geq 2$. Therefore, $q + 1 \equiv 2 \bmod 4$. This implies that a Sylow 2-subgroup of H either has at least two fixed points on PQ or at least one orbit of length 2 on PQ. As a Sylow 2-subgroup S_0 of $\Delta(P, l)$ has exactly two fixed points on PQ, namely P and P', whereas the remaining points of PQ are split by S_0 into orbits of length $|S_0| \geq 4$, we have that there exists a Sylow 2-subgroup S of H such that $\{P, P'\}^S = \{P, P'\}$, whereas the remaining points of PQ are split by S into orbits of length at least 4. The same reasoning as in case 1 now yields an $\eta \in H$ with $\{P, P'\}^\eta = \{Q, Q'\}$. □

14.9 Theorem. *Let* $\mathfrak{A}$ *be a finite generalized André plane with kernel* $\mathrm{GF}(q)$ *and of order* q^d. *Furthermore, let* P, Q, P', Q' *be four distinct points on* l_∞ *and* O *an affine point of* $\mathfrak{A}$. *If each of the groups* $\Delta(P, OP')$ *and* $\Delta(Q, OQ')$ *contains a cyclic subgroup of order* $u^{-1}(q^d - 1)$ *where* $u = \mathrm{lcm}(q^i - 1 \,|\, 1 \leqslant i,$ $i < d, i$ *divides* $d)$, *then either* $d = 1$ *and* $\mathfrak{A}$ *is desarguesian or* $d = 2$ *and* $\mathfrak{A}$ *is a Hall plane.*

PROOF. We may assume that $\mathfrak{A}$ is non-desarguesian. Then $d \geqslant 2$. Assume $u^{-1}(q^d - 1) = 2$. Then $q^d - 1 = p^{rd} - 1$ has no p-primitive prime divisor. By Zsigmondy's theorem $r = 1$ and $d = 2$ and hence $p + 1 = 2$: a contradiction. Thus we may use 14.8, 11.4, 9.2 and 2.1 to assume that $\mathfrak{A}$ has the form $\pi(V)$ with $V = L \oplus L$ where $L = \mathrm{GF}(q^d)$, that $P = V(0) \cap l_\infty$, $P' = V(\infty) \cap l_\infty$, $Q = V(1) \cap l_\infty$ and that the components of π which are distinct from $V(0)$ and $V(\infty)$ have the form $\{(x, x^{\alpha(m)}m) \,|\, x \in L\}$ with $m \in L^*$ and $\alpha(m) \in \mathrm{Aut}(L)$.

Let $Q' = V(a) \cap l_\infty$, let Γ_0 be the cyclic group of order $u^{-1}(q^d - 1)$ which is contained in $\Delta((0), V(\infty))$ and let Γ_1 be the cyclic group of the same order which is contained in $\Delta((1), V(a))$. As we know, $\Delta((\infty), V(0))$ also contains a cyclic subgroup Γ_∞ of order $u^{-1}(q^d - 1)$. It follows from 14.8 and 11.7 that Γ_0, Γ_1 and Γ_∞ are the only cyclic subgroups of order $u^{-1}(q^d - 1)$ of $\Delta((0), V(\infty))$ etc. unless $q = 3$ and $d = 2$. In the latter case, $\mathfrak{A}$ is the Hall plane of order 9 by 8.4. Thus we may assume that $q^d > 9$, as there are no non-desarguesian generalized André planes of order 4 and 8 by 10.2.

Put $A = \Gamma_0\Gamma_\infty$ and denote by A_i ($i = 1, a$) the stabilizer of $V(i)$ in A. By 13.9 and 9.12, we obtain $A_1 = A_{1,a} = A_a$. Hence A_1 normalizes Γ_1. We want to show that A_1 centralizes Γ_1. Let $q = p^r$ and assume that there is a p-primitive divisor of $p^{rd} - 1$. Using 10.5 and Schur's lemma, we see that A_1 and Γ_1 induce the same group of automorphisms on $V(1)$. Thus A_1 and Γ_1 centralize each other on $V(1)$. As Γ_1 induces the identity on $V(a)$, they also centralize each other on $V(a)$. Therefore, A_1 and Γ_1 centralize each other in this case, since $V = V(1) \oplus V(a)$. As $L = \mathrm{GF}(p)[A_1]$, we infer that Γ_1 consists only of L-linear mappings.

If there is no p-primitive prime divisor of $p^{rd} - 1$, then, by 6.2, either $rd = 2$ and hence $d = 2$ and $p + 1 = 2^s$ or $p = 2$ and $rd = 6$. In the latter case, $r > 1$ by 10.3. Therefore, there exists a q-primitive prime divisor of $q^d - 1$. This yields again that Γ_1 consists of L-linear mappings. Thus we may assume $r = 1$, $d = 2$ and $p + 1 = 2^s$. Then $p \equiv 3 \bmod 4$. By 14.7, the Sylow 2-subgroups of $\mathrm{GL}(2, p)$ are semidihedral of order 2^{s+2}. Furthermore, $A_1\Gamma_1$ is contained in a Sylow 2-subgroup S of $\mathrm{GL}(2, p)$. As $2^s > 4$, the group S contains only one cyclic subgroup of order 2^s. Hence A_1 and Γ_1 centralize each other on $V(1)$ and therefore on V. This shows that Γ_1 consists in all cases of L-linear mappings only.

As $\Gamma_0 \subseteq A$, the group Γ_0 is L-linear as well. Let Γ_0^* and Γ_1^* be the groups induced in $\mathrm{PGL}(V, L) = \mathrm{PGL}(2, q^d)$ by Γ_0 and Γ_1 respectively. Then $|\Gamma_0^*| = |\Gamma_1^*| = u^{-1}(q^d - 1)$. Let H be the subgroup of $\mathrm{PGL}(V, L)$ generated by Γ_0^* and Γ_1^* and put $H_0 = H \cap \mathrm{PSL}(2, q^d)$. By 13.9 and 9.12, $H \neq \mathrm{PGL}(2, q^d)$. Furthermore, $u^{-1}(q^d - 1) > 2$ and $|\mathrm{PGL}(2, q^d) : \mathrm{PSL}(2, q^d)| \leqslant 2$ imply $H_0 \neq \{1\}$. Using 13.9, 9.12, 14.1 and $\Gamma_0^* \neq \Gamma_1^*$ we obtain one of the following cases:

a) $u^{-1}(q^d - 1) = 4$.
b) $H_0 \cong A_4$.
c) $H_0 \cong S_4$.
d) $H_0 \cong A_5$.
e) $H_0 \cong \mathrm{PSL}(2, p^m)$ with m a divisor of dr where $q = p^r$.
f) $H_0 \cong \mathrm{PGL}(2, p^m)$ with $2m$ a divisor of dr and $p > 2$.

Assume $u^{-1}(q^d - 1) = 4$. Then there are no p-primitive prime divisors of $p^{rd} - 1$. Hence $r = 1$ and $d = 2$. This yields $q + 1 = 4$, yet $q^d > 3^2$. Therefore, a) does not occur.

Case b) is the same as case e), if q is even. In case c), q is always odd. Case d) is the same as case e), if q is a power of 2 or a power of 5. Thus we assume that we are in one of the cases b), c) or d) but not e). By 14.4, 14.5 resp. 14.6, we obtain $H \cong A_4$, S_4 or A_5. This yields $u^{-1}(q^d - 1) = 3$ or 5. Therefore, 3 resp. 5 is a q-primitive prime divisor of $q^d - 1$. Assume $u^{-1}(q^d - 1) = 3$. As $q^2 - 1 \equiv 0 \bmod 3$, we have $d = 2$ and hence $q + 1 = 3$: a contradiction. If $u^{-1}(q^d - 1) = 5$, then $d = 2$ or 4. If $d = 4$, then $u = q^2 - 1$ and thus $q^2 + 1 = 5$. Thus $q = 2$: a contradiction. If $d = 2$, then $q + 1 = 5$ and $q = 4$: again a contradiction.

It remains to consider the cases e) and f). In case f) we have $H = H_0$, as $\mathrm{PGL}(2, p^m)$ is its own normalizer in $\mathrm{PGL}(2, q^d)$. Thus we have that H is a subgroup of $\mathrm{PGL}(2, p^m)$ which contains $\mathrm{PSL}(2, p^m)$. Furthermore, m is a proper divisor of dr, since H is a proper subgroup of $\mathrm{PGL}(2, p^{dr})$. We infer from 14.3 that $u^{-1}(q^d - 1)$ is a divisor of $p^m - 1$ or of $p^m + 1$.

Every p-primitive divisor of $q^d - 1$ divides $u^{-1}(q^d - 1)$ and hence $p^m - 1$ or $p^m + 1$. If it divides $p^m - 1$, then $p^m - 1 = q^d - 1$: a contradiction. Therefore, it divides $p^m + 1$. This implies that it divides $p^{2m} - 1$, whence $p^{2m} - 1 = q^d - 1 = p^{rd} - 1$. Thus $2m = rd$ in this case.

If $q = p$ and $d = 2$, then $m = 1$ and $2m = rd$. Therefore, $2m \neq rd$ implies $q^d = 64$ and $d = 2$ or 3 by 10.3. If $d = 2$, then $u^{-1}(q^d - 1) = 9$ divides $2^m - 1$ or $2^m + 1$. As m is a proper divisor of $dr = 6$, we have that 9 divides $2^m + 1$, whence $m = 3$ and $2m = 6 = dr$: a contradiction. If $d = 3$, then $u^{-1}(q^d - 1) = 21$ divides $2^m - 1$ or $2^m + 1$: again a contradiction. Thus $2m = rd$ in all cases.

Put $q^d = t^2$. Then H contains a subgroup isomorphic to $\mathrm{PSL}(2, t)$ and is contained in a subgroup of $\mathrm{PGL}(2, t^2)$ which is isomorphic to $\mathrm{PGL}(2, t)$. This yields $|V(0)^H| = t(t - 1)$. Furthermore, $V(0)^H$ is contained in the desarguesian spread π_0 over L. Moreover, H splits π_0 into two orbits. Hence, we obtain $\pi = \pi_0$, if $|\pi \cap \pi_0| > t(t - 1)$: a contradiction. Therefore, $\pi \cap \pi_0 = V(0)^H$. Let $X \in \pi_0 \backslash \pi$. Then there exists a Sylow p-subgroup Σ of H which fixes X pointwise. As $|X| = t^2$ and $X \notin \pi$, we infer that X is a Baer subplane of $\pi(V)$. If $Y \in \pi$ and $|X \cap Y| \geqslant 2$, then $Y \in \pi \backslash \pi_0$. Let D

be the set of points at infinity of $\pi(V)$ which are on lines in $\pi\backslash\pi_0$. It follows from $|\pi\backslash\pi_0| = t + 1$ that $\pi(V)$ is derivable via D and that $\pi_0(V)$ is the derived plane. Thus, by the development of section 13, $\pi(V)$ is the Hall plane of order t^2. Moreover, $q = t$ and $d = 2$.   □

14.10 Corollary (Foulser 1969). *Let $\mathfrak{A}$ be a finite generalized André plane and G the collineation group of $\mathfrak{A}$. If $\mathfrak{A}$ is neither desarguesian nor a Hall plane, then there are two distinct points P and Q on l_∞ with $\{P,Q\}^G = \{P,Q\}$. If $\mathfrak{B} \subseteq l_\infty$ is an orbit of G with $\mathfrak{B} \cap \{P,Q\} = \emptyset$, then $|\mathfrak{B}| > 2$.*

14.11 Corollary (Foulser 1969). *Let $L(+, \circ)$ be a finite generalized André system and $L(+, \cdot)$ be the field belonging to it. Let $V = L \oplus L$ be the vector space of rank 2 over $L(+, \cdot)$ and $\pi(V)$ the generalized André plane defined by $L(+, \circ)$. Assume that $\pi(V)$ is neither desarguesian nor a Hall plane. If G is the stabilizer of $(0,0)$ in the collineation group of $\pi(V)$, then G is a subgroup of $(\Gamma L(1, L) \times \Gamma L(1, L))\langle\rho\rangle$, where ρ is the mapping defined by $(x, y)^\rho = (y, x)$.*

This follows from 14.10, 11.7, 9.5 and 9.1.

14.12 Corollary (Foulser 1969). *Let $L(+, \circ)$ be a finite generalized André system. Then $\mathfrak{A}(L)$ is an André plane, if and only if $L(+, \circ)$ is an André system.*

PROOF. If $L(+, \circ)$ is an André system, then we are done. Assume that $\mathfrak{A}(L)$ is an André plane. If $\mathfrak{A}(L)$ is desarguesian, there is nothing to prove. Thus we may assume $d \geqslant 2$. The case $d = 2$ is taken care of by 12.5. Hence we may assume $d \geqslant 3$. This yields that $\mathfrak{A}(L)$ is not a Hall plane. The Corollary now follows from 14.10 and 12.4.   □

The only finite generalized André systems we have seen so far are the André systems and the Dickson nearfields. If F is a Dickson nearfield of type $\{q, n\}$, then $v = n$, where v is defined as in section 10. Therefore, by 12.4, F is an André system, if and only if n divides $q - 1$. This shows that there are Dickson nearfields which are not André systems. Hence, by 14.12, there exist generalized André planes which are not André planes. We shall exhibit now generalized André systems all of which are due to D. Foulser and some of which are neither André systems nor nearfields thus proving that there are generalized André planes which are neither André planes nor nearfield planes. More examples of the type described below and more information about them can be found in Foulser 1967b.

Let q be a power of a prime and let d be a positive integer all of whose prime divisors divide $q - 1$. Define the mapping λ from I_{q^d-1} onto I_d by $\lambda(i) \equiv i \bmod d$. Let w be a generator of the multiplicative group of $F = \mathrm{GF}(q^d)$ and define $x \circ w^i$ by $x \circ w^i = x^{q^{\lambda(i)}} w^i$. As $\lambda(i) \equiv i \bmod d$, we have $x \circ w^i = x^{q^i} w^i$.

14.13 Lemma. F_λ *is a generalized André system with* $k(F_\lambda) = GF(q)$.

PROOF. Let $i, j \in I_{q^d-1}$ and assume $i \equiv j$ mod $q^t - 1$, where $t = (d, \lambda(i) - \lambda(j))$. Let r be a prime dividing d and let r^a be the highest power of r which divides d. Let r^b be a divisor of $i - j$ and assume $b \leqslant a$. Then r^b divides $\lambda(i) - \lambda(j)$, as $\lambda(i) - \lambda(j) \equiv i - j$ mod d. Hence r^b divides t. By 6.3 a), r^{b+1} divides $q^t - 1$ and hence $i - j$, as r divides $q - 1$ and $i \equiv j$ mod $q^t - 1$. It follows that r^a divides $i - j$. As this is true for all primes dividing d, we have $i \equiv j$ mod d. This yields $\lambda(i) = \lambda(j)$, whence $t = d$. Therefore, $i \equiv j$ mod $q^d - 1$ and hence $i = j$.

As λ is a mapping from I_{q^d-1} onto I_d, the group Γ generated by all the $\rho^{\lambda(i)}$ with $i \in I_{q^d-1}$, where ρ is the mapping defined by $x^\rho = x^q$ is the Galois group of F over $GF(q)$. By 10.7, $k(F_\lambda) = GF(q)$. □

14.14 Lemma. *Let* F_λ *be as above. Then*:

a) $n_r(F_\lambda) = \{w^i \mid i \equiv 0 \text{ mod } ord_d(q)\}$, *where* $ord_d(q)$ *denotes the order of* q mod d.
b) $n_m(F_\lambda) = \{w^i \mid i \equiv 0 \text{ mod } d(d, q-1)^{-1}\}$.
c) $n_r(F_\lambda) = n_m(F_\lambda)$ *unless* $q \equiv 3$ mod 4 *and* $d \equiv 0$ mod 8.

PROOF. a) By 11.1, $w^i \in n_r(F_\lambda)$, if and only if $kq^i \equiv k$ mod d for all k. This is equivalent to $q^i \equiv 1$ mod d, i.e. $i \equiv 0$ mod $ord_d(q)$.

b) By 11.1, $w^i \in n_m(F_\lambda)$, if and only if $iq^k \equiv i$ mod d for all k. This is equivalent to $i(q - 1) \equiv 0$ mod d, i.e. $i \equiv 0$ mod $d(d, q - 1)^{-1}$.

c) It follows easily from 6.3 that $ord_d(q) = d(d, q - 1)^{-1}$ unless $q \equiv 3$ mod 4 and $d \equiv 0$ mod 8. In the latter case $q^A \equiv 1$ mod d, if $A = d(2(d, q - 1))^{-1}$. This establishes c). □

REMARK. If $q \equiv 3$ mod 4 and $d \equiv 0$ mod 8, then $ord_d(q) = 2d(d, q^2 - 1)^{-1}$.

14.15 Lemma. *Let* F_λ *be as above. Then the following statements are equivalent*:

a) F_λ *is nearfield.*
b) d *divides* $q - 1$.
c) F_λ *is an André system.*

PROOF. As F_λ is a nearfield, if and only if $F_\lambda^* = n_r(F_\lambda)$, we infer from 14.14 that F_λ is a nearfield, if and only if $ord_d(q) = 1$. This proves the equivalence of a) and b).

As $\lambda(i + j) = \lambda(i)$, if and only if $j \equiv 0$ mod d, we have $v = d$. Hence b) and c) are equivalent by 12.4. This proves 14.15. □

By 14.15 and 14.12, it is now easy to see that there exist generalized André planes which are neither André planes nor nearfield planes.

CHAPTER III

Rank-3-Planes

This chapter starts with Wagner's celebrated theorem that finite line transitive affine planes are translation planes. This theorem is used in the proof of Kallaher's & Liebler's theorem on affine planes of rank 3. Finally, rank-3-planes with an orbit of length 2 on l_∞ are investigated.

15. Line Transitive Affine Planes

In this section we shall prove Wagner's celebrated theorem that a finite affine plane $\mathfrak{A}$ admitting a group of collineations acting transitively on the set of lines of $\mathfrak{A}$ is a translation plane. On our way to establishing this theorem we shall prove a lot of other results which will also be useful in the sequel.

15.1 Lemma (Gleason 1956). *Let G be a finite group operating on a set Ω and let p be a prime. If Ψ is a subset of Ω such that for every $\alpha \in \Psi$ there is a p-subgroup Π_α of G fixing α but no other point of Ω, then Ψ is contained in an orbit of G.*

PROOF. Let $\alpha \in \Psi$ and let $\Phi = \alpha^G$. Then Π_α splits Φ into orbits the length of each of them excepting $\{\alpha\}$ being divisible by p. Hence $|\Phi| \equiv 1 \bmod p$. Let $\beta \in \Psi \backslash \Phi$. Then Φ is split by Π_β into orbits the length of each of which is divisible by p. Hence $|\Phi| \equiv 0 \bmod p$: a contradiction. Therefore $\Psi \subseteq \Phi$. $\qquad\square$

15.2 Lemma (Gleason 1956). *Let $\mathfrak{A}$ be a finite affine plane and H a collineation group of $\mathfrak{A}$. Put $H(P) = H \cap \mathrm{T}(P)$. If h is an integer greater* *than* 1 *such that* $|H(P)| = h$ *for all* $P \, I \, l_\infty$, *then* $\mathfrak{A}$ *is a translation plane and* H *contains the translation group of* $\mathfrak{A}$.

PROOF. Put $T_0 = T \cap H$. Then $|T_0| = 1 + (n + 1)(h - 1)$, if n is the order of $\mathfrak{A}$. Hence $|T_0| > n$, as $h \geqslant 2$. On the other hand, $|T_0|$ divides n^2. Therefore, $n^2 = |T_0|r$ with $1 \leqslant r < n$. Thus $n^2 = r(1 + (n + 1)(h - 1))$. This yields $1 \equiv r \bmod n + 1$. As $1 \leqslant r < n$, we have $r = 1$ and therefore $|T_0| = n^2$. $\square$

15.3 Lemma (Ostrom 1964). *Let* $\mathfrak{A}$ *be a finite affine plane of order* n. *If* $|T| > n$, *then* n *is a power of a prime*.

PROOF. Let $P \, I \, l_\infty$, l where l is a line distinct from l_∞. Then

$$n|T(P)| \geqslant |l^T||T(P)| = |l^T||T_l| = |T| > n.$$

Therefore, $T(P) \neq \{1\}$ for all $P \, I \, l_\infty$. By 1.1, 1.6a) and the finiteness of T, it follows that T is an elementary abelian p-group.

Let l be a line with $|l^T|$ maximal and let t be the number of point orbits of T in $\mathfrak{A}$. Then $n^2 = |T|t$. Furthermore, $t + n + 1$ is the number of point orbits of T in the projective closure $\mathfrak{A} \vee l_\infty$ of $\mathfrak{A}$. By the Theorem of Dembowski-Hughes-Parker, $t + n + 1$ is also the number of line orbits of T in $\mathfrak{A} \vee l_\infty$. Hence $t + n$ is the number of line orbits of T in $\mathfrak{A}$. By the maximality of $|l^T|$, we get the inequality $|l^T|(t + n) \geqslant n(n + 1)$. From $n^2 = |T|t$, we obtain $t < n$. Therefore, $2|l^T|n > n(n + 1)$ which implies $2|l^T| > n + 1 > n$. As $|l^T|$ is a divisor of n, this yields $|l^T| = n$. As n is a divisor of $|T| = p^r$, the lemma is proved. $\square$

15.4 Lemma. *Let* G *be a finite permutation group acting transitively on* Ω *and assume that* $n = |\Omega|$ *is even. If* Σ *is a Sylow 2-subgroup of* G *and* $\sigma \in \mathfrak{Z}(\Sigma)$, *then the number of elements left fixed by* σ *is distinct from* $\sqrt{n}$ *and* $\sqrt{n + 1} - 1$.

PROOF. Let $\Phi = \{w, | w \in \Omega, w^\sigma = w\}$. As $\sigma \in \mathfrak{Z}(\Sigma)$, the set Φ is invariant under Σ. Therefore, Φ is the union of orbits of Σ. Let 2^a be the highest power of 2 dividing n. As G operates transitively on Ω, we have that 2^a divides the length of each orbit of Σ (Wielandt [1964, Theorem 3.4]). Thus 2^a divides $|\Phi|$. If $|\Phi| = \sqrt{n}$, then 2^{2a} divides n. Hence $a = 0$, i.e. n is odd: a contradiction. If $|\Phi| = \sqrt{n + 1} - 1$, then 2^{a+1} divides $(\sqrt{n + 1} - 1) \cdot (\sqrt{n + 1} + 1) = n$: again a contradiction. $\square$

Next we formulate two lemmas which we shall prove simultaneously.

15.5 Lemma. *Let* $\mathfrak{P}$ *be a finite projective plane of even order, let* (P, l) *be a flag of* $\mathfrak{P}$ *and* G *a group of collineations of* $\mathfrak{P}$ *fixing* (P, l). *Assume that* G *operates transitively on the set of points on* l *which are distinct from* P. *Let* N *be an elementary abelian 2-subgroup of the Sylow 2-subgroup* Σ *of* G. *Denote* *by* * *the restriction map of G onto* $l\backslash\{P\}$. *If* $\nu \in \mathrm{N}\backslash\{1\}$ *and* $\nu^* \in \mathfrak{Z}(\Sigma^*)$, *then* ν *is an elation of* $\mathfrak{P}$. *In particular, every involution in* $\mathfrak{Z}(\Sigma)$ *is an elation.*

15.6 Lemma. *Let* $\mathfrak{P}$ *be a finite projective plane of odd order, let* P *and* Q *be distinct points of* $\mathfrak{P}$ *and* G *a group of collineations fixing* P *and* Q *and acting transitively on* $PQ\backslash\{P,Q\}$. *Let* N *be an elementary abelian 2-subgroup of the Sylow 2-subgroup* Σ *of* G *and denote by* * *the restriction map of* G *onto* $PQ\backslash\{P,Q\}$. *If* $\nu \in \mathrm{N}\backslash\{1\}$ *and* $\nu^* \in \mathfrak{Z}(\Sigma^*)$, *then* ν *is an involutory homology of* $\mathfrak{P}$. *In particular, every involution in* $\mathfrak{Z}(\Sigma)$ *is a homology.*

PROOF. Let $\nu \in \mathrm{N}\backslash\{1\}$ and $\nu^* \in \mathfrak{Z}(\Sigma)$. If $\nu^* = 1$, then ν is a perspectivity with axis l or PQ. Thus we may assume $\nu^* \neq 1$. Assume that ν is not a perspectivity. Then ν is a Baer involution by 4.6, i.e. ν fixes a subplane of order m with $m^2 = n$ pointwise. As $l^\nu = l$ resp. $(PQ)^\nu = PQ$, we see that ν fixes exactly $m + 1$ points of l resp. PQ, i.e. ν has $m = \sqrt{n}$ fixed points in the case of 15.5 or $m - 1 = \sqrt{n-1+1} - 1$ fixed points in the case of 15.6 contrary to 15.4. $\qquad\square$

15.7 Lemma (Wagner 1965). *Let* $\mathfrak{A}$ *be a finite affine plane and* G *a group of collineations of* $\mathfrak{A}$. *Then* G *operates transitively on the set of points of* $\mathfrak{A}$, *if and only if* G_P *operates transitively on the set of affine lines through* P *for all* $P \mathbin{\mathrm{I}} l_\infty$.

PROOF. Let $\mathfrak{p}_1, \ldots, \mathfrak{p}_s$ be the orbits of G which are contained in l_∞ and put $\mathfrak{L}_i = \{l \mid l \cap l_\infty \in \mathfrak{p}_i\}$, $i = 1, 2, \ldots, s$. Then $\mathfrak{L}_i$ is invariant under G for all i.

Assume that G operates transitively on the set of points of $\mathfrak{A}$. Then G has exactly $s + 1$ point orbits in $\mathfrak{A} \vee l_\infty$. By the Theorem of Dembowski-Hughes-Parker, G has exactly $s + 1$ line orbits in $\mathfrak{A} \vee l_\infty$. As $\{l_\infty\}$ is one such orbit and as the $\mathfrak{L}_i$ are invariant under G, we see that each $\mathfrak{L}_i$ is an orbit of G. Let $P \mathbin{\mathrm{I}} l_\infty$ and let l and m be affine lines through P. Then there exists an i with $P \in \mathfrak{p}_i$. Hence $l, m \in \mathfrak{L}_i$. As $\mathfrak{L}_i$ is an orbit of G, there exists $\gamma \in G$ with $l^\gamma = m$. This implies $(l \cap l_\infty)^\gamma = m \cap l_\infty = l \cap l_\infty$, i.e. $\gamma \in G_P$. This establishes the first part of 15.7.

Conversely, assume that G_P operates transitively on the set of affine lines through P for all $P \mathbin{\mathrm{I}} l_\infty$. Then $\mathfrak{L}_i$ is an orbit of G for all i. Thus G has exactly $s + 1$ line orbits in $\mathfrak{A} \vee l_\infty$. Therefore, G has exactly $s + 1$ point orbits in $\mathfrak{A} \vee l_\infty$. As G splits l_∞ into s point orbits, G operates transitively on the set of points of $\mathfrak{A}$. $\qquad\square$

15.8 Lemma. *Let* $\mathfrak{A}$ *be a finite affine plane of even order and* G *a group of collineations of* $\mathfrak{A}$. *If* G *operates transitively on the set of points of* $\mathfrak{A}$, *then* G *contains a non-trivial translation of* $\mathfrak{A}$. *In particular, if* G *operates primitively on the set of points of* $\mathfrak{A}$, *then* $\mathfrak{A}$ *is a translation plane and* G *contains the translation group of* $\mathfrak{A}$.

PROOF. Let $P \mathrel{I} l_\infty$. By 15.7, G_P operates transitively on the set of affine lines through P. By the dual of 15.5, G_P contains an involutory elation τ. If τ is not a translation, then there is a line l through P with $l \neq l_\infty$ and $\tau \in G(P,l)$. As G_P permutes the affine lines through P transitively, there exists an integer $h \in \mathbb{N}\setminus\{1\}$ with $|G(P,x)| = h$ for all affine lines x through P. Denote by E the group of all elations with centre P which are contained in G. Then $|E| = |G(P,l_\infty)| + n(h-1)$. As n and $|E|$ are divisible by h, the order of $G(P,l_\infty)$ is divisible by h. Hence $G(P,l_\infty) \neq \{1\}$. The remaining statement is now trivial, as the group of all elations which are contained in G is a normal subgroup of G. □

15.9 Lemma. *Let $\mathfrak{A}$ be a finite plane of even order and G a group of collineations of $\mathfrak{A}$. If G_l operates transitively on the set of affine points on l for all affine lines l of $\mathfrak{A}$, then $G(P) \neq \{1\}$ for all $P \mathrel{I} l_\infty$.*

PROOF. Let $l_1, \ldots, l_n$ be all the affine lines through P. Then $G(P,l_i) \neq \{1\}$ for at least one $i \in \{1,2,\ldots,n,\infty\}$ by 15.5. If $i = \infty$, then we are done. Let $i \neq \infty$. Obviously, G operates transitively on the set of points of $\mathfrak{A}$. Hence G_P operates transitively on $\{l_1, \ldots, l_n\}$ by 15.7. Therefore, there exists $h \in \mathbb{N}\setminus\{1\}$ with $|G(P,l_i)| = h$ for all $i \in \{1,2,\ldots,n\}$. This yields that h divides $|G(P)| = |G(P,l_\infty)|$ as in the proof of 15.8. □

15.10 Lemma. *Let n be an odd integer and G a permutation group of order $2n$ which operates transitively on a set Ω of length n. If $G_{a,b} = \{1\}$ for all distinct $a,b \in \Omega$, then for each pair $a,b \in \Omega$ with $a \neq b$ there exists an involution $\sigma \in G$ with $a^\sigma = b$.*

PROOF. As n is odd, every involution in G has exactly one fixed point. Therefore, G contains exactly n involutions.

Let $1 \neq \gamma \in G_a$. Then $|\{\{x,x^\gamma\} \mid x \in \Omega\setminus\{a\}\}| = \frac{1}{2}(n-1)$. Let $1 \neq \delta \in G_b$ and assume

$$\{\{x,x^\gamma\} \mid x \in \Omega\setminus\{a\}\} \cap \{\{x,x^\delta\} \mid x \in \Omega\setminus\{b\}\} \neq \varnothing.$$

Then there exists $x \in \Omega\setminus\{a,b\}$ with $x^\gamma = x^\delta$. This yields $x^{\gamma\delta} = x$ and hence $(\gamma\delta)^2 = 1$. From this we infer $\gamma\delta = \delta\gamma$. This implies $a^\delta = a$ and therefore, $a = b$. Thus

$$\left| \bigcup_{a \in \Omega} \{\{x,x^{\gamma(a)}\} \mid x \in \Omega\setminus\{a\}\} \right| = \sum_{a \in \Omega} |\{\{x,x^{\gamma(a)}\} \mid x \in \Omega\setminus\{a\}\}|$$
$$= \tfrac{1}{2}n(n-1),$$

where $\gamma(a)$ is defined by $G_a = \{1,\gamma(a)\}$. As $\frac{1}{2}n(n-1)$ is the number of 2-subsets of Ω, the lemma is proved. □

15.11 Lemma (Wagner 1964). *Let Ω be a set of even length and put $|\Omega| = n + 1$. Let G be a group of permutations of Ω with the property: For all $a \in \Omega$, the stabilizer G_a contains a subgroup F_a of even order which operates* *transitively on* $\Omega\backslash\{a\}$. *Furthermore,* $F_{a,b,c} = \{1\}$ *for all* $b,c \in \Omega\backslash\{a\}$ *with* $b \neq c$. *Then* G_a *operates primitively on* $\Omega\backslash\{a\}$.

PROOF. F_a operates as a Frobenius group on $\Omega\backslash\{a\}$. Hence F_a has a Frobenius kernel K_a, i.e. a normal subgroup K_a which operates transitively and regularly on $\Omega\backslash\{a\}$. Furthermore, F_a contains an involution γ. As $|K_a| = n$ and n is odd, $\gamma \notin K_a$. Hence $K_a\langle\gamma\rangle$ is a subgroup of F_a which satisfies the hypotheses of 15.10. We may assume $F_a = K_a\langle\gamma\rangle$.

Assume that G_a operates imprimitively on $\Omega\backslash\{a\}$ and let $I = \{a_1, \ldots, a_t\}$ be a system of imprimitivity of G_a. Then $1 < t < n$ and t divides n. Hence t is odd. Put $T = I \cup \{a\}$. Then $T \neq \Omega$. Therefore, there exists $b \in \Omega\backslash T$. By 15.9, there exists an involution $\gamma_i \in F_b$ with $a^{\gamma_i} = a_i$. We show $T^{\gamma_i} = T$. Obviously $a^{\gamma_i} \in T$ for all i. Let $c = a_j^{\gamma_i} \notin T$. Then $i \neq j$, as $a_i^{\gamma_i} = a \in T$. Since c is not in T, there exists an involution $\delta \in F_c$ with $a_i^\delta = a$. From this it follows that $a^{\gamma_i\delta} = a_i^\delta = a$. Hence $\gamma_i\delta \in G_a$. Furthermore, $a_i^{\gamma_i\delta} = a^\delta = a_i \in I \cap I^{\gamma_i\delta}$. As I is a system of imprimitivity of G_a, we obtain $I = I^{\gamma_i\delta}$. Therefore, $c = c^\delta = a_j^{\gamma_i\delta} \in I \cap (\Omega\backslash T) = \emptyset$: a contradiction. Hence T is invariant under all γ_i.

Let $D = \langle\gamma_1, \ldots, \gamma_t\rangle$. Then D leaves T invariant and operates transitively on T. Furthermore, D is a subgroup of F_b. Hence $|D|$ is not divisible by 4. As $|D| = (t + 1)|D_a|$, and $D_a \subseteq F_{b,a}$, and as $t + 1$ is even, we have $D_a = \{1\}$, i.e. $|D| = t + 1$. Hence $D = \{1, \gamma_1, \ldots, \gamma_t\}$. Therefore, D is an elementary abelian 2-group, a contradiction. $\square$

15.12 Corollary (Wagner 1964). *If in addition to the assumption of* 15.11 *the Frobenius kernel* K_a *of* F_a *is normal in* G_a, *then* K_a *is an elementary abelian p-group and n is a power of p.*

PROOF. K_a is abelian, as $F_{a,b}$ contains an involution. Since G_a operates primitively on $\Omega\backslash\{a\}$, it follows that K_a is characteristically simple. Hence K_a is an elementary abelian p-group. $\square$

15.13 Corollary (Wagner 1965). *Let* $\mathfrak{P}$ *be a finite projective plane and* (O, l) *a non-incident point-line-pair of* $\mathfrak{P}$. *If there exists an involutory* (P, OQ)-*homology for all pairs of distinct points* P *and* Q *on* l, *then the order of* $\mathfrak{P}$ *is a power of a prime.*

This follows from 13.9 and 15.12.

15.14 Lemma (Ostrom & Wagner 1959). *Let* $\mathfrak{P}$ *be a projective plane of order* n *and let* Γ *be a 2-group of collineations of* $\mathfrak{P}$. *If the fixed points and fixed lines of* Γ *form a subplane of order* m, *then there exists an integer* g *with* $n = m^{2^g}$.

PROOF. We make induction on $|\Gamma|$. If $\Gamma = \{1\}$, then $m = n$ and $g = 0$. Assume $\Gamma \neq \{1\}$. Let γ be an involution in $\mathfrak{Z}(\Gamma)$. As Γ fixes a subplane pointwise, γ is a Baer involution, i.e. γ fixes a subplane $\mathfrak{F}$ of order s with $s^2 = n$ elementwise. Since $\gamma \in \mathfrak{Z}(\Gamma)$, the group Γ induces a 2-group Γ^* of collineations in $\mathfrak{F}$. Furthermore, the subplane fixed elementwise by Γ is a subplane of $\mathfrak{F}$. Since $\gamma^* = 1$, we have $|\Gamma^*| < |\Gamma|$. By induction, there exists an integer g such that $m^{2^{g-1}} = s$. Hence $m^{2^g} = s^2 = n$. □

15.15 Theorem (Wagner 1965). *Let $\mathfrak{A}$ be a finite affine plane and Γ a group of collineations of $\mathfrak{A}$. Then the following statements are equivalent:*

a) Γ *operates transitively on the set of lines of $\mathfrak{A}$.*
b) Γ *operates transitively on the set of affine points of $\mathfrak{A}$ and also on the set of points on l_∞.*
c) Γ *acts transitively on the set of flags of $\mathfrak{A}$.*

PROOF. Assume a). Let $P \mathrel{I} l_\infty$. Then Γ_P acts transitively on the set of affine lines through P. Hence Γ acts transitively on the set of points of $\mathfrak{A}$ by 15.7. It follows immediately from the line transitivity that Γ is transitive on l_∞. Thus b) is a consequence of a).

Assume b). By assumption, Γ has a point orbit of length n^2 and a point orbit of length $n + 1$ in $\mathfrak{A} \vee l_\infty$. As $(n^2, n + 1) = 1$, the stabilizer Γ_P of an affine point P is still transitive on l_∞. Hence Γ is flag transitive. This establishes c).

a) is obviously a consequence of c). □

15.16 Theorem (Wagner 1965). *Let $\mathfrak{A}$ be a finite affine plane. If G is a group of collineations which acts transitively on the set of lines of $\mathfrak{A}$, then $\mathfrak{A}$ is a translation plane and G contains the translation group of $\mathfrak{A}$.*

PROOF. If G contains a translation $\neq 1$, then $|G(P)| = |T(P) \cap G| = h > 1$ for a fixed $h \in \mathbb{N}$ and all $P \mathrel{I} l_\infty$, since G operates transitively on l_∞. This yields the assertions of 15.16 by 15.2. Hence it suffices to prove that G contains a non-identity translation. If the order n of $\mathfrak{A}$ is even, then G contains such a translation by 15.15 and 15.8. Therefore, we may assume that n is odd.

a) If (P, l) is a flag of $\mathfrak{A}$ and $k = |G_{P,l}|$, then $|G| = n^2(n + 1)k$.

This follows from 15.15 and the remark that $n^2(n + 1)$ is the number of flags of $\mathfrak{A}$.

As $n(n + 1)$ is the number of lines of $\mathfrak{A}$, we obtain by a):
b) $|G_l| = nk$.

As n^2 is the number of points of $\mathfrak{A}$, we have by a):
c) $|G_P| = nk$. Moreover, if $A \mathrel{I} l_\infty$, then $|G_{P,A}| = k$.

The last assertion follows from $G_{P,A} = G_{P,PA}$ and a).

Since G_A permutes transitively the points of $\mathfrak{A}$, we have:
d) $|G_A| = n^2 k$ for $A \mathrel{I} l_\infty$.

As G_A acts transitively on the set of points of $\mathfrak{A}$, it follows from 15.7 that $G_{A,B}$ operates transitively on the set of affine lines through B for all $B \mathrel{I} l_\infty$. Hence we have proved:

e) Let $A, B \mathrel{I} l_\infty$. If l and m are two affine lines through B, then $|G_{A,B,l}| = |G_{A,B,m}|$.

Next we prove that n is a power of a prime. Put $\mathfrak{P} = \mathfrak{A} \vee l_\infty$ and let $\mathfrak{S}$ be a subplane of $\mathfrak{P}$ with the properties:

(1) There exists a 2-subgroup of G the fixed points and fixed lines of which are the points and lines of $\mathfrak{S}$.

(2) There does not exist a proper subplane of $\mathfrak{S}$ with the property (1).

As $\mathfrak{P}$ and $\{1\}$ satisfy (1) and as $\mathfrak{P}$ is finite, there exists such a subplane $\mathfrak{S}$.

From $l_\infty^G = l_\infty$ we infer that l_∞ is a line of $\mathfrak{S}$. Let Σ be a 2-subgroup of G and assume that Σ is maximal with respect to the property that the fixed points and fixed lines of Σ are the points and lines of $\mathfrak{S}$. Let $\mathfrak{B}$ be the affine plane one obtains from $\mathfrak{S}$ by deleting l_∞ and put $\Lambda = G_{\mathfrak{B}}$. Furthermore, let 2^a be the highest power of 2 dividing $n + 1$ and 2^b the highest power of 2 dividing k. Put $|\Sigma| = 2^c$.

f) Each flag of $\mathfrak{B}$ is left fixed by an element of Λ which induces in $\mathfrak{B}$ an involutory homology.

Let (P, l) be a flag of $\mathfrak{B}$. Then $\Sigma \subseteq G_{P,l}$. Hence, by a), 2^c divides 2^b. As n is odd, $a + b \geqslant b + 1$. Therefore there exists a 2-group Σ^* of G with $\Sigma \subseteq \Sigma^*$ and $|\Sigma^* : \Sigma| = 2$. This implies that Σ is normal in Σ^* and hence $\Sigma^* \subseteq \Lambda$. By (2), Σ^* induces an involutory homology in $\mathfrak{B}$. Thus Σ^* fixes a flag (Q, m) of $\mathfrak{B}$. Since $|G_{Q,m}| = k$ and $\Sigma \subseteq \Sigma^* \subseteq G_{Q,m}$, we obtain $c < b$. Therefore, Σ is not a Sylow 2-subgroup of $G_{P,l}$. Hence, there exists a 2-group Σ^{**} with $\Sigma \subseteq \Sigma^{**} \subseteq G_{P,l}$ and $|\Sigma^{**} : \Sigma| = 2$. This Σ^{**} induces an involutory homology in $\mathfrak{B}$ which fixes (P, l).

g) Let A, B I l_∞ and assume that A, B belong to $\mathfrak{S}$. Furthermore, let l and h be lines of $\mathfrak{B}$ through B. If there exists a collineation in Λ inducing an involutory homology in $\mathfrak{B}$ and fixing A, B and l, then there exists a collineation in Λ inducing an involutory homology in $\mathfrak{B}$ and fixing A, B and h. Let $\overline{\Lambda}$ be the subgroup of Λ which fixes $\mathfrak{S}$ pointwise. Then $\Sigma \subseteq \overline{\Lambda}$ and $\overline{\Lambda} \lhd \Lambda$. Hence $\overline{\Lambda} \lhd \Lambda_{A,B,l}$. By assumption $\Lambda_{A,B,l}/\overline{\Lambda}$ contains an involution. This shows that Σ is not a Sylow 2-subgroup of $G_{A,B,l}$. By e), $|G_{A,B,l}| = |G_{A,B,h}|$. Hence Σ is not a Sylow 2-subgroup of $G_{A,B,h}$. Therefore, there exists a 2-group Σ^* with $\Sigma \subseteq \Sigma^* \subseteq G_{A,B,h}$ and $|\Sigma^* : \Sigma| = 2$. Then Σ^* induces an involutory homology in $\mathfrak{S}$ which fixes A, B and h.

h) The order of $\mathfrak{A}$ is a power of a prime.

By 15.14, it suffices to show that $\mathfrak{B}$ has prime power order. As $\Lambda/\overline{\Lambda}$ is isomorphic to a group of collineations of $\mathfrak{S}$, we may assume $\overline{\Lambda} = \{1\}$ and we may identify Λ with the group of collineations induced on $\mathfrak{S}$ by Λ.

Case 1: Λ does not contain an involutory homology with centre a point of $\mathfrak{B}$. By f), Λ contains an involutory homology σ. Let l be the axis and A the centre of σ. The A I l_∞. Using g) and 4.8, we see that each affine line through $l \cap l_\infty$ is the axis of an involutory homology in Λ with centre A. Therefore, $\mathfrak{S}$ is (A, l_∞)-transitive by 13.10. We may assume that $\mathfrak{S}$ is not a translation plane. Then A is the centre of every involutory homology in Λ. By this and f), each line of $\mathfrak{B}$ which does not pass through A is the axis of an involutory homology. This implies by 13.10 that $\mathfrak{S}$ is the dual of a translation plane. Hence the order of $\mathfrak{B}$ is a power of a prime.

Case 2: There is exactly one point O in $\mathfrak{B}$ which is the centre of an involutory homology in Λ. In this case, f) and g) together with 15.13 yield that the order of $\mathfrak{B}$ is a power of a prime.

Case 3: There exist two distinct points of $\mathfrak{B}$ which are centres of involutory homologies of Λ. Assume that every line of $\mathfrak{B}$ carries a centre, and let v be the number of centres. Furthermore, denote by k_l the number of centres on l. Then $v(m+1) = \sum_l k_l$, where m is the order of $\mathfrak{B}$. As $k_l \geqslant 1$ for all l and since there exists l with $k_l \geqslant 2$, we have $v(m+1) \geqslant m(m+1)$. Hence $v > m$. Therefore, by 13.9 and 15.3, m is a power of a prime. We may thus assume that there exists a line l of $\mathfrak{B}$ which does not carry a centre. Put $B = l \cap l_\infty$ and let $B \neq A \mathrel{I} l_\infty$ and suppose P is an affine centre. As there exists an involutory homology fixing A, B and AP, there is an involutory homology ρ fixing A, B and l by g). If l is the axis of ρ, then A is its centre. If l is not the axis, then B is the centre, as l does not carry an affine centre. In either case, A is the only fixed point of ρ in $l_\infty \setminus \{B\}$. Hence $\Lambda_{B,l}$ acts transitively on $l_\infty \setminus \{B\}$ by 15.1.

As there are two centres, Λ contains a non-trivial translation τ by 13.9. If B is not the centre of τ, then the translation group of $\mathfrak{B}$ has order greater than m by the transitivity of Λ_B on $l_\infty \setminus \{B\}$. Hence m is a power of a prime by 15.3. Therefore, we may assume that all translations in Λ have centre B. This implies that all affine centres lie on a line h through B. Thus $h^\Lambda = h$. Moreover, if σ is an involutory homology in Λ fixing l, then B is the centre of σ, as $h^\sigma = h$. Using the transitivity of $\Lambda_{B,l}$ on $l_\infty \setminus \{B\}$, we see by 13.9 that there exist at least $m-1$ non-trivial elations with centre B and axis distinct from l_∞. As $\Lambda(B, l_\infty) \neq \{1\}$, the group of all elations with centre B has order greater than m. Hence, by the dual of 15.3, m is a power of a prime.

By h), we have $n = p^r$ for some prime p. Let Π be a Sylow p-subgroup of G. Then Π acts transitively on the set of p^{2r} points of $\mathfrak{A}$. As p does not divide $n + 1$, there exists $A \mathrel{I} l_\infty$ with $A^\Pi = A$. Let B be a point on l_∞ distinct from A. Then $|B^\Pi|$ divides $|\Pi|$ and $|B^\Pi| \leqslant n = p^r$. As p^{2r} divides $|\Pi|$, we have that p^r divides $|\Pi_B|$. Choose $B \neq A$ such that $|\Pi_B| \geqslant |\Pi_C|$ for all $C \mathrel{I} l_\infty$ with $C \neq A$. Let F be a fixed point of Π_B with $A \neq F \mathrel{I} l_\infty$. Then $\Pi_B = \Pi_{B,F} \subseteq \Pi_F$. As $|\Pi_B| \geqslant |\Pi_F|$, we have $\Pi_B = \Pi_F$. From the transitivity of Π on the set of points of $\mathfrak{A}$, we infer from 15.7 that Π_F and hence Π_B operate transitively on the set of affine lines through F. Furthermore, Π_B also acts transitively on the set of affine lines through A, as we shall see now. Let Σ be a Sylow p-subgroup of G_B which contains Π_B. By d), $|G_B| = n^2 k$. Hence Σ is a Sylow p-subgroup of G, as $|G| = n^2(n+1)k$. Furthermore, $\Pi_B \subseteq \Sigma_A$. This implies $\Pi_B = \Sigma_A$, since Σ and Π are conjugate. By 15.7, Σ_A and hence Π_B have the required transitivity property.

Finally, let $1 \neq \tau \in \mathfrak{Z}(\Pi_B)$. Then τ fixes the points A and B. Thus τ fixes at least one line l of $\mathfrak{A}$. If $l \cap l_\infty$ is not a fixed point of Π_B, then there exists $\mu \in \Pi_B$ such that $l^\mu \cap l$ is an affine point which is obviously fixed by

τ. Thus $(l^\mu \cap l)A$ is a fixed line of τ. As $((l^\mu \cap l)A) \cap l_\infty = A$, we infer that there is always a fixed line l of τ such that $l \cap l_\mu$ is a fixed point of Π_B. By what we have seen, $|l^{\Pi_B}| = n$. Hence all the lines through $l \cap l_\infty$ are fixed by τ. As $o(\tau) = p^s$, we infer that τ is an elation. This and $A^\tau = A$ and $B^\tau = B$ yield that τ is a translation.   □

15.17 Theorem. *Let $\mathfrak{A}$ be a finite affine plane. If G is a group of collineations which acts transitively on the set of lines of $\mathfrak{A}$, then G acts primitively on the set of points of $\mathfrak{A}$.*

PROOF. By 15.16, $\mathfrak{A}$ is a translation plane and G contains the translation group T of $\mathfrak{A}$. Furthermore, G_P acts transitively on the set of lines through P for all points P of $\mathfrak{A}$. Let $\mathfrak{B}$ be a set of points of $\mathfrak{A}$ such that $|\mathfrak{B}| \geqslant 2$ and $\mathfrak{B}^\gamma = \mathfrak{B}$ for all $\gamma \in G$ with $\mathfrak{B}^\gamma \cap \mathfrak{B} \neq \emptyset$. We have to show that $\mathfrak{B}$ consists of all the points of $\mathfrak{A}$. In order to prove this, we note first that $|T_\mathfrak{B}| = |\mathfrak{B}|$. Pick a point $P \in \mathfrak{B}$. As G_P operates transitively on l_∞, there exists an $h \in \mathbb{N}\backslash\{1\}$ such that $|T_\mathfrak{B} \cap T(A)| = h$ for all $A \, I \, l_\infty$. It follows from 15.2 that $T(A) \subseteq T_\mathfrak{B}$ and hence $T = T_\mathfrak{B}$.   □

16. Affine Planes of Rank 3

Let G be a permutation group on the set Ω. We shall call G a *rank-3-group* on Ω, if it acts transitively on Ω and if G_α for $\alpha \in \Omega$ splits Ω into exactly 3 orbits. An affine plane $\mathfrak{A}$ will be called a *rank-3-plane*, if it admits a group of collineations acting as a rank-3-group on the set of points of $\mathfrak{A}$. Any such collineation group will be called a rank-3-collineation group of $\mathfrak{A}$.

Every nearfield plane is a rank-3-plane. For, if P and Q are the two "special points" on l_∞, then $G = T\Delta(P, OQ)\Delta(Q, OP)$ permutes the points of $\mathfrak{A}$ transitively. Furthermore, $G_O = \Delta(P, OQ)\Delta(Q, OP)$ has exactly four orbits, namely $\{O\}$, $OP\backslash\{O\}$, $OQ\backslash\{O\}$ and the set of the remaining points. As there exists a collineation ρ fixing O and interchanging P and Q by 3.11, we see that $G\langle\rho\rangle$ is a rank-3-collineation group of the given nearfield plane.

Every Hall-plane is a rank-3-plane: Let the Hall plane $H(q)$ be derived from the desarguesian plane $A(q^2)$ via D. Consider the collineation group G of $A(q^2)$ which fixes D globally. As we have seen, G splits the points of l_∞ into two orbits. Furthermore, $|G(O, l_\infty)| = q^2 - 1$. Hence the stabilizer of every line l in G acts doubly transitively on l. Hence G operates as a rank-3-group on the set of points of $A(q^2)$. As G is also a collineation group of $H(q)$, we see that $H(q)$ is a rank-3-plane.

The Lüneburg planes which we shall construct later on are also rank-3-planes.

The next theorem was proved independently by Johnson & Kallaher 1974 and by myself 1973.

16.1 Theorem. *Let $\mathfrak{A}$ be a finite affine plane and let G be a group of collineations of $\mathfrak{A}$. If G_l acts doubly transitively on the set of points on l for all lines l of $\mathfrak{A}$, then $\mathfrak{A}$ is a translation plane and G contains the translation group of $\mathfrak{A}$.*

PROOF. Let n be the order of $\mathfrak{A}$.

a) For $U \text{ I } l_\infty$ either $G(U) = \{1\}$ or $|G(U)| = n$.

Assume $G(U) \neq \{1\}$ and let l be an affine line through U. As $G(U)$ is normal in G_l and as $G(U)$ operates faithfully on l, we have $|G(U)| = n$ by the 2-transitivity of G_l on l.

Case 1: n is even. Then $G(U) \neq \{1\}$ for all $U \text{ I } l_\infty$ by 15.9. Hence we are done by a).

Case 2: n is odd. Obviously, we have:

b) G is point transitive.

Next we prove:

c) If G contains an involutory homology with affine centre, then 16.1 holds.

This follows from b) and 13.9.

d) If $G(U) = \{1\}$ for all $U \text{ I } l_\infty$, then there exists $V \text{ I } l_\infty$ such that the axis of every involutory homology in G passes through V. In particular, $V^G = V$.

Let σ be an involutory homology in G and let l be the axis of σ. By c), $l \neq l_\infty$. Then $V = l \cap l_\infty$. Let h be a second line through V. By b) and 15.7, there exists $\gamma \in G_V$ with $l^\gamma = h$. Hence there exists an involutory homology ρ with axis h. Let U be the centre of σ and W be the centre of ρ. If $U = W$, then 13.9 implies $G(U) \neq \{1\}$. Hence $U \neq W$. As there are exactly n affine lines through V, each point on l_∞ which is distinct from V is the centre of a homology with axis through V. By 15.1, G_V acts transitively on $l_\infty \setminus \{V\}$. As G cannot be transitive on l_∞ by 15.16, $G = G_V$. Let τ be an involutory homology in G. If V is the centre of τ, then G contains an involutory homology with affine centre by 4.9 contradicting c). Hence V is on the axis of τ.

e) If G has no fixed points, then 16.1 holds.

This follows from d) and a).

Therefore, we may assume that G has a fixed point $V \text{ I } l_\infty$.

f) If the axis of every involutory homology in G passes through V, then 16.1 holds.

If l_∞ is an axis, then we are done by c). Assume that l_∞ is not an axis. Let $U \text{ I } l_\infty$ and $U \neq V$. Furthermore, let h be an affine line through U. As h is fixed by an involutory homology in G_h by 15.6, U is the centre of an involutory homology in G. As $G_h = G_{V,h}$, the group G_h acts transitively on the set of affine lines through V. (Remember that G_h acts doubly transitively on l.) Using 13.9, we obtain $|G(U)| = n$. As this is true for all $U \neq V$, the plane $\mathfrak{A}$ is a translation plane and G contains the translation group.

By f), d) and a), there exists a point U I l_∞ with $|G(U)| = n$. If U is not a fixed point of G, then 16.1 holds. Hence we may assume $V = U$. Furthermore, we may assume that there exists an involutory homology in G with centre V. Let X I l_∞, $X \neq V$. Choose an affine line h through X. Then X is fixed by an involutory homology ρ. We may assume that l_∞ is not the axis of ρ. Then V and X are the only fixed points of ρ on l_∞. Hence, by 15.1, $G = G_V$ acts transitively on $l_\infty \setminus \{V\}$. Therefore, every point U I l_∞ with $U \neq V$ is on the axis of an involutory homology with centre V. Therefore, we may assume by 4.8 that V is the centre of every involutory homology in G. Applying 15.6 again, we see that every line h not through V is the axis of an involutory homology in G. This yields by 13.9 that the dual of $\mathfrak{A} \vee l_\infty$ is a translation plane with respect to V. This implies by 1.12 that there is exactly one involutory (V, h)-homology for all lines h not through V.

Again let h be a line not through V and let Σ be a Sylow 2-subgroup of G_h. Furthermore, let A be an abelian normal subgroup of Σ and put $N = \{\alpha \,|\, \alpha \in A, \ \alpha^2 = 1\}$. Then N is a characteristic subgroup of A and hence a normal subgroup of Σ. Let * be the restriction of G_h onto h. Let $1 \neq \sigma \in N$ with $\sigma^* \in \mathfrak{Z}(\Sigma^*)$. By 15.6, σ^* is a homology. Therefore, h is the axis of σ. Hence $\sigma^* = 1$, i.e. $N^* \cap \mathfrak{Z}(\Sigma^*) = \{1\}$. This implies $N^* = \{1\}$, as Σ^* is a 2-group. Then N consists of (V, h)-homologies only. This yields $|N| \leqslant 2$ by the above remark. Therefore, A is cyclic. We infer from this by Gorenstein [1968, 5.4.10] that Σ is either cyclic, or dihedral, or semidihedral, or a generalized quaternion group. In either case, Σ^* contains a cyclic subgroup Z^* of index 1 or 2. Let Z be the preimage of Z^*. Then $|\Sigma : Z| \leqslant 2$. Hence $Z \cap \mathfrak{Z}(\Sigma) \neq \{1\}$, whence Z contains the only central involution σ of Σ. (Recall that $\mathfrak{Z}(\Sigma)$ is cyclic.) Let τ be an involution in Z. Then $\tau^{*2} = 1$. Hence $\tau^* \in Z^* \cap \mathfrak{Z}(\Sigma^*)$. This yields $\tau^* = 1$ by 15.6. Therefore, $\tau = \sigma$ and Z contains exactly one involution.

As n is odd, Σ has an affine fixed point P on h. Let Q and R be affine points distinct from P with Q I PV, R I h and

$$|\Sigma_Q| = \max\{|\Sigma_X| \,\big|\, P \neq X \text{ I } PV\}$$

and

$$|\Sigma_R| = \max\{|\Sigma_Y| \,\big|\, P \neq Y \text{ I } h\}.$$

We show that Σ_Q is a Sylow 2-subgroup of $G_{P, Q, h}$. Let Δ be a Sylow 2-subgroup of $G_{P, Q, h}$ which contains Σ_Q. Then there exists $\gamma \in G_{P, h}$ with $\Delta^\gamma \subseteq \Sigma$. As $\Delta \subseteq G_{P, Q, h}$, we have $\Delta^\gamma \subseteq \Sigma_{Q^\gamma}$. Hence

$$|\Sigma_Q| \geqslant |\Sigma_{Q^\gamma}| \geqslant |\Delta^\gamma| = |\Delta| \geqslant |\Sigma_Q|.$$

This establishes $\Delta = \Sigma_Q$. Similarly, Σ_R is a Sylow 2-subgroup of $G_{P, R}$.

Q is contained in an orbit of length $n - 1$ of G_P, as G_{PV} is 2-transitive on PV. Furthermore, R lies in an orbit of length $n(n - 1)$ of G_P, as $|G(V, PV)| = n$ and as $G_{P, l}$ is transitive on $l \setminus \{P\}$. Therefore, we have:

(*) $|G_P| = (n - 1)|G_{P, Q}| = n(n - 1)|G_{P, R}|.$

As n is odd, (*) yields $|\Sigma_Q| = |\Sigma_R|$. Since the involution σ which lies in Z has no fixed points on $\overline{PV}$ other than P, we have $\Sigma_Q \cap Z = \{1\}$. This implies $|\Sigma_Q| \leqslant 2$. On the other hand $\sigma \in \Sigma_R$. Thus $|\Sigma_R| \geqslant 2$. Hence $|\Sigma_Q| = |\Sigma_R| = 2$. Let $1 \neq \tau \in \Sigma_Q$, then τ has at least 2 affine fixed points on $\overline{PV}$. As $\overline{PV}$ cannot be the axis of τ, we infer that τ is a Baer involution. Hence τ fixes a point F on h which is distinct from P and $l \cap l_\infty$. As $\sigma \in \Sigma_F$, we obtain the contradiction $2 = |\Sigma_R| \geqslant |\Sigma_F| \geqslant 4$. This final contradiction proves 16.1. $\qquad\square$

16.2 Lemma (Kallaher 1969b). *Let $\mathfrak{A}$ be a finite affine plane and G a rank-3-collineation group of $\mathfrak{A}$, then one of the following holds:*

a) *G acts transitively on the set of lines of $\mathfrak{A}$ and G_l acts as a rank-3-group on l for all lines l of $\mathfrak{A}$.*

b) *G has exactly two orbits on l_∞ and G_l operates doubly transitively on l for all lines l of $\mathfrak{A}$.*

PROOF. Let P be a point of $\mathfrak{A}$ and let $\mathfrak{p}$ and $\mathfrak{q}$ be the orbits of G_P other than $\{P\}$. Furthermore, let l and m be lines through P and assume that l carries points of $\mathfrak{p}$ as well as points of $\mathfrak{q}$. As m meets at least one of $\mathfrak{p}$ and $\mathfrak{q}$, we may assume that m carries a point R of $\mathfrak{p}$. Let Q be a point of $\mathfrak{p}$ which is on l. Then there exists $\gamma \in G_P$ with $Q^\gamma = R$. Hence $l^\gamma = m$. This implies that G is flag transitive. Furthermore, G_l acts as a rank-3-group on l for the particular l under consideration and hence for all lines, as G acts flag transitively on $\mathfrak{A}$. This is case a). Thus we may assume that for all lines l through P either $l\backslash\{P\} \subseteq \mathfrak{p}$ or $l\backslash\{P\} \subseteq \mathfrak{q}$. Therefore, the set of lines through P is split into two orbits, one of them consisting of all the lines l with $l\backslash\{P\} \subseteq \mathfrak{p}$ and the other one consisting of all the lines l with $l\backslash\{P\} \subseteq \mathfrak{q}$. Therefore, G has at most two orbits on l_∞. But G cannot be transitive on l_∞, as in this case G would be flag transitive by 15.15. Hence G splits l_∞ into two orbits. As $G_{P,l}$ operates transitively on $l\backslash\{P\}$, we see finally that G_l acts doubly transitively on l. This is case b). $\qquad\square$

16.3 Corollary (Kallaher 1969b, Liebler 1970). *Let $\mathfrak{A}$ be a finite affine plane. If G is a rank-3-group of collineations of $\mathfrak{A}$, then $\mathfrak{A}$ is a translation plane and G contains the translation group of $\mathfrak{A}$.*

This follows from 16.2, 16.1 and 15.16.

16.4 Theorem (Kallaher 1969b). *Let $\mathfrak{A}$ be a finite affine plane and let G be a rank-3-collineation group of $\mathfrak{A}$. Then one of the following holds:*

a) *G has no fixed point on l_∞ and G operates primitively on the set of points of $\mathfrak{A}$.*

b) *G has a fixed point on l_∞ and G operates imprimitively on the set of points of $\mathfrak{A}$.*

This follows from 16.3, 16.2, 15.17 and 1.3.

17. Rank-3-Planes with an Orbit of Length 2 on the Line at Infinity

Our aim in this section is to prove:

17.1 Theorem (Kallaher & Ostrom 1971, Kallaher 1974, Lüneburg 1974, 1976 a + b, Lüneburg & Ostrom 1975). *Let $\mathfrak{A}$ be a finite affine plane of order n. If G is a rank-3-collineation group of $\mathfrak{A}$ such that G has an orbit of length 2 on l_∞, then $\mathfrak{A}$ is a generalized André plane or $n = 5^2, 7^2, 11^2, 23^2, 29^2, 59^2$.*

The numbers listed in this theorem are really exceptions, as the nearfield planes over the seven exceptional nearfields admit collineation groups satisfying the hypotheses of the theorem. But it is an easy exercise after all that we have done so far to prove that these planes are not generalized André planes.

PROOF. According to 16.3, $\mathfrak{A}$ is a translation plane and G contains the translation group T of $\mathfrak{A}$. Let O be an affine point of $\mathfrak{A}$ and put $H = G_O$. Then $G = TH$. As T operates trivially on l_∞, the groups G and H have the same orbits on l_∞. Let $\{P, Q\}$ be the orbit of length 2 of H on l_∞.

H has three point orbits on $\mathfrak{A}$. One is $\{O\}$. The other two consist of the points on OP and OQ other than O, resp. the points off the lines OP and OQ. The length of the first of these is $2(n - 1)$ and the length of the second $(n - 1)^2$. Furthermore, $|H : H_{P,Q}| = 2$.

Since $\mathfrak{A}$ is a translation plane, $n = p^r$ where p is a prime. Let π be a p-primitive prime divisor of $p^r - 1$ and let π^s be the highest power of π dividing $p^r - 1$. If $\pi = 2$, then p is odd and hence $r = 1$. In this case $\mathfrak{A}$ is desarguesian and hence a generalized André plane. Therefore, we may assume $r > 1$ whence $\pi > 2$. As $(n - 1)^2$ divides $|H|$, the highest power of π dividing $|H|$ is π^{2s+t} for some non-negative integer t. Since $\pi \neq 2$ and $|H : H_{P,Q}| = 2$, we see that π^{2s+t} divides $|H_{P,Q}|$. Put $K = H_{P,Q}$ and denote by $K(P, OQ)$ the group of all (P, OQ)-homologies contained in K. Define $K(Q, OP)$ similarly. Let Σ be a Sylow π-subgroup of K. Then $\Sigma \cap K(P, OQ)$ is a Sylow π-subgroup of $K(P, OQ)$, as $K(P, OQ)$ is normal in K. The order of $K(P, OQ)$ divides $n - 1$. Hence $|\Sigma \cap K(P, OQ)| \leqslant \pi^s$. Let Σ^* be the centralizer of $T(Q)$ in Σ. Then $\Sigma^* \subseteq K(P, OQ)$, since Σ^* fixes O and $T(Q)$ acts transitively on OQ. In particular, $\Sigma^* \subseteq K(P, OQ) \cap \Sigma$. By 3.8, $K(P, OQ) \cap \Sigma \subseteq \Sigma^*$. Hence $\Sigma^* = K(P, OQ) \cap \Sigma$.

As Σ^* is the centralizer of $T(Q)$ in Σ, the quotient group Σ/Σ^* is isomorphic to a subgroup of $\mathrm{GL}(r, p)$. Hence Σ/Σ^* is cyclic and $|\Sigma/\Sigma^*| \leqslant \pi^s$ by 10.5. Since

$$\pi^{2s+t} = |\Sigma| = |\Sigma/\Sigma^*||\Sigma^*|,$$

the inequalities $|\Sigma/\Sigma^*| \leqslant \pi^s$ and $|\Sigma^*| \leqslant \pi^s$ imply $t = 0$ and $|\Sigma/\Sigma^*| = |\Sigma^*| = \pi^s$.

Similarly $|\Sigma^{**}| = \pi^s$ where $\Sigma^{**} = \Sigma \cap K(Q, OP)$. Hence $\Sigma = \Sigma^*\Sigma^{**}$, as $\Sigma^* \cap \Sigma^{**} = \{1\}$. Furthermore, $\Sigma^{**} \cong \Sigma/\Sigma^*$. Hence Σ^{**} is cyclic. Likewise, Σ^* is cyclic. As Σ^* and Σ^{**} centralize each other, Σ is abelian.

Assume that Σ^* is normal in $K(P, OQ)$. As Σ^* is a Sylow π-subgroup of $K(P, OQ)$, it is the only Sylow π-subgroup of $K(P, OQ)$. Furthermore, $K(P, OQ)$ and $K(Q, OP)$ are conjugate in H. Therefore, Σ^{**} is the only Sylow π-subgroup of $K(Q, OP)$. Therefore, Σ is normal in $K(P, OQ)K(Q, OP)$ and hence in H. The group H acts transitively on $l_\infty \backslash \{P, Q\}$. Therefore, Σ splits $l_\infty \backslash \{P, Q\}$ into orbits of equal length. As this length is a divisor of $n - 1$ and is, at the same time, greater than or equal to π^s, it is equal to π^s. Therefore, $|\Sigma_W| = \pi^s$ for all $W \mathrel{I} l_\infty$ with $W \neq P, Q$. Let $\sigma \in \Sigma_W$ centralize $T(W)$. Then σ fixes each point on OW, as it fixes O and as $T(W)$ acts transitively on the set of affine points on OW. This together with $P^\sigma = P$ and $Q^\sigma = Q$ implies $\sigma = 1$. Thus Σ_W operates faithfully on $T(W)$. It follows therefore from 10.5 that Σ_W operates irreducibly on $T(W)$. Hence $\mathfrak{A}$ is a generalized André plane in this case by 9.3.

Our next aim is to show that Σ is almost always normal in $K(P, OQ)$.

17.2 Lemma (Hering). *Let V be an elementary abelian p-group of order p^r and let π be a p-primitive divisor of $p^r - 1$. Furthermore, let Σ be a non-trivial π-subgroup and A an abelian subgroup of $GL(r, p)$. If Σ normalizes A, then ΣA is cyclic.*

PROOF. Σ and hence ΣA act irreducibly on V by 10.5. Therefore, by Clifford's Theorem (see e.g. Huppert [1967, V.17.3, p. 565]), V is a completely reducible A-module. Let $V_1, \ldots, V_s$ be the homogeneous components of the A-module V. Then $A\Sigma$ and hence Σ operate transitively on $\{V_1, \ldots, V_s\}$. Thus s divides $|\Sigma|$. This implies that π divides s provided $s \neq 1$. Let $|V_1| = p^t$. Then $p^r = p^{ts}$, as $V = \bigoplus_{i=1}^s V_i$ and $|V_i| = p^t$ for all i. As π is a p-primitive divisor of $p^r - 1$, we see that r is the order of p modulo π. Hence $r = st$ divides $\pi - 1$. Thus s is not divisible by π, whence $s = 1$. Therefore, $V = V_1$ is the direct sum of isomorphic irreducible A-modules. We infer from 9.1 that $|A\Sigma : \mathfrak{C}_{A\Sigma}(A)|$ divides r. As A is abelian, $|A\Sigma : \mathfrak{C}_{A\Sigma}(A)|$ is a power of π. On the other hand r divides $\pi - 1$, as we have seen above. Thus $A\Sigma \subseteq \mathfrak{C}_{A\Sigma}(A)$. Since Σ is cyclic by 10.5, $A\Sigma$ is abelian. Schur's Lemma then implies that $A\Sigma$ is cyclic. $\qquad\square$

17.3 Corollary. *Let V be an elementary abelian p-group of order p^r and let π be a p-primitive prime divisor of $p^r - 1$. Let N be a subgroup of $GL(r, p)$ all Sylow subgroups of which are cyclic. Then N contains exactly one Sylow π-subgroup.*

PROOF. By Passman [1968, Prop. 12.11], there exists a cyclic normal subgroup M of N such that N/M is also cyclic. Let Σ be a Sylow π-subgroup of N. Then $M\Sigma$ is cyclic by 17.2. Furthermore, $N' \subseteq M$, as N/M is abelian. This yields $N' \subseteq M\Sigma$. Hence $M\Sigma$ is normal in N. As Σ is characteristic in $M\Sigma$, it is normal in N. $\qquad\square$

We now return to the proof of 17.2. By 3.5, we have:

(A) If $K(P,OQ)$ is soluble, then $K(P,OQ)$ contains a normal subgroup N all Sylow subgroups of which are cyclic and such that $K(P,OQ)/N$ is isomorphic to a subgroup of S_4.

(B) If $K(P,OQ)$ is not soluble, then $K(P,OQ)$ contains a normal subgroup L of index 1 or 2 which is the direct product of a group $M \cong SL(2,5)$ and a group N with cyclic Sylow subgroups. Furthermore, $(|N|,30) = 1$.

If $\pi \neq 3$ and if $K(P,OQ)$ is soluble in case $\pi = 5$, then Σ^* is normal in $K(P,OQ)$ by (A) and 17.3. Therefore we may assume $\pi \in \{3,5\}$ for all p-primitive prime divisors π of $p^r - 1$ and furthermore that $K(P,OQ)$ is not soluble, if 5 is a p-primitive prime divisor of $p^r - 1$.

If $\pi = 3$, then $r = 2$ as $p^2 \equiv 1 \bmod 3$. If $\pi = 5$, then $p^2 \equiv 1 \bmod 5$ or $p^4 \equiv 1 \bmod 5$. Thus $r = 2$ or 4 in this case. If $r = 4$, then 5 divides $p^2 + 1$. As $|K(P,OQ)|$ divides $n - 1 = p^4 - 1$ and $SL(2,5) \subseteq K(P,OQ)$, we have that 120 divides $p^4 - 1$. Consequently $p \neq 2,3$. This yields $p^2 - 1 \equiv 0 \bmod 12$ and hence $p^2 + 1 = 2 \cdot 5^t$, since every odd prime divisor of $p^2 + 1$ is a p-primitive prime divisor of $p^4 - 1$. Furthermore $5^t = |\Sigma^*|$. It follows from (B) that $|\Sigma^*| = 5$. Hence $t = 1$, i.e. $p^2 + 1 = 10$. As a consequence $p = 3$: a contradiction. Thus $r = 2$ in either case.

17.4 Lemma (Lüneburg 1974). *Let $\mathfrak{A}$ be a finite affine plane and let l be an affine line of $\mathfrak{A}$ and P a point on l_∞ with $P \nmid l$. If G is the collineation group of $\mathfrak{A}$ and if $l \cap l_\infty$ is not fixed by G_P, then either $\Delta(P,l)$ is cyclic or $T(P) = \{1\}$.*

PROOF. Assume $T(P) \neq \{1\}$. If $\Delta(P,l) = \{1\}$, there is nothing to prove. Hence we may assume $\Delta(P,l) \neq \{1\}$. Let $\gamma \in G_P$ be such that l^γ is not parallel to l. Then $\Delta(P,l^\gamma) = \Delta(P,l)^\gamma \neq \{1\}$. Therefore, by 13.9, there exists an elation ϵ with centre P and $l^\epsilon = l^\gamma$. As l and l^γ are not parallel, we have $\epsilon \notin T(P)$. It follows that the group E of all elations with centre P has a non-trivial partition π, all components of which are normal in E. Hence E is abelian by 1.1 and $K(E,\pi)$ is a Galoisfield by 1.6 a). As $\Delta(P,l)$ is isomorphic to a subgroup of the multiplicative group of $K(E,\pi)$, we see that $\Delta(P,l)$ is cyclic. □

17.5 Lemma. *Let p be a prime and let $\mathfrak{A}$ be a translation plane of order p^2. Furthermore, let O,P,Q be three distinct points of $\mathfrak{A}$ with $P,Q \mathrel{\text{I}} l_\infty$ and $O \nmid l_\infty$. Denote by G the collineation group of $\mathfrak{A}$. If $G(P,OQ)$ contains a subgroup isomorphic to $SL(2,5)$ and if $G_{O,P}$ operates transitively on $OP \setminus \{O,P\}$, then $p = 11, 19, 29, 59$.*

PROOF. As $T(P) \neq \{1\}$ and $G(P,OQ)$ is not cyclic, 17.4 yields $G_P = G_{P,Q}$. Furthermore, by 3.5, $G(P,OQ)$ contains a subgroup L of index 1 or 2 and $L = S \times N$ where $S \cong SL(2,5)$ and $(|N|,30) = 1$. This yields that S is the only subgroup of L which is isomorphic to $SL(2,5)$. As $SL(2,5)$ does not contain a subgroup of index 2, the group S is the only subgroup of $G(P,OQ)$ which is isomorphic to $SL(2,5)$. Hence S is normal in $G_{O,P,Q} = G_{O,P}$. Since $G_{O,P}$ operates transitively on $OP\setminus\{P,O\}$, it induces a group of automorphisms in $T(P)$ which acts transitively on $T(P)\setminus\{1\}$. Hence $G_{O,P}$ induces a group $\overline{G}_{O,P}$ in $PGL(2,p)$ which acts transitively on the projective line over $GF(p)$. Hence $p + 1$ divides $|\overline{G}_{O,P}|$. Obviously $\overline{S} \cong A_5$. Hence $\overline{S} = \overline{G}_{O,P}$ by 14.6. Therefore, $p + 1$ divides 60. This yields $p \in \{2,3,5,11,19,23,29,59\}$. On the other hand 120 divides $|G(P,OQ)|$ which is a divisor of $p^2 - 1$. Thus $p \geqslant 11$ and $p \neq 23$. $\square$

Hence, by 17.5, if $\pi \in \{3,5\}$ and if $K(P,OQ)$ is not soluble, then $n = 11^2$, 19^2, 29^2 or 59^2. As we have seen, the numbers 11^2, 29^2 and 59^2 are really exceptions. We shall prove in section 20 that 19^2 does not occur among the exceptions.

We may now assume that 3 is the only p-primitive prime divisor of $p^r - 1$ and that $K(P,OQ)$ is soluble. We may further assume that Σ^* is not normal in $K(P,OQ)$. The next lemma takes care of this situation.

17.6 Lemma. *Let p be a prime and let $\mathfrak{A}$ be a translation plane of order p^2. Furthermore, let O, P, Q be three distinct points of $\mathfrak{A}$ with P, $Q \text{ I } l_\infty$ and $O \nmid l_\infty$. Denote by G the collineation group of $\mathfrak{A}$. If $G(P,OQ)$ is soluble and contains more than one Sylow 3-subgroup and if furthermore $G_{O,P}$ operates transitively on $OP\setminus\{O,P\}$, then $p \in \{5,7,11,23\}$.*

PROOF. $G_{O,P} = G_{O,P,Q}$ by 17.4. Hence $G(P,OQ)$ is normal in $G_{O,P}$. By assumption, $G_{O,P}$ induces a group $\overline{G}_{O,P}$ in $PGL(2,p)$ which acts transitively on the set of points of the projective line over $GF(p)$. Thus $p + 1$ divides $|\overline{G}_{O,P}|$. The group $G(P,OQ)$ is mapped by $^-$ onto a normal subgroup M of $\overline{G}_{O,P}$. As $G(P,OQ)$ contains more than one Sylow 3-subgroup and as $G(P,OQ)$ is soluble, we infer from 14.1, using the fact that p does not divide $|G(P,OQ)|$, that M contains a subgroup A with $A \cong A_4$ and $|M:A| = 2$. Since A is generated by its Sylow 3-subgroups, A is characteristic in M and hence normal in $\overline{G}_{O,P}$. Therefore, by 14.4 and 14.5, $|\overline{G}_{O,P}:A| \leqslant 2$. Hence $p + 1$ divides 24. This yields $p \in \{2,3,5,7,11, 23\}$. As 12 divides $p^2 - 1$, we find $p \in \{5,7,11,23\}$. $\square$

By 17.6, $n \in \{5^2, 11^2, 23^2\}$, if 3 is the only p-primitive divisor of $p^r - 1$ and if $K(P,OQ)$ is soluble.

We may assume from now on that there does not exist a p-primitive prime divisor of $p^r - 1$. Applying 6.2 yields $p^r = 2^6$ or $r = 2$ and $p + 1 = 2^s$ for some integer s.

Assume $p^r = 2^6$. As H acts transitively on the set of 63^2 points off the lines OP and OQ and since $|H:K| = 2$, the group K also permutes this set of 63^2 points transitively. Assume that 3 does not divide $|K(P,OQ)|$. Then the Sylow 3-subgroup of K operates faithfully on $T(Q)$. As 3^4 divides $|K|$ and since 3^4 is the highest power of 3 dividing $|GL(6,2)|$, each Sylow 3-subgroup of K is isomorphic to a Sylow 3-subgroup of $GL(6,2)$. Let Σ be a Sylow 3-subgroup of K. As K permutes the 63^2 points off the lines OP and OQ transitively, Σ operates regularly on the set of these points. Furthermore, Σ has an orbit $\mathfrak{S}$ of length 9 on OP. Let $X \in \mathfrak{S}$. Then $|\Sigma_X| = 9$. Moreover, Σ_X is elementary abelian, as we shall now prove. Let $\tau \in T(P)$ with $O^\tau = X$. Then $\Sigma_X = \mathfrak{C}_\Sigma(\tau)$. Thus it suffices to show that the Sylow 3-subgroup of $GL(6,2)_v$ is elementary abelian for a non-zero vector v in the vector space V of rank 6 over $GF(2)$. Let $V = U_1 \oplus U_2 \oplus U_3$ with $|U_i| = 4$. Let σ_i $(i = 1, 2)$ be elements of order 3 in $GL(U_i)$ and define ρ_i by

$$(u_1 + u_2 + u_3)^{\rho_1} = u_1^{\sigma_1} + u_2 + u_3$$

resp.

$$(u_1 + u_2 + u_3)^{\rho_2} = u_1 + u_2^{\sigma_2} + u_3 \, .$$

Then $\langle \rho_1, \rho_2 \rangle$ is an elementary abelian 3-subgroup of $GL(6,2)$ which fixes every vector in U_3. Thus Σ_X is elementary abelian. Therefore, according to Huppert [1967, V8.15 b)], Σ_X cannot operate regularly on $OQ \setminus \{O, Q\}$. Hence there exist $\sigma \in \Sigma_X$ and $Y I OQ$ with $\sigma \neq 1$ and $Y \neq O, Q$ and $Y^\sigma = Y$. Putting $W = QX \cap PY$, we see $\sigma \in \Sigma_W = \{1\}$: a contradiction. Hence 3 does divide $|K(P, OQ)|$.

Denote by K^* the group of automorphisms induced by K on $T(Q)$. Then $K^* \cong K/K(P, OQ)$. From

$$K(Q, OP) \cong K(Q, OP)K(P, OQ)/K(P, OQ)$$

we infer that K^* contains a normal subgroup Z which is isomorphic to $K(Q, OP)$. As $K(Q, OP)$ is conjugate to $K(P, OQ)$ in H, the order of $K(Q, OP)$ is divisible by 3. Thus $|Z|$ is divisible by 3. Assume $|Z| = 3$. As K^* operates transitively on $T(Q) \setminus \{1\}$, the Z-irreducible subgroups of $T(Q)$ form a partition π of $T(Q)$ the components of which all have the same order, namely 4. Using 1.6 a) and the finiteness of $T(Q)$, we find that $K^* \subseteq \Gamma L(3, 4)$. As 7^2 divides $|K|$ and as $|K(P, OQ)| = |Z| = 3$, the order of K^* is divisible by 7^2. Hence 7^2 divides $|\Gamma L(3, 4)| = 2^7 \cdot 3^4 \cdot 5 \cdot 7$: a contradiction. Therefore $|K(P, OQ)| > 3$, i.e. $|K(P, OQ)| \in \{9, 21, 63\}$.

If $|K(P, OQ)| = 63$, then $\mathfrak{A}$ is either desarguesian or a nearfield plane. In the first case, $\mathfrak{A}$ is a generalized André plane. Let $\mathfrak{A}$ be a nearfield plane. As $3, 9 \not\equiv 1 \bmod 7$, there is exactly one Sylow 7-subgroup Z in $K(P, OQ)$. Furthermore, $|\mathrm{Aut}(Z)| = 2 \cdot 3$ and all Sylow 3-subgroups of $K(P, OQ)$ are cyclic. Hence the centre of $K(P, OQ)$ has an order divisible by 3. Thus there exists a normal subgroup N of order 21 in $K(P, OQ)$. As $K(P, OQ)$ is isomorphic to the multiplicative group of a coordinatizing nearfield F, the multiplicative group of F contains a normal subgroup A of order 21. Obviously, A operates irreducibly on $F(+)$. Therefore, $F^* \subseteq \Gamma L(1, 64)$ by 9.1. This shows that F is a generalized André system.

If $|K(P,OQ)| = 9$, then $|K(P,OQ)K(Q,OP)| = 81$ and

$$A = K(P,OQ)K(Q,OP)$$

is normal in H. This implies that all the orbits of A in $l_\infty \backslash \{P,Q\}$ have the same length λ. As λ divides 63 and 81 and is divisible by 9, we have $\lambda = 9$. Therefore, $|A_W| = 9$ for all W I l_∞ with $W \neq P,Q$. By 3.5 a), $K(P,OQ)$ and $K(Q,OP)$ are cyclic. Moreover A is abelian and A_W is a diagonal of $A = K(P,OQ)K(Q,OP)$. Hence A_W is likewise cyclic. From this we infer that A_W operates irreducibly on $T(W)$. By 9.3, $\mathfrak{A}$ is a generalized André plane in this case.

Finally if $|K(P,OQ)| = 21$, then $K(P,OQ)$ is cyclic by 3.5 c). Hence $A = K(P,OQ)K(Q,OP)$ is abelian. Again all orbits of A which are contained in $l_\infty \backslash \{P,Q\}$ have the same length λ. Hence $\lambda = 21$ or 63. If $\lambda = 21$, then it follows as above that $\mathfrak{A}$ is a generalized André plane. Thus assume $\lambda = 63$. If W I l_∞ and $W \neq P,Q$, then $|A_W| = 7$. Let S be a Sylow 3-subgroup of A. Then $S \cap A_W = \{1\}$ for all such W. Hence S acts regularly on $l_\infty \backslash \{P,Q\}$. As the group K^* of automorphisms induced on $T(Q)$ by K contains a cyclic normal subgroup of order 21, namely $K(Q,OP)^*$, we deduce, with the aid of 9.1, that $K^* \subseteq \Gamma L(1,64)$. Therefore $|K^*|$ divides $2 \cdot 3^3 \cdot 7$. On the other hand, 63^2 divides $|K|$. Moreover, $|K| = |K(P,OQ)||K^*| = 21|K^*|$. Thus $3 \cdot 63 = 3^3 \cdot 7$ divides $|K^*|$. Therefore, $|K| = 63^2$ or $|K| = 2 \cdot 63^2$. In either case, K contains a normal subgroup B of order 63^2. As K permutes the 63^2 points off the lines OP and OQ transitively, B does also. Moreover B acts regularly on the set of these points. In particular, B_W acts sharply transitively on $OW \backslash \{O,W\}$. Let Σ be a Sylow 3-subgroup of B_W. Then Σ has order 9. Hence Σ is abelian. Moreover, $S \cap \Sigma = \{1\}$. Let S_1 be a Sylow 3-subgroup of $K(P,OQ)$ and S_2 a Sylow 3-subgroup of $K(Q,OP)$. Then $S = S_1 S_2$. Furthermore Σ normalizes each S_i. As $|S_i| = 3$ and $|\Sigma| = 9$, it follows that Σ centralizes S_1 and S_2. Hence $S\Sigma$ is an abelian group of order 3^4. In particular, $S\Sigma$ is a Sylow 3-subgroup of K. This yields that $(S\Sigma)^*$ is a Sylow 3-subgroup of K^*. As $|(S\Sigma)^*| = 3^3$ and $K^* \subseteq \Gamma L(1,64)$, we see that $(S\Sigma)^*$ is a Sylow 3-subgroup of $\Gamma L(1,64)$. This is a contradiction, since $(S\Sigma)^*$ is abelian but the Sylow 3-subgroups of $\Gamma L(1,64)$ are not. This settles the case $n = 2^6$.

It remains to consider the case $n = p^2$ where p is a prime with $p + 1 = 2^s$.

17.7 Lemma (Lüneburg 1976 a + b). *Let p be a prime with $p \equiv 3$ mod 4 and let $\mathfrak{A}$ be a translation plane of order p^2. Furthermore, let O, P, Q be three distinct points of $\mathfrak{A}$ with P, Q I l_∞ and $O \nmid l_\infty$. If G is a group of collineations of $\mathfrak{A}$ fixing O, P and Q and permuting transitively the affine points off the lines OP and OQ, then the order of a Sylow 2-subgroup of $G(P,OQ)$ is either 2^s or 2^{s+1}, where 2^s is the highest power of 2 dividing $p + 1$. Moreover* *if the order of a Sylow 2-subgroup of $G(P, OQ)$ is 2^s, then the Sylow 2-subgroups of $G(P, OQ)$ are cyclic, unless 2^{s+1} divides $|G(Q, OP)|$.*

PROOF. Let Σ be a Sylow 2-subgroup of G. By our assumption, $(p^2 - 1)^2$ divides $|G|$. Hence $|\Sigma| = 2^{2s+2+b}$ with $b \geqslant 0$. From $P^\Sigma = P$ we infer that Σ normalizes $T(P)$. Put $\Sigma^* = \mathfrak{C}_\Sigma(T(P))$. Then Σ/Σ^* is a subgroup of $GL(2, p)$. As $|GL(2, p)| = p(p^2 - 1)(p - 1)$, we have $|\Sigma/\Sigma^*| \leqslant 2^{s+2}$. On the other hand, $O^{\Sigma^*} = O$ implies $\Sigma^* \subseteq G(Q, OP)$. Since $G(Q, OP)$ centralizes $T(P)$, we obtain finally $\Sigma^* = \Sigma \cap G(Q, OP)$. Hence Σ^* is a Sylow 2-subgroup of $G(Q, OP)$. If $|\Sigma^*| = 2^t$, then $t \leqslant s + 1$, as $|G(Q, OP)|$ divides $p^2 - 1$. Furthermore $2^{2s+2+b-t} = |\Sigma/\Sigma^*| \leqslant 2^{s+2}$ and hence $s + b - t \leqslant 0$. This yields $s \leqslant s + b \leqslant t \leqslant s + 1$. Hence $|\Sigma^*| = 2^s$ or 2^{s+1}.

Assume $|\Sigma^*| = 2^s$. Then $b = 0$ and $|\Sigma/\Sigma^*| = 2^{s+2}$. Therefore, Σ/Σ^* is a Sylow 2-subgroup of $GL(2, p)$. Hence, by 14.7, Σ/Σ^* is semidihedral. Put $\Sigma^{**} = G(P, OQ) \cap \Sigma$. Then $|\Sigma^{**}| = 2^s$ or 2^{s+1}, as our assumptions are symmetric in P and Q. Assume $|\Sigma^{**}| = 2^s$. As $\Sigma^{**} \cong \Sigma^{**}\Sigma^*/\Sigma^*$, we have that Σ^{**} is isomorphic to a normal subgroup of index 4 in Σ/Σ^*. But a semidihedral group has only one normal subgroup of index 4 and this normal subgroup is cyclic. □

17.8 Lemma (Lüneburg 1976 a + b). *Let p be a prime with $p \equiv 3$ mod 4 and let $\mathfrak{A}$ be a translation plane of order p^2. Furthermore let O, P, Q be three distinct points of $\mathfrak{A}$ with P, Q I l_∞ and O ł l_∞. If G is a group of collineations of $\mathfrak{A}$ fixing O, P and Q and if G splits the set of affine points off the lines OP and OQ into two orbits of length $\frac{1}{2}(p^2 - 1)^2$, then the order of a Sylow 2-subgroup of $G(P, OQ)$ is either 2^{s-1}, 2^s or 2^{s+1}, where 2^s is the highest power of 2 dividing $p + 1$. Moreover if the order of a Sylow 2-subgroup of $G(P, OQ)$ is 2^{s-1}, then the Sylow 2-subgroups of $G(P, OQ)$ are cyclic unless 2^s divides $|G(Q, OP)|$.*

The proof of 17.8 is similar to the proof of 17.7 and is left as an exercise to the reader.

Now we are able to finish the proof of 17.1 in the case where there are no p-primitive prime divisors of $p^r - 1$. We were left with the case, where the order of $\mathfrak{A}$ is p^2 with $p + 1 = 2^s \geqslant 4$. Using (B) and Lemma 17.5, we see that $K(P, OQ)$ is soluble. By 17.7 and 17.8, the Sylow 2-subgroups of $K(P, OQ)$ have order either 2^{s-1} or 2^s or 2^{s+1}. If they have order 2^{s-1}, then they are cyclic, as $K(P, OQ)$ and $K(Q, OP)$ are conjugate in H. If $s = 2$, then $p = 3$. In this case we are done by 8.4. If $s = 3$, then $p = 7$. This is one of the exceptions. Therefore we may assume $s \geqslant 4$. This yields $s \geqslant 5$, as $2^4 - 1 = 15$ is not a prime.

As $K(P, OQ)$ is soluble, $K(P, OQ)$ contains a normal Z-subgroup N such that $K(P, OQ)/N$ is isomorphic to a subgroup of S_4. We show that all Sylow subgroups of odd order of N are in the centre of $K(P, OQ)$. Let a be the largest prime dividing $|N|$. If $a = 2$, then there is nothing to prove. Let $a > 2$. As N is a Z-group, the Sylow a-subgroup A of N is normal in N

by a theorem of Burnside (see e.g. Gorenstein [1968, Theorem 7.4.3]). As $|N|$ divides $p^2 - 1 = 2^s(p - 1)$, we see that $|A|$ divides $p - 1$. Therefore, A operates reducibly on $T(P)$. Let Σ be a Sylow 2-subgroup of $K(P, OQ)$. Then $|\Sigma| \geq 2^{s-1} \geq 16$. Furthermore, Σ normalizes A, as A is a characteristic subgroup of N. The centralizer of a subgroup of order p of $T(P)$ in Σ has order 2, as $p \equiv 3 \bmod 4$. Therefore, A leaves invariant at least 8 subgroups of order p of $T(P)$. This is more than enough to assure that A is contained in the centre of $GL(2, p)$ and hence in the centre of $K(P, OQ)$. As a normal Hall subgroup A has a complement B in N. As $A \subseteq \mathfrak{Z}(N)$, the complement B is normal in N. Since normal Hall subgroups are characteristic subgroups, B is also normal in $K(P, OQ)$. The assertion now follows by induction.

Put $\Sigma_0 = \Sigma \cap N$, where, Σ is a Sylow 2-subgroup of $K(P, OQ)$. Then Σ_0 is a Sylow 2-subgroup of N. Furthermore Σ_0 is normal in N as all Sylow subgroups of odd order of N are in the centre of $K(P, OQ)$. Hence Σ_0 is normal in $K(P, OQ)$. Moreover Σ is normal in ΣN.

Assume that Σ is not normal in $K(P, OQ)$. Then 3 divides $K(P, OQ)/N$. For otherwise $K(P, OQ) = \Sigma N$. But Σ is normal in ΣN. Let M/N be the largest normal 2-subgroup of $K(P, OQ)/N$ and let Σ_1 be a Sylow 2-subgroup of M. Furthermore let C be the complement of Σ_0 in N. Then $M = \Sigma_1 C$ and $\Sigma_1 \supseteq \Sigma_0$. As C centralizes Σ_1, we have that Σ_1 is normal in M and hence normal in $K(P, OQ)$. If $N \neq M$, then $|\Sigma_1/\Sigma_0| = 4$, as $K(P, OQ)/N$ is isomorphic to a subgroup of S_4 and as $\Sigma N/N$ is not normal in $K(P, OQ)/N$. Moreover Σ_1/Σ_0 is elementary abelian. This implies that Σ_1 is a generalized quaternion group. Since there exists an element in $K(P, OQ)$ which induces an automorphism of order 3 in Σ_1 (recall that Σ_1 is normal in $K(P, OQ)$), we deduce that Σ_1 is the quaternion group of order 8, as this is the only generalized quaternion group admitting an automorphism of order 3. Hence $|\Sigma_1| = 8$ and $|\Sigma| = 16$. As Σ is not cyclic, we obtain $16 = |\Sigma| \geq 2^s$, i.e. $s \leq 4$: a contradiction. This proves $M = N$ and $\Sigma_0 = \Sigma_1$. Moreover $|\Sigma : \Sigma_0| = 2$, as Σ is not normal, and therefore $|\Sigma_0| \geq 2^{s-1}$. As Σ_0 is the intersection of all the Sylow 2-subgroups of $K(P, OQ)$, we have that Σ_0 is the only normal subgroup of order $|\Sigma_0|$ in $K(P, OQ)$. As Σ_0 is cyclic of order $\geq 2^{s-1}$, it contains a cyclic subgroup Σ^* of order 2^{s-1}. Obviously, Σ^* is characteristic in $K(P, OQ)$. We infer from $2^{s-1} \geq 16$ that Σ^* is the only cyclic normal subgroup of order 2^{s-1}.

Assume that Σ is normal in $K(P, OQ)$. Then Σ contains a unique normal cyclic subgroup Σ^* of order 2^{s-1}: This is true by 17.8 in the case that $|\Sigma| = 2^{s-1}$. If $|\Sigma| \geq 2^s$, then Σ is cyclic or a generalized quaternion group. In either case, Σ contains such a unique normal cyclic subgroup, as $s \geq 5$. It follows that Σ^* is the only cyclic normal subgroup of order 2^{s-1} of $K(P, OQ)$.

Let Σ^* be the only normal cyclic subgroup of order 2^{s-1} of $K(P, OQ)$. As $K(Q, OP)$ is conjugate to $K(P, OQ)$, it also contains exactly one cyclic normal subgroup Σ^{**} of order 2^{s-1}. Therefore $A = \Sigma^* \Sigma^{**}$ is an abelian normal subgroup of H of order 2^{2s-2}. As H acts transitively on $l_\infty \setminus \{P, Q\}$, the orbits of A contained in $l_\infty \setminus \{P, Q\}$ all have the same length λ. As λ is a divisor of $p^2 - 1$, we have $\lambda \leqslant 2^{s+1}$. Let W I l_∞ with $W \neq P, Q$. Then

$$2^{2s-2} = |A| = |W^A||A_W| \leqslant 2^{s+1}|A_W|.$$

Hence $|A_W| \geqslant 2^{s-3} \geqslant 4$. One proves as usual that $\mathfrak{C}_{A_W}(T(W)) = \{1\}$. Hence A_W is isomorphic to a subgroup of the group $GL(2, p)$. Furthermore A is the direct product of the two cyclic groups Σ^* and Σ^{**}. Therefore A contains exactly three involutions. This implies that A_W contains but one involution. Hence A_W is cyclic. This yields finally that A_W acts irreducibly on $T(W)$. Therefore, $\mathfrak{A}$ is a generalized André plane by 9.3. Thus the proof of 17.1 will be finished if we can show that a plane of order 19^2 satisfying the hypotheses of 17.1 is a generalized André plane. This will be proved in section 20.

18. The Planes of Type $R * p$

In this section we follow Lüneburg 1974.

Let p be a prime and let $\mathfrak{A}$ be an affine plane of order p^2. We shall call $\mathfrak{A}$ of type $R * p$, if $\mathfrak{A}$ possesses the following properties:

(j) If G is the collineation group of $\mathfrak{A}$, then G_l acts doubly transitively on the set of affine points on l for all affine lines l of $\mathfrak{A}$.

(ij) There exist three distinct points P, Q and O of $\mathfrak{A}$ with P, Q I l_∞ and $O \nparallel l_\infty$ such that $G(P, OQ)$ contains a subgroup S_0 and $G(Q, OP)$ contains a subgroup S_∞ with $S_0 \cong S_\infty \cong SL(2, 5)$. Moreover, the set of orbits of S_0 on l_∞ is equal to the set of orbits of S_∞ on l_∞.

We shall determine all planes of type $R * p$ in this section.

18.1 Theorem. *If $\mathfrak{A}$ is an affine plane of type $R * p$, then $\mathfrak{A}$ is a translation plane and p is one of the primes $11, 19, 29, 59$.*

PROOF. $\mathfrak{A}$ is a translation plane by 16.1. Furthermore, $p \in \{11, 19, 29, 59\}$ by 17.5. $\square$

Let $p \in \{11, 19, 29, 59\}$ and let S be a subgroup of $SL(2, p)$ which is isomorphic to $SL(2, 5)$. As $p^2 - 1 \equiv 0 \bmod 10$, there exists such an S. Denote by V the vector space of rank 2 over $GF(p)$. Let $\sigma \in S$ and assume that 1 is an eigenvalue of σ. As $\det(\sigma) = 1$, the multiplicity of 1 as an eigenvalue of σ is 2. Hence σ is a transvection. Therefore $\sigma^p = 1$. As p does not divide $|S| = 120$, we obtain $\sigma = 1$. Thus S operates regularly on $V \setminus \{0\}$, i.e. all the orbits of S in $V \setminus \{0\}$ have the length 120.

Let Z_0 be the maximal subgroup of odd order of $\mathfrak{Z}(GL(2, p))$. Then $|Z_0| = 5, 3^2, 7, 29$ for $p = 11, 19, 29, 59$ resp. The group SZ_0 acts transitively on $V \setminus \{0\}$: This is certainly true for $p = 11$, as $11^2 - 1 = 120$. Let $p = 19$,

29 or 59. In these cases, the number of orbits of S is 3, 7 resp. 29. Furthermore Z_0 permutes the orbits of S. As Z_0 operates regularly on $V \setminus \{0\}$ and since neither 9 nor 7 nor 29 divides 120, we see that Z_0 permutes the orbits of S transitively. For $p = 29$ or 59, the group SZ_0 operates regularly on $V \setminus \{0\}$. In the case $p = 19$, the subgroup of order 3 of Z_0 fixes all the orbits of S.

Put $W = V \oplus V$, $V(0) = \{(x, 0) \mid x \in V\}$, $V(\infty) = \{(0, x) \mid x \in V\}$ and $W^* = W \setminus (V(0) \cup V(\infty))$. The group $G = Z_0 S \times Z_0 S$ operates on W by $(x, y)^{(\alpha, \beta)} = (x^\alpha, y^\beta)$. Put $H = S \times S$ and $A = \{(\zeta, \zeta) \mid \zeta \in Z_0\}$. Then AH is a subgroup of order $120^2 |Z_0|$ of G. Furthermore, G fixes $V(0)$ and $V(\infty)$.

18.2 Lemma. *AH decomposes W^* in case*

$p = 11$ *into one orbit of length 120^2,*
$p = 19$ *into 3 orbits of length $3 \cdot 120^2$,*
$p = 29$ *into 7 orbits of length $7 \cdot 120^2$,*
$p = 59$ *into 29 orbits of length $29 \cdot 120^2$.*

PROOF. Let $(x, y) \in W^*$. Furthermore let $\alpha, \beta \in S$ and $\zeta \in Z_0$ be such that $(x, y)^{(\zeta, \zeta)(\alpha, \beta)} = (x, y)$. Then $x^{\zeta\alpha} = x$ and $y^{\zeta\beta} = y$. If $p = 29$ or 59, we obtain $\zeta\alpha = 1 = \zeta\beta$, as SZ_0 acts regularly on $V \setminus \{0\}$. This proves 18.2 in these two cases.

Let $p = 19$. Then x and x^ζ are in the same orbit of S. Therefore $\zeta^3 = 1$. As α is uniquely determined by ζ and x and as β is uniquely determined by ζ and y, we have consequently $|(AH)_{(x, y)}| \leqslant 3$. On the other hand, if $\zeta^3 = 1$, then ζ fixes all the orbits of S. Therefore there exist $\alpha, \beta \in S$ with $x^{\zeta\alpha} = x$ and $y^{\zeta\beta} = y$. Thus $|(AH)_{(x, y)}| = 3$.

Finally, if $p = 11$, then S acts transitively on $V \setminus \{0\}$. Therefore H and hence AH operate transitively on W^*. This proves 18.2.   □

Put $B = Z_0 \times Z_0$. Then $G = BH$. Furthermore $|B| = |Z_0|^2$. As B centralizes AH, it permutes the orbits of AH. Since SZ_0 acts transitively on $V \setminus \{0\}$, the group $G = BH$ acts transitively on W^*. Therefore, we have:

18.3 Lemma. *B permutes transitively the orbits of AH which are contained in W^*.*

Put $Z = \mathfrak{Z}(GL(2, p))$ and define $V(\zeta)$ for all $\zeta \in Z$ by

$$V(\zeta) = \{(x, x^\zeta) \mid x \in V\}.$$

If $S_1 = \{(\alpha, \alpha) \mid \alpha \in S\}$, then S_1 is the stabilizer of $V(1)$ in H. Moreover AS_1 operates transitively on $V(1) \setminus \{(0, 0)\}$. Therefore $V(1) \setminus \{(0, 0)\}$ is completely contained in an orbit B_1 of AH. Moreover, $B_1 \subseteq W^*$ and $B_1 = \bigcup_{\eta \in H}(V(1)^\eta \setminus \{(0, 0)\})$. Let Z_0 be generated by ω and put $B_{i+1} = B_1^{(1, \omega^i)}$. Then we have:

18.4 Lemma. *The orbits of AH which are contained in W^* are in case*

$p = 11$ *the set* B_1,
$p = 19$ *the sets* B_1, B_2, B_3,
$p = 29$ *the sets* $B_1, \ldots, B_7$,
$p = 59$ *the sets* $B_1, \ldots, B_{29}$.

PROOF. This is trivial for $p = 11$. Let $p > 11$. By 18.3, B acts transitively on the set of the orbits in question. Put $C = \{(1, \omega^i) \mid i = 0, 1, \ldots, |Z_0| - 1\}$. Then $B = AC$. Hence C acts transitively on the set of orbits. This proves the assertion in case $p = 29$ and $p = 59$. Therefore assume $p = 19$ and let C_0 be the subgroup of order 3 of C. Then C_0 fixes all the orbits in question. As $C = C_0 \cup C_0(1, \omega) \cup C_0(1, \omega^2)$, we obtain finally that B_1, B_2 and B_3 are the orbits of AH which are contained in W^*. $\square$

18.5 Lemma. *If ρ is the mapping defined by $(x, y)^\rho = (y, x)$, then we have:*

(j) ρ *normalizes* AH.
(ij) $\{B_1, B_2, \ldots\}^\rho = \{B_1, B_2, \ldots\}$.
(iij) $B_1^\rho = B_1$.
(iv) *If $i \neq 1$, then $B_i^\rho \neq B_i$.*

PROOF. (j) is trivial and (ij) is a consequence of (j). (iij) follows from (ij) and $V(1)^\rho = V(1)$. In order to prove (iv) assume that $B_{i+1}^\rho = B_{i+1}$. By 18.4, $V(\omega^i) \setminus \{(0,0)\} \subseteq B_{i+1}$. Therefore, $(x^{\omega^i}, x) = (x, x^{\omega^i})^\rho \in B_{i+1}$ for all $x \in V \setminus \{0\}$. Because of $B_1 = \bigcup_{\eta \in H}(V(1) \setminus \{(0,0)\})^\eta$, there exist, for every $x \in V \setminus \{0\}$, an element $y \in V$ and elements $\alpha, \beta \in S$ with $(x^{\omega^i}, x) = (y^\alpha, y^{\beta \omega^i})$. This yields $y = x^{\omega^i \alpha^{-1}}$ and therefore $x = x^{\omega^{2i} \alpha^{-1} \beta}$. This proves that x and $x^{\omega^{2i}}$ are in the same orbit of S. Hence $2i \equiv 0 \bmod 3$, if $p = 19$, and $2i = 0$, if $p = 29$ or 59. In either case $i = 0$. $\square$

18.6 Lemma. *Let V be a vector space of rank 2 over $\mathrm{GF}(q)$ and let $W, Z, V(0), V(\infty), V(\zeta)$ be as before. If S is a non-cyclic subgroup of $\mathrm{GL}(2, q)$ acting irreducibly on V, then $\Re = \{V(0), V(\infty)\} \cup \{V(\zeta) \mid \zeta \in Z\}$ is the set of subspaces of rank 2 of W which are left invariant by $S_1 = \{(\alpha, \alpha) \mid \alpha \in S\}$.*

PROOF. Let $X \in \Re$. Then X is left invariant by S_1. Thus assume that X is a subspace of rank 2 of W which is left invariant by S_1. We may assume $X \neq V(0), V(\infty)$. From $X^{S_1} = X$ and the irreducibility of S, we infer $X \cap V(0) = \{(0,0)\} = X \cap V(\infty)$. Hence X is a diagonal of $W = V(0) \oplus V(\infty)$, i.e. there exists $\sigma \in \mathrm{GL}(2, q)$ with $X = \{(x, x^\sigma) \mid x \in V\}$. In particular $X = \{(x^\alpha, x^{\alpha\sigma}) \mid x \in V\}$ for all $\alpha \in S$. On the other hand $X = \{(x^\alpha, x^{\sigma\alpha}) \mid x \in V\}$ for all $\alpha \in S$, as $X^{S_1} = X$. Hence $\alpha\sigma = \sigma\alpha$ for all $\alpha \in S$. Thus σ lies in the centralizer C of S. By Schur's Lemma $|C| = q - 1$ or $|C| = q^2 - 1$, as $Z \subseteq C$ and $|Z| = q - 1$. If $|C| = q^2 - 1$, then C would be its own centralizer, as C is cyclic. This is impossible, because S is not cyclic. Hence $|C| = q - 1$, i.e. $Z = C$, whence $\sigma \in Z$. $\square$

Let a_i be the number of $X \in \mathfrak{R} \backslash \{ V(0), V(\infty)\}$ with $X \backslash \{(0,0)\} \subseteq B_i$. As B centralizes AS_1 and since B permutes the B_i's transitively by 18.3, we have $a_1 = a_2 = \cdots = a$. Because of $|\mathfrak{R}| = p + 1$ we then have:

18.7 Lemma. *We have*

$a = 10$ *if* $p = 11$,
$a = 6$ *if* $p = 19$,
$a = 4$ *if* $p = 29$,
$a = 2$ *if* $p = 59$.

Let V be a vector space and let σ be a set of subspaces of V. We shall call σ a *partial spread* of V if σ contains more than one element and if $V = X \oplus Y$ for all $X, Y \in \sigma$ with $X \neq Y$.

18.8 Lemma. *If* $X \in \mathfrak{R} \backslash \{ V(0), V(\infty)\}$ *and* $\pi = \{ X^\gamma | \gamma \in AH \}$, *then we have*:

(a) $\bigcup_{Y \in \pi}(Y \backslash \{(0,0)\})$ *is an orbit of* AH *contained in* W^*.
(b) π *is a partial spread of* W.

PROOF. a) follows from 18.4.

(b) Let $\bigcup_{Y \in \pi}(Y \backslash \{(0,0)\}) = B_i$. Consider the incidence structure $(B_i, \pi, \in)$. This is a tactical configuration, as B_i and π are orbits of AH. The parameters of $(B_i, \pi, \in)$ are $v = |B_i|$, $b = |\pi|$, $k = |X \cap B_i|$ and r. We have to show $r = 1$. As $AS_1 \subseteq (AH)_X$, we have $b \leqslant 120$. Furthermore, $k = p^2 - 1$. Hence $vr = bk \leqslant 120(p^2 - 1)$. From 18.2 we infer

$$120^2 r \leqslant 120(11^2 - 1) = 120^2 \qquad \text{if} \quad p = 11,$$
$$3 \cdot 120^2 r \leqslant 120(19^2 - 1) = 3 \cdot 120^2 \qquad \text{if} \quad p = 19,$$
$$7 \cdot 120^2 r \leqslant 120(29^2 - 1) = 7 \cdot 120^2 \qquad \text{if} \quad p = 29,$$
$$29 \cdot 120^2 r \leqslant 120(59^2 - 1) = 29 \cdot 120^2 \qquad \text{if} \quad p = 59.$$

Thus $r \leqslant 1$ in all cases. As $k = p^2 - 1 \neq 0$, we find $r \geqslant 1$. Hence $r = 1$ and $b = 120$ in all cases. $\square$

18.9 Corollary. $(AH)_X = AS_1$.

We call π as constructed in 18.8 an *i-partial spread*, if $\bigcup_{Y \in \pi} Y = B_i \cup \{(0,0)\}$.

18.10 Lemma. *For each i there exist*

exactly five i-partial spreads if $p = 11$,
exactly three i-partial spreads if $p = 19$,
exactly two i-partial spreads if $p = 29$,
exactly one i-partial spread if $p = 59$.

PROOF. By 18.7, it suffices to show that $|\pi \cap \Re| = 2$ for all i-partial spreads π. As B operates transitively on the set of all π's in question, we may assume $\pi = \{V(1)^\gamma | \gamma \in AH\}$. Since A fixes $V(1)$ we have $\pi = \{V(1)^\gamma | \gamma \in H\}$. Let $1 \neq \sigma \in \mathfrak{Z}(S)$. Then $\sigma = -1$ and $V(1), V(-1) \in \pi \cap \Re$. Hence $|\pi \cap \Re| \geqslant 2$. Assume $X \in \pi \cap \Re$. Obviously, $X \neq V(0), V(\infty)$. Thus $X = V(\zeta)$ for some $\zeta \in Z$. On the other hand there are $\alpha, \beta \in S$ with $V(1)^{(\alpha, \beta)} = V(\zeta)$. As

$$V(1)^{(\alpha, \beta)} = V(1)^{(\alpha, \alpha)(1, \alpha^{-1}\beta)} = V(1)^{(1, \alpha^{-1}\beta)},$$

we may assume $\alpha = 1$. This yields $\{(x, x^\beta) | x \in V\} = \{(x, x^\zeta) | x \in V\}$. Therefore $\beta = \zeta \in Z \cap S = \mathfrak{Z}(S) = \{1, -1\}$. □

For each i pick an i-partial spread φ_i. Then $\kappa = \{V(0), V(\infty)\} \cup \varphi_1 \cup \varphi_2 \cup \cdots$ is a spread of W by 18.8. Hence $\kappa(W)$ is a translation plane. If $\eta \in Z \times Z$ and $\kappa^\eta = \{V(0), V(\infty)\} \cup \varphi_1^\eta \cup \varphi_2^\eta \cup \cdots$, then $\kappa^\eta(W)$ and $\kappa(W)$ are isomorphic. Hence we may always assume that $\varphi_1 = \{V(1)^\gamma | \gamma \in H\}$. Thus we obtain up to isomorphism only one plane $\kappa(W)$, if $p = 11$. If $p = 59$, then there is also only one plane $\kappa(W)$ by 18.10. In all cases, AH is a subgroup of the collineation group of $\kappa(W)$. Moreover, $S_0 = \{(\alpha, 1) | \alpha \in S\}$ and $S_\infty = \{(1, \alpha) | \alpha \in S\}$ are groups of homologies which are isomorphic to $SL(2, 5)$ and the set of orbits of S_0 on l_∞ is the same as the set of orbits of S_∞ on l_∞. Hence $\kappa(W)$ is always of type $R * p$.

18.11 Lemma. *Let N be a subgroup of H which is isomorphic to $SL(2, 5)$. If N leaves a subspace X of rank 2 of W invariant and if $X \neq V(0), V(\infty)$, then N and S_1 are conjugate in H.*

PROOF. As $H \cong SL(2, 5) \times SL(2, 5)$, it contains only three involutions, namely the involutions $\sigma_1, \sigma_2, \sigma_3$ defined by

$$(x, y)^{\sigma_1} = (-x, y), \qquad (x, y)^{\sigma_2} = (x, -y), \qquad (x, y)^{\sigma_3} = (-x, -y).$$

Let σ be the involution in N. If N did not act faithfully on X, we would have that σ were the identity on X, as σ is contained in all non-trivial normal subgroups of N. But then $\sigma = \sigma_1$ or $\sigma = \sigma_2$ and hence $X = V(0)$ or $X = V(\infty)$: a contradiction. Hence N acts faithfully and therefore irreducibly on X. This yields $X \cap V(0) = \{(0, 0)\} = X \cap V(\infty)$. Therefore there exists $\beta \in GL(2, p)$ with $X = \{(x, x^\beta) | x \in V\}$. Let $\sigma \in S$ and $X^{(\sigma, 1)} = X$. Then, for all $x \in V$, there exists $y \in V$ with $(x^\sigma, x^\beta) = (y, y^\beta)$. As a consequence $x = y$ and $x^\sigma = x$ for all x. Therefore $S_0 \cap N = \{(1, 1)\}$. Similarly $S_\infty \cap N = \{(1, 1)\}$. Hence N is a diagonal of H. So there exists $\gamma \in \text{Aut}(S)$ with $N = \{(\alpha, \alpha^\gamma) | \alpha \in S\}$. From this we infer

$$\{(x^\alpha, x^{\alpha\beta}) | x \in V\} = X = X^{(\alpha, \alpha^\gamma)} = \{(x^\alpha, x^{\beta\alpha^\gamma}) | x \in V\}$$

for all $\alpha \in S$, whence $\alpha\beta = \beta\alpha^\gamma$. This proves $\beta \in \Re_{GL(2, p)}(S)$. By 14.6, $\Re_{GL(2, p)}(S) = ZS$. Therefore we may assume $\beta \in S$. But then $(\beta, 1) \in H$. Moreover $X^{(\beta, 1)} = V(1)$ and $(\beta, 1)^{-1}N(\beta, 1) \subseteq H_{V(1)} = S_1$. □

18.12 Theorem. *If $\mathfrak{A}$ is a plane of type $R * p$, then $\mathfrak{A}$ is isomorphic to one of the planes $\kappa(W)$.*

PROOF. By 18.1, $\mathfrak{A}$ is a translation plane of order 11^2, 19^2, 29^2 or 59^2. Thus we may identify the points of $\mathfrak{A}$ with the elements of W. By 2.1, we may further assume that $OP = V(0)$ and $OQ = V(\infty)$. Applying 14.6, we see that we may identify $S_0 \times S_\infty$ with H, where H is the group used in the construction of the $\kappa(W)$'s. Let σ be the spread of W defining the lines of $\mathfrak{A}$ and let $X \in \sigma\setminus\{V(0), V(\infty)\}$. Then $|H_X| = 120$ by the assumption made on the orbits of S_0 and S_∞. This yields that H_X is a diagonal of $H = S \times S$. Hence $H_X \cong \mathrm{SL}(2,5)$. Therefore H_X is conjugate to S_1 in H. Thus X^H is an i-partial spread. This proves that σ is the union of i-partial spreads and the set $\{V(0), V(\infty)\}$. $\square$

18.13 Lemma. *If $\kappa(W)$ and $\kappa'(W)$ are isomorphic, then there exists an isomorphism σ from $\kappa(W)$ onto $\kappa'(W)$ with $\{V(0), V(\infty)\}^\sigma = \{V(0), V(\infty)\}$.*

This follows immediately from 14.8.

18.14 Lemma. *If σ is an isomorphism of $\kappa(W)$ onto $\kappa'(W)$ with $\{V(0), V(\infty)\}^\sigma = \{V(0), V(\infty)\}$, then $\sigma \in (Z \times Z)H\langle\rho\rangle$, where $(x, y)^\rho = (y, x)$.*

PROOF. $\{V(0), V(\infty)\}^\sigma = \{V(0), V(\infty)\}$ implies $(0, 0)^\sigma = (0, 0)$. Hence, by 1.10, $\sigma \in \mathrm{GL}(W)$. As we have seen several times, S_0 is a characteristic subgroup of $\Delta((0), V(\infty))$ and S_∞ is a characteristic subgroup of $\Delta((\infty), V(0))$. Therefore $\{S_0^\sigma, S_\infty^\sigma\} = \{S_0, S_\infty\}$. Replacing κ' by κ'^ρ and σ by $\sigma\rho$ if necessary, we may assume $S_0^\sigma = S_0$ and $S_\infty^\sigma = S_\infty$. This yields that σ induces a mapping in $V(0)$ which is in the normalizer of S_0 in $\mathrm{GL}(V(0))$. Hence, by 14.6, there exist $\zeta \in Z$ and $\alpha \in S$ with $(x, 0)^\sigma = (x^{\zeta\alpha}, 0)$. Similarly $(0, x)^\sigma = (0, x^{\eta\beta})$ with $\beta \in S$ and $\eta \in Z$. Therefore $\sigma \in (Z \times Z)H$. $\square$

Put $I = (Z \times Z)H\langle\rho\rangle$ and $D = \{(\zeta, \zeta) \mid \zeta \in Z\}$ and $\pi = \{V(1)^\gamma \mid \gamma \in AH\}$.

18.15 Lemma.

(a) $I_\pi = DH\langle\rho\rangle$.
(b) $((Z \times Z)H)_\pi = DH$. Moreover $((Z \times Z)H)_\pi$ fixes all i-partial spreads for all i.

PROOF. (a) $DH\langle\rho\rangle \subseteq I_\pi$ by 18.5. Let $\eta \in I_\pi$. As $H\langle\rho\rangle \subseteq I_\pi$, we may assume $\eta \in Z \times Z$. Then $\mathfrak{R}^\eta = \mathfrak{R}$ and hence $V(1)^\eta = V(1)$ or $V(1)^\eta = V(-1)$. Put $\eta = (\alpha, \beta)$. Then

$$V(1)^{(\alpha, \beta)} = V(1)^{(\alpha, \alpha)(1, \alpha^{-1}\beta)} = V(1)^{(1, \alpha^{-1}\beta)} = V(\alpha^{-1}\beta).$$

Therefore $\alpha = \beta$ or $\alpha = -\beta$. This implies $\eta \in DH$ proving (a). (b) follows immediately from (a). $\square$

18.16 Theorem. *There exists up to isomorphism exactly one plane of type* $R * 11$, *namely the plane over the exceptional nearfield of order* 11^2 *the multiplicative group of which is isomorphic to* $SL(2,5)$.

This follows from 18.12 and the remark after 18.10.

18.17 Theorem. *There exist up to isomorphism exactly three planes of type* $R * 19$. *If* G *is the collineation group of any of these planes, then* $G_{\{V(0), V(\infty)\}}$ *is conjugate to* $DH\langle\rho\rangle$ *in* I.

PROOF. By 18.10, there are exactly $3^3 = 27$ planes $\kappa(W)$. By 18.11, 18.12 and 18.14, the number of orbits of I on the set of the $\kappa(W)$'s is precisely the number of isomorphism types in question.

Let $\kappa(W)$ be one of these planes and K its collineation group and let $G = K_{\{V(0), V(\infty)\}}$. By 18.14 and 18.15, we have $DH \subseteq G \subseteq I$. Put $k = |G : DH|$. Then

$$a = |I|(|DH|k)^{-1}$$

is the number of planes in the orbit $\kappa(W)^I$. Now

$$|I| = |(Z \times Z)H\langle\rho\rangle| = 2|Z|^2|H||(Z \times Z) \cap H|^{-1} = \tfrac{1}{2}18^2|H|$$

and

$$|DH| = |D||H||D \cap H|^{-1} = \tfrac{1}{2}18|H|.$$

Thus $ak = 18$.

Assume that G acts transitively on $\{B_1, B_2, B_3\}$. As DH induces the identity on this set, there exists $\zeta \in Z$ with $(1, \zeta) \in G$ such that $(1, \zeta)$ induces a 3-cycle on $\{B_1, B_2, B_3\}$. We may assume that ζ is a 3-element. Furthermore we may assume $\pi \subseteq \kappa$. Then $\pi^{(1, \zeta^3)} = \pi$ and hence $V(\zeta^3) \in \pi$. Therefore $\zeta^3 = 1$ or $\zeta^3 = -1$. As ζ is a 3-element, we have $\zeta^3 = 1$. By 18.3, $Z_0 \times Z_0$ and hence $\{1\} \times Z_0$ act transitively on $\{B_1, B_2, B_3\}$. As Z_0 is cyclic of order 9, we have consequently that $(1, \zeta)$ fixes each B_i: a contradiction. Thus G induces a 2-group on $\{B_1, B_2, B_3\}$. Hence there exists an i with $B_i^G = B_i$. We may assume $i = 1$. Let $\varphi_{21}, \varphi_{22}, \varphi_{23}$ be the three 2-partial spreads of W. By 18.5, φ_{2i}^ρ ($i = 1, 2, 3$) are the three 3-partial spreads of W. Put

$$\kappa_i = \{V(0), V(\infty)\} \cup \pi \cup \varphi_{2i} \cup \varphi_{2i}^\rho \qquad (i = 1, 2, 3).$$

If G_i has the meaning for $\kappa_i(W)$ that G has for $\kappa(W)$, we have $G_i = I_\pi$ for all i. Hence $k_i = 2$ for all i by 18.15. Therefore the orbit of $\kappa_i(W)$ under I contains exactly 9 planes. If $i \neq j$, then $\kappa_i(W)$ and $\kappa_j(W)$ are non-isomorphic, as every isomorphism from $\kappa_i(W)$ onto $\kappa_j(W)$ fixes π. Therefore $|\bigcup_{i=1}^3 \kappa_i(W)^I| = 3 \cdot 9 = 27$. □

18.18 Theorem. *There exist up to isomorphism exactly nine planes of type* $R * 29$, *one being the nearfield plane over the exceptional nearfield of order*

29^2. *If G is the collineation group fixing $\{V(0), V(\infty)\}$, then $G = DH\langle\rho\rangle$ for seven of the remaining eight planes and $G = DH$ for the last one.*

PROOF. We have

$$|I| = |(Z \times Z)H\langle\rho\rangle| = 2|Z|^2|H||(Z \times Z) \cap H|^{-1} = \tfrac{1}{2}|Z|^2|H| = \tfrac{1}{2}28^2|H|.$$

Moreover $DH \subseteq G \subseteq I$. Let $k = |G : DH|$. As $|DH| = \tfrac{1}{2}28|H|$, we have $|I : G| = 28k^{-1}$. This is the number of planes in $\kappa(W)^I$.

By 18.3, B acts transitively on $\{B_1, \ldots, B_7\}$. Put $|E| = \{(\zeta, \zeta^{-1}) | \zeta \in Z_0\}$. Then $B = AE$ and $A \cap E = \{(1, 1)\}$. Therefore E acts sharply transitively on $\{B_1, \ldots, B_7\}$. Let $\kappa_0 = \{V(0), V(\infty)\} \cup \bigcup_{\xi \in E} \pi^\xi$. Then $\kappa_0(W)$ is a plane of type $R*29$ which admits E as a group of collineations. As ρ normalizes E and $\pi^\rho = \pi$, we see that ρ is also a collineation of $\kappa_0(W)$. Hence, by 18.15, $G_\pi = I_\pi$, whence $|G| = 7|I_\pi| = 14|DH|$. Thus $k = 14$ in this case and $|I : G| = 2$. On the other hand, if G permutes $\{B_1, \ldots, B_7\}$ transitively, then we infer from $DH \subseteq G \subseteq I$ that $E \subseteq G$ and that $\kappa(W) \in \kappa_0(W)^I$.

If G does not act transitively on $\{B_1, \ldots, B_7\}$, then G/DH is a 2-group as $|I : DH| = 28$. Therefore we may assume $B_1^G = B_1$ and also $\pi \subseteq \kappa$. Therefore $G \subseteq I_\pi = DH\langle\rho\rangle$.

By 18.10 there are 2^6 planes $\kappa(W)$ with $\pi \subseteq \kappa$. Of these $2^3 = 8$ admit ρ as a collineation and for exactly one of them $G \nsubseteq DH\langle\rho\rangle$. The remaining seven are pairwise non-isomorphic as $G = I_\pi$ in these cases. The corresponding orbits of I contain $28 : 2 = 14$ planes each. Hence if n is the number of orbits which contain those planes for which $G = DH$, then

$$128 = 2^7 = 28n + 7 \cdot 14 + 2 = 28n + 100,$$

whence $n = 1$.   □

Finally, we have by the remark after 18.10 and by 18.12:

18.19 Theorem. *There exists up to isomorphism exactly one plane of type $R*59$, namely the plane over the exceptional nearfield of order 59^2.*

19. The Planes of Type $F*p$

In this section we follow Lüneburg 1974.

Let p be a prime and let $\mathfrak{A}$ be an affine plane of order p^2. We shall call $\mathfrak{A}$ of type $F*p$ if $\mathfrak{A}$ possesses the following properties:

(j) If G is the collineation group of $\mathfrak{A}$, then G_l acts doubly transitively on the set of affine points on l for all affine lines l of $\mathfrak{A}$.

(ij) There exist three distinct points P, Q and O of $\mathfrak{A}$ with $P, Q \mathrel{I} l_\infty$ and $O \mathrel{\not I} l_\infty$ such that $G(P, OQ)$ contains a subgroup S_0 and $G(Q, OP)$ contains a subgroup S_∞ with $S_0 \cong S_\infty \cong \mathrm{SL}(2, 3)$ and such that the orbits of S_0 on l_∞ are the same as the orbits of S_∞ on l_∞.

As a consequence of 3.2, 3.3 and 3.9, we have that S_0 is normal in $G(P, OQ)$ and that S_∞ is normal in $G(Q, OP)$. Furthermore, by using 3.5, we see that S_0 is the only subgroup of $G(P, OQ)$ which is isomorphic to SL(2,3) and that the analogous statement for S_∞ is also true.

Applying these remarks, 16.1, 17.4 and 17.6 yields:

19.1 Theorem. *If $\mathfrak{A}$ is a plane of type $F * p$, then $\mathfrak{A}$ is a translation plane and $p \in \{5, 7, 11, 23\}$.*

Let $p \in \{5, 7, 11, 23\}$ and let V be a vector space of rank 2 over GF(p). Let S be a subgroup of SL(2, p) which is isomorphic to SL(2,3). (Such a subgroup exists.) As p does not divide $|SL(2,3)| = 24$, the group S acts regularly on $V \setminus \{0\}$. We let $S \times S$ operate on $W = V \oplus V$ by defining $(x, y)^{(\alpha, \beta)} = (x^\alpha, y^\beta)$. Denote by S_1 the stabilizer of $V(1) = \{(x, x) \mid x \in V\}$ in $S \times S$. Then $S_1 = \{(\alpha, \alpha) \mid \alpha \in S\}$. Put $Z = \mathfrak{Z}(GL(2, p))$ and let $V(0), V(\infty), V(\zeta)$ for $\zeta \in Z$ have the usual meaning. Finally, put $\mathfrak{R} = \{V(0), V(\infty)\} \cup \{V(\zeta) \mid \zeta \in Z\}$. By 18.16, $\mathfrak{R}$ is the set of all subspaces of rank 2 of W which are left invariant by S_1.

Denote by N the normalizer of S in GL(2, p). By 14.4 and 14.5, $|N : SZ| = 2$.

19.2 Lemma. *The set of subgroups of $S \times S$ which are isomorphic to SL(2,3) and which fix exactly $p + 1$ subspaces of rank 2 of W split into two conjugacy classes. These two classes fuse under $N \times N$ into one conjugacy class. Moreover, if H is a subgroup of $S \times S$ which is isomorphic to SL(2,3) and if H fixes a subspace of rank 2 which is distinct from $V(0)$ and $V(\infty)$, then H is in one of these classes.*

PROOF. Pick $\rho \in N \setminus SZ$ and put $S_2 = \{(\alpha, \alpha^\rho) \mid \alpha \in S\}$. Then S_2 is a subgroup of $S \times S$ which is conjugate under $N \times N$ to S_1. Hence S_2 fixes exactly $p + 1$ subspaces of rank 2. Assume that S_1 and S_2 are conjugate under $S \times S$. Then there exist $\sigma, \tau \in S$ with $S_1^{(\sigma, \tau)} = S_2$, i.e. $\{(\beta^\sigma, \beta^\tau) \mid \beta \in S\} = \{(\alpha, \alpha^\rho) \mid \alpha \in S\}$. This yields $\beta^\tau = \beta^{\sigma\rho}$ for all $\beta \in S$. Hence $\sigma\rho\tau^{-1}$ centralizes S. Therefore $\sigma\rho\tau^{-1} \in Z$, i.e. $\rho \in ZS$: a contradiction. Thus we have at least two such conjugacy classes.

Let X be a subspace of rank 2 of V with $X \neq V(0), V(\infty)$ and let H be a subgroup of $S \times S$ isomorphic to SL(2,3) and assume $X^H = X$. The argument using the involutions we have made in the proof of 18.11 yields also in the present situation that H acts faithfully and hence irreducibly on X. Therefore $X \cap V(0) = \{(0,0)\} = X \cap V(\infty)$. It follows that there exists $\delta \in GL(2, p)$ with $X = \{(x, x^\delta) \mid x \in V\}$. Likewise we obtain that H is a diagonal of $S \times S$. Hence there exists $\gamma \in \text{Aut}(S)$ with $H = \{(\alpha, \alpha^\gamma) \mid \alpha \in S\}$. From this we infer

$$\{(x^\alpha, x^{\alpha\delta}) \mid x \in V\} = X = X^{(\alpha, \alpha^\gamma)} = \{(x^\alpha, x^{\delta\alpha^\gamma}) \mid x \in V\}.$$

Therefore $\alpha^{\gamma} = \delta^{-1}\alpha\delta$ for all $\alpha \in S$, i.e. $\delta \in N$. This yields that H is conjugate to S_1 or to S_2. $\square$

19.3 Lemma. *If* $\zeta \in Z$, *then* $V(\zeta)^{S \times S} \cap \Re = \{V(\zeta), V(-\zeta)\}$.

PROOF. Put $S_{\infty} = \{1\} \times S$. Then $S \times S = S_1 S_{\infty}$. Hence $V(\zeta)^{S \times S} = V(\zeta)^{S_{\infty}}$. Let $V(\eta) \in V(\zeta)^{S \times S}$. Then there exists $\alpha \in S$ with $V(\eta) = V(\zeta\alpha)$. Therefore $\eta = \zeta\alpha$. This yields $\alpha \in S \cap Z$. Hence $\eta = \zeta$ or $\eta = -\zeta$. $\square$

19.4 Lemma. *If* $\mathfrak{A}$ *is a plane of type* $F * p$, *then we may identify the points of* $\mathfrak{A}$ *with the elements of* W *such that* $V(0)$, $V(1)$ *and* $V(\infty)$ *are lines of* $\mathfrak{A}$ *and such that* $S_0 = S \times \{1\}$ *and* $S_{\infty} = \{1\} \times S$.

PROOF. By 19.1 and 2.1, we may identify the points of $\mathfrak{A}$ with the elements of W such that $OP = V(0)$ and $OQ = V(\infty)$. By 14.4 and 14.5, we may assume $S_0 = S \times \{1\}$ and $S_{\infty} = \{1\} \times S$, and, by 19.2, we may assume that $S_1 = \{(\alpha, \alpha) \mid \alpha \in S\}$ fixes a line X through O distinct from $V(0)$ and $V(\infty)$. It follows from 18.6 that $X = V(\zeta)$ with $\zeta \in Z$. As $\{1\} \times Z$ acts transitively on $\Re \setminus \{V(0), V(\infty)\}$, we may assume finally that $\zeta = 1$. $\square$

19.5 Theorem. *There exists up to isomorphism exactly one plane of type* $F * 5$, *namely the plane over the exceptional nearfield of order* 25.

PROOF. Let $\mathfrak{A}$ be of type $F * 5$. By 19.4, $\mathfrak{A} \cong \kappa(W)$ with $\kappa = \{V(0), V(\infty)\} \cup V(1)^{S \times S}$. This establishes the uniqueness part of the theorem. The existence part is trivial. $\square$

19.6 Theorem. *There exists up to isomorphism exactly one plane of type* $F * 23$, *namely the plane over the exceptional nearfield of order* 23^2.

PROOF. By 19.4, we may assume $\mathfrak{A} = \kappa(W)$ with $V(0), V(1), V(\infty) \in \kappa$ and $S_0 \times S_{\infty} = S \times S$. As $23^2 - 1 = 22 \cdot 24$, the group $S \times S$ decomposes $\kappa \setminus \{V(0), V(\infty)\}$ into 22 orbits each of length 24. Denote by S_1 the stabilizer of $V(1)$ in $S \times S$ and denote by S_2 a subgroup of $S \times S$ which is isomorphic to SL(2, 3), which fixes a subspace of rank 2 distinct from $V(0)$ and $V(\infty)$ and which is not conjugate to S_1. Such an S_2 exists by 19.2. Let $X \in \kappa \setminus \{V(0), V(\infty)\}$ and put $H = (S \times S)_X$. Then H is a diagonal of $S \times S$, whence $H \cong$ SL(2, 3). By 19.2, H is conjugate to S_1 or S_2. By 18.6, S_1 fixes exactly 24 subspaces of rank 2, namely the subspaces belonging to $\Re$. Therefore, by 19.3, there exist at most 11 orbits of the form $X^{S \times S}$ with H conjugate to S_1. As the same assertion holds for S_2 and since there are exactly 22 orbits of the form $X^{S \times S}$, we see that $\mathfrak{A}$ is unique up to isomorphism. The existence is clear. $\square$

19.7 Lemma. *Let* $\mathfrak{A}$ *and* $\mathfrak{A}'$ *be two planes of type* $F * p$ *and identify the points of* $\mathfrak{A}$ *as well as the points of* $\mathfrak{A}'$ *with the elements of* W *according to 19.4. If* $\mathfrak{A}$

and $\mathfrak{A}'$ *are isomorphic, then there exists an isomorphism* σ *from* $\mathfrak{A}$ *onto* $\mathfrak{A}'$ *with* $\{V(0), V(\infty)\}^\sigma = \{V(0), V(\infty)\}$.

This follows from 14.8.

19.8 Lemma. *Let* $\mathfrak{A}$ *and* $\mathfrak{A}'$ *be two planes of type* $F * p$. *The points of* $\mathfrak{A}$ *as well as the points of* $\mathfrak{A}'$ *are identified with the elements of* W *as in 19.4. If* σ *is an isomorphism from* $\mathfrak{A}$ *onto* $\mathfrak{A}'$ *with* $\{V(0), V(\infty)\}^\sigma = \{V(0), V(\infty)\}$, *then* $\sigma \in (N \times N)\langle\rho\rangle$, *where* ρ *is defined by* $(x, y)^\rho = (y, x)$.

PROOF. Replacing σ by $\sigma\rho$ and $\mathfrak{A}'$ by $\mathfrak{A}'^\rho$, if necessary, we may assume $V(0)^\sigma = V(0)$ and $V(\infty)^\sigma = V(\infty)$. By the uniqueness remark made at the beginning of this section, $S_0^\sigma = S_0$ and $S_\infty^\sigma = S_\infty$. Furthermore $(0,0)^\sigma = (0,0)$. Hence σ is linear by 1.10. Using all this information, we see that there exist $\alpha, \beta \in N$ with $\sigma = (\alpha, \beta)$. $\qquad\qquad\square$

Put $\pi_\zeta = V(\zeta)^{S \times S}$ and $M = \{(\nu, \nu) \mid \nu \in N\}$. Then $\pi_\zeta^M = \pi_\zeta$. If $D = \{(\zeta, \zeta) \mid \zeta \in Z\}$, then $D \subseteq M$. Furthermore,

$$|N \times N| = |N|^2 = |N : ZS|^2 |ZS|^2 = 4|Z|^2|S|^2|Z \cap S|^{-2} = (p-1)^2 24^2.$$

The group $D(S \times S)$ is normal in $D(S \times S)M$. Hence

$$|D(S \times S)M : D(S \times S)| = |M : M \cap (D(S \times S))|$$
$$= |M : D(M \cap (S \times S))| = |M : DS_1|$$
$$= |M||DS_1|^{-1} = |N||DS_1|^{-1}$$
$$= 2|ZS||DS_1|^{-1} = 2.$$

This yields

$$|D(S \times S)M| = 2|D(S \times S)| = 2 \cdot \tfrac{1}{2}(p-1)|S|^2 = (p-1)24^2.$$

Therefore $|(N \times N) : D(S \times S)M| = p - 1$.

Let $(\sigma, \tau) \in (N \times N)_{\pi_1}$. As $N \times N = D(S \times S)M(\{1\} \times N)$, we may assume $\sigma = 1$. Since $S \times S = S_1(\{1\} \times S)$ there exists $\alpha \in S$ with

$$\{(x, x^\tau) \mid x \in V\} = \{(x, x^\alpha) \mid x \in V\}.$$

Hence $\tau = \alpha \in S$, i.e. $(N \times N)_{\pi_1} = D(S \times S)M$. Thus $|\pi_1^{N \times N}| = p - 1$.

By 19.3, $|\pi_\zeta \cap \mathfrak{R}| = 2$. Hence there are exactly $\tfrac{1}{2}(p-1)$ such π_ζ. All of them belong to $\pi_1^{N \times N}$.

Put $G = (N \times N)\langle\rho\rangle$, where $(x, y)^\rho = (y, x)$, and set $\pi_1^{N \times N} = \mathfrak{F}$.

19.9 Lemma. *Assume* $p \neq 5$. *Then we have*:

a) $G_{\pi_1} = D(S \times S)M\langle\rho\rangle$.
b) $\bigcap_{\pi \in \mathfrak{F}} G_\pi = D(S \times S)M$.

PROOF. a) As we have already seen $D(S \times S)M \subseteq G_{\pi_1}$. Furthermore $\rho \in G_{\pi_1}$. Since $N \times N$ is normal in G and since ρ fixes π_1, we see that $\mathfrak{F}$ is invariant under G. Furthermore,

$$p - 1 = |\mathfrak{F}| = |G : G_{\pi_1}| \leqslant |G : D(S \cap S)M\langle\rho\rangle| = p - 1,$$

whence $G_{\pi_1} = D(S \times S)M\langle\rho\rangle$.

   b) Assume $\pi_\zeta^\rho = \pi_\zeta$. Since

$$V(\zeta)^\rho = \{(x^\zeta, x) \mid x \in V\} = \{(x^\zeta, x^{\zeta\zeta^{-1}}) \mid x \in V\} = V(\zeta^{-1}),$$

we infer from 19.3 that $V(\zeta^{-1}) \in \{V(\zeta), V(-\zeta)\}$. If $V(\zeta^{-1}) = V(\zeta)$, then $\zeta^2 = 1$ and hence $\pi_\zeta = \pi_1$, as $\pi_1 = \pi_{-1}$. If $V(\zeta^{-1}) = V(-\zeta)$, then $\zeta^2 = -1$ which is impossible, as $p \equiv 3 \bmod 4$. (Here we have used the assumption $p \neq 5$.) This proves b). $\qquad\square$

19.10 Theorem. *There exist up to isomorphism two planes of type $F * 7$. One of them is the plane over the exceptional nearfield of order 7^2. If H is the collineation group of the second plane, then, up to conjugacy,*

$$H_{\{V(0), V(\infty)\}} = D(S \times S)M\langle(1, \nu)\rho\rangle$$

where $\nu \in (N \cap \mathrm{SL}(2, 7)) \setminus S$.

PROOF. We infer from $7^2 - 1 = 2 \cdot 24$ that S has exactly two orbits on $V \setminus \{0\}$, since S acts regularly on $V \setminus \{0\}$. It follows from $|Z| = 6$ and $|Z \cap S| = 2$ that Z fixes each of these orbits. Using 14.4, we see $N \subseteq \mathrm{SL}(2, 7)Z$. Hence

$$N = N \cap (\mathrm{SL}(2, 7)Z) = (N \cap \mathrm{SL}(2, 7))Z.$$

As $|N : ZS| = 2$, the order of N is $2 \cdot \frac{1}{2}(7 - 1)24 = 3 \cdot 48$. Therefore

$$3 \cdot 48 = |Z| |N \cap \mathrm{SL}(2, 7)| |Z \cap N \cap \mathrm{SL}(2, 7)|^{-1} = 3 |N \cap \mathrm{SL}(2, 7)|.$$

Thus $N^* = N \cap \mathrm{SL}(2, 7)$ is a group of order 48. As 7 does not divide 48, we have consequently that N^* operates transitively on $V \setminus \{0\}$. In particular, N acts transitively on $V \setminus \{0\}$.

   Put $B_1 = (\bigcup_{X \in \pi_1} X) \setminus \{(0, 0)\}$.

   (a) If $\zeta \in Z$, then $V(\zeta) \subseteq B_1 \cup \{(0, 0)\}$: Obviously B_1 is invariant under $S \times S$. Hence $(x, x^\zeta) \in B_1$ for all $x \in V \setminus \{0\}$ and all $\xi \in S$. If $0 \neq x \in V$, then there exists $\xi \in S$ with $x^\zeta = x^\xi$, as S and SZ have the same orbits on V. Thus $(x, x^\zeta) = (x, x^\xi) \in B_1$. This proves (a).

   The group $D(S \times S)M(\{1\} \times Z)$ permutes $\{\pi_\zeta \mid \zeta \in Z\}$ transitively. Therefore $D(S \times S)M(\{1\} \times Z) \subseteq (N \times N)_{B_1}$. Moreover $|(N \times N) : D(S \times S)M(\{1\} \times Z)| = 2$. Let $(1, \nu) \in (N \times N) \setminus (D(S \times S)M(\{1\} \times Z))$ and put $B_2 = B_1^{(1, \nu)}$.

   (b) $B_1 \cap B_2 = \emptyset$: Let $(x, y) \in B_1 \cap B_2$. Then there exist $u, v \in V \setminus \{0\}$ and $\alpha, \beta, \gamma, \delta \in S$ with $(u^\alpha, u^\beta) = (x, y) = (v^\gamma, v^{\delta\nu})$. This yields $u^\beta = u^{\alpha\gamma^{-1}\delta\nu}$. Now u^β and $u^{\alpha\gamma^{-1}\delta}$ lie in the same orbit of S in $V \setminus \{0\}$. As N^* and hence N operate transitively on $V \setminus \{0\}$ and since $\nu \in N \setminus ZS$, the elements $u^{\alpha\gamma^{-1}\delta}$ and $u^{\alpha\gamma^{-1}\delta\nu} = u^\beta$ are in different orbits of S. This contradiction proves (b).

(c) $B_1 \cup B_2 = W^* = W \setminus \{V(0) \cup V(\infty)\}$: Let $(x, y) \in W^*$. As N acts transitively on $V \setminus \{0\}$, there exist $\zeta \in Z$, $\alpha \in S$ and $i \in \{0, 1\}$ with $y = x^{\zeta \alpha \nu^i}$. Furthermore there exists $\beta \in S$ with $x^\beta = x^\zeta$. Hence

$$(x, y) = (x, x^{\beta \alpha \nu^i}) \in B_1 \cup B_2.$$

From $B_2 = B_1^{(1, \nu)}$, (b) and (c) we infer:

(d) $|B_i| = \frac{1}{2}|W^*| = \frac{1}{2}(7^2 - 1)^2$.

Consider the incidence structure $(B_1, \pi_\zeta, \in)$. This is a tactical configuration with parameters $v = \frac{1}{2}(7^2 - 1)^2$, $b = |S \times S||S_1|^{-1} = 24$, $k = 7^2 - 1$ and r. As $\frac{1}{2}(7^2 - 1)^2 r = vr = bk = 24(7^2 - 1)$, we have $r = 1$. Thus π_ζ is a 1-partial spread of W. As $\pi_\zeta = \pi_{-\zeta}$, we have a total of three 1-partial spreads and a total of three 2-partial spreads. The union of $\{V(0), V(\infty)\}$ with a 1- and a 2-partial spread yields a spread κ of W. Thus we obtain $3^2 = 9$ translation planes $\kappa(W)$. It follows from 19.1 that $D(S \times S)M$ is contained in the collineation group of each of these planes. Therefore these planes are all of type $F * 7$.

We have to determine the number of isomorphism types of planes of type $F * 7$. In order to do this, it suffices by 19.4, 19.7 and 19.8 to determine the number of orbits of $(N \times N)\langle \rho \rangle$ on the set of planes $\kappa(W)$.

Let H be the collineation group of $\kappa(W)$ and put $K = H_{\{V(0), V(\infty)\}}$. As $\{1\} \times Z$ acts transitively on $\{\pi_\zeta | \zeta \in Z\}$, we have that 3 divides $|(N \times N)\langle \rho \rangle : K|$. This together with $|(N \times N)\langle \rho \rangle : D(S \times S)M| = 2(7 - 1) = 4 \cdot 3$ and $D(S \times S)M \subseteq K$ yields $k = |K : D(S \times S)M| \in \{1, 2, 4\}$. Let a_i be the number of isomorphism types with $k = 2^i$. Then $9 = a_0 12 + a_1 6 + a_2 3$. Therefore $a_0 = 0$. As 6 does not divide 9, we obtain $a_2 \neq 0$. The Sylow 2-subgroups of $(N \times N)\langle \rho \rangle / D(S \times S)M$ are all conjugate. Therefore the planes with $k = 4$ are all isomorphic. Hence $a_2 = 1$ and then $a_1 = 1$. This proves that there exist up to isomorphism exactly two planes of type $F * 7$.

We have $|N^* : S| = 2$, where $N^* = N \cap \mathrm{SL}(2, 7)$. This implies $|D(S \times S)M(\{1\} \times N^*) : D(S \times S)M| = 2$. From this we obtain $|D(S \times S)M(\{1\} \times N^*)\langle \rho \rangle : D(S \times S)M| = 4$. .

If $K = D(S \times S)(\{1\} \times N^*)\langle \rho \rangle$, then $|\{\pi_1^\lambda | \lambda \in K\}| = 2$. As K permutes B_1, B_2 transitively, $\kappa_2 = \{V(0), V(\infty)\} \cup \pi_1 \cup \pi_1^\lambda$ is a spread of W, provided λ is an element in K which switches B_1 and B_2. As $\{1\} \times N^* \subseteq K$, the plane $\kappa_2(W)$ is the plane over the exceptional nearfield of order 7^2.

Let $\kappa_1(W)$ be a plane of type $F * 7$ which is not isomorphic to $\kappa_2(W)$. Let L be the stabilizer of $\{V(0), V(\infty)\}$ in its collineation group. Then $|L : D(S \times S)M| = 2$. Hence L is conjugate to a subgroup of index 2 in K. We may assume $L \subseteq K$. As $\kappa_1(W)$ is not isomorphic to $\kappa_2(W)$, we see $\{1\} \times N^* \not\subseteq L$. If $\rho \in L$, then $B_1^\rho = B_1$. Furthermore $\pi_1 \subseteq \kappa_1$ by 19.9. Lemma 19.9 also yields that ρ fixes exactly one 2-partial spread. Hence $\kappa_1 = \kappa_2$: a contradiction. Hence $\rho \notin L$.

Let $\nu \in N^* \setminus S$. Then $D(S \times S)M(\{1\} \times N^*)$, $D(S \times S)M\langle \rho \rangle$ and

$D(S \times S)M\langle(1,\nu)\rho\rangle$ are the only subgroups of index 2 in K which contain $D(S \times S)M$. Therefore $L = D(S \times S)M\langle(1,\nu)\rho\rangle$.   □

As L acts transitively on $\{V(0), V(\infty)\}$ and $\kappa_1 \setminus \{V(0), V(\infty)\}$ and since the same is true for K and κ_2, we have:

19.11 Corollary. *If $\mathfrak{A}$ is of type $F * 7$ and if G is its collineation group, then $TG_{\{V(0), V(\infty)\}}$ is a rank-3-collineation group of $\mathfrak{A}$.*

Finally, we have to determine the planes of type $F * 11$.

19.12 Theorem. *There exist up to isomorphism exactly four planes of type $F * 11$, one of them being the plane over the exceptional nearfield of order 11^2 with soluble multiplicative group. If G is the collineation group of one of the other planes, then $G_{\{V(0), V(\infty)\}}$ is conjugate to $D(S \times S)\langle\rho\rangle$.*

PROOF. Put $B_\zeta = (\bigcup_{x \in \pi_\zeta} X) \setminus \{(0,0)\}$. Then $B_\zeta = B_{-\zeta}$ as $\pi_\zeta = \pi_{-\zeta}$. The group S has five orbits of length 24 on $V \setminus \{0\}$. As 5 does not divide 24, the group SZ acts transitively on $V \setminus \{0\}$.

(a) $W^* = \bigcup_{\zeta \in Z} B_\zeta$: Let $(x, y) \in W^*$. Then there exist $\zeta \in Z$ and $\alpha \in S$ with $y = x^{\alpha\zeta}$. Therefore $(x, y) = (x, x^\zeta)^{(1, \alpha)} \in B_\zeta$.

(b) If $B_\zeta \cap B_\eta \neq \emptyset$, then $\zeta = \eta$ or $\zeta = -\eta$: Let $(x, y) \in B_\zeta \cap B_\eta$. Then $y = x^{\alpha\zeta} = x^{\beta\eta}$ with $\alpha, \beta \in S$. Hence $\beta^{-1}\alpha = \eta\zeta^{-1} \in S \cap Z = \{1, -1\}$.

(a) and (b) yield:

(c) $|B_\zeta| = \frac{1}{5}(11^2 - 1)^2$.

Let $\nu \in N \setminus ZS$. Then $Z\langle\nu\rangle$ is abelian. As Z operates transitively on the set of orbits of S in $V \setminus \{0\}$, we may assume that ν fixes each of these orbits.

(d) If $\zeta \in Z$, then $V(\zeta)^{(1, \nu)} \subseteq B_\zeta \cup \{(0,0)\}$: Let $0 \neq x \in V$. Then x^ζ and $x^{\zeta\nu}$ are in the same orbit of S. Hence there exists $\alpha \in S$ with $x^{\zeta\nu} = x^{\zeta\alpha}$. Therefore $(x, x^{\zeta\nu}) = (x, x^{\zeta\alpha}) \in B_\zeta$ proving (d).

(e) π_ζ and $\pi_\zeta^{(1, \nu)}$ are partial spreads of W. This is proved similarly to corresponding earlier statements.

(f) Define ρ by $(x, y)^\rho = (y, x)$. Then $B_\zeta^\rho = B_\zeta$ implies $\zeta = 1$ or $\zeta = -1$: From $B_\zeta^\rho = B_\zeta$ and $S_1^\rho = S_1$ we infer $V(\zeta)^\rho = V(\zeta)$ or $V(\zeta)^\rho = V(-\zeta)$. As $V(\zeta)^\rho = \{(x, x^{\rho^{-1}}) \mid x \in V\}$, we obtain $\zeta^2 = 1$ and hence $\zeta = 1$ or -1 or $\zeta^2 = -1$. But $11 \equiv 3 \bmod 4$ implies $\zeta^2 \neq -1$ proving (f).

We shall denote the five partial spreads π_ζ by $\pi_1, \ldots, \pi_5$ in such a way that $\pi_1 = V(1)^{S \times S}$. Furthermore we put $\pi_i' = \pi_i^{(1, \nu)}$. Choose $\varphi_i \in \{\pi_i, \pi_i'\}$. Then $\kappa = \{V(0), V(\infty)\} \cup \bigcup_{i=1}^5 \varphi_i$ is a spread of W. Hence $\kappa(W)$ is a translation plane. As $(N \times N)\langle\rho\rangle$ induces S_2 on $\{\{\pi_1, \ldots, \pi_5\}, \{\pi_1', \ldots, \pi_5'\}\}$, the index of H in $(N \times N)\langle\rho\rangle$ is divisible by 2, where H is the stabilizer of $\{V(0), V(\infty)\}$ in the full collineation group of $\kappa(W)$.

Assume that 5 divides $k = |H : D(S \times S)M|$. Then $\{1\} \times Z \subseteq H$. Thus $\kappa = \{V(0), V(\infty)\} \cup \bigcup_{i=1}^5 \pi_i$ or $\kappa = \{V(0), V(\infty)\} \cup \bigcup_{i=1}^5 \pi_i'$. In these cases, ρ is also a collineation of $\kappa(W)$. Hence $k = 10$, i.e. there are exactly two planes $\kappa(W)$ where 5 divides k and these two planes are isomorphic.

If 5 does not divide k, then $k = 1$ or 2. Let a_0 be the number of isomorphism types with $k = 1$ and a_1 the number of isomorphism types with $k = 2$. Then

$$32 = 2^5 = a_0 20 + a_1 10 + 2.$$

There are exactly $2^2 = 4$ planes with $\pi_1 \subseteq \kappa$ and $\rho \in H$. For one of them, 5 divides k. All these planes are pairwise nonisomorphic. Hence $a_1 \geqslant 3$. This yields $a_0 = 0$ and $a_1 = 3$. $\square$

20. Exceptional Rank-3-Planes

We are now able to finish the proof of 17.1 and to say more about the exceptions which occur.

20.1 Theorem. *Let $\mathfrak{A}$ be a finite affine plane of order n. If G is a non-soluble rank-3-collineation group of $\mathfrak{A}$ with an orbit of length 2 on l_∞, then $n = 11^2$, 29^2, or 59^2 and $\mathfrak{A}$ is the plane over the exceptional nearfield of order n, where in the case $n = 11^2$ the nearfield with non-soluble multiplicative group is meant.*

PROOF. It follows from 14.11 that $\mathfrak{A}$ is not a generalized André-plane. Let $\{P, Q\}$ be the orbit of length 2 on l_∞ and let O be an affine point. The work of section 17 shows that $G(P, OQ)$ is non-soluble. Hence $G(P, OQ)$ contains a subgroup S_0 isomorphic to $\mathrm{SL}(2, 5)$. Likewise $G(Q, OP)$ contains a subgroup S_∞ isomorphic to $\mathrm{SL}(2, 5)$. Furthermore $n = 11^2$, 19^2, 29^2, or 59^2.

The group $S_0 S_\infty$ is normal in G_0. Thus the orbits of $S_0 S_\infty$ on $l_\infty \setminus \{P, Q\}$ all have the same length λ. As $n - 1 = 120s$ with $s = 1, 3, 7$, or 29, either $\lambda = 120$ or $\lambda = n - 1$. As 7 and 29 do not divide 120, we have $\lambda = 120$ or $n = 19^2$ and $\lambda = 3 \cdot 120$. If the second possibility occurs, then S_0 permutes the three orbits of S_∞ in $l_\infty \setminus \{P, Q\}$ transitively. If follows that S_0 contains a normal subgroup of index 3 or 6 which is not the case. Hence $\lambda = 120$. This proves that $\mathfrak{A}$ is of type $R * p$. Theorem 20.1 follows now from 18.16, 18.17, 18.18 and 18.19. $\square$

This finishes also the proof of 17.1.

20.2 Theorem. *Let $\mathfrak{A}$ be a finite affine plane of order n and let G be a soluble rank-3-collineation group of $\mathfrak{A}$ with an orbit of length 2 on l_∞. If $\mathfrak{A}$ is not a generalized André-plane, then $n = 5^2$, 7^2, 11^2 or 23^2. If $n = 5^2$, 11^2 or 23^2, then $\mathfrak{A}$ is the plane over the exceptional nearfield F of order n, where F is the* *exceptional nearfield with soluble F^* in the case $n = 11^2$. Moreover, the two planes of type $F * 7$ satisfy the assumptions of this theorem.*

PROOF. By 17.1, 20.1, and the remarks after 17.3, we know that $n = 5^2$, 7^2, 11^2 or 23^2. If $n = 7^2$, then the planes of type $F * 7$ satisfy the hypotheses of the theorem. (It is easily seen by a look at their collineation groups that they are not generalized André-planes.)

Assume that $n = 5^2$, 11^2 or 23^2. Then 3 is a p-primitive divisor of $n - 1$ and 3 divides $|G(P, OQ)| = |G(Q, OP)|$ where $\{P, Q\}$ is the orbit of length 2 on l_∞ of the group G and O is some affine point. Moreover, if Σ is a Sylow 3-subgroup of $G(P, OQ)$, then Σ is not normal in $G(P, OQ)$.

By 3.5, the group $G(P, OQ)$ contains a normal Z-group N such that $G(P, OQ)/N$ is isomorphic to a subgroup of S_4. As Σ is not normal in $G(P, OQ)$, we have by 17.3 that Σ is not contained in N. Furthermore, $|\Sigma| = 3$, as 9 does not divide $n - 1$. Hence $N\Sigma$ is also a Z-group. By 17.3, the group Σ is normal in $N\Sigma$. Since Σ is not normal in $G(P, OQ)$, we deduce that $N\Sigma/N$ is not normal in $G(P, OQ)/N$. Thus $G(P, OQ)/N$ contains a subgroup isomorphic to A_4. This implies that the Sylow 2-subgroups of $G(P, OQ)$ are generalized quaternion groups of order 8 or 16. This yields that 4 does not divide $|N|$, since the Sylow 2-subgroups of N are cyclic and Σ permutes the subgroups of order 4 of a quaternion group of order 8 transitively. We infer therefore from $n - 1 = 2^3 \cdot 3$, $5 \cdot 2^3 \cdot 3$, $11 \cdot 2^4 \cdot 3$ that N is abelian. Let C be the maximal subgroup of odd order of N. Then C is a normal Hall subgroup of N and hence a normal Hall subgroup of $G(P, OQ)$. Therefore C has a complement D in $G(P, OQ)$ by the Schur-Zassenhaus Theorem. It follows that $|D| = 2^3 \cdot 3$ or $|D| = 2^4 \cdot 3$. It is now easily seen that D contains a subgroup $S_0 \cong \mathrm{SL}(2, 3)$. As $|C| = 1$, 5 or 11, we have that $\mathrm{Aut}(C)$ is cyclic. This implies that S_0 centralizes C. We obtain therefore from 3.6 that S_0 is the only subgroup of CS_0 which is isomorphic to $\mathrm{SL}(2, 3)$. As $\mathrm{SL}(2, 3)$ does not contain a subgroup of index 2, we finally see that S_0 is the only subgroup of $G(P, OQ)$ which is isomorphic to $\mathrm{SL}(2, 3)$. Likewise we obtain that $G(Q, OP)$ contains exactly one subgroup S_∞ which is isomorphic to $\mathrm{SL}(2, 3)$.

$S_0 S_\infty$ is a normal subgroup of G_0. Hence all the orbits of $S_0 S_\infty$ in $l_\infty \backslash \{P, Q\}$ have the same length λ and λ is divisible by 24. As 5 and 11 do not divide 24, we find $\lambda = 24$ or $n = 23^2$ and $\lambda = 2 \cdot 24$. In the latter case, S_0 permutes the orbits of S_∞ in cycles of length 2. It follows that S_0 contains a subgroup of index 2 which is not the case. Thus $\lambda = 24$ in all cases and $\mathfrak{A}$ is of type $F * p$. Theorem 20.2 now follows from 19.5, 19.6 and 19.12.

CHAPTER IV

The Suzuki Groups and Their Geometries

In this chapter we investigate the Suzuki groups as well as the Möbius planes and translation planes belonging to them. The investigations culminate in R. Liebler's characterization of the Lüneburg planes. This chapter is a blending of my set of lecture notes 1965b and an unpublished set of lecture notes by A. Cronheim. I would like to thank him very much indeed for allowing me to incorporate his notes into this chapter.

21. The Suzuki Groups $S(K,\sigma)$

Let K be a commutative field of characteristic 2 with $|K| > 2$. Moreover let σ be an automorphism of K such that $x^{\sigma^2} = x^2$ for all $x \in K$. We denote by $\mathfrak{P}$ the 3-dimensional projective space over K and we let (x_0, x_1, x_2, x_3) be ths coordinates of the points of $\mathfrak{P}$.

Let E be the plane defined by the equation $x_0 = 0$ and put $U = (0,1,0,0)K$. We introduce coordinates x, y, z in the affine space $\mathfrak{P}_E$ by

$$x = x_2 x_0^{-1}, \qquad y = x_3 x_0^{-1}, \qquad z = x_1 x_0^{-1}.$$

Finally let $\mathfrak{O}$ be the set of points of $\mathfrak{P}$ consisting of U and all those points of $\mathfrak{P}_E$ whose coordinates (x, y, z) satisfy the equation

$$z = xy + x^{\sigma+2} + y^{\sigma}.$$

First of all we note:

21.1 Lemma. $\sigma + 1$ and $\sigma + 2$ are automorphisms of K^*.

PROOF. We have $1 = 2 - 1 = \sigma^2 - 1 = (\sigma + 1)(\sigma - 1) = (\sigma - 1)(\sigma + 1)$. Thus $\sigma + 1$ is an automorphism of K^*. This and $\sigma + 2 = (1 + \sigma)\sigma$ yield that $\sigma + 2$ is also an automorphism of K^*.   □

21.2 Lemma. *If* $K \cong \mathrm{GF}(q)$, *then* $|\mathfrak{D}| = q^2 + 1$.

The proof is trivial.

For $a, b \in K$ we denote by $\tau(a, b)$ the mapping defined by

$$(x, y, z)^{\tau(a, b)} = (x + a, y + b + a^\sigma x, z + ab$$
$$+ a^{\sigma + 2} + b^\sigma + ay + a^{\sigma + 1} x + bx).$$

Obviously, $\tau(a, b)$ is a collineation of $\mathfrak{P}_E$ and hence extends to a collineation of $\mathfrak{P}$ which we also denote by $\tau(a, b)$.

21.3 Lemma. $T = \{\tau(a, b) \,|\, a, b \in K\}$ *is a group which fixes* $\mathfrak{D}$ *and* U, *and* T *acts transitively and regularly on* $\mathfrak{D} \backslash \{U\}$. *If* $a \neq 0$, *then* $o(\tau(a, b)) = 4$; *if* $b \neq 0$ *then* $o(\tau(0, b)) = 2$. *Moreover* $\mathfrak{Z}(T) = \{\tau(0, b) \,|\, b \in K\}$.

PROOF. We have

$$(x, y, z)^{\tau(a, 0)} = (x + a, y + a^\sigma x, z + a^{\sigma + 2} + ay + a^{\sigma + 1} x)$$

and

$$(x, y, z)^{\tau(0, b)} = (x, y + b, z + b^\sigma + bx).$$

Hence

$$(x, y, z)^{\tau(a, 0)\tau(0, b)} = (x + a, y + a^\sigma x, z + a^{\sigma + 2} + ay + a^{\sigma + 1} x)^{\tau(0, b)}$$
$$= (x + a, y + b + a^\sigma x, z + a^{\sigma + 2} + ay$$
$$+ a^{\sigma + 1}(x + a) + b^\sigma + b(x + a))$$
$$= (x, y, z)^{\tau(a, b)}.$$

Therefore $\tau(a, 0)\tau(0, b) = \tau(a, b)$.

On the other hand

$$(x, y, z)^{\tau(0, b)\tau(a, 0)} = (x, y + b, z + b^\sigma + bx)^{\tau(a, 0)}$$
$$= (x + a, y + b + a^\sigma x, z + b^\sigma + bx$$
$$+ a^{\sigma + 2} + a(y + b) + a^{\sigma + 1} x)$$
$$= (x, y, z)^{\tau(a, b)}.$$

Thus $\tau(0, b)\tau(a, 0) = \tau(a, b)$.

Furthermore

$$(x, y, z)^{\tau(0, b)\tau(0, c)} = (x, y + b, z + b^{\sigma} + bx)^{\tau(0, c)}$$
$$= (x, y + b + c, z + b^{\sigma} + bx + c^{\sigma} + cx)$$
$$= (x, y, z)^{\tau(0, b+c)}.$$

Hence $\tau(0, b)\tau(0, c) = \tau(0, b + c)$. This yields in particular $\tau(0, b)^2 = 1$.

$$\tau(0, c)\tau(a, b) = \tau(0, c)\tau(a, 0)\tau(0, b) = \tau(a, 0)\tau(0, c)\tau(0, b)$$
$$= \tau(a, 0)\tau(0, b)\tau(0, c) = \tau(a, b)\tau(0, c).$$

Therefore $\tau(0, c) \in \mathfrak{Z}(T)$ for all $c \in K$.

We have

$$(x, y, z)^{\tau(a, 0)\tau(b, 0)} = (x + a, y + a^{\sigma}x, z + a^{\sigma+2} + ay + a^{\sigma+1}x)^{\tau(b, 0)}$$
$$= (x + a + b, y + a^{\sigma}x + b^{\sigma}x + b^{\sigma}a, z')$$

where $z' = z + a^{\sigma+2} + ay + a^{\sigma+1}x + b^{\sigma+2} + by + ba^{\sigma}x + b^{\sigma+1}x + b^{\sigma+1}a$. On the other hand we have

$$(x, y, z)^{\tau(a+b, b^{\sigma}a)} = (x + a + b, y + b^{\sigma}a + a^{\sigma}x + b^{\sigma}x, z'')$$

where

$$z'' = z + (a + b)b^{\sigma}a + (a + b)^{\sigma+2} + b^{\sigma^2}a + (a + b)y$$
$$+ (a + b)^{\sigma+1}x + b^{\sigma}ax.$$

Now

$$(a + b)b^{\sigma}a + b^{\sigma^2}a + (a + b)^{\sigma+2} = b^{\sigma}a^2 + b^{\sigma+1}a + b^2a + (a^{\sigma} + b^{\sigma})(a + b)^2$$
$$= b^{\sigma+1}a + b^{\sigma}a^2 + b^2a^{\sigma} + a^{\sigma}a^2 + a^{\sigma}b^2$$
$$+ b^{\sigma}a^2 + b^{\sigma}b^2$$
$$= b^{\sigma+1}a + a^{\sigma+2} + b^{\sigma+2}$$

and

$$(a + b)^{\sigma+1} + b^{\sigma}a = (a^{\sigma} + b^{\sigma})(a + b) + b^{\sigma}a$$
$$= a^{\sigma+1} + a^{\sigma}b + b^{\sigma}a + b^{\sigma+1} + b^{\sigma}a$$
$$= a^{\sigma+1} + a^{\sigma}b + b^{\sigma+1}.$$

Hence $z' = z''$ and thus $\tau(a, 0)\tau(b, 0) = \tau(a + b, ab^{\sigma})$. Therefore we obtain

$$\tau(a, b)\tau(c, d) = \tau(a, 0)\tau(c, 0)\tau(0, b + d) = \tau(a + c, ac^{\sigma})\tau(0, b + d)$$
$$= \tau(a + c, 0)\tau(0, ac^{\sigma})\tau(0, b + d)$$
$$= \tau(a + c, 0)\tau(0, ac^{\sigma} + b + d)$$
$$= \tau(a + c, ac^{\sigma} + b + d).$$

This proves that T is a group.

From $(0,0,0)^{\tau(x, y)} = (x, y, xy + x^{\sigma+2} + y^{\sigma})$ we infer that $\tau(x, y) = \tau(x', y')$ if and only if $x = x'$ and $y = y'$. Moreover $\tau(x, y) = 1 = \tau(0, 0)$ if and only if $x = 0$ and $y = 0$. Thus, if $a \neq 0$, we have $\tau(a, 0)^2 = \tau(a + a, a^{1+\sigma}) = \tau(0, a^{1+\sigma}) \neq 1$ and hence $o(\tau(a, 0)) = 4$. This yields $o(\tau(a, b)) = 4$ for all $a, b \in K$ with $a \neq 0$ because of $\tau(a, b) = \tau(a, 0)\tau(0, b)$ and $\tau(0, b) \in \mathfrak{Z}(T)$.

Let $\tau(a, b) \in \mathfrak{Z}(T)$. Then $\tau(a, 0) = \tau(a, b)\tau(0, b) \in \mathfrak{Z}(T)$. Hence

$$\tau(a + c, ac^{\sigma}) = \tau(a, 0)\tau(c, 0) = \tau(c, 0)\tau(a, 0) = \tau(c + a, ca^{\sigma}).$$

Therefore $ac^{\sigma} = ca^{\sigma}$ for all c by the remark made above. Putting $c = 1$ yields $a = a^{\sigma}$. Therefore $a(c^{\sigma} - c) = 0$ for all c. As $\sigma \neq 1$, we infer $a = 0$. Thus $\mathfrak{Z}(T) = \{\tau(0, b) \mid b \in K\}$.

Let $\mathfrak{O}^*$ be the orbit of $(0,0,0)$ under T. Then $\mathfrak{O}^* = \mathfrak{O} \setminus \{U\}$. Moreover T acts transitively and regularly on $\mathfrak{O}^*$. It remains to show that U is fixed by T.

Let l be the affine line which is defined by the equations $x = 0 = y$ and let $(0, x_1, x_2, x_3)K$ be the intersection of l and E. Then $x_2 = x_3 = 0$ and therefore $U = l \cap E$. The line l is mapped by $\tau(a, b)$ onto the line defined by the equations $x = a, y = b$ which is parallel to l. Hence $U^T = U$.   □

21.4 Corollary. *For all* $a, b, c, d \in K$, *we have* $\tau(a, b)\tau(c, d) = \tau(a + c, b + d + ac^{\sigma})$.

For $k \in K^*$ we define the collineation $\eta(k)$ by

$$(x, y, z)^{\eta(k)} = (kx, k^{\sigma+1}y, k^{\sigma+2}z).$$

It follows from 21.1 that $Z = \{\eta(k) \mid k \in K^*\}$ is a group isomorphic to K^*.

21.5 Lemma. *For all* $a, b \in K$ *and all* $k \in K^*$ *we have* $\eta(k)^{-1}\tau(a, b)\eta(k) = \tau(ka, k^{\sigma+1}b)$, *i.e.,* Z *normalizes* T. *Also* $U^Z = U$ *and* $O^Z = O$, *where* O *is the affine point with coordinates* $(0, 0, 0)$. *Moreover,* Z *leaves* $\mathfrak{O}$ *invariant and acts (via inner automorphisms) transitively on* $\mathfrak{Z}(T) \setminus \{1\}$.

PROOF. It follows immediately from the definition of Z that Z fixes O and the line l defined by the equations $x = 0 = y$. Hence Z fixes U.

A straightforward computation shows $\tau(a, b)\eta(k) = \eta(k)\tau(ka, k^{\sigma+1}b)$. Hence $\eta(k)^{-1}\tau(a, b)\eta(k) = \tau(ka, k^{\sigma+1}b)$.

As $\eta(k)$ normalizes T, the set $\mathfrak{O}^{*\eta(k)}$ is an orbit of T. From $O = O^{\eta(k)}$ we infer $\mathfrak{O}^{*\eta(k)} = \mathfrak{O}^*$. Hence Z fixes $\mathfrak{O}$.

Finally, it follows from $\tau(0, 1)^{\eta(k)} = \tau(0, k^{\sigma+1})$ as well as 21.1 and 21.3 that Z acts transitively on $\mathfrak{Z}(T) \setminus \{1\}$.   □

Let ω be the collineation of $\mathfrak{P}$ defined by $(x_0, x_1, x_2, x_3)^{\omega} = (x_1, x_0, x_3, x_2)$. Obviously $\omega^2 = 1$ and $U^{\omega} = O$ and $O^{\omega} = U$.

Let $X = (x, y, z) \in \mathfrak{D} \setminus \{O, U\}$. Then $z = xy + x^{\sigma+2} + y^{\sigma}$. This yields

$$
\begin{aligned}
xz^{\sigma} + z(x^{\sigma+1} + y) &= x(x^{\sigma}y^{\sigma} + x^{2+2\sigma} + y^2) + (xy + x^{\sigma+2} + y^{\sigma})(x^{\sigma+1} + y) \\
&= x^{\sigma+1}y^{\sigma} + x^{2\sigma+3} + xy^2 + x^{\sigma+2}y + xy^2 + x^{2\sigma+3} \\
&\quad + x^{\sigma+2}y + y^{\sigma}x^{\sigma+1} + y^{\sigma+1} \\
&= y^{\sigma+1}.
\end{aligned}
$$

If $z = 0$, then this yields $y = 0$ and hence $x = 0$, as $x^{\sigma+2} = z + xy + y^{\sigma}$. Thus $z \neq 0$, as $X \neq O$. Let $X = (x_0, x_1, x_2, x_3)K$. We may assume $x_0 = z^{-1}$. Then $x = x_2 z$, $y = x_3 z$ and $z = x_1 z$. Therefore $X = (z^{-1}, 1, xz^{-1}, yz^{-1})K$. This yields $X^{\omega} = (1, z^{-1}, yz^{-1}, xz^{-1})K$. This is the affine point with the coordinates $x' = yz^{-1}$, $y' = xz^{-1}$, $z' = z^{-1}$ Therefore

$$
\begin{aligned}
x'y' + x'^{\sigma+2} + y'^{\sigma} &= xyz^{-2} + y^{\sigma+2}z^{-\sigma-2} + x^{\sigma}z^{-\sigma} \\
&= z^{-\sigma-2}(xyz^{\sigma} + y^{\sigma+2} + x^{\sigma}z^2).
\end{aligned}
$$

As $xy = z + x^{\sigma+2} + y^{\sigma}$, we find

$$
xyz^{\sigma} + y^{\sigma+2} + x^{\sigma}z^2 = z^{\sigma+1} + x^{\sigma+2}z^{\sigma} + y^{\sigma}z^{\sigma} + y^{\sigma+2} + x^{\sigma}z^2.
$$

Furthermore

$$
x^{\sigma+2}z^{\sigma} = x^{\sigma+2}(x^{\sigma}y^{\sigma} + x^{2+2\sigma} + y^2) = x^{2\sigma+2}y + x^{4+3\sigma} + x^{\sigma+2}y^2.
$$

Also

$$
y^{\sigma}z^{\sigma} = x^{\sigma}y^{2\sigma} + x^{2+2\sigma}y^{\sigma} + y^{\sigma+2}
$$

and

$$
x^{\sigma}z^2 = x^{\sigma+2}y^2 + x^{3\sigma+4} + x^{\sigma}y^{2\sigma}.
$$

Hence

$$
xyz^{\sigma} + y^{\sigma+2} + x^{\sigma}z^2 = z^{\sigma+1} + x^{\sigma+2}z^{\sigma} + y^{\sigma}z^{\sigma} + y^{\sigma+2} + x^{\sigma}z^2 = z^{\sigma+1}.
$$

Therefore we obtain $x'y' + x'^{\sigma+2} + y'^{\sigma} = z^{-\sigma-2}z^{\sigma+1} = z^{-1} = z'$. Thus we have proved

21.6 Lemma. ω *leaves $\mathfrak{D}$ invariant.*

We denote by $S(K, \sigma)$ the group of all projective collineations of $\mathfrak{P}$ which leave $\mathfrak{D}$ invariant. Then $TZ \subseteq S(K, \sigma)$ and $\omega \in S(K, \sigma)$. As T is transitive on $\mathfrak{D} \setminus \{U\}$ and ω switches O and U, we deduce that $S(K, \sigma)$ acts doubly transitively on $\mathfrak{D}$.

21.7 Theorem (Tits 1962). *If $P \in \mathfrak{D}$, then there exists exactly one plane E_P of $\mathfrak{P}$ with $\mathfrak{D} \cap E_P = \{P\}$. If l is a line through P which is not contained in E_P, then l carries exactly one more point of $\mathfrak{D}$.*

PROOF. Because of the transitivity of $S(K, \sigma)$ on $\mathfrak{D}$, we may assume that $P = U$. Clearly $E \cap \mathfrak{D} = \{U\}$. Let l be a line through U which does not lie in E. Then there exists $a, b \in K$ such that l consists just of U and the affine points having coordinates (a, b, z) with $z \in K$. Thus there exists exactly one point other than U of $\mathfrak{D}$ which is on l, namely the point with the coordinates $(a, b, ab + a^{\sigma+2} + b^\sigma)$. $\square$

21.8 Theorem (Tits 1962). *Let K be a commutative field of characteristic 2 with $|K| > 2$ and assume that K admits an automorphism σ such that $x^{\sigma^2} = x^2$ for all $x \in K$. If $\mathfrak{D}$ is the point set defined above in the projective space of dimension 3 over K, then S(K, σ) acts doubly transitively on $\mathfrak{D}$. Moreover, if $\gamma \in S(K, \sigma)_U$, then there exists exactly one triple $(k, a, b) \in K^* \times K \times K$ with $\gamma = \eta(k)\tau(a, b)$, and if $\gamma \in S(K, \sigma) \backslash S(K, \sigma)_U$, then there exists exactly one quintuple $(k, a, b, c, d) \in K^* \times K \times K \times K \times K$ with $\gamma = \eta(k)\tau(a, b)\omega\tau(c, d)$.*

PROOF. We remarked already that S(K, σ) acts 2-transitively on $\mathfrak{D}$. Put $G = S(K, \sigma)$ and let $\eta \in G_{U, O}$. Then $E^\eta = E$ and $E_O^\eta = E_O$ by 21.7. From $U^\omega = O$ we infer $E^\omega = E_O$. Hence E_O is defined by the equation $z = 0$. Therefore $(x, y, z)^\eta = (x', y', fz)$ with $f \in K^*$.

The line OU is invariant under η. Hence $(0, 0, z)^\eta = (0, 0, fz)$. Therefore $x' = ax + by$ and $y' = cx + dy$. The point with coordinates $(0, y, y^\sigma)$ is on $\mathfrak{D}$. Therefore the point with coordinates (by, dy, fy^σ) also belongs to $\mathfrak{D}$. Hence we have

$$(*) \qquad fy^\sigma = bdy^2 + b^{\sigma+2}y^{\sigma+2} + d^\sigma y^\sigma$$

for all $y \in K$. As $|K| > 2$, there are two elements $y, z \in K^*$ with $y \neq z$. If $(y^2, y^{\sigma+2}, y^\sigma) = k(z^2, z^{\sigma+2}, z^\sigma)$ for some $k \in K$, then $k = (yz^{-1})^2$ and hence $y^{\sigma+2} = (yz^{-1})^2 z^{\sigma+2} = y^2 z^\sigma$. This yields $y^\sigma = z^\sigma$ and thus $y = z$: a contradiction. This proves $(y^2, y^{\sigma+2}, y^\sigma) \notin (z^2, z^{\sigma+2}, z^\sigma)K$. This shows that $(*)$ has the two linearly independent solutions $(y^2, y^{\sigma+2}, y^\sigma)$ and $(z^2, z^{\sigma+2}, z^\sigma)$.

From $(*)$ we deduce

$$(**) \quad 0 = (f + d^\sigma)k^\sigma y^\sigma + bdk^2 y^2 + b^{\sigma+2}k^{\sigma+2}y^{\sigma+2} \qquad \text{for all } k, y \in K.$$

As every solution of $(*)$ also solves $(**)$, we see that $(**)$ has two linearly independent solutions. Thus $(**)$ defines for all $k \in K^*$ the same plane in the vector space of rank 3 over K. Therefore, given $k \in K$, there exists an $l \in K$ with $(f + d^\sigma)k^\sigma = (f + d^\sigma)l$, $bdk^2 = bdl$ and $b^{\sigma+2}k^{\sigma+2} = b^{\sigma+2}l$.

Assume $b \neq 0$. Then $k^{\sigma+2} = l$ and hence $bdk^2 = bdk^{\sigma+2}$; thus $bd(k^\sigma - 1) = 0$. Choosing $k \neq 1$ we obtain $d = 0$ and therefore $fk^\sigma = fl$ yielding $k^\sigma = l$, as $f \neq 0$. Thus $k^{\sigma+2} = l = k^\sigma$. From this we deduce $k^2 = 1$ and hence $k = 1$, a contradiction. This proves $b = 0$. As a result, we have $0 = (f + d^\sigma)y^\sigma$ and thus $f = d^\sigma$.

The point $(x, 0, x^{\sigma+2})$ is on $\mathfrak{D}$. Therefore $(ax, cx, fx^{\sigma+2})$ is on $\mathfrak{D}$. Hence $fx^{\sigma+2} = acx^2 + a^{\sigma+2}x^{\sigma+2} + c^\sigma x^\sigma$. Similar arguments as above yield $c = 0$ and $f = a^{\sigma+2}$.

Now $(a^{\sigma+1})^\sigma = a^{\sigma+2} = f = d^\sigma$ and hence $d = a^{\sigma+1}$. Therefore

$$(x, y, z)^\eta = (ax, a^{\sigma+1}y, a^{\sigma+2}z) = (x, y, z)^{\eta(a)}.$$

Let $\gamma \in G_U$. Then there exists exactly one $\tau(a,b) \in T$ with $O^\gamma = O^{\tau(a,b)}$. Hence there exists $k \in K^*$ with $\gamma\tau(a,b)^{-1} = \eta(k)$, whence $\gamma = \eta(k)\tau(a,b)$. This decomposition is unique, as ZT is the semidirect product of T with Z.

Let $\gamma \notin G_U$. Then $U \neq U^\gamma$. Therefore there exists exactly one $\tau(c,d) \in T$ with $U^\gamma = O^{\tau(c,d)}$. This yields $U^{\gamma\tau(c,d)^{-1}\omega} = U$ and consequently $\gamma\tau(c,d)^{-1}\omega = \eta(k)\tau(a,b)$ for some k, a, b. Thus

$$\gamma = \eta(k)\tau(a,b)\omega\tau(c,d).$$

Assume $\gamma = \eta(k)\tau(a,b)\omega\tau(c,d) = \eta(k')\tau(a',b')\omega\tau(c',d')$. Then $O^{\tau(c,d)} = U^\gamma = O^{\tau(c',d')}$. Hence $c = c'$, $d = d'$ and therefore

$$\eta(k)\tau(a,b) = \eta(k')\tau(a',b'),$$

whence $k = k'$, $a = a'$, and $b = b'$. $\square$

Let K be a commutative field of characteristic 2 with $\mathrm{GF}(4) \subseteq K$. If $\sigma \in \mathrm{Aut}(K)$, then $\mathrm{GF}(4) = \mathrm{GF}(4)^\sigma$, since the multiplicative group of $\mathrm{GF}(4)$ is just the set of all cube roots of unity of K. As $|\mathrm{Aut}(\mathrm{GF}(4))| = 2$, we deduce $x^{\sigma^2} = x$ for all $x \in \mathrm{GF}(4)$. Hence $\sigma^2 \neq 2$. Therefore, if $K \cong \mathrm{GF}(2^n)$ and if K admits an automorphism σ with $x^{\sigma^2} = x^2$ for all $x \in K$, then $n = 2r + 1 > 1$. (Recall that we are always assuming $|K| \geqslant 4$.)

Let $K \cong \mathrm{GF}(2^{2r+1})$ with $2r + 1 > 1$. Put $\sigma = 2^{r+1}$. Then $x^{\sigma^2} = x^{2^{2r+2}} = x^2$. Hence K admits an automorphism of the required type.

Finally, let $\alpha \in \mathrm{Aut}(K)$ be such that $x^{\alpha^2} = x^2$ for all x. Then $x^\alpha = x^{2^t}$ with $1 \leqslant t < 2r + 1$. This implies $x^2 = x^{2^{2t}}$ for all x. This yields $2t \equiv 1 \bmod 2r + 1$. Hence $2t = 2r + 1 + 1$ and $t = r + 1$. Thus we have proved

21.9 Theorem.

a) If $K \cong \mathrm{GF}(2^{2r+1})$ with $r \geqslant 1$, then K admits exactly one automorphism σ with $x^{\sigma^2} = x^2$ for all $x \in K$.
b) If $K \cong \mathrm{GF}(2^{2r})$, then K does not possess an automorphism σ with $x^{\sigma^2} = x^2$ for all $x \in K$.

From 21.9 we obtain immediately

21.10 Theorem. Let K be an algebraic extension of $\mathrm{GF}(2)$ with $|K| \geqslant 4$.

a) If K does not possess a subfield isomorphic to $\mathrm{GF}(4)$, then K admits exactly one automorphism σ with $x^{\sigma^2} = x^2$ for all $x \in K$.
b) If K has a subfield isomorphic to $\mathrm{GF}(4)$, then K does not admit an automorphism σ with $x^{\sigma^2} = x^2$ for all $x \in K$.

Because of 21.9 we may write $S(q)$ instead of $S(K,\sigma)$ if $K \cong GF(q)$. The groups $S(q)$ are the Suzuki groups named after M. Suzuki who found them in 1960. The generalizations $S(K,\sigma)$ are due to J. Tits 1962.

21.11 Theorem. $|S(q)| = (q^2 + 1)q^2(q - 1)$. *Moreover* 3 *does not divide* $|S(q)|$.

PROOF. The first assertion follows from 21.2 and 21.8 and the second from the fact that $q = 2^{2r+1} \equiv 2 \mod 3$.   □

22. The Simplicity of the Suzuki Groups

In this section we shall prove that the Suzuki groups are simple.

22.1 Lemma. a) *Let* $\gamma \in S(K,\sigma)$. *If* γ *has three fixed points on* $\mathfrak{D}$, *then* $\gamma = 1$.
b) *If* $\gamma \in ZT \backslash T$, *then* γ *has a fixed point other than* U *on* $\mathfrak{D}$.

PROOF. a) We may assume wlog that $U^\gamma = U$ and $O^\gamma = O$, since $S(K,\sigma)$ is doubly transitive on $\mathfrak{D}$. Hence $(x, y, z)^\gamma = (kx, k^{\sigma+1}y, k^{\sigma+2}z)$. By assumption there is $(x, y, z) \in \mathfrak{D}\backslash\{U, O\}$ with $(x, y, z) = (kx, k^{\sigma+1}y, k^{\sigma+2}z)$. As $(x, y, z) \neq (0, 0, 0)$, at least one of the equations $1 = k$, $1 = k^{\sigma+1}$, $1 = k^{\sigma+2}$ holds. It follows from 21.1 that $k = 1$. This proves a).
  b) We have $\gamma = \eta(k)\tau(a, b)$ with $k \neq 1$. Hence

$$(x, y, z)^\gamma = (kx + a, k^{\sigma+1}y + b + a^\sigma kx, \ldots).$$

Put $x = (k + 1)^{-1}a$ and $y = (k^{\sigma+1} + 1)^{-1}(b + ak(k + 1)^{-1}a)$. This is possible by 21.1, as $k \neq 1$. An easy computation then shows that

$$(x, y, xy + x^{\sigma+2} + y^\sigma)^\gamma = (x, y, xy + x^{\sigma+2} + y^\sigma).$$

(Recall that one need not compute the third coordinate of the image of $(x, y, xy + x^{\sigma+2} + y^\sigma)$, since $\mathfrak{D}$ is fixed by γ.)   □

22.2 Lemma. $S(K,\sigma)$ *is generated by all its involutions which have a fixed point on* $\mathfrak{D}$.

PROOF. Put $G = S(K,\sigma)$. Then ω normalizes $G_{O,U}$. Furthermore $(x, y, z)^\omega = (yz^{-1}, xz^{-1}, z^{-1})$, if $z \neq 0$. Therefore

$$(1, 1, 1)^{\omega\eta(k)\omega} = (k, k^{\sigma+1}, k^{\sigma+2})^\omega = (k^{\sigma+1-\sigma-2}, k^{1-\sigma-2}, k^{-\sigma-2}) = (1, 1, 1)^{\eta(k^{-1})}.$$

As $\omega\eta(k)\omega\eta(k^{-1})^{-1}$ fixes O, U and $(1, 1, 1)$, we obtain by 22.1 that $\omega\eta(k)\omega = \eta(k^{-1})$. This yields that $\omega\eta(k)$ is an involution.

The mapping $x \to x^2$ is bijective. Hence there exists $l \in K$ with $l^2 = k$. Consequently

$$(l, l^{\sigma+1}, l^{\sigma+2})^{\omega\eta(k)} = (kl^{-1}, k^{\sigma+1}l^{-\sigma-1}, k^{\sigma+2}l^{-\sigma-2}) = (l, l^{\sigma+1}, l^{\sigma+2}).$$

Therefore $\omega\eta(k)$ has a fixed point. From this we infer that $\eta(k) = \omega\omega\eta(k)$ is the product of two involutions with fixed point.

Let $\tau \in T$ and $1 \neq k \in K^*$. Then $\eta(k)\tau$ has a fixed point other than U on $\mathfrak{D}$ by 22.1. Therefore $\eta(k)\tau$ is conjugate to some element in $G_{O,U}$ whence $\eta(k)\tau$ is a product of two involutions with fixed points. Thus $\tau = \eta(k)^{-1}\eta(k)\omega$ is the product of four involutions with fixed points. The lemma now follows from 21.8. $\qquad\square$

22.3 Lemma. *All involutions of* $S(K,\sigma)$ *which have a fixed point on* $\mathfrak{D}$ *are conjugate. If* $S(K,\sigma)$ *is finite, then all involutions on* $S(K,\sigma)$ *are conjugate.*

This follows from 21.5 and the transitivity of $S(K,\sigma)$ on $\mathfrak{D}$ and the fact that $|\mathfrak{D}|$ is odd if K is finite.

22.4 Lemma. $S(K,\sigma)' = S(K,\sigma)$.

PROOF. Let $a \in K$ with $a \neq a^{\sigma}$. Then

$$\tau(a,0)^{-1}\tau(1,0)^{-1}\tau(a,0)\tau(1,0) = \tau(0, a + a^{\sigma})$$

is an involution with fixed point. 22.4 now follows from 22.2 and 22.3. $\qquad\square$

22.5 Theorem (Iwasawa). *Let* G *be a finite or infinite permutation group on the set* $\mathfrak{S}$ *with the following properties:*

a) G *acts primitively on* $\mathfrak{S}$.
b) *There exists* $P \in \mathfrak{S}$ *such that* G_P *contains a normal soluble subgroup* A *which together with its conjugates generates* G.
c) $G = G'$.

Then G *is simple.*

PROOF. Let $1 \neq N \trianglelefteq G$. Then N acts transitively on $\mathfrak{S}$, as G is primitive. Therefore NA contains all the conjugates of A whence $G = NA$. From $G/N = NA/N \cong A/(N \cap A)$ we infer that G/N is soluble. This yields $G = N$, since $G = G'$. $\qquad\square$

22.6 Theorem (Suzuki 1962, Tits 1962). *The Suzuki group* $S(K,\sigma)$ *is simple.*

PROOF. In 22.5, set $\mathfrak{S} = \mathfrak{D}$, $P = U$ and $A = \mathfrak{Z}(T)$. Then 22.6 is a consequence of 22.2, 22.3, 22.4, and 22.5. $\qquad\square$

22.7 Theorem. $S(K,\sigma) = \langle \mathfrak{Z}(T), \omega \rangle$.

PROOF. From $\mathfrak{Z}(T) = \{\tau(0,b) \,|\, b \in K\}$ we infer $(0,0,0)^{\mathfrak{Z}(T)} = \{(0,b,b^{\sigma}) \,|\, b$

$\in K$ }. If $b \neq 0$, then $(0, b, b^{\sigma})^{\omega} = (b^{1-\sigma}, 0, b^{-\sigma})$. Moreover

$$(b^{1-\sigma}, 0, b^{-\sigma})^{\tau(0, c)} = (b^{1-\sigma}, c, b^{-\sigma} + c^{\sigma} + cb^{1-\sigma}).$$

As $-1 + \sigma$ is the inverse of $1 + \sigma$ and hence an automorphism of K^*, we see that

$$\mathfrak{D} \setminus \{ U \} = \{ (b^{1-\sigma}, c, b^{-\sigma} + c^{\sigma} + cb^{1-\sigma}) \mid b, c \in K \}.$$

Therefore $\mathfrak{D} = \{ U \} \cup U^{\omega 3(T)} \cup U^{\omega 3(T) \omega 3(T)}$. This yields $\langle 3(T), \omega \rangle \supseteq 3(T^g)$ for all $g \in S(K, \sigma)$. Thus $\langle 3(T), \omega \rangle = S(K, \sigma)$ by 22.2 and 22.3. $\square$

22.8 Corollary. *If* $g, h \in S(K, \sigma)$ *with* $U^g \neq U^h$, *then* $S(K, \sigma) = \langle 3(T^g), 3(T^h) \rangle$.

22.9 Corollary. *If* $g \in S(K, \sigma)$ *and if* τ *is an involution of* $S(K, \sigma)$ *which has a fixed point on* $\mathfrak{D}$ *and if* $\tau \notin 3(T^g)$, *then* $S(K, \sigma) = \langle 3(T^g), \tau \rangle$.

23. The Lüneburg Planes

First we want to determine the action of $\tau(a, b)$ on the plane E. We have

$$(\xi, \eta, \zeta)^{\tau(a, b)} = (\xi + a, \eta + b + a^{\sigma}\xi, \zeta + ab$$
$$+ a^{\sigma+2} + b^{\sigma} + a\eta + a^{\sigma+1}\xi + b\xi).$$

Pick $(\xi, \eta, \zeta) \neq (0, 0, 0)$ and let G be the line whose affine points have coordinates $(\lambda\xi, \lambda\eta, \lambda\zeta)$. Then the affine points on the line $G^{\tau(a, b)}$ are the points with coordinates

$$(\lambda\xi + a, \lambda\eta + b + a^{\sigma}\lambda\xi, \lambda\zeta + ab + a^{\sigma+2} + b^{\sigma} + a\lambda\eta + a^{\sigma+1}\lambda\xi + b\lambda\xi).$$

Thus the line $G^{\tau(a, b)}$ is parallel to the line H consisting of the points with coordinates

$$(\lambda\xi, \lambda\eta + a^{\sigma}\lambda\xi, \lambda\zeta + a\lambda\eta + a^{\sigma+1}\lambda\xi + b\lambda\xi).$$

The point (ξ, η, ζ) is projectively represented by $(1, \zeta, \xi, \eta)K$. Hence $G = (1, 0, 0, 0)K + (1, \zeta, \xi, \eta)K$. Therefore $G \cap E = (0, \zeta, \xi, \eta)K$. Similarly we obtain

$$H \cap E = (0, \zeta + a\eta + (a^{\sigma+1} + b)\xi, \xi, \eta + a^{\sigma}\xi)K.$$

Consequently

$$((0, x_1, x_2, x_3)K)^{\tau(a, b)} = (0, x_1 + (a^{\sigma+1} + b)x_2 + ax_3, x_2, a^{\sigma}x_2 + x_3)K.$$

Consider the mapping $\tau^*(a, b)$ defined by

$$(x_1, x_2, x_3)^{\tau^*(a, b)} = (x_1 + (a^{\sigma+1} + b)x_2 + ax_3, x_2, a^{\sigma}x_2 + x_3).$$

Then the restriction of $\tau(a, b)$ to E is induced by $\tau^*(a, b)$.

If I is the line defined by the equations $x_0 = 0 = x_2$, then $I^{\tau(a,b)} = I$. Furthermore $U \subseteq I$ and $(0,1,0)^{\tau^*(a,b)} = (a^{\sigma+1} + b, 1, a^\sigma)$. Given x_1 and x_3, there exist $a, b \in K$ such that $a^\sigma = x_3$ and $a^{\sigma+1} + b = x_1$. Therefore, T acts transitively on the set of points of E which are not on I. Consequently, I is the only fixed line of T in E.

We have

$$I^\omega = \{(0, x_1, 0, x_3) \,|\, x_1, x_3 \in K\}^\omega = \{(x_1, 0, x_3, 0) \,|\, x_1, x_3 \in K\}.$$

Hence the coordinates of the affine points on I^ω satisfy the equations $y = z = 0$. Moreover

$$(x, 0, 0)^{\tau(a,b)} = (x + a, b + a^\sigma x, ab + a^{\sigma+2} + b^\sigma + a^{\sigma+1}x + bx).$$

Therefore

$$I^{\omega\tau(a,b)} = \big\{ \big(x_0, abx_0 + a^{\sigma+2}x_0 + b^\sigma x_0 + a^{\sigma+1}x_3 + bx_3, \\ x_3 + ax_0, bx_0 + a^\sigma x_3 \big) \,|\, x_0, x_3 \in K \big\}.$$

Put $g_\infty = \{(0, s, 0, t) \,|\, s, t \in K\}$ and

$$g_{a,b} = \big\{ \big(s, [ab + a^{\sigma+2} + b^\sigma]s + [a^{\sigma+1} + b]t, t + as, bs + a^\sigma t\big) \,|\, s, t \in K \big\}.$$

We want to prove that $\pi = \{g_\infty\} \cup \{g_{a,b} \,|\, a, b \in K\}$ is a spread. In order to do this we first prove

23.1 Lemma. $S(K, \sigma)_U$ *operates transitively on the set of points* $\mathfrak{P}_E$ *which do not belong to* $\mathfrak{D}$.

PROOF. Let (x, y, z) be a point of $\mathfrak{P}_E$ which is not in $\mathfrak{D}$. Then $z + xy + x^{\sigma+2} + y^\sigma \neq 0$. By 21.1 there exists $k \in K^*$ with $k^{\sigma+2} = z + xy + x^{\sigma+2} + y^\sigma$. Furthermore there exist $a, b \in K$ with $k + a = x$ and $b + a^\sigma k = y$. From this it follows

$$(1,0,0)^{\eta(k)\tau(a,b)} = (k + a, b + a^\sigma k, ab + a^{\sigma+2} + b^\sigma + a^{\sigma+1}k + bk)$$
$$= (x, y, ab + a^{\sigma+2} + b^\sigma + a^{\sigma+1}k + bk).$$

Now we have

$$\begin{aligned}
ab + a^{\sigma+2} + b^\sigma + a^{\sigma+1}k + bk &= a(a^\sigma k + y) + a^{\sigma+2} + a^2 k^\sigma \\
&\quad + y^\sigma + a^{\sigma+1}k + a^\sigma k^2 + yk \\
&= ay + a^{\sigma+2} + a^2 k^\sigma + y^\sigma + a^\sigma k^2 + yk \\
&= xy + (k + x)^{\sigma+2} + (k + x)^2 k^\sigma \\
&\quad + y^\sigma + (k + x)^\sigma k^2 \\
&= xy + (k + x)^\sigma x^2 + k^{\sigma+2} + k^\sigma x^2 + y^\sigma \\
&= xy + x^{\sigma+2} + y^\sigma + k^{\sigma+2} \\
&= z.
\end{aligned}$$

Hence $(1,0,0)^{\eta(k)\tau(a,b)} = (x, y, x)$. □

23.2 Theorem. $\pi = \{ g_\infty \} \cup \{ g_{a,b} \mid a, b \in K \}$ *is a spread of* $V = K \oplus K \oplus K \oplus K$. *Moreover there exists a subgroup of* $\mathrm{SL}(V)$ *isomorphic to* $\mathrm{S}(K, \sigma)$ *which acts doubly transitively on* π. *Finally* $\mathrm{K}(V, \pi) \cong K$ *and* $\mathrm{rank}_{\mathrm{K}(V,\pi)}(V) = 4$.

PROOF. We first consider the elements of π as lines of the projective 3-space $\mathfrak{P}$ over K. Then $\mathrm{S}(K, \sigma)$ permutes π doubly transitively. It follows from 23.1 that $\mathrm{S}(K, \sigma)$ operates transitively on the set of points of $\mathfrak{P}$ which do not belong to $\mathfrak{D}$. Hence every point of $\mathfrak{P}$ is on a line belonging to π. Consequently $V = \bigcup_{X \in \pi} X$. Also

$$g_\infty + g_{0,0} = \{ (0, s, 0, t) \mid s, t \in K \} + \{ (s, 0, t, 0) \mid s, t \in K \} = V.$$

We deduce from this and the double transitivity of $\mathrm{S}(K, \sigma)$ on π that $V = X \oplus Y$ for all $X, Y \in \pi$ with $X \neq Y$. Thus π is a spread of V.

The determinant of ω is 1. Hence $\omega \in \mathrm{SL}(V)$. As $\mathrm{SL}(V)$ is a normal subgroup of $\mathrm{GL}(V)$, we deduce from 22.3 that every involution of $\mathrm{S}(K, \sigma)$ having a fixed point is induced by an involution in $\mathrm{SL}(V)$. Let G be the group generated by all involutions of $\mathrm{SL}(V)$ which induce an involution in $\mathrm{S}(K, \sigma)$ having a fixed point in $\mathfrak{D}$. Let $\overline{G}$ be the group induced by G on $\mathfrak{P}$. By 22.2 we have $\overline{G} = \mathrm{S}(K, \sigma)$. Moreover, by the general theory of projective spaces, $\overline{G} \cong G/\mathfrak{Z}(G)$ and $\mathfrak{Z}(G) = G \cap \mathfrak{Z}(\mathrm{SL}(V))$. As the centre of $\mathrm{SL}(V)$ consists of all the mappings $v \to kv$ of V onto itself with $k^4 = 1$, we deduce $\mathfrak{Z}(\mathrm{SL}(V)) = \{1\}$. Hence $G \cong \overline{G} = \mathrm{S}(K, \sigma)$.

Certainly $\mathrm{K}(V, \pi)$ contains a subfield K_0 isomorphic to K such that V is of rank 4 over K_0. Therefore $\mathrm{rank}_{\mathrm{K}(V,\pi)}(V) = 2$ or 4. If it is 2, then $\pi(V)$ is desarguesian. It then follows that G is isomorphic to a subgroup of $\mathrm{GL}(2, \mathrm{K}(V, \pi))$. Therefore the stabilizer of g_∞ in $\mathrm{GL}(2, \mathrm{K}(V, \pi))$ contains elements of order 4, a contradiction. Hence $\mathrm{rank}_{\mathrm{K}(V,\pi)}(V) = 4$. □

The planes $\pi(V)$ constructed in this section are called the *Lüneburg planes*.

24. The Subgroups of the Suzuki Groups

We start with a lemma that will be useful in the sequel.

24.1 Lemma. *Let* G *be a finite group and* H *a subgroup of* G. *If* $\mathfrak{C}_G(h) \subseteq H$ *for all* $h \in H$ *having prime order, then* H *is a Hall subgroup of* G.

PROOF. Let S be a Sylow subgroup of H. We have to show that S is also a Sylow subgroup of G. Assume by way of contradiction that T is a Sylow subgroup of G with $S \subseteq T$ and $S \neq T$. Then there is a subgroup S^* of T which properly contains S with S normal in S^*. Since S is normal in S^*, we have $S \cap 3(S^*) \neq \{1\}$. Hence there exists an element s of prime order in $S \cap 3(S^*)$. Consequently $S^* \subseteq \mathfrak{C}_G(s) \subseteq H$, a contradiction. $\square$

Let $q = 2^{2\lambda+1}$ and $G = S(q)$. Put $r = 2^{\lambda+1}$. Then $r^2 = 2q$ and hence $x^{r^2} = x^2$ for all $x \in GF(q)$.

The element x of the group X is called *real*, if there exists $y \in X$ with $x^y = x^{-1}$.

24.2 Theorem (Suzuki 1962). *Let S be a Sylow 2-subgroup of G, then we have*:

a) $\exp(S) = 4$.
b) $3(S) = \{\sigma \mid \sigma \in S, \sigma^2 = 1\}$ *and* $|3(S)| = q$.
c) $\mathfrak{N}_G(S)$ *is a Frobenius group of order $q^2(q-1)$. The Frobenius kernel is S and each Frobenius complement is cyclic of order $q-1$.*
d) *If $1 \neq s \in S$, then $\mathfrak{C}_G(s) \subseteq S$.*
e) *The element $s \in S$ is real, if and only if $s \in 3(S)$.*
f) *S does not contain a quaternion group.*
g) *If Z is a Frobenius complement of $\mathfrak{N}_G(S)$, then Z acts transitively on $3(S) \setminus \{1\}$.*

PROOF. a), b), c), and g) have been proved earlier. In order to prove d), e), and f), we may assume $S = T$.

d) Let $1 \neq \tau(a,b) \in T$ and $\gamma \in \mathfrak{C}_G(\tau(a,b))$. As U is the only fixed point of $\tau(a,b)$ on $\mathfrak{D}$, we have $U^\gamma = U$. Since G_U is a Frobenius group and $\gamma \in G_U$, we deduce $\gamma \in T$. Thus $\mathfrak{C}_G(\tau(a,b)) \subseteq T$ proving d).

e) Let $\tau(a,b) \in T$ be real and assume $\tau(a,b)^2 \neq 1$. Using Lemma 21.5, we see that we may assume $a = 1$, since $\tau(a,b)^2 \neq 1$ implies $a \neq 0$. As $\tau(1,b)$ is real, there exists $\gamma \in G$ with $\tau(1,b)\gamma = \gamma\tau(1,b)^{-1}$. Obviously, γ^2 centralizes $\tau(1,b)$. Hence $\gamma^2 \in T$ by d) and thus $\gamma \in T$. Therefore $\gamma = \tau(c,d)$ for some $c,d \in K$. We thus have

$$\tau(1,b)\tau(c,d) = \tau(c,d)\tau(1,b)^{-1} = \tau(c,d)\tau(1,b+1).$$

From 21.4 we then obtain $b + d + c^\sigma = b + d + 1 + c$. Thus $c^\sigma = 1 + c$. This implies $c^2 = c^{\sigma^2} = 1 + c^\sigma = 1 + 1 + c = c$. Therefore $c = 0$ or $c = 1$. This yields $c^\sigma = c$ and hence $1 + c = c$, a contradiction. This contradiction shows that $\tau(a,b)^2 = 1$ and hence $\tau(a,b) \in 3(T)$.

Conversely, if $\tau \in 3(T)$, then $\tau = \tau^{-1}$ and τ is real.

f) is implied by e), as the elements of order 4 in the quaternion group of order 8 are real. This finishes the proof of 24.2. $\square$

For the element g of the group X we put $\mathfrak{C}_X^*(g) = \{x \mid x \in X, g^x \in \{g, g^{-1}\}\}$.

We call the element g of the group X *strongly real*, if g is the product of two involutions of X.

24.3 Lemma. *Let* $1 \neq g \in G$ *be strongly real. Then* g *is an involution or* g *is conjugate to* $\tau\omega$ *where* τ *is some element in* $\mathfrak{Z}(T)\backslash\{1\}$. *If* $\tau, \tau' \in \mathfrak{Z}(T)\backslash\{1\}$ *are such that* $\tau\omega$ *and* $\tau'\omega$ *are conjugate, then* $\tau = \tau'$. *Hence there are exactly* $q - 1$ *conjugacy classes of strongly real elements which are not involutions. If* g *is not an involution, then* $|\mathfrak{C}_G(g)|$ *is odd.*

PROOF. There exist involutions h_1 and h_2 such that $g = h_1 h_2$. Let S_i be the Sylow 2-subgroup containing h_i. If $S_1 = S_2$, then g is an involution. Assume $S_1 \neq S_2$. Let S_3 be the Sylow 2-subgroup containing ω and let R be the fixed point of S_3. Then $G_U \cap G_R$ is a Frobenius complement of G_R which by 24.2 operates transitively on $\mathfrak{Z}(S_3)\backslash\{1\}$. As G is 2-transitive on $\mathfrak{D}$, there exists $\eta \in G$ with $S_1^\eta = T$ and $S_2^\eta = S_3$. Furthermore, there exists $\delta \in G_U \cap G_R$ with $h_2^{\eta\delta} = \omega$. It follows from $U^\delta = U$ that $T^\delta = T$. Hence $\tau = h_1^{\eta\delta} \in \mathfrak{Z}(T)\backslash\{1\}$. Thus $g^{\eta\delta} = \tau\omega$ is of the required form.

Let $a = \tau\omega$ and $b = \tau'\omega$. Assume that $o(a)$ is even. Then a centralizes a 2-element which is different from 1. Hence a is a 2-element by 24.2 d). Thus $o(a) = 2$ by 24.2 e). This yields that τ centralizes ω. Therefore τ fixes U and R whence $\tau = 1$, a contradiction. Hence $o(a)$ is odd. Similarly, we see that $o(b)$ is odd. As a centralizes no 2-element other than 1, we have that $|\mathfrak{C}_G(a)|$ is odd. Therefore $\omega \notin \mathfrak{C}_G(a)$, but $\omega \in \mathfrak{C}_G^*(a)$. Thus $\langle\omega\rangle$ is a Sylow 2-subgroup of $\mathfrak{C}_G^*(a)$, since $|\mathfrak{C}_G^*(a) : \mathfrak{C}_G(a)| \leqslant 2$. Similarly, $\langle\omega\rangle$ is a Sylow 2-subgroup of $\mathfrak{C}_G^*(b)$.

Let $g \in G$ and assume $a^g = b$. Then there exists $h \in \mathfrak{C}_G(b)$ with $\langle\omega\rangle^{gh} = \langle\omega\rangle$ whence $gh \in \mathfrak{C}_G(\omega)$. Furthermore $a^{gh} = b^h = b$. This yields $\tau^{gh} = \tau'$. From this we infer $U^{gh} = U$, as U is the only fixed point of τ and τ'. Therefore $gh \in G_U \cap \mathfrak{C}_G(\omega) \subseteq G_U \cap S_3 = \{1\}$ whence $\tau = \tau'$. $\square$

24.4 Lemma. *Let* $g \in G$ *be strongly real and assume* $g^2 \neq 1$. *Then* $A = \mathfrak{C}_G(g)$ *is abelian. If* $1 \neq g' \in A$, *then* g' *is strongly real and* $\mathfrak{C}_G(g') = A$. *Moreover* A *is a Hall subgroup of* G *and* $\mathfrak{C}_G(A)/A$ *is cyclic of order 2 or 4.*

PROOF. $|A|$ is odd by 24.3. As g is strongly real, there exist involutions ρ and τ with $g = \rho\tau$. Therefore $\rho^{-1}g\rho = g^{-1}$ whence $\rho \in \mathfrak{N}_G(A)$. We infer that $|\mathfrak{N}_G(A)/A|$ is even.

Let $a, b \in A$, let β be an involution in $\mathfrak{N}_G(A)$ and assume that $a^{-1}\beta a\beta = b^{-1}\beta b\beta$. Then $ab^{-1}\beta = \beta ab^{-1}$. Hence $ab^{-1} \in \mathfrak{N}_G(\beta) \cap A = \{1\}$ by 24.2 d). Thus $a = b$. As A is finite, we obtain $A = \{a^{-1}\beta a\beta \,|\, a \in A\}$. Now $\beta(a^{-1}\beta a\beta)\beta = \beta a^{-1}\beta a = (a^{-1}\beta a\beta)^{-1}$. Thus $a^\beta = a^{-1}$ for all $a \in A$. This implies that A is abelian. Moreover $(a\beta)^2 = 1$ whence $a\beta$ is an involution for all $a \in A$. As $a = (\beta a^{-1})\beta$, we have that a is strongly real for all $a \in A$.

Let $1 \neq g' \in A$. Then $A \subseteq \mathfrak{C}_G(g')$. As g' is strongly real and $o(g')$ is odd, $\mathfrak{C}_G(g')$ is abelian. Hence $\mathfrak{C}_G(g') \subseteq \mathfrak{C}_G(g) = A$. Therefore $\mathfrak{C}_G(g') = A$. From this and 24.1 we infer that A is a Hall subgroup of G.

It remains to show that $\mathfrak{N}_G(A)/A$ is cyclic. As A is a normal abelian Hall subgroup of $\mathfrak{N}_G(A)$, there exists a complement C of A in $\mathfrak{N}_G(A)$ by the Schur-Zassenhaus theorem. Moreover, since A is its own centralizer, $\mathfrak{N}_G(A)/A$ is isomorphic to a group of automorphisms of A. Since $a^\beta = a^{-1}$ for all involutions β in $\mathfrak{N}_G(A)$ and all $a \in A$, the group $\mathfrak{N}_G(A)/A$ contains exactly one involution. We infer from $C \cong \mathfrak{N}_G(A)/A$ that C contains exactly one involution β. Therefore $C \subseteq \mathfrak{C}_G(\beta)$ whence C is a 2-group by 24.2 d). This implies that C is either cyclic or a generalized quaternion group. Now 24.2 a) and f) yield that C is cyclic of order 2 or 4. $\qquad\square$

24.5 Corollary. *If* g, $g' \in G \setminus \{1\}$ *are strongly real and if* $\mathfrak{C}_G(g) \neq \mathfrak{C}_G(g')$, *then* $\mathfrak{C}_G(g) \cap \mathfrak{C}_G(g') = \{1\}$.

PROOF. Assume $\mathfrak{C}_G(g) \cap \mathfrak{C}_G(g') \neq \{1\}$. If g is involutorial, then $\mathfrak{C}_G(g)$ is a Sylow 2-subgroup of G. We infer from $\mathfrak{C}_G(g) \cap \mathfrak{C}_G(g') \neq \{1\}$ that $\mathfrak{C}_G(g')$ is also a Sylow 2-subgroup of G. Hence $\mathfrak{C}_G(g) = \mathfrak{C}_G(g')$ in that case.

If g is not an involution, then g' is not an involution. Hence $|\mathfrak{C}_G(g)|$ and $|\mathfrak{C}_G(g')|$ are odd. Pick $h \in \mathfrak{C}_G(g) \cap \mathfrak{C}_G(g')$ with $h \neq 1$. Then 24.4 implies $\mathfrak{C}_G(g) = \mathfrak{C}_G(h) = \mathfrak{C}_G(g')$. $\qquad\square$

24.6 Lemma. *If* $g \in G$ *is real, then* g *is strongly real.*

PROOF. This is certainly true, if $g^2 = 1$. We may thus assume that $g^2 \neq 1$. Then $\mathfrak{C}_G(g)$ has odd order by 24.2 d). Let $h \in G$ such that $g^h = g^{-1}$. Then $h^2 \in \mathfrak{C}_G(g)$ and $h \neq 1$. Thus h is a 2-element by 24.2 d) whence $h^2 = 1$, as $|\mathfrak{C}_G(g)|$ is odd. Therefore $g = hg^{-1}h$ is the product of the involutions hg^{-1} and h. $\qquad\square$

24.7 Theorem (Suzuki 1962). *If* $G \cong S(q)$, *then all elements of odd order of* G *are real. If* π *is the set of centralizers of the real elements of* G *which are different from* 1, *then* π *is a normal partition of* G. *The partition* π *consists of four classes of conjugate subgroups of* G: *the Sylow 2-subgroups of* G, *one class of cyclic groups of order* $q - 1$, *one class of cyclic groups of order* $q + r + 1$, *and one class of cyclic groups of order* $q - r + 1$, *where* $r^2 = 2q$.

PROOF. The normalizer of $G_{O,U}$ is a dihedral group of order $2(q-1)$. Therefore all the elements of $G_{O,U}$ are strongly real. Let $\mathfrak{R}_0, \mathfrak{R}_1, \ldots, \mathfrak{R}_s$ be all the conjugacy classes of centralizers of strongly real elements of odd order and let the labelling be such that $G_{O,U} \in \mathfrak{R}_0$. Let $A_0, A_1, \ldots, A_s$ be a transversal of $\mathfrak{R}_0, \mathfrak{R}_1, \ldots, \mathfrak{R}_s$. Put $a_i = |A_i|$ and $b_i = |\mathfrak{N}_G(A_i):A_i|$. Then $a_0 = q - 1$, $b_0 = 2$ and $b_i = 2$ or 4 by 24.4.

(j) $(a_i, a_k) = 1$ for $i \neq k$. In particular a_i divides $q^2 + 1$ for $i \geqslant 1$.

This follows, using 24.4, from a theorem of Wielandt (see e.g. Huppert [1967, III.5.8, p. 285]).

We infer from 24.5 that for $X, Y \in \mathfrak{R}_i$ either $X = Y$ or $X \cap Y = \{1\}$. Hence $\bigcup_{X \in \mathfrak{R}_i} X$ consists of $b_i^{-1}(a_i - 1)$ classes of conjugate elements, each class consisting of $|G|a_i^{-1}$ elements. Therefore it follows from 24.3 that

(ij)
$$q - 1 = \sum_{i=0}^{s} b_i^{-1}(a_i - 1).$$

Let d be the number of non-real elements of odd order. Then we have

(iij) $$g = |G| = 1 + (q^2 + 1)(q^2 - 1) + \sum_{i=0}^{s}(a_i - 1)g b_i^{-1} a_i^{-1} + d.$$

By 21.10 we know that 3 does not divide a_i. Hence $a_i \geqslant 5$. Therefore $(a_i - 1)a_i^{-1} = 1 - a_i^{-1} \geqslant 1 - \frac{1}{5} = \frac{4}{5}$. Let u be the number of i's with $b_i = 2$ and put $v = s + 1 - u$. Then

$$1 = g^{-1} + g^{-1}(q^2 + 1)(q^2 - 1) + \sum_{i=0}^{s}(a_i - 1)b_i^{-1}a_i^{-1} + dg^{-1}$$

$$> \frac{4}{5}u\frac{1}{2} + \frac{4}{5}v\frac{1}{4} = \frac{2}{5}u + \frac{1}{5}v.$$

Hence $2u + v \leqslant 4$. As $b_0 = 2$, we have $u \geqslant 1$. Thus the only solutions of $2u + v \leqslant 4$ are $(u, v) = (2, 0), (1, 0), (1, 1), (1, 2)$.

Assume $u = 2$ and $v = 0$. Then $s = 1$. By (j), we have $2(q - 1) = q - 2 + a_1 - 1$ whence $a_1 = q + 1$. But this contradicts the fact that a_1 divides $q^2 + 1$.

Hence $u = 1$. Assume $v = 0$. Then $2(q - 1) = q - 2$ by (j) implying $q = 0$. Assume $v = 1$. Then $4(q - 1) = 2(q - 2) + a_1 - 1$ which yields $a_1 = 2q + 1$. By (j), we have $q^2 + 1 = k(2q + 1)$ for some integer k. We deduce $k \equiv 1 \bmod q$. As $q > 2$, we obtain $k > 1$ and hence $k \geqslant q + 1$. Thus $q^2 + 1 \geqslant (q + 1)(2q + 1)$, a contradiction. Therefore $u = 1$ and $v = 2$. This yields $4(q - 1) = 2(q - 2) + a_1 - 1 + a_2 - 1$ and hence

(iv) $$a_1 + a_2 = 2(q + 1).$$

Now

$$(a_1 - 1)g(4a_1)^{-1} + (a_2 - 1)g(4a_2)^{-1} = g(4a_1a_2)^{-1}[2a_1a_2 - a_1 - a_2]$$

$$= g(4a_1a_2)^{-1}[2a_1a_2 - 2(q + 1)]$$

by (iv). Using this and (iij) we obtain

$$g = 1 + (q^2 + 1)(q^2 - 1) + \frac{g(q - 2)}{2(q - 1)} + \frac{g(2a_1a_2 - 2(q + 1))}{4a_1a_2} + d.$$

Since $g = (q^2 + 1)q^2(q - 1)$, we get

$$(q^2 + 1)q^2(q - 1) = 1 + (q^2 + 1)(q^2 - 1) + \frac{1}{2}(q^2 + 1)q^2(q - 2)$$

$$+ (q^2 + 1)q^2(q - 1)\frac{a_1a_2 - q - 1}{2a_1a_2} + d$$

$$= 1 + (q^2 + 1)(q^2 - 1) + \frac{1}{2}(q^2 + 1)q^2(q - 2)$$

$$+ \frac{1}{2}(q^2 + 1)q^2(q - 1) - \frac{(q^2 + 1)q^2(q^2 - 1)}{2a_1a_2} + d.$$

A straightforward computation shows that

$$-\tfrac{1}{2}q^2(q^2-1) = -\tfrac{1}{2}(q^2+1)q^2(q^2-1)a_1^{-1}a_2^{-1} + d.$$

Put $d = ca_1a_2q^2(q-1)$. Then we obtain

(v) $$\qquad\qquad (q^2+1)(q+1) = a_1a_2(q+1) + 2ca_1^2a_2^2.$$

Let x be non-real of odd order. Then $\mathfrak{C}_G(x)$ does not contain a real element other than 1. As the A_i's are nilpotent Hall subgroups of G, it follows by Wielandt's theorem (Huppert loc. cit.) that $|\mathfrak{C}_G(x)|$ is relatively prime to $2(q-1)a_1a_2$. As $2, q-1, a_1, a_2$ are relatively prime, $q^2(q-1)a_1a_2$ divides $|G : \mathfrak{C}_G(x)|$. Hence d is divisible by $q^2(q-1)a_1a_2$ whence c is an integer. Since a_1a_2 divides q^2+1 and $(q+1, q^2+1) = 1$, we see that $q+1$ divides c.

Assume $c > 0$. Then $(q^2+1)(q+1) \geqslant a_1a_2(q+1) + 2(q+1)a_1^2a_2^2$ and hence $q^2+1 \geqslant a_1a_2(1 + 2a_1a_2)$. We deduce from (iv) that $a_1a_2 \geqslant q+1$. Hence $q^2+1 \geqslant (q+1)(2q+3) > q^2+1$, a contradiction. Thus $c = 0$ and therefore $d = 0$. This proves that π is a normal partition.

We have $a_1 + a_2 = 2(q+1)$ by (iv) and $a_1a_2 = q^2+1$ by (v). Thus $a_1 = 2(q+1) - a_2$ and hence $q^2+1 = 2(q+1)a_2 - a_2^2$. Solving this quadratic equation yields $a_2 = q + r + 1$ or $a_2 = q - r + 1$ where $r^2 = 2q$. In the first case $a_1 = q - r + 1$ and in the second $a_1 = q + r + 1$. Thus we may assume $a_1 = q + r + 1$ and $a_2 = q - r + 1$.

The group $\mathrm{PGL}(4,q)$ contains a cyclic subgroup Z of order $(q+1)(q^2+1)$ (see e.g. Huppert [1967, II.7.3 c], p. 188]). As q is even and $|\mathrm{PGL}(4,q)| = q^6(q^2+1)(q-1)^3(q^2+q+1)(q+1)^2$, the subgroup U of Z of order q^2+1 is a Hall subgroup of $\mathrm{PGL}(4,q)$. By Wielandt's theorem, A_1 and A_2 are conjugate to subgroups of U. Hence A_1 and A_2 are cyclic. $\qquad\square$

24.8 Corollary (Suzuki 1962). *If $1 \neq \gamma \in S(q)$, then $\mathfrak{C}_{S(q)}(\gamma)$ is nilpotent.*

PROOF. If γ is a 2-element, then $\mathfrak{C}_{S(q)}(\gamma)$ is a 2-group by 24.2 d). If γ is not a 2-element, then γ has odd order and is strongly real by 24.7 and 24.6. Hence $\mathfrak{C}_{S(q)}(\gamma)$ is abelian in that case by 24.4. $\qquad\square$

24.9 Corollary (Suzuki 1962). *The Sylow subgroups of odd order of $S(q)$ are cyclic.*

PROOF. Let Σ be a Sylow subgroup of odd order of $S(q)$. Pick $\zeta \in \mathfrak{Z}(\Sigma) \setminus \{1\}$. Then $\Sigma \subseteq \mathfrak{C}_{S(q)}(\zeta)$. Moreover $\mathfrak{C}_{S(q)}(\zeta) \in \pi$. Hence $\mathfrak{C}_{S(q)}(\zeta)$, and therefore Σ, is cyclic by 24.7.

24.10 Theorem (Suzuki 1962). *Let π be the partition of $S(q)$ described in 24.7. If W is a subgroup of $S(q)$, then there exists $X \in \pi$ with $W \subseteq \mathfrak{N}_{S(q)}(X)$ or W is isomorphic to $S(q_0)$ with $q_0 \leqslant q$.*

PROOF. We may assume $W \neq \{1\}$.

Case 1: All Sylow subgroups of W are cyclic. By a theorem of Burnside, W contains a cyclic normal subgroup V distinct from $\{1\}$. As V is cyclic, there exists $X \in \pi$ with $V \subseteq X$. Let $y \in W$, then $X^y \in \pi$, since π is a normal partition. On the other hand $\{1\} \neq V = V^y \subseteq X \cap X^y$. Hence $X = X^y$, i.e., $W \subseteq \mathfrak{N}_{S(q)}(X)$.

Case 2: The group W contains a normal 2-subgroup Σ_0 distinct from $\{1\}$. Then the same argument works by taking for X a Sylow 2-subgroup of $S(q)$ containing Σ_0.

By 24.9, it remains to consider

Case 3: The group W contains more than one Sylow 2-subgroup. Let Σ_0 be a Sylow 2-subgroup of W. As π is a normal partition, we may assume $\Sigma_0 \subseteq T$. If Σ_0 is cyclic, then by 24.9 and Burnside's theorem, Σ_0 is normal. Hence Σ_0 is not cyclic. Also Σ_0 is not a quaternion group by 24.2 f). Thus Σ_0 contains more than one involution. As Σ_0 is not normal in W, there exists an involution in $W \backslash \Sigma_0$. Using that $S(q)$ acts doubly transitively on $\mathfrak{O}$, we may as well assume that $\omega \in W$.

Put $I = \Sigma_0 \cap \mathfrak{Z}(T)$ and $|I| = q_0$. Then $q_0 \geqslant 4$. Furthermore put $N = \mathfrak{N}_W(\Sigma_0)$. Obviously $N = W_U$. Put $|W:N| = k + 1$. Then $k + 1$ is the number of Sylow 2-subgroups of W.

(a) $|N : \Sigma_0| = q_0 - 1$.

In order to prove this let ρ and σ be involutions in W with $\rho\Sigma_0 = \sigma\Sigma_0$. Then $\rho\sigma \in \Sigma_0$. If $\rho \neq \sigma$, then $\rho\sigma \neq 1$ and U is the only fixed point of $\langle\rho\sigma\rangle$. As $\langle\rho\sigma\rangle$ is normalized by ρ and σ, we have $U^\rho = U = U^\sigma$. Hence $\rho, \sigma \in \Sigma_0$. Therefore, if $x \in W\backslash\Sigma_0$ then $x\Sigma_0$ contains at most one involution.

The number of involutions in W is $(k+1)(q_0 - 1)$ and $q_0 - 1$ is the number of involutions in Σ_0. The number of left cosets of Σ_0 being distinct from Σ_0 is $|W:\Sigma_0| - 1$. Hence

$$(k + 1)(q_0 - 1) \leqslant q_0 - 1 + |W:\Sigma_0| - 1 = q_0 - 2 + |W:N||N:\Sigma_0|$$

$$= q_0 - 2 + (k + 1)|N:\Sigma_0|.$$

This yields $q_0 - 1 \leqslant (q_0 - 2)(k + 1)^{-1} + |N:\Sigma_0|$.

Let Σ_1 be a Sylow 2-subgroup of W which is distinct from Σ_0. Then $|\{\Sigma_0^\xi | \xi \in \mathfrak{Z}(\Sigma_1)\}| \geqslant q_0$. Hence $k + 1 \geqslant q_0$ and thus $(q_0 - 2)(k + 1)^{-1} \leqslant 1$ and $q_0 - 1 \leqslant |N:\Sigma_0|$.

As $N = W_U$, there exists a cyclic subgroup A with $N = \Sigma_0 A$ and $\Sigma_0 \cap A = \{1\}$. Also, the order of A divides $q - 1$. This implies that A operates regularly on $I\backslash\{1\}$. Hence $|N:\Sigma_0| = |A| \leqslant q_0 - 1$. Therefore $|N:\Sigma_0| = q_0 - 1$.

From $q - 1 \equiv 0 \bmod |A|$ and $|A| = q_0 - 1$ we deduce:

(b) $q = q_0^s$ for some integer s.

Next we prove:

(c)   If $x \in W\backslash N$, then $x\Sigma_0$ contains exactly one involution.

We have $|W:\Sigma_0| = (k+1)(q_0-1)$ and $|N:\Sigma_0| = q_0 - 1$. Also, all involutions of N belong to Σ_0. Therefore, the $(k+1)(q_0-1) - (q_0-1) = k(q_0-1)$ involutions outside of N lie in the union of the $k(q_0-1)$ cosets of Σ_0 which are not contained in N. As each of these cosets contains at most one involution, (c) is proved.

(d) $|\Sigma_0| = k = q_0^2$.

The points $U, U^{\omega\tau}$ with $\tau \in I$ are pairwise distinct. If $U = U^{\omega\tau}$, then $\tau = 1$, since $O = U^\omega = U^{\omega\tau} = O^\tau$.

Assume $U^{\omega\tau} = U^{\omega\tau_1\omega\tau_2}$. Then $O^{\tau'} = O^{\tau_1\omega}$, where $\tau' = \tau\tau_2$. Now

$$O^{\tau_1\omega} = (0,0,0)^{\tau(0,a_1)\omega} = (0,a_1,a_1^\sigma)^\omega = \left[(1,a_1^\sigma,0,a_1)\mathrm{GF}(q)\right]^\omega$$
$$= (a_1^\sigma,1,a_1,0,)\mathrm{GF}(q)$$

and

$$O^{\tau'} = (0,0,0)^{\tau(0,a)} = (0,a,a^\sigma) = (1,a^\sigma,0,a)\mathrm{GF}(q).$$

This implies $a_1 = 0$ and hence $1 = 0$, a contradiction. Thus the orbit of U under W contains at least $1 + q_0 + (q_0-1)q_0 = q_0^2 + 1$ points. Hence $k + 1 \geqslant q_0^2 + 1$, i.e., $k \geqslant q_0^2$.

Each left coset of N contains exactly $q_0 - 1$ left cosets of Σ_0. Therefore, any left coset of N, including N itself, contains exactly $q_0 - 1$ involutions of W. If ρ is an involution in $W \setminus N$, then $U^\rho \neq U$. Put $P = U^\rho$. Then, of course, $P^\rho = U$. Let $\rho = \rho_1, \ldots, \rho_t$, where $t = q_0 - 1$, be all the involutions in ρN. Then $\rho_i = \rho\nu_i$ for some $\nu_i \in N = W_U$. Hence $P^{\rho_i} = P^{\rho\nu_i} = U$. Thus $U^{\rho_i} = P$ for all i. Consequently $\rho_1\rho_i \in W_{U,P}$ for $i = 2, \ldots, t$. Hence $|W_{U,P}| \geqslant q_0 - 1$ and therefore $|W_{U,P}| = q_0 - 1$. Since there are k left cosets of N other than N, we obtain in this way at least k complements of Σ_0 in N. As N is a Frobenius group, we therefore obtain $|\Sigma_0| + k(q_0 - 2) \leqslant |N| = (q_0-1)|\Sigma_0|$. Thus $k \leqslant |\Sigma_0|$. Therefore $q_0^2 \leqslant k \leqslant |\Sigma_0|$. As $|I| = q_0$ we have in particular $\Sigma_0 \setminus I \neq \varnothing$.

Pick $\sigma \in \Sigma_0 \setminus I$. Then $\sigma = \tau(a,b)$ with $a \neq 0$. Also $\tau(a,b)^2 = \tau(0,a^{1+\alpha})$ where α is the automorphism of $\mathrm{GF}(q)$ with $x^{\alpha^2} = x^2$ for all $x \in \mathrm{GF}(q)$. If $\rho \in \Sigma_0 \setminus I$ is such that $\rho^2 = \sigma^2$ and if $\rho = \tau(c,d)$, then $c^{1+\alpha} = a^{1+\alpha}$ and hence $c = a$ by 21.1. This implies $\rho\sigma \in I$. Hence there are at most q_0 such ρ's with $\rho^2 = \sigma^2$. Therefore $|\Sigma_0| \leqslant q_0^2$ whence $q_0^2 \leqslant k \leqslant |\Sigma_0| \leqslant q_0^2$. This proves (d).

As an immediate consequence we have:

(e) $|W| = (q_0^2 + 1)q_0^2(q_0 - 1).$

The group $A = W_{O,U}$ is cyclic of order $q_0 - 1$. If $\eta(k) \in A$, then $k^{q_0-1} = 1$. Hence:

(f) $A = \{\eta(k) \mid k \in \mathrm{GF}(q_0)^*\}.$

Next we prove:

(g) $\tau(0,b) \in I$, if and only if $b \in GF(q_0)$.

Let $\tau(0,g)$ be a fixed involution in I. By 21.5 we get $\tau(0,g)^{\eta(k)} =$

$\tau(0, k^{\alpha+1}g)$, where α is again the automorphism of $\mathrm{GF}(q)$ satisfying $x^{\alpha^2} = x^2$ for all $x \in \mathrm{GF}(q)$. From this and 21.1 we infer

$$I = \left\{ \tau(0, k^{\alpha+1}g) \mid k \in \mathrm{GF}(q_0) \right\}.$$

Now $U^{\omega\tau(0,b)} = (0, b, b^\alpha)$ and hence, if $b \neq 0$,

$$U^{\omega\tau(0,b)\omega\tau(0,c)} = (0, b, b^\alpha)^{\omega\tau(0,c)} = \left[(b^\alpha, 1, b, 0)\mathrm{GF}(q) \right]^{\tau(0,c)}$$

$$= \left[(1, b^{-\alpha}, b^{1-\alpha}, 0)\mathrm{GF}(q) \right]^{\tau(0,c)} = (b^{1-\alpha}, 0, b^{-\alpha})^{\tau(0,c)}$$

$$= (b^{1-\alpha}, c, b^{-\alpha} + c^\alpha + cb^{1-\alpha}).$$

This implies in particular that the y-coordinates of the points in $U^W \setminus \{U\}$ are in $g\,\mathrm{GF}(q_0)$.

Putting $b = c = g$ we obtain

$$U^{\omega\tau(0,\,g)\omega\tau(0,\,g)\omega} = \left(g^{1-\alpha},\, g,\, g^{-\alpha} + g^\alpha + g^{2-\alpha} \right)^\omega$$

$$= \left[(1,\, g^{-\alpha} + g^\alpha + g^{2-\alpha},\, g^{1-\alpha},\, g)\mathrm{GF}(q) \right]^\omega$$

$$= \left(g^{-\alpha} + g^\alpha + g^{2-\alpha},\, 1,\, g,\, g^{1-\alpha} \right)\mathrm{GF}(q)$$

$$= (1, A,\, gA,\, g^{1-\alpha}A)\mathrm{GF}(q)$$

$$= \left(gA,\, g^{1-\alpha}A, A \right),$$

where $A = (g^{-\alpha} + g^\alpha + g^{2-\alpha})^{-1}$. (It is easily seen that $g^{-\alpha} + g^\alpha + g^{2-\alpha} \neq 0$.)

By the remark made above $g^{1-\alpha}A \in g\,\mathrm{GF}(q_0)$. Hence $g^{-\alpha}A \in \mathrm{GF}(q_0)$ whence $g^\alpha(g^{-\alpha} + g^\alpha + g^{2-\alpha}) \in \mathrm{GF}(q_0)$. This yields $1 + g^2 + g^{2\alpha} \in \mathrm{GF}(q_0)$ and therefore $(g + g^\alpha)^2 = g^2 + g^{2\alpha} \in \mathrm{GF}(q_0)$. This implies $g + g^\alpha \in \mathrm{GF}(q_0)$. Consequently $g^\alpha + g^2 = (g + g^\alpha)^\alpha \in \mathrm{GF}(q_0)$ and hence $g + g^2 \in \mathrm{GF}(q_0)$. From this we infer $[\mathrm{GF}(q_0)(g) : \mathrm{GF}(q_0)] \leq 2$. As $q = 2^{2\lambda+1}$, we obtain $\mathrm{GF}(q_0)(g) = \mathrm{GF}(q_0)$, i.e., $g \in \mathrm{GF}(q_0)$. Therefore $I = \{\tau(0, b) \mid b \in \mathrm{GF}(q_0)\}$. From this, 22.7 and the fact that $\omega \in W$, we finally see that $W \cong \mathrm{S}(q_0)$. $\qquad\square$

24.11 Corollary (Suzuki 1962). *If q_0 is a divisor of q and if $q_0 - 1$ divides $q - 1$, then $\mathrm{S}(q)$ contains a subgroup S_0 isomorphic to $\mathrm{S}(q_0)$. Moreover, if $S_1 \subseteq \mathrm{S}(q)$ and $S_1 \cong \mathrm{S}(q_0)$, then S_1 and S_0 are conjugate in $\mathrm{S}(q)$.*

It should be mentioned that this character-free proof of 24.10 is due to A. Cronheim.

25. Möbius Planes

An incidence structure $\mathfrak{M} = (\mathfrak{P}, \mathfrak{C}, \mathrm{I})$ is called a *Möbius plane*, if the following hold:

(1) If A, B, C are three distinct *points* in $\mathfrak{P}$, then there exists exactly one *circle* k in $\mathfrak{C}$ with $A, B, C\,\mathrm{I}\,k$.

(2) Let $A, B \in \mathfrak{P}$ and $k \in \mathfrak{C}$. If $A \mathrel{I} k$ and $B \mathrel{\not I} k$ then there exists exactly one $k' \in \mathfrak{C}$ with $A, B \mathrel{I} k'$ and $k \cap k' = \{A\}$.

(3) Each $k \in \mathfrak{C}$ is incident with at least one $A \in \mathfrak{P}$ and there exist four points in $\mathfrak{P}$ which are not concircular.

For $P \in \mathfrak{P}$ define $\mathfrak{P}_P$, $\mathfrak{C}_P$, and I_P by $\mathfrak{P}_P = \mathfrak{P} \backslash \{P\}$, $\mathfrak{C}_P = \{k \mid k \in \mathfrak{C}, P \mathrel{I} k\}$ and $I_P = I \cap (\mathfrak{P}_p \times \mathfrak{C}_P)$. Furthermore put $\mathfrak{M}(P) = (\mathfrak{P}_P, \mathfrak{C}_P, I_P)$.

25.1 Lemma. *If $\mathfrak{M}$ is a Möbius plane, then $\mathfrak{M}(P)$ is an affine plane for all points P of $\mathfrak{M}$.*

PROOF. Call the circles of $\mathfrak{C}_P$ lines of $\mathfrak{M}(P)$. Let Q and R be two distinct points of $\mathfrak{M}(P)$. By (1), there exists exactly one line of $\mathfrak{M}(P)$ joining Q and R. Condition (2) gives us Euclid's axiom of parallelism for $\mathfrak{M}(P)$ and (3) guarantees non-degeneracy.   □

We shall say that the circle k *avoids, touches,* respectively *intersects,* the circle k', if $|k \cap k'| = 0$, 1, respectively 2.

A set $\mathfrak{B}$ of circles is called a *pencil* with *carrier* P if $\mathfrak{B}$ is a parallel class in $\mathfrak{M}(P)$.

25.2 Theorem. *Let $\mathfrak{M}$ be a finite Möbius plane. Then there exists a positive integer $n \geqslant 2$ with:*

a) *The number of points of $\mathfrak{M}$ is $n^2 + 1$.*

b) *The number of circles of $\mathfrak{M}$ is $n(n^2 + 1)$.*

c) *Each circle carries exactly $n + 1$ points.*

d) *Each point is on exactly $n(n + 1)$ circles.*

e) *Given distinct points A and B, then there exist exactly $n + 1$ circles k with $A, B \mathrel{I} k$.*

f) *Every pencil contains exactly n circles.*

g) *There exist exactly $n^2 - 1$ circles touching a given circle.*

h) *There exist exactly $\frac{1}{2} n^2(n + 1)$ circles intersecting a given circle.*

i) *There exist exactly $\frac{1}{2} n(n - 1)(n - 2)$ circles avoiding a given circle.*

The integer n is called the *order* of $\mathfrak{M}$.

PROOF. Pick a point P of $\mathfrak{M}$. Then $\mathfrak{M}(P)$ is a finite affine plane. Let n be its order. Then $\mathfrak{M}(P)$ has n^2 points. Therefore a) is true. This implies that $\mathfrak{M}(P)$ is an affine plane of order n for all points P of $\mathfrak{M}$. Thus c), d), e), and f) are true.

As $\mathfrak{M}$ is obviously a tactical configuration, we have $(n^2 + 1)n(n + 1) = b(n + 1)$, where b is the number of circles. Hence $b = n(n^2 + 1)$ proving b).

By c), each circle belongs to $n + 1$ pencils. This and f) imply g).

Given a point-pair on k, then by e) there exist exactly n circles intersecting k in the given point-pair. As three points determine a circle uniquely and as there are $\frac{1}{2} n(n + 1)$ point-pairs on k, there are exactly $\frac{1}{2} n^2(n + 1)$ circles intersecting k.

Finally, i) follows from b), g), and h). This finishes the proof of the theorem.   □

The automorphism σ of the Möbius plane $\mathfrak{M}$ is called a *reflection* of $\mathfrak{M}$ if $\sigma \neq 1$ and if σ fixes a circle pointwise.

25.3 Lemma. *If σ is a reflection of a Möbius plane, then $\sigma^2 = 1$.*

PROOF. Let k be the circle fixed pointwise by σ and pick P on k. Then σ induces a perspectivity with affine axis in $\mathfrak{M}(P)$. Hence there exists a pencil $\mathfrak{B}(P, \sigma)$ with carrier P which is fixed elementwise by σ. Let Q be on k and distinct from P. Then we obtain similarly a pencil $\mathfrak{B}(Q, \sigma)$ with carrier Q which is fixed elementwise by σ. Let R be a point off k. Then there exist two circles l and m with $l \in \mathfrak{B}(P, \sigma)$ and $m \in \mathfrak{B}(Q, \sigma)$ and $R \mathrel{I} l, m$. As $R \mathrel{\not{I}} k$, we have $P \mathrel{\not{I}} m$ and hence $l \neq m$. It follows from $l^{\sigma} = l$ and $m^{\sigma} = m$ that $\{R, R^{\sigma}\} \subseteq l \cap m$. Furthermore $R \neq R^{\sigma}$, as $\sigma \neq 1$. Thus $\{R, R^{\sigma}\} = l \cap m$ and $\{R^{\sigma}, R^{\sigma^2}\} = l^{\sigma} \cap m^{\sigma} = l \cap m$. Hence $R = R^{\sigma^2}$ whence $\sigma^2 = 1$.   □

25.4 Lemma. *If k is a circle of a Möbius plane, then there exists at most one reflection fixing k pointwise.*

PROOF. Let σ and τ be reflections fixing k pointwise. Then $\sigma^2 = \tau^2 = (\sigma\tau)^2 = 1$ by 25.3. Hence $\sigma\tau = \tau\sigma$. If $P \mathrel{I} k$ then $\mathfrak{B}(P, \sigma) = \mathfrak{B}(P, \tau)$, as σ and τ centralize each other. Likewise $\mathfrak{B}(Q, \sigma) = \mathfrak{B}(Q, \tau)$ for $P \neq Q \mathrel{I} k$. Repeating the argument of the proof of 25.3, we then see that $\{R, R^{\sigma}\} = l \cap m = \{R, R^{\tau}\}$, whence $\sigma = \tau$.   □

25.5 Lemma. *Let $\mathfrak{M}$ be a finite Möbius plane and let σ be an involutory automorphism of $\mathfrak{M}$. Denote by $\mathfrak{F}$ the set of fixed points of σ. Then one of the following holds:*

(1) $\mathfrak{F} = \varnothing$.
(2) $|\mathfrak{F}| = 1$.
(3) $|\mathfrak{F}| = 2$.
(4) *There exists a circle k with $\mathfrak{F} = \{P \mid P \mathrel{I} k\}$.*

PROOF. Let $\mathfrak{F} \neq \varnothing$ and $P \in \mathfrak{F}$. Then σ induces an involutory collineation in $\mathfrak{M}(P)$. If σ is a perspectivity in $\mathfrak{M}(P)$, then (2), (3) or (4) holds.

Assume that σ is not a perspectivity. Then σ induces a Baer involution in the projective closure of $\mathfrak{M}(P)$. As l_{∞} is fixed by σ, we infer that $\mathfrak{F} \setminus \{P\}$ is the set of points of an affine Baer subplane $\mathfrak{U}$ of $\mathfrak{M}(P)$. In particular $|\mathfrak{F}| = m^2 + 1$ where m^2 is the order of $\mathfrak{M}$. Let $S \notin \mathfrak{F}$. Then there exists exactly one line of $\mathfrak{U}$ which passes through S; i.e., there exists exactly one circle k through S and P with $|k \cap \mathfrak{F}| = m + 1$. As P was arbitrary, we conclude that $m + 1$ is a divisor of $m^2 + 1$, a contradiction.   □

A non-empty point set $\mathfrak{D}$ of a projective 3-space $\mathfrak{P}$ is called an *ovoid* if it satisfies the following conditions:

(A) No three points of $\mathfrak{D}$ are collinear.
(B) If $P \in \mathfrak{D}$, then there exists a plane E of $\mathfrak{P}$ with $\mathfrak{D} \cap E = \{P\}$.
(C) If $P \in \mathfrak{D}$ and if E is a plane of $\mathfrak{P}$ with $E \cap \mathfrak{D} = \{P\}$, then all lines l through P which are not contained in E carry a point of $\mathfrak{D}$ distinct from P.

If $\mathfrak{D}$ is an ovoid, then given $P \in \mathfrak{D}$ there is exactly one plane E with $E \cap \mathfrak{D} = \{P\}$. This plane is called the *tangent plane* at $\mathfrak{D}$ in P.

25.6 Theorem. *Let $\mathfrak{D}$ be an ovoid in the projective 3-space $\mathfrak{P}$. Define $\mathfrak{M}(\mathfrak{D})$ as follows:*

1° *The points of $\mathfrak{M}(\mathfrak{D})$ are the points of $\mathfrak{D}$.*
2° *The circles of $\mathfrak{M}(\mathfrak{D})$ are the planes of $\mathfrak{P}$ which carry at least two points of $\mathfrak{D}$.*
3° *Incidence in $\mathfrak{M}(\mathfrak{D})$ is equivalent to incidence in $\mathfrak{P}$.*

Then $\mathfrak{M}(\mathfrak{D})$ is a Möbius plane.

PROOF. a) Three distinct points of $\mathfrak{M}(\mathfrak{D})$ are on exactly one circle of $\mathfrak{M}(\mathfrak{D})$, as no three points of $\mathfrak{D}$ are collinear.

b) Let P, Q be points and k a circle of $\mathfrak{M}(\mathfrak{D})$ with $P \mathrel{I} k$ and $Q \mathrel{\not I} k$. Let T be the tangent plane at $\mathfrak{D}$ in P. Then $T \cap k$ is a tangent line of $\mathfrak{D}$ which lies in k. Let l be the plane determined by $T \cap k$ and Q. As $P, Q \mathrel{I} l$ and $P \neq Q$, we have that l is a circle of $\mathfrak{M}(\mathfrak{D})$. As $Q \mathrel{\not I} k$, we see that $k \cap l = T \cap k$. Hence $k \cap l \cap \mathfrak{D} = \{P\}$ whence the circle l touches the circle k in the point P. Let l' be a second circle through Q touching k in P. Then $k \cap l'$ is a tangent line of $\mathfrak{D}$. It follows from (C) that $k \cap l' = k \cap T$ whence $l' = l$.

c) The points of $\mathfrak{D}$ are not complanar by (B) and (C). Hence there exist four non-concyclic points in $\mathfrak{M}(\mathfrak{D})$.   □

A Möbius plane $\mathfrak{M}$ is called *egglike*, if there exists an ovoid $\mathfrak{D}$ in a projective 3-space such that $\mathfrak{M}$ and $\mathfrak{M}(\mathfrak{D})$ are isomorphic.

25.7 Theorem. *If $\mathfrak{M}$ is an egglike Möbius plane, then $\mathfrak{M}(P)$ is desarguesian for all points P of $\mathfrak{M}$.*

The proof is obvious.

25.8 Theorem. *If $\mathfrak{D}$ is an ovoid in* $PG(3,q)$, *then* $|\mathfrak{D}| = q^2 + 1$. *If E is a plane of* $PG(3,q)$, *then E is either a tangent plane of $\mathfrak{D}$ or $E \cap \mathfrak{D}$ consists of the $q + 1$ points of an oval of E.*

PROOF. Let $P \in \mathfrak{D}$. There are $q^2 + q + 1$ lines through P. As exactly $q + 1$ of them are tangent lines of $\mathfrak{D}$, it follows from (A) and (C) that $|\mathfrak{D}| = q^2 + 1$.

Using 25.6 and 25.2, we see that exactly $q(q^2 + 1)$ planes of $PG(3,q)$ intersect $\mathfrak{D}$ in the $q + 1$ points of an oval. Moreover there are exactly $q^2 + 1$ tangent planes at $\mathfrak{D}$. As $q(q^2 + 1) + q^2 + 1 = (q + 1)(q^2 + 1)$ is the total number of planes in $PG(3,q)$, the theorem is proved.   □

25.9 Theorem (Qvist 1952). *Let $\mathfrak{o}$ be an oval in a finite projective plane $\mathfrak{P}$ of even order. Then there exists a point K in $\mathfrak{P}$ not on $\mathfrak{o}$ such that the lines through K are just the tangent lines of $\mathfrak{o}$.*

PROOF. Let $P \in \mathfrak{o}$. Then P is on exactly $n + 1$ lines, where n is the order of $\mathfrak{P}$. We infer from $|\mathfrak{o}| = n + 1$ that there is exactly one tangent through P at $\mathfrak{o}$. Hence the number of tangents of $\mathfrak{o}$ is $n + 1$.

Let P and Q be two distinct points on $\mathfrak{o}$ and let $X \text{ I } PQ$ with $X \neq P, Q$. As $|\mathfrak{o} \setminus \{P, Q\}| = n - 1$ is odd, there exists at least one tangent of $\mathfrak{o}$ through X. Hence every point of a secant is on a tangent of $\mathfrak{o}$. As there are exactly $n + 1$ tangents and as every secant carries $n + 1$ points, we deduce that every point X of $\mathfrak{P}$ which is on a secant is on exactly one tangent. Hence if K is the intersection of two tangents, each line through K is either a tangent or does not meet $\mathfrak{o}$ at all. Since every point of $\mathfrak{o}$ is joined by a line to K, all the lines through K are tangents.   □

K is called the *knot* of $\mathfrak{o}$.

25.10 Theorem (Segre 1959). *Let $\mathfrak{D}$ be an ovoid in* $PG(3,q)$ *where q is a power of 2. Define the mapping φ from the set of planes of* $PG(3,q)$ *into the set of points as follows: If E is the tangent plane of $\mathfrak{D}$ at P, then $E^\varphi = P$. If E is not a tangent plane, then E^φ is the knot of $E \cap \mathfrak{D}$ in E. Then φ is bijective and $\pi = (\varphi^{-1}, \varphi)$ is a symplectic polarity of* $PG(3,q)$. *Moreover, each collineation of* $PG(3,q)$ *which fixes $\mathfrak{D}$ centralizes π.*

PROOF. Let E be a plane which is not a tangent plane of $\mathfrak{D}$ and put $P = E^\varphi$. Let $l_1, \ldots, l_{q+1}$ be all the lines of E which pass through P. Then $l_1, \ldots, l_{q+1}$ are tangents of $\mathfrak{D}$ by 25.9 and 25.8. Hence l_i is contained in exactly one tangent plane E_i of $\mathfrak{D}$. Thus there are at least $(q + 1)(q - 1) + 1 = q^2$ planes through P which are not tangent planes. As there are exactly $q^2 + q + 1$ planes containing P, there are exactly $q + 1$

tangent planes and hence exactly $q + 1$ tangents through P. This shows that φ is injective and therefore bijective.

Let P be a point and E a plane of $\mathrm{PG}(3, q)$ and assume $P \mathrel{I} E$. If $P = E^\varphi$, then $P^{\varphi^{-1}} = E$ and hence $E^\varphi \mathrel{I} P^{\varphi^{-1}}$. If $P \neq E^\varphi$, then PE^φ is a tangent line and hence $E^\varphi \mathrel{I} P^{\varphi^{-1}}$. This proves that π is a symplectic polarity. The remaining statement of 25.10 is now trivial. $\qquad\square$

26. The Möbius Planes Belonging to the Suzuki Groups

Let $q = 2^{2r+1} \geqslant 8$ and assume that $\mathfrak{M}$ is a Möbius plane of order q admitting a group G of automorphisms isomorphic to $S(q)$.

26.1 Lemma. *Let Π be a Sylow 2-subgroup of G. Then Π has a fixed point and acts sharply transitively on the set of the remaining points.*

PROOF. As $q^2 + 1$ is odd, Π has a fixed point P. Let $\sigma \in \Pi$ be an involution. If σ has more than one fixed point, then σ is a reflection by 25.5, because $q^2 - 1$ is odd. Let $k(\sigma)$ be the circle fixed pointwise by σ. As $\sigma \in \mathfrak{Z}(\Pi)$, we have that $k(\sigma)$ is fixed by Π. Also $P \mathrel{I} k(\sigma)$, since σ has no fixed points not on $k(\sigma)$. Let τ be another involution in Π. Since all involutions of G are conjugate by 22.3, τ also is a reflection. If $|k(\sigma) \cap k(\tau)| \geqslant 2$, then τ fixes more than two points of $k(\sigma)$, since $q - 1$ is odd. (Recall that $k(\sigma)$ is fixed by Π.) Therefore $|k(\sigma) \cap k(\tau)| \geqslant 3$ and thus $k(\sigma) = k(\tau)$. Hence $\sigma = \tau$ by 25.4, a contradiction. Therefore we have $k(\sigma) \cap k(\tau) = \{P\}$. As Π contains exactly $q - 1$ involutions, there exist exactly $q - 1$ circles through P which touch each other mutually and each one of which is fixed pointwise by an involution in Π. Let $\mathfrak{p}$ be the pencil with carrier P to which all these circles belong. Then $\mathfrak{p}$ is fixed by Π. Moreover $|\mathfrak{p}| = q$ by 25.2 f). Hence there exists a circle $k \in \mathfrak{p}$ with $k^\Pi = k$ and $k \neq k(\sigma)$ for all involutions $\sigma \in \Pi$. Let $Q \mathrel{I} k$ and $Q \neq P$. Then $|\Pi_Q| \geqslant q$, as $|\Pi| = q^2$. Therefore $\Pi_Q \cap \mathfrak{Z}(\Pi) \neq \{1\}$, a contradiction. Therefore, if $\sigma \in \mathfrak{Z}(\Pi) \backslash \{1\}$, then P is the only fixed point of σ. Thus $\Pi_Q = \{1\}$ for all $Q \neq P$ whence the orbit of Q under Π has length q^2. $\qquad\square$

26.2 Lemma. *G acts doubly transitively on the set of points of $\mathfrak{M}$. Moreover, each element of $G \backslash \{1\}$ has at most two fixed points.*

PROOF. Let Π and Σ be distinct Sylow 2-subgroups of G and assume that the point P is fixed by Π and Σ. As $\Pi \cap \Sigma = \{1\}$, there are exactly $2(q - 1)$ involutions in $\Pi \cup \Sigma$. It then follows from 26.1 that $\Pi \cup \Sigma$ contains translations of $\mathfrak{M}(P)$ with distinct centres. Therefore $\mathfrak{Z}(\Pi)$ and $\mathfrak{Z}(\Sigma)$ centralize each other in contradiction to 24.2 d). This contradiction shows that Π and Σ have distinct fixed points. 26.2 now follows from 26.1, 21.8, and 22.1.

26.3 Lemma. *G acts transitively on the set of circles of $\mathfrak{M}$. The stabilizer of a circle has order $q(q-1)$.*

PROOF. Let P be a point of $\mathfrak{M}$ and put $H = G_P$. Then $H = \mathfrak{N}_G(\Pi)$ for some Sylow 2-subgroup Π of G. The number of pencils with carrier P is $q + 1$. Hence there is a pencil $\mathfrak{p}$ with carrier P which is fixed by Π. If $k \in \mathfrak{p}$, then $|\Pi_k| = q$ by 26.1. Therefore $|H_k| = qt$ where t divides $q - 1$.

Assume $G_k \ne H_k$. Then G_k acts transitively on k, as H_k acts transitively on $k \backslash \{P\}$. Thus $q + 1$ divides $|G_k|$ and therefore $|G| = (q^2 + 1)q^2(q - 1)$, a contradiction. This proves $G_k = H_k$.

Let $\mathfrak{R} = \{k^\sigma | \sigma \in G\}$. Then we have

$$(q^2 + 1)q^2(q - 1) = |G| = |\mathfrak{R}| |G_k| = |\mathfrak{R}| qt.$$

Therefore $|\mathfrak{R}| \geqslant q(q^2 + 1)$, as $t \leqslant q - 1$. $\qquad\square$

26.4 Theorem (Hughes 1962, Lüneburg 1964). *If $q = 2^{2r+1} \geqslant 8$, then there exists up to isomorphisms exactly one Möbius plane of order q admitting a group of automorphisms isomorphic to $S(q)$.*

PROOF. The existence follows from section 21 and 25.6. To prove uniqueness let $\mathfrak{M}$ be a Möbius plane of order q and let G be a group of automorphisms of $\mathfrak{M}$ which is isomorphic to $S(q)$. Pick a circle k of $\mathfrak{M}$. Then G_k has a fixed point P on k. Also $|G_k| = q(q - 1)$. Hence G_k acts sharply 2-transitively on $k \backslash \{P\}$ by 26.2. Let Π be the Sylow 2-subgroup of G whose fixed point is P. Then Π_k is the Frobenius kernel of the Frobenius group G_k. Hence Π_k is elementary abelian. Therefore $\Pi_k = \mathfrak{Z}(\Pi)$. Put $\Pi_k = Z$ and $G_K = \Lambda$ and let $P_0 = P, P_1, \ldots, P_q$ be all the points on k. Furthermore put $G_i = G_{P_i}$. Finally let $\mathfrak{H} = \{G_i | i = 0, 1, \ldots, q\}$ and $\Delta = \{\gamma | \gamma \in G, G_0^\gamma \in \mathfrak{H}\}$. If $i > 0$, then, obviously, $\mathfrak{H} = \{G_0\} \cup \{G_i^\zeta | \zeta \in Z\}$.

Define the mappings φ and ψ by

$$Q^\varphi = G_0\sigma \quad \text{if and only if} \quad P^\sigma = Q,$$

$$l^\psi = \Lambda\tau \quad \text{if and only if} \quad k^\tau = l.$$

We infer from 26.2 that φ is a bijection of the set of points of $\mathfrak{M}$ onto the set of right cosets of G_0 and 26.3 implies that ψ is a bijection of the set of circles onto the set of right cosets of Λ. Let $G_0\sigma$ be incident with $\Lambda\tau$ if and only if $P^\sigma I k^\tau$. Obviously, $G_0\sigma$ is incident with $\Lambda\tau$, if and only if $\sigma\tau^{-1} \in \Delta$.

Let $\mathfrak{M}^*$ be a second Möbius plane of order q admitting a group G^* of automorphisms isomorphic to $S(q)$. Let G_0^*, Λ^*, $\mathfrak{H}^*$, Δ^* and Z^* have the corresponding meaning to G_0, Δ, $\mathfrak{H}$, Δ and Z. If $G_1^* \in \mathfrak{H}^* \backslash \{G_0^*\}$, then there

exists an isomorphism ρ from G onto G^* with $G_0^\rho = G_0^*$ and $G_1^\rho = G_1^*$. This implies $Z^\rho = Z^*$. This again implies $\mathfrak{H}^\rho = \mathfrak{H}^*$ and hence $\Delta^\rho = \Delta^*$. As $\Lambda = Z(G_0 \cap G_1)$, we have $\Lambda^\rho = \Lambda^*$. Therefore $\mathfrak{M}$ and $\mathfrak{M}^*$ are isomorphic.
□

27. S(q) as a Collineation Group of PG(3, q)

27.1 Lemma. *Let $\mathfrak{P}$ be a desarguesian projective plane. Then $\mathfrak{P}$ does not admit a collineation group G isomorphic to $S(K, \sigma)$ such that all involutions of G are central collineations.*

PROOF. Assume by way of contradiction that G is a collineation group of $\mathfrak{P}$ isomorphic to $S(K, \sigma)$ for some K and σ and that all involutions of G are central collineations. Let Σ be a 2-subgroup of G which is mapped by an isomorphism of G onto $S(K, \sigma)$ onto the group T. Let $1 \ne \sigma \in \mathfrak{Z}(\Sigma)$. Then σ is an involution. Thus σ is a (C, l)-perspectivity for some C and l. Let L be the coordinatizing field of $\mathfrak{P}$.

Case 1: The characteristic of L is distinct from 2. Then $C \nmid l$. As $\Delta(C, l)$ is isomorphic to the multiplicative group of L, we have that σ is the only involutory (C, l)-perspectivity of $\mathfrak{P}$. Let σ_1 be a second involution in $\mathfrak{Z}(\Sigma)$. Then σ_1 is a (D, m)-homology. As $\sigma\sigma_1 = \sigma_1\sigma$, we have $C^{\sigma_1} = C$ and $l^{\sigma_1} = l$. It follows therefore from $\sigma \ne \sigma_1$ that $D \mathbin{I} l$ and $C \mathbin{I} m$. Let σ_3 be a third involution in $\mathfrak{Z}(\Sigma)$. Then similar arguments show that σ_3 is an $(l \cap m, DC)$-homology. Hence $|\mathfrak{Z}(\Sigma)| \le 4$, a contradiction.

Case 2: The characteristic of L is 2. Then $C \mathbin{I} l$. Let σ_1 be an involution in $\mathfrak{Z}(\Sigma)$ distinct from σ. Let D be the centre and m the axis of σ_1. If $m \ne l$, then $C = l \cap m$, as $C^{\sigma_1} = C$. Likewise $D = l \cap m$. This shows that all involutions in $\mathfrak{Z}(\Sigma)$ either have the same centre or the same axis. We may assume wlog that they all have the same axis. Let $\gamma \in G$ and $\Sigma_1 = \Sigma^\gamma \ne \Sigma$. Then all the involutions in $\mathfrak{Z}(\Sigma_1)$ have axis l^γ. If $l^\gamma \ne l$, then $l \cap l^\gamma$ is a fixed point of $\langle \mathfrak{Z}(\Sigma), \mathfrak{Z}(\Sigma_1) \rangle = G$. If $l^\gamma = l$, then l is a fixed line of G. We may therefore assume wlog that G has a fixed line l. Let E be the group of all elations with axis l. Then $E \cap G$ is normal in G. As G is a simple, non-abelian group by 22.6, we obtain as a result $E \cap G = \{1\}$. Now E acts transitively on the points of $\mathfrak{P}_l$ by 1.14. We may therefore assume that G fixes a point P not on l. As G is generated by its involutions with fixed points, we conclude $G \subseteq SL(2, L)$.

Again let $1 \ne \rho \in \mathfrak{Z}(\Sigma)$. Then the centre C of ρ is on l. Furthermore, Σ fixes C, whence $\Sigma \subseteq SL(2, L)_C$. But this contradicts the fact that $SL(2, L)_C$ does not contain an element of order 4. This proves 27.1.
□

27.1 is wrong without the assumption made about the involutions of G, as one can easily construct commutative fields of any characteristic whose automorphism groups contain subgroups isomorphic to $S(q)$.

27.2 Corollary. *If $\mathfrak{P}$ is a finite desarguesian projective plane, then the collineation group of $\mathfrak{P}$ does not contain a subgroup which is isomorphic to a Suzuki group.*

PROOF. This follows at once from 27.1 using the simplicity of the Suzuki groups and the fact that $P\Gamma L(3, q)/PSL(3, q)$ is always soluble.   □

27.3 Theorem (Lüneburg 1965a). *Let $q = 2^{2r+1} \geqslant 8$ and let $\mathfrak{P}$ be the 3-dimensional projective space over $GF(q)$. Then the full collineation group of $\mathfrak{P}$ contains subgroups isomorphic to S(q). Moreover all these groups are contained in the projective group of $\mathfrak{P}$ and are conjugate under this group. Let G be a subgroup of the collineation group of $\mathfrak{P}$ which is isomorphic to S(q). Then G splits the set of points of $\mathfrak{P}$ into two orbits. One of these orbits is an ovoid $\mathfrak{O}$. The set of planes is also split by G into two orbits: the set of tangent planes of $\mathfrak{O}$ and the set of the remaining planes. The set of secants of $\mathfrak{O}$ as well as the set of exterior lines of $\mathfrak{O}$ are line orbits of G. The set of tangent lines of $\mathfrak{O}$ is split into two orbits, one of them being a spread of $\mathfrak{P}$.*

PROOF. Let G be a collineation group of $\mathfrak{P}$ isomorphic to S(q).

(1) G has no fixed point.

Assume that G has a fixed point P. The lines and planes through P together with the inclusion relation as incidence form a desarguesian plane. Hence by 27.2, the group G fixes all the lines through P individually. Thus G consists only of perspectivities with centre P. As the group of all perspectivities with centre P has order $q^3(q - 1)$, we arrive at the contradiction that $(q^2 + 1)q^2(q - 1)$ divides $q^3(q - 1)$.

(2) G is contained in the projective group of $\mathfrak{P}$.

This follows from the fact that G is simple and that $P\Gamma L(4, q)/PGL(4, q)$ is cyclic.

(3) If Π is a Sylow 2-subgroup of G, then Π has exactly one fixed point. The set of fixed points of the Sylow 2-subgroups of G is an ovoid of $\mathfrak{P}$.

The centre of Π cannot fix a plane pointwise by (1), as two planes always have a point in common and since G is generated by the centres of two distinct Sylow 2-subgroups.

Let P and Q be distinct points fixed by Π. As $q - 1$ is odd, Π has a third fixed point on PQ. It follows from (2) and Baer [1952, p. 66–69] that Π fixes PQ pointwise. There are $q + 1$ planes containing PQ. Hence there exists a plane E with $PQ \subseteq E = E^{\Pi}$. As PQ is fixed pointwise by Π, the group Π induces a group Π^* of elations with axis PQ in E. As E is desarguesian, all elations of E are involutions. Furthermore, each involution of Π is contained in a cyclic group of order 4. Therefore $\mathfrak{Z}(\Pi)$ induces the identity on E, a contradiction. This proves that Π has at most one fixed point. As the number of points of $\mathfrak{P}$, being $(q^2 + 1)(q + 1)$, is odd, Π has exactly one fixed point.

Let $\mathfrak{O}$ be the set of fixed points of the Sylow 2-subgroups of G. Then by (1) and the remark just made, $|\mathfrak{O}| = q^2 + 1$. It follows by 24.10 that G acts in its natural representation on $\mathfrak{O}$. We consider the following incidence structure $\mathfrak{B}$:

(a) The points of $\mathfrak{B}$ are the points of $\mathfrak{O}$.
(b) The blocks of $\mathfrak{B}$ are the lines of $\mathfrak{P}$ which meet $\mathfrak{O}$ in more than one point.
(c) Incidence is inherited by incidence in $\mathfrak{P}$.

As G is twofold transitive on $\mathfrak{O}$, we have that $\mathfrak{B}$ is a block design with $v = q^2 + 1$ points, b blocks, k points per block and r blocks through each point. Moreover, any two points are joint by exactly $\lambda = 1$ block. It follows that $r(k - 1) = \lambda(v - 1) = q^2$.

Assume $k \geqslant 3$ and let P, Q, R be three distinct collinear points of $\mathfrak{O}$. As G acts in its natural representation on $\mathfrak{O}$, we have $|G_{P,Q}| = q - 1$ and $|G_{P,Q,R}| = 1$. Therefore all the points on the line PQ belong to $\mathfrak{O}$. Thus $k = q + 1$ whence $r = q$. Now $(q^2 + 1)q = vr = bk = b(q + 1)$, a contradiction. Hence no three points of $\mathfrak{O}$ are collinear.

The dual arguments show that each Sylow 2-subgroup Π of G fixes exactly one plane E. As the number of points on E is odd, the fixed point P of Π is on E. If $|E \cap \mathfrak{O}| \geqslant 2$, then $\mathfrak{O} \subseteq E$, since Π acts transitively on $\mathfrak{O}\setminus\{P\}$. Hence $E^G = E$ in contradiction to the dual of (1). Hence $E \cap \mathfrak{O} = \{P\}$. This proves that $\mathfrak{O}$ is an ovoid.

(4) G has exactly two point- and two plane-orbits. One of the point orbits is $\mathfrak{O}$ and one of the plane orbits consists of the tangent planes of $\mathfrak{O}$.

G acts transitively on the set of tangent planes of $\mathfrak{O}$, as the above arguments show. Furthermore, by (3), 25.6, 25.8, and 26.3, the group G acts transitively on the set of the remaining planes. Moreover G operates transitively on $\mathfrak{O}$. Let X and Y be two points not on $\mathfrak{O}$ and let π be the polarity defined in 25.10. Then X^π, Y^π are planes which are not tangent to $\mathfrak{O}$. Hence there exists $\gamma \in G$ with $X^{\pi\gamma} = Y^\pi$. Using 25.10 again, we obtain $X^\gamma = X^{\gamma\pi^2} = X^{\pi\gamma\pi} = Y^{\pi^2} = Y$. This proves (4).

(5) The secants as well as the exterior lines of $\mathfrak{O}$ each form a line orbit of G.

The secants form a line orbit, since G acts doubly transitively on $\mathfrak{O}$. From this and 25.10, it follows that the exterior lines also form an orbit.

Now let $\mathfrak{O}^*$ be the ovoid defined in section 21. Then $\mathfrak{M}(\mathfrak{O})$ and $\mathfrak{M}(\mathfrak{O}^*)$ are isomorphic by 26.4. Let φ be an isomorphism of $\mathfrak{M}(\mathfrak{O})$ onto $\mathfrak{M}(\mathfrak{O}^*)$. We have to show that we can extend φ to a collineation of $\mathfrak{P}$. Let π be the polarity defined by 25.10 with respect to $\mathfrak{O}$, let π^* be the polarity defined with respect to $\mathfrak{O}^*$, and define the collineation κ as follows: If X is a point of $\mathfrak{O}$, then $X^\kappa = X^\varphi$. If X is not on $\mathfrak{O}$, then $X^\kappa = X^{\pi\varphi\pi^*}$. If E is a tangent plane, then $E^\kappa = E^{\pi\varphi\pi^*}$, and if E is not a tangent plane, then $E^\kappa = E^\varphi$. It is easily seen that κ is a collineation of $\mathfrak{P}$ which maps $\mathfrak{O}$ onto $\mathfrak{O}^*$. As G is contained in $\mathrm{PGL}(4, q)$, we also get the result that $G^\gamma = S(q)$.

We still have to show that G and $S(q)$ are conjugate within $\mathrm{PGL}(4, q)$. But this follows from the fact that $\gamma \in \mathrm{P\Gamma L}(4, q)$ and that all the automorphisms of $\mathrm{GF}(q)$ induce collineations in $\mathfrak{P}$ leaving $\mathfrak{O}^*$ invariant, as is easily seen from the definition of $\mathfrak{O}^*$.

The remaining statement that G splits the set of tangent lines into two orbits one of them being a spread now follows from 23.2 and 23.1. $\qquad\square$

28. S(q) as a Collineation Group of a Plane of Order q^2

We shall assume throughout this section that $\mathfrak{P}$ is a projective plane of order q^2 where $q = 2^{2\alpha+1} \geqslant 8$ and that G is a group of collineations of $\mathfrak{P}$ which is isomorphic to $S(q)$.

28.1 Lemma (Dembowski 1966). *If G has a point orbit $\mathfrak{p}$ of length $q^2 + 1$, then either $\mathfrak{p}$ is an oval of $\mathfrak{P}$ or all the points of $\mathfrak{p}$ are collinear.*

PROOF. It follows from 24.10 that G acts in its natural representation on $\mathfrak{p}$. Consider the following incidence structure $\mathfrak{B}$: The points of $\mathfrak{B}$ are the points of $\mathfrak{p}$ and the blocks of $\mathfrak{B}$ are the lines of $\mathfrak{P}$ carrying more than one point of $\mathfrak{p}$. Then $\mathfrak{B}$ is a block design with parameters $v = q^2 + 1, b, k, r$, and $\lambda = 1$. Hence $r(k-1) = \lambda(v-1) = q^2$. Therefore $r = 2^s$ with $s \leqslant 2(2\alpha + 1)$. If $s = 0$, then $r = 1$ and all the points in $\mathfrak{p}$ are collinear. Thus assume $s \geqslant 1$. As G operates in its natural action on $\mathfrak{p}$, we have $k = 2 + t(q-1)$ with $t \geqslant 0$. Therefore

$$1 + t(q-1) = q^2 r^{-1} = 2^{2(2\alpha+1)-s}.$$

This implies that $2^{2\alpha+1} - 1$ divides $2^{2(2\alpha+1)-s} - 1$ which yields that $2\alpha + 1$ divides $2(2\alpha + 1) - s$ and hence s. Therefore $s = 2\alpha + 1$ or $s = 2(2\alpha + 1)$, i.e., $r = q$ or $r = q^2$. If $r = q$, then $k = q + 1$ and thus $(q^2 + 1)q = vr = bk = b(q + 1)$, a contradiction. Thus $r = q^2$ and $k = 2$. $\qquad\square$

28.2 Lemma (Dembowski 1966). *G has a fixed element, if and only if all involutions in G are elations.*

PROOF. Assume that G has a fixed element. We may then assume that G fixes a line f. If G fixes f pointwise, then all involutions of G are elations. Therefore we may assume that G operates non-trivially on f. In this case, G acts faithfully on f, as G is simple. Using 24.10, we see that G operates in its natural action on the set of points incident with f. Hence all involutions of G have exactly one fixed point on f. Therefore they cannot be Baer involutions.

Conversely, assume that all involutions of G are elations. Let Π be a Sylow 2-subgroup of G and $1 \neq \sigma \in \mathfrak{Z}(\Pi)$. Then σ is an involution and hence an elation. Let C be the centre and a the axis of σ. Then C and a are both fixed by Π. Pick $\tau \in \mathfrak{Z}(\Pi) \setminus \{1, \sigma\}$. Let D be the centre and b the axis of τ. If $b \neq a$, then $D = C$, as $\sigma\tau = \tau\sigma$. In other words: If there exist two involutions in Π with distinct axes, then all involutions in Π have the same centre. Thus all involutions of Π either have the same centre or the same axis. We may assume that they all have the same axis, a say.

Let Σ be a Sylow 2-subgroup of G distinct from Π. Then all involutions in Σ also have the same axis, b say. If $a = b$, then a is fixed by G, as G is generated by $\mathfrak{Z}(\Sigma) \cup \mathfrak{Z}(\Pi)$. If $a \neq b$, then $a \cap b$ is a fixed point of G for the same reason. $\square$

28.3 Lemma (Dembowski 1966). *All involutions of G are elations.*

PROOF. Assume that G contains a Baer involution. Then all involutions of G are Baer involutions by 22.3. If σ is an involution, we denote by $\mathfrak{E}(\sigma)$ the Baer subplane of $\mathfrak{P}$ fixed pointwise by σ.

Let Π be a Sylow 2-subgroup of G and assume $\mathfrak{E}(\sigma) = \mathfrak{E}(\tau)$ for all $\sigma, \tau \in \mathfrak{Z}(\Pi) \setminus \{1\}$. Then Π acts regularly outside of $\mathfrak{E}(\sigma)$. Hence q^2 divides $q^4 + q^2 + 1 - q^2 - q - 1 = q(q^3 - 1)$, a contradiction. Thus there exist involutions $\sigma, \tau \in \Pi$ with $\mathfrak{E}(\sigma) \neq \mathfrak{E}(\tau)$. As $\sigma\tau = \tau\sigma$, we then have that τ induces an involutory collineation τ^* in $\mathfrak{E}(\sigma)$. Since $q = 2^{2\alpha+1}$ is not a square, τ^* is an elation. Let C be the centre and l the axis of τ^*. As all the fixed lines of Π are fixed by σ and τ, they all belong to $\mathfrak{E}(\sigma)$ and hence pass through C. Similarly, all the fixed points of Π belong to $\mathfrak{E}(\sigma)$ and are on l. Therefore, if Π has more than one fixed point, then l is fixed by the normalizer of Π. If Π has just one fixed point, then this fixed point is fixed by the normalizer of Π. As this normalizer is a maximal subgroup of G, we see that Π has a fixed element Φ such that the orbit of Φ under G has length 1 or $q^2 + 1$. The first case cannot occur by 28.2. Hence G has an orbit of length $q^2 + 1$. We may assume that it is a point orbit. As G has no fixed elements, it follows from 28.1 that it is an oval. But this again contradicts 28.2, as the knot of the oval is fixed by G. This final contradiction proves the lemma. $\square$

It follows from 28.3 and 28.2 that G has a fixed element. We may assume that G has a fixed point P.

28.4 Lemma. *G acts in its natural representation on the set of lines through P.*

PROOF. As $|G| = (q^2 + 1)q^2(q - 1)$ does not divide $q^4(q^2 - 1)$, the group G cannot consist only of perspectivities with centre P. Therefore G acts faithfully on the set of lines through P since it is simple. 28.4 then follows from 24.10. $\square$

28.5 Lemma. *P is not the centre of an involution in G.*

PROOF. 28.4 yields that every involution fixes exactly one line through P. Therefore P cannot be the centre of such an involution. $\square$

28.6 Lemma. *Let l and m be two distinct lines through P. Then $G_{l,m}$ has exactly two fixed points on l. The remaining points are split by $G_{l,m}$ into $q + 1$ orbits of length $q - 1$.*

PROOF. $G_{l,m}$ is cyclic of order $q - 1$. Let $\delta \in G_{l,m}$ and let Q and R be points on l with $P \ne Q \ne R \ne P$ and $Q^\delta = Q$ and $R^\delta = R$. Furthermore, let H be the group generated by δ. By 28.4, there exists $\sigma \in G$ with $l^\sigma = m$ and $m^\sigma = l$. We have $G_{l,m}^\sigma = G_{l,m}$ and therefore $H^\sigma = H$, as $G_{l,m}$ is cyclic. Thus Q, R, Q^σ, and R^σ are fixed points of H no three of which are collinear. This yields that H fixes the three lines l, m, and $P(QR^\sigma \cap Q^\sigma R)$. Therefore $H = \{1\}$ by 28.4. This shows that any non-identity element of $G_{l,m}$ has at most two fixed points on l.

Let γ be an element of prime order in $G_{l,m}$. As $(q - 1, q^2) = 1$, we have that γ fixes a point Q on l other than P. As this is the only fixed point of γ on $l \setminus \{P\}$ by the remark made above and as $G_{l,m}$ is abelian, we see that Q is a fixed point of $G_{l,m}$. The remaining statement is now trivial. $\qquad \square$

28.7 Lemma. *Let Π be a Sylow 2-subgroup of G. If $P \mathrel{I} l = l^\Pi$, then Π has at least $q + 1$ fixed points on l.*

PROOF. P is a fixed point of Π on l. Let σ be an involution in Π. Then σ is an elation with axis l. Let Q be the centre of σ. Then $P \ne Q$ by 28.6. As $\sigma \in \mathfrak{Z}(\Pi)$, we also have $Q^\Pi = Q$. As $q^2 - 1$ is odd, Π has a third fixed point R on l. Let m be a line through P which is distinct from l. Then $G_{l,m}$ normalizes Π. It follows from 28.6 that at least one of the points Q and R lies in an orbit of length $q - 1$ of $G_{l,m}$. Thus Π has at least $q + 1$ fixed points on l. $\qquad \square$

28.8 Lemma. *Let Π be a Sylow 2-subgroup of G and assume that there exist involutions in Π with distinct centres. Then G has, besides $\{P\}$, three point orbits of length $q^2 + 1$, $(q^2 + 1)(q - 1)$ and $(q^2 + 1)q(q - 1)$ respectively. The point orbit $\mathfrak{o}$ of length $q^2 + 1$ is an oval whose knot is P. The tangent lines and the secants of $\mathfrak{o}$ each form a line orbit of G. The exterior lines of $\mathfrak{o}$ are split into two orbits of length $\frac{1}{4}(q - r + 1)q^2(q - 1)$ and $\frac{1}{4}(q + r + 1) q^2(q - 1)$ respectively, where $r^2 = 2q$.*

PROOF. Let l be the line through P fixed by Π. By assumption, there exist distinct points Q and R on l which are centres of involutions. Futhermore, $Q, R \ne P$ by 28.5. We then deduce from 28.6 that no two involutions of Π have the same centre. Thus there exist exactly $q - 1$ points on l which are centres of involutions in Π.

Let m be a line distinct from l and assume $|m^\Pi| \leqslant q^2$. Then $m \cap l$ is a centre. As there are $q^2 + 1 - (q - 1) = q^2 - q + 2$ points on l which are not centres, there exist $q^2(q^2 - q + 2)$ lines distinct from l and meeting l in a point which is not a centre. The set of all these lines is split by Π into $q^2 - q + 2$ orbits of length q^2.

If $m \cap l$ is the centre of an involution, then Π_m is cyclic of order 2 or 4, as Π is of exponent 4 and does not contain a quaternion group by 24.2 and as Π_m contains exactly one involution. Therefore $|m^\Pi| = \frac{1}{2}q^2$ or $|m^\Pi| = \frac{1}{4}q^2$.

Assume that $|m^\Pi| = \frac{1}{2}q^2$ for all lines m such that $m \cap l$ is a centre. As there are $q - 1$ centres on l, there are $2(q - 1)$ line orbits of Π whose length is $\frac{1}{2}q^2$. As $\{l\}$ is a line orbit of Π, we thus have that Π possesses $1 + 2(q - 1) + q^2 - q + 2 = q^2 + q + 1$ line orbits.

Now we count the point orbits of Π. Since Π operates regularly on the set of points off l, there are q^2 point orbits of length q^2. Moreover, Π has at least $q + 1$ point orbits of length 1. As there are still $q(q - 1)$ points left, the number of point orbits of Π is greater than $q^2 + q + 1$ which contradicts the Dembowski-Hughes-Parker theorem.

Hence there exists a line m with $|m^\Pi| = \frac{1}{4}q^2$.

Assume that $|h^\Pi| = \frac{1}{4}q^2$ for all lines h with $\Pi_h \neq \{1\}$. Then Π has exactly $1 + 4(q - 1) + q^2 - q + 2 = q^2 + 3q - 1$ line orbits. As Π has q^2 point orbits of length q^2 and at least $q + 1$ point orbits of length 1, there exist $q^2 + 3q - 1 - q^2 - q - 1 = 2(q - 1)$ further point orbits all of which are contained in l.

Let j be a line through P distinct from l and let Q be a fixed point of $G_{l, j}$ on l other than P. If $Q^\Pi = Q$, then $|Q^G| = q^2 + 1$. Therefore, by 28.1, either Q^G is a line or Q^G is an oval. If Q^G is a line, then all the involutions in Π have the same centre $l \cap Q^G$, a contradiction. Thus Q^G is an oval. Let R be a centre on l and s a secant through R. Then $|s^\Pi| = \frac{1}{2}q^2$, a contradiction. Thus Q is not fixed by Π. As $G_{l, j}$ normalizes Π, it fixes Q^Π as a set. Therefore $|Q^\Pi| \geqslant q$ and hence $|Q^\Pi| = q$, as the centre of Π fixes all the points on l.

Now Π has at least $q + 1$ fixed points on l and P is the only common fixed point of Π and $G_{l, j}$; hence Π has at least $2(q - 1) + 1 = 2q - 1$ fixed points on l, since $G_{l, j}$ permutes the fixed points of Π in orbits of length $q - 1$. Of these, $q + 1$ have been taken into account already. Thus among the remaining $2(q - 1)$ point orbits, there are $2q - 1 - (q + 1) = q - 2$ of length 1. There remain $2q - 2 - q + 2 = q$ orbits, one of which has length q as we have seen. The remaining $q - 1$ orbits are permuted transitively by $G_{l, j}$, as is easily seen using 28.6 and the fact that the only fixed point of $G_{l, j}$ on $l \backslash \{P\}$ lies in the orbit of length q already counted. Thus the remaining orbits all have length 2^a. Consequently

$$2q - 1 + q + 2^a(q - 1) = q^2 + 1$$

which yields $(2^a + 1)(q - 1) = (q - 1)^2$, a contradiction.

Hence there also is a line n with $|n^\Pi| = \frac{1}{2}q^2$.

As G_l normalizes Π and acts transitively on the set of centres, there are $q - 1$ line orbits of length $\frac{1}{2}q^2$ and $2(q - 1)$ line orbits of length $\frac{1}{4}q^2$. Thus Π has a total of $1 + q^2 - q + 2 + q - 1 + 2(q - 1) = q^2 + 2q$ line orbits.

Π splits the points off l into q^2 orbits of length q^2. Using the Dembowski-Hughes-Parker theorem, we see that the points on l are split by Π into $2q$ orbits, $q + 1$ of them having length 1. Let $\lambda_1, \ldots, \lambda_{q-1}$ be the lengths of the remaining orbits. Then $1 \leqslant \lambda_i \leqslant q$ and $\sum_{i=1}^{q-1}\lambda_i = q(q - 1)$. Hence $\lambda_i = q$ for all i. This implies that $G_{l,\,j}$ permutes these $q - 1$ orbits transitively. Hence G_l splits the set of points on l into four orbits. Two of these have length 1. The other two have length $q - 1$ and $q(q - 1)$ respectively. This yields that G splits the set of points of $\mathfrak{P}$ into four orbits, one being $\{P\}$. The others have length $q^2 + 1$, $(q^2 + 1)(q - 1)$ and $(q^2 + 1)q(q - 1)$ respectively.

By the Dembowski-Hughes-Parker theorem, G splits the set of lines of $\mathfrak{P}$ into four orbits. Let $\mathfrak{o}$ be the point orbit of length $q^2 + 1$. By 28.1, $\mathfrak{o}$ is a line or an oval. As the points of $\mathfrak{o}$ are not centres, $\mathfrak{o}$ must be an oval whose knot obviously is P. Again it follows from 24.10 that G operates in its natural action on $\mathfrak{o}$. Hence the secants of $\mathfrak{o}$ form a line orbit. Therefore the set of exterior lines of $\mathfrak{o}$ is split into two orbits $\mathfrak{E}_1$ and $\mathfrak{E}_2$. Put $|\mathfrak{E}_i| = \lambda_i$. Then

$$\lambda_1 + \lambda_2 = q^4 + q^2 + 1 - q^2 - 1 - \tfrac{1}{2}(q^2 + 1)q^2 = \tfrac{1}{2}q^2(q^2 - 1).$$

Let C be a centre on the line l and let $\{Q\} = \mathfrak{o} \cap l$. Then Π acts transitively on $\mathfrak{o}\backslash\{Q\}$. Therefore, the line orbit of length $\tfrac{1}{2}q^2$ of Π all of whose lines pass through C consists of secants. Therefore the other two line orbits of Π consisting of lines through C contain only exterior lines. Therefore 4 divides $|G_e|$ for some exterior line e. Moreover $|G_e| = 4m$ with m odd. We may assume $e \in \mathfrak{E}_1$. Then λ_1 is even and therefore λ_2 is. Thus 4 divides $|G_f|$ for all $f \in \mathfrak{E}_1 \cup \mathfrak{E}_2$. Let G_i be the stabilizer of a line in $\mathfrak{E}_i$. Then $|G_i| = 4m_i$, where m_i is odd. Moreover, it follows from 24.10 that m_i divides $q - r + 1$ or $q + r + 1$, where $r^2 = 2p$. Now $\lambda_i = (q^2 + 1)q^2(q - 1)/4m_i$ and thus

$$\frac{1}{4}(q^2 + 1)q^2(q - 1)\left[\frac{1}{m_1} + \frac{1}{m_2}\right] = \frac{1}{2}q^2(q^2 - 1).$$

This implies $(q^2 + 1)(m_1 + m_2) = 2(q + 1)m_1m_2$. From $(q^2 + 1, 2(q + 1)) = 1$ we infer that $q^2 + 1$ divides m_1m_2 and $2(q + 1)$ divides $m_1 + m_2$. Furthermore

$$m_1 + m_2 \leqslant 2(q + r + 1) = 2(q + 1) + 2r < 2(q + 1) + 2q,$$

whence $m_1 + m_2 = 2(q + 1)$ and $m_1m_2 = q^2 + 1$. This implies $\{m_1, m_2\} = \{q - r + 1, q + r + 1\}$. $\square$

28.9 Lemma. *Let Π be a Sylow 2-subgoup of G and assume that all involutions of Π have the same centre. Then G fixes a line l not through P. The points of l form an orbit of length $q^2 + 1$. The set of points of $\mathfrak{P}$ which are not on l and distinct from P is split into two orbits of length $(q^2 + 1)(q - 1)$ and $(q^2 + 1)q(q - 1)$ respectively. The set of lines of $\mathfrak{P}$ which* *do not pass through P and are distinct from l is split into two orbits of length $(q^2 + 1)(q - 1)$ and $(q^2 + 1)q(q - 1)$ respectively.*

PROOF. Let Σ and T be two distinct Sylow 2-subgroups of G and let h and m be the two lines through P fixed by Σ and T respectively. Let Q be the centre of the involutions in Σ and R the centre of the involutions in T. Then $Q \text{ I } h$ and $R \text{ I } m$. From 28.5 we get $P \neq Q, R$. From this and $\Sigma \neq T$ we obtain $P \not{I} l = QR$. As l is fixed by $\mathfrak{Z}(\Sigma)$ and $\mathfrak{Z}(T)$, it is fixed by $\langle \mathfrak{Z}(\Sigma), \mathfrak{Z}(T) \rangle = G$.

Let X be a point with $P \neq X \not{I} l$ and $X^{\Pi} = X$. It follows from 28.6 that $G_X = \Pi$. Hence $|X^G| = (q^2 + 1)(q - 1)$. As there is always such an X by 28.7, G has a point orbit of length $(q^2 + 1)(q - 1)$. As l is a fixed line of G, dual arguments show that G also has a line orbit of length $(q^2 + 1)(q - 1)$.

Let $\mathfrak{p}$ be a point orbit of length $(q^2 + 1)(q - 1)$ and let m and n be lines such that $m \cap \mathfrak{p} \neq \emptyset \neq n \cap \mathfrak{p}$ and $P \not{I} m, n$. In order to prove the existence of $\gamma \in G$ with $m^\gamma = n$, we may assume that m and n have a point X of $\mathfrak{p}$ in common. Put $\Pi = G_X$. Then Π acts transitively on $l \setminus \{PX \cap l\}$. Hence there exists $\gamma \in \Pi$ with $m^\gamma = n$. Therefore, $\mathfrak{g} = \{m \mid P \not{I} m, m \cap \mathfrak{p} \neq \emptyset\}$ is a line orbit of G. Now $(\mathfrak{p}, \mathfrak{g})$ is a tactical configuration with parameters $v = (q^2 + 1)(q - 1)$, b, k, and $r = q^2$. Moreover the points X and Y of $\mathfrak{p}$ are not joined by a line of $\mathfrak{g}$, if and only if P, X, and Y are collinear. Hence

$$q^2(k - 1) = (q^2 + 1)(q - 1) - (q - 1) = q^2(q - 1),$$

whence $k = q$. Therefore

$$q^2(q^2 + 1)(q - 1) = vr = bk = bq.$$

Thus $b = q(q^2 + 1)(q - 1)$. Therefore, we have found four line orbits of length 1, $q^2 + 1$, $(q^2 + 1)(q - 1)$ and $(q^2 + 1)q(q - 1)$ respectively. Since this accounts for all the lines, G has exactly four line orbits and hence also four point orbits. As three of them have length 1, $q^2 + 1$, and $(q^2 + 1)(q - 1)$ respectively, the last one must have length $(q^2 + 1)q(q - 1)$. $\quad\square$

28.10 Corollary. *Same assumptions as in 28.9. Let $\mathfrak{p}_1, \mathfrak{p}_2, \mathfrak{g}_1, \mathfrak{g}_2$ be the point and line orbits with $|\mathfrak{p}_1| = |\mathfrak{g}_1| = (q^2 + 1)(q - 1)$ and $|\mathfrak{p}_2| = |\mathfrak{g}_2| = (q^2 + 1)q(q - 1)$. Then we have:*

a) *$(\mathfrak{p}_1, \mathfrak{g}_1)$ is a tactical configuration with parameters $v = b = (q^2 + 1)(q - 1)$ and $r = k = 0$.*

b) *$(\mathfrak{p}_1, \mathfrak{g}_2)$ is a tactical configuration with parameters $v = (q^2 + 1)(q - 1)$, $b = (q^2 + 1)q(q - 1)$, $r = q^2$ and $k = q$.*

c) *$(\mathfrak{p}_2, \mathfrak{g}_1)$ is a tactical configuration with parameters $v = (q^2 + 1)q(q - 1)$, $b = (q^2 + 1)(q - 1)$, $r = q$ and $k = q^2$.*

d) *$(\mathfrak{p}_2, \mathfrak{g}_2)$ is a tactical configuration with parameters $v = b = (q^2 + 1)q(q - 1)$ and $r = k = q(q - 1)$.*

PROOF. We already proved a) and b). c) is the dual of b) and d) is a consequence of a), b) and c). $\qquad\square$

By now we have proved:

28.11 Theorem (Lüneburg 1965a). *Let $\mathfrak{P}$ be a projective plane of order q^2, where $q = 2^{2\alpha+1} \geqslant 8$ and let G be a group of collineations of $\mathfrak{P}$ which is isomorphic to $S(q)$. Then one of the following holds:*

a) *G fixes a non-incident point line pair (P, l) of $\mathfrak{P}$ and G acts doubly transitively on the set of lines through P as well as on the set of points on l. Moreover, G has two further non-trivial point orbits $\mathfrak{p}_1$ and $\mathfrak{p}_2$ and two further line orbits $\mathfrak{g}_1$ and $\mathfrak{g}_2$ such that $|\mathfrak{p}_1| = |\mathfrak{g}_1| = (q^2 + 1)(q - 1)$ and $|\mathfrak{p}_2| = |\mathfrak{g}_2| = (q^2 + 1)q(q - 1)$.*

b) *G fixes an oval $\mathfrak{o}$ and its knot P. It operates doubly transitively on the set of lines through P and on $\mathfrak{o}$. It also acts transitively on the set of secants of $\mathfrak{o}$ and splits the set of exterior lines into two orbits of length $\frac{1}{4}(q - r + 1)q^2(q - 1)$ and $\frac{1}{4}(q + r + 1)q^2(q - 1)$ respectively, where $r^2 = 2q$. Furthermore, G has two point orbits of length $(q^2 + 1)(q - 1)$ and $(q^2 + 1)q(q - 1)$ respectively.*

c) *The dual to b).*

Remark. All three possibilities really occur. We have seen that a) occurs in the Lüneburg planes and it follows from results of Pollatsek 1971 that c) also occurs in these planes. Hence b) occurs in the dual planes.

29. Geometric Partitions

Let π be a partition of the additively written abelian group A. The subgroup U of A is called π-*admissible*, if for all $X \in \pi$ we have that $X \cap U \neq \{0\}$ implies $X \subseteq U$. A non-trivial partition π of A is called *geometric*, if $X + Y$ is π-admissible for all $X, Y \in \pi$.

29.1 Lemma (Baer 1963). *Let π be a geometric partition of the abelian group A. Then π is a spread, if and only if there exist $X, Y \in \pi$ such that $A = X + Y$.*

PROOF. If π is a spread, then, of course, $A = X + Y$ for all $X, Y \in \pi$ with $X \neq Y$. (This also shows that every spread is a geometric partition.)

In order to prove the converse, let $X, Y \in \pi$ be such that $A = X + Y$. Furthermore, let $Z \in \pi \setminus \{X\}$. As the lattice of subgroups of A is modular, we have

$$X + Z = (X + Z) \cap A = (X + Z) \cap (X + Y) = X + ((X + Z) \cap Y).$$

Since X is a proper subgroup of $X + Z$, it follows that $(X + Z) \cap Y \neq \{0\}$. Therefore $Y \subseteq X + Z$, as π is geometric. This proves $X + Z \supseteq X + Y = A$, i.e., $A = X + Z$.

Let U and W be distinct components of π. We may assume that $U \neq X$. Then applying the above argument twice yields $A = X + U = W + U$. $\square$

If π is a geometric partition which is not a spread, then we call π a *spatial* partition. As we did in section 1, denote by $K(A,\pi)$ the kernel of the partition π.

29.2 Theorem (Baer 1963). *Let π be a spatial partition of the abelian group A. Then $K(A,\pi)$ is a skewfield. Moreover, the π-admissible subgroups of A are exactly the subspaces of the $K(A,\pi)$-vector space A. In particular, π consists of all the subspaces of rank 1.*

PROOF. Let $0 \neq \kappa \in K(A,\pi)$. Then κ is injective by 1.6 a). To prove that κ is surjective, it suffices to show that $X^\kappa = X$ for all $X \in \pi$. Let $X \in \pi$ and pick $Y \in \pi \setminus \{X\}$. Furthermore, let $\pi_0 = \{Z \mid Z \in \pi, Z \subseteq X + Y\}$. Then π_0 is a geometric partition of $X + Y$, as π is geometric. We infer from 29.1 that π_0 is in fact a spread of $X + Y$. It follows therefore from 1.6 b) that the restriction of κ to $X + Y$ is surjective. Hence $X^\kappa = X$. Therefore, κ is an automorphism of A. Let $Z \in \pi$. Then $Z^{\kappa^{-1}} = Z^{\kappa\kappa^{-1}} = Z$. Thus $\kappa^{-1} \in K(A, \pi)$ whence $K(A,\pi)$ is a skewfield.

Denote by σ the set of all $X + Y$ with $X, Y \in \pi$ and $X \neq Y$. We consider the geometry $\mathfrak{G}$ whose points are the elements of A, whose lines are the cosets $X + a$ with $X \in \pi$ and $a \in A$, and whose planes are the cosets $S + a$ with $S \in \sigma$ and $a \in A$. We show now that $\mathfrak{G}$ is an affine space:

1) Let a and b be distinct points of $\mathfrak{G}$. Then there is exactly one line of $\mathfrak{G}$ containing a and b.

Let $X \in \pi$ and $c \in A$ be such that $a,b \in X + c$. Then $a - b \in X$. As $a - b \neq 0$, the component X is uniquely determined. Moreover $X + c = X + b$. Hence there is at most one line joining a and b. Conversely, there exists $X \in \pi$ such that $a - b \in X$. This yields $a,b \in X + b$. This proves 1).

2) Each line of $\mathfrak{G}$ contains at least two points.

This follows from the fact that $|X| \geqslant 2$ for all $X \in \pi$.

3) Let a and b be two distinct points of $\mathfrak{G}$ and let P be a plane such that $a,b \in P$. Then the line joining a and b is contained in P.

There are $X, Y, Z \in \pi$ such that $a,b \in P = X + Y + b$ and $a \in Z + b$. Therefore $0 \neq a - b \in Z \cap (X + Y)$ whence $Z \subseteq X + Y$, as π is geometric. This yields $Z + b \subseteq X + Y + b$.

4) If a, b, c are three non-collinear points of $\mathfrak{G}$, then there exists exactly one plane P with $a, b, c \in P$.

As a, b, c are not collinear, we have $a - c \neq 0 \neq b - c$. Hence there exist $X, Y \in \pi$ with $a - c \in X$ and $b - c \in Y$. If $X = Y$, then $a, b, c \in X + c$, a contradiction. Thus $X + Y \in \sigma$ and a, b, c are on the plane $X + Y + c$. To prove uniqueness, let $S \in \sigma$ be such that $a, b, c \in S + c$. Then $a - c, b - c \in S$ whence $X + Y \subseteq S$ by 3). It follows from 29.1 that $X + Y = S$. This proves 4).

5) Each plane of $\mathfrak{G}$ contains three non-collinear points.

This is trivial.

6) There exist four points not contained in a plane.

As π is spatial, there exist $X, Y, Z \in \pi$ such that $Z \not\subseteq X + Y$. Hence $Z \cap (X + Y) = \{0\}$. Pick $x \in X$, $y \in Y$, $z \in Z$ with $x, y, z \neq 0$. Assume that $0, x, y, z$ are contained in the plane P. Then $P \in \sigma$, as $0 \in P$. It then follows from 5) that $X, Y, Z \subseteq P$. We infer from 29.1 that $X + Y = P \supseteq Z$, a contradiction.

Two lines $X + a$ and $Y + b$ of $\mathfrak{G}$ are said to be *parallel*, if they are contained in a plane and if either $X + a = Y + b$ or $(X + a) \cap (Y + b) = \emptyset$. The proof of 1.4 shows that $X + a$ and $Y + b$ are parallel, if and only if $X = Y$.

7) Let L and M be parallel lines and P and Q be planes of $\mathfrak{G}$ such that $L \subseteq P$ and $M \subseteq Q$. If P and Q have a point in common, then $P \cap Q$ contains a line.

There are $a, b \in A$ and $X \in \pi$ such that $L = X + a$ and $M = X + b$ by the remark made above. Moreover there exist $S, T \in \sigma$ with $P = S + a$ and $Q = T + b$. It follows that $X \subseteq S$ and $X \subseteq T$. Let c be a point in $P \cap Q$. Then $P = S + c$ and $Q = T + c$. Therefore $X + c \subseteq P \cap Q$. This proves 7).

8) Let L, M, N be lines of $\mathfrak{G}$ and assume that M and N are parallel to L. Then M and N are parallel to each other.

This is trivial.

9) If L is a line of $\mathfrak{G}$, then there exists a line parallel to and distinct from L.

This is trivial, too.

From 1) to 9) we conclude that $\mathfrak{G}$ is an affine space which is not an affine plane (see e.g. Lenz [1967, p. 136]). Therefore the group Δ of all homologies with centre 0 acts transitively on $X \setminus \{0\}$ for all $X \in \pi$ (see e.g. Lenz [1967, p. 141]).

The mapping τ defined by $x^{\tau(a)} = x + a$ is an isomorphism of A onto the translation group T of $\mathfrak{G}$. If $\delta \in \Delta$, then

$$0^{\delta^{-1}\tau(a)\delta} = 0^{\tau(a)\delta} = a^\delta = 0^{\tau(a^\delta)}.$$

Hence $\delta^{-1}\tau(a)\delta = \tau(a^\delta)$. This yields

$$\tau\big((a+b)^\delta\big) = \delta^{-1}\tau(a+b)\delta = \delta^{-1}\tau(a)\delta\delta^{-1}\tau(b)\delta = \tau(a^\delta + b^\delta).$$

Therefore $(a+b)^\delta = a^\delta + b^\delta$. This proves $\Delta \subseteq \mathrm{K}(A,\pi)$. As Δ acts transitively on $X\setminus\{0\}$ for all $X \in \pi$, we see that each $X \in \pi$ is a subspace of rank 1 of the $\mathrm{K}(A,\pi)$-vector space A. This yields that the π-admissible subgroups of A are exactly the subspaces of the $\mathrm{K}(A,\pi)$-vector space A. $\quad\square$

30. Rank-3-Groups

In this section we follow D. G. Higman 1964.

Let G be a permutation group which acts transitively on a set Ω. If $a \in \Omega$, then $\mathfrak{D}(a)$ denotes the set of orbits of G_a. For $\Delta \in \mathfrak{D}(a)$ we put

$$\Delta^{\pi(a)} = \{a^\gamma \mid \gamma \in G, a \in \Delta^\gamma\}.$$

30.1 Lemma. $\pi(a)$ is a permutation of $\mathfrak{D}(a)$ with $\pi(a)^2 = 1$.

PROOF. Let $\pi = \pi(a)$ and let $\Delta \in \mathfrak{D}(a)$. As G is transitive, there exists $\delta \in G$ with $a^\delta \in \Delta$. Therefore $a \in \Delta^{\delta^{-1}}$ and hence $a^{\delta^{-1}} \in \Delta^\pi$. This proves $\Delta^\pi \neq \varnothing$.

Next we prove that Δ^π is an orbit of G_a. Let $x, y \in \Delta^\pi$. Then there exist $\gamma, \delta \in G$ with $x = a^\gamma$ and $y = a^\delta$ and $a^{\gamma^{-1}}, a^{\delta^{-1}} \in \Delta$. Hence there is $\eta \in G_a$ with $a^{\gamma^{-1}} = a^{\delta^{-1}\eta}$. This implies $\gamma^{-1}\eta^{-1}\delta \in G_a$. Furthermore

$$x^{\gamma^{-1}\eta^{-1}\delta} = a^{\eta^{-1}\delta} = a^\delta = y.$$

This proves that there exists $\Delta' \in \mathfrak{D}(a)$ with $\Delta^\pi \subseteq \Delta'$.

Let $z \in \Delta'$. By the definition of Δ^π and the remark that Δ^π is not empty, there exists $\gamma \in G$ with $a^\gamma \in \Delta^\pi$ and $a \in \Delta^\gamma$. Furthermore there is $\sigma \in G_a$ with $z = a^{\gamma\sigma}$, and $a^{(\gamma\sigma)^{-1}} = a^{\sigma^{-1}\gamma^{-1}} = a^{\gamma^{-1}} \in \Delta$. Thus $z \in \Delta^\pi$ whence $\Delta^\pi = \Delta' \in \mathfrak{D}(a)$.

Pick $x \in \Delta^{\pi^2}$. Then there exists $\gamma \in G$ with $x = a^\gamma$ and $a^{\gamma^{-1}} \in \Delta^\pi$. This yields a $\delta \in G$ with $a^{\gamma^{-1}} = a^\delta$ and $a^{\delta^{-1}} \in \Delta$. We infer that $\delta\gamma \in G_a$ and

$$x = a^\gamma = a^{\delta^{-1}\delta\gamma} \in \Delta^{\delta\gamma} = \Delta.$$

This proves that $\Delta^{\pi^2} \cap \Delta \neq \varnothing$ and hence $\Delta^{\pi^2} = \Delta$, as Δ^{π^2} and Δ are orbits of G_a. $\quad\square$

30.2 Lemma. It is $\Delta^{\pi(x)\gamma} = \Delta^{\gamma\pi(x^\gamma)}$ for all $x \in \Omega$, all $\Delta \in \mathfrak{D}(x)$, and all $\gamma \in G$.

PROOF. As $\Delta^{\pi(x)\gamma}$ and $\Delta^{\gamma\pi(x^\gamma)}$ both belong to $\mathfrak{D}(x^\gamma)$, it suffices to show that their intersection is not empty. Pick $z \in \Delta^{\pi(x)\gamma}$. Then $z^{\gamma^{-1}} \in \Delta^{\pi(x)}$. There exists $\delta \in G$ with $z = x^\delta$. Hence $x^{\delta\gamma^{-1}} \in \Delta^{\pi(x)}$ whence $x^{\gamma\delta^{-1}} \in \Delta$ by 30.1. This implies $x^{\gamma\delta^{-1}\gamma} \in \Delta^\gamma$. Now $\Delta^\gamma \in \mathfrak{D}(x^\gamma)$. Therefore $z = x^\delta = (x^\gamma)^{\gamma^{-1}\delta} \in \Delta^{\gamma\pi(x^\gamma)}$. $\qquad\square$

30.3 Lemma. *If G is finite, then $|\Delta| = |\Delta^{\pi(a)}|$ for all $\Delta \in \mathfrak{D}(a)$.*

PROOF. Let $\pi = \pi(a)$ and let $\Delta = \{x_1, \ldots, x_n\}$. There exist $\gamma_1, \ldots, \gamma_n \in G$ such that γ_i^{-1} maps a onto x_i. Denote by y_i the image of a under γ_i. Then $y_i \in \Delta^\pi$ for all i. If $y_i = y_j$, then $\gamma_i \gamma_j^{-1} \in G_a$. This yields $x_j = x_i$ whence $i = j$. This proves $|\Delta| \leqslant |\Delta^\pi|$. From this and 30.1 we infer $|\Delta^\pi| \leqslant |\Delta^{\pi^2}| = |\Delta|$. $\qquad\square$

30.4 Lemma. *Let G be finite. Then $\pi(a)$ fixes an orbit of G_a other than $\{a\}$, if and only if $|G|$ is even.*

PROOF. Let $\pi = \pi(a)$, let $\{a\} \neq \Delta \in \mathfrak{D}(a)$ and assume $\Delta = \Delta^\pi$. Then there exists $\gamma \in G$ with $a^\gamma \in \Delta^\pi = \Delta$ and $a^{\gamma^{-1}} \in \Delta$. Therefore there exists $\sigma \in G_a$ with $a^{\gamma\sigma} = a^{\gamma^{-1}}$. This yields $\gamma\sigma \notin G_a$, but $(\gamma\sigma)^2 \in G_a$, as $a^{(\gamma\sigma)^2} = a^\sigma = a$. Therefore the order of $\gamma\sigma$ is even, whence $|G|$ is even.

Assume conversely that $|G|$ is even. Then there exists $\gamma \in G$ with $\gamma \neq 1 = \gamma^2$. Pick $b \in \Omega$ with $b^\gamma \neq b$. As G acts transitively on Ω, we may assume $b = a$. There exists $\Delta \in \mathfrak{D}(a)$ with $a^{\gamma^{-1}} \in \Delta$. Hence $a^{\gamma^{-1}} = a^\gamma \in \Delta \cap \Delta^\pi$, whence $\Delta = \Delta^\pi$. Finally $\Delta \neq \{a\}$, since $a \neq a^\gamma$. $\qquad\square$

From now on, we shall assume that G is finite and that $|\mathfrak{D}(a)| = 3$ for one and therefore all $a \in \Omega$, i.e., we shall assume that G is a rank-3-group.

Let $\mathfrak{D}(a) = \{\{a\}, \Delta, \Gamma\}$ and pick $\gamma \in G$. Then $\mathfrak{D}(a^\gamma) = \{\{a^\gamma\}, \Delta^\gamma, \Gamma^\gamma\}$. Assume that there exists $\delta \in G$ with $\Delta^\delta = \Gamma^\gamma$. Then Γ^γ is an orbit of G_b and of G_c where $b = a^\gamma$ and $c = a^\delta$. Moreover $|\Delta| = |\Delta^\delta| = |\Gamma^\gamma| = |\Gamma|$, whence $|\Omega| = 1 + 2|\Delta|$. The group G cannot act imprimitively, since in that case either $\{a\} \cup \Delta$ or $\{a\} \cup \Gamma$ would be a set of imprimitivity implying that $1 + |\Delta|$ divides $1 + 2|\Delta|$ which is impossible. Hence G_b and G_c are maximal subgroups of G. If $G_b \neq G_c$, then $G = \langle G_b, G_c \rangle$. As Γ^γ is an orbit of G_b and of G_c, this would imply G being intransitive. Thus $G_b = G_c$, whence $b = c$ and $\gamma\delta^{-1} \in G_a$. Moreover, $\Gamma^{\gamma\delta^{-1}} = \Delta$ and this is a contradiction. This shows that there are mappings Δ and Γ from Ω into $\bigcup_{a \in \Omega} \mathfrak{D}(a)$ such that $\mathfrak{D}(a) = \{\{a\}, \Delta(a), \Gamma(a)\}$ and $\Delta(a)^\gamma = \Delta(a^\gamma)$ as well as $\Gamma(a)^\gamma = \Gamma(a^\gamma)$ for all $a \in \Omega$ and all $\gamma \in G$.

Put $|\Omega| = n$, $|\Delta(a)| = k$ and $|\Gamma(a)| = l$. Then $n = k + l + 1$.

30.5 Lemma. a) *There exist two non-negative integers λ and μ with the property: If $a, b \in \Omega$, then $|\Delta(a) \cap \Delta(b)| = \lambda$ if $b \in \Delta(a)$, and $|\Delta(a) \cap \Delta(b)| = \mu$ if $b \in \Gamma(a)$.*

b) *There exist two non-negative integers λ_1 and μ_1 with the property*: *If* $a, b \in \Omega$, *then* $|\Gamma(a) \cap \Gamma(b)| = \lambda_1$ *if* $b \in \Gamma(a)$, *and* $|\Gamma(a) \cap \Gamma(b)| = \mu_1$ *if* $b \in \Delta(a)$.

PROOF. a) Let $a, b, a', b' \in \Omega$ and assume $b \in \Delta(a)$ and $b' \in \Delta(a')$. There exists $\gamma \in G$ with $a^\gamma = a'$. Hence $b^\gamma \in \Delta(a)^\gamma = \Delta(a^\gamma) = \Delta(a')$. Therefore there is an $\eta \in G_{a'}$ with $b^{\gamma\eta} = b'$. This implies

$$|\Delta(a) \cap \Delta(b)| = |\Delta(a)^{\gamma\eta} \cap \Delta(b)^{\gamma\eta}| = |\Delta(a^{\gamma\eta}) \cap \Delta(b^{\gamma\eta})|$$
$$= |\Delta(a') \cap \Delta(b')|.$$

This proves the existence of λ. The remaining statements are proved similarly. □

30.6 Lemma. *Let $\pi(a)$ be defined as at the beginning of this section. Then we have*:

a) *If $|G|$ is odd, then $\Delta(a)^{\pi(a)} = \Gamma(a)$ for all $a \in \Omega$.*
b) *If $|G|$ is even, then $\Delta(a)^{\pi(a)} = \Delta(a)$ and $\Gamma(a)^{\pi(a)} = \Gamma(a)$ for all $a \in \Omega$.*

PROOF. This follows immediately from 30.4. □

30.7 Lemma. *Let $a, b \in \Omega$.*

a) *If $|G|$ is odd, then $b \in \Delta(a)$ if and only if $a \in \Gamma(b)$.*
b) *If $|G|$ is even, then $b \in \Delta(a)$ if and only if $a \in \Delta(b)$.*

PROOF. a) Let $b \in \Delta(a)$. There exists $\gamma \in G$ with $a^{\gamma^{-1}} = b$. Hence $a^\gamma \in \Delta(a)^{\pi(a)} = \Gamma(a)$ by 30.6 a). Therefore $a \in \Gamma(a)^{\gamma^{-1}} = \Gamma(a^{\gamma^{-1}}) = \Gamma(b)$.
If $a \in \Gamma(b)$, then $b^\gamma = a \in \Gamma(b)$. Therefore $b^{\gamma^{-1}} \in \Gamma(b)^{\pi(b)} = \Delta(b)$ and hence $b \in \Delta(b)^\gamma = \Delta(b^\gamma) = \Delta(a)$.
b) is proved similarly.

30.8 Lemma. *If $|G|$ is even, then $\lambda_1 = l - k + \mu - 1$ and $\mu_1 = l - k + \lambda + 1$.*

PROOF. Let $a, b \in \Omega$ and assume $b \in \Gamma(a)$. By 30.7, we have $a \in \Gamma(b)$. Now $\Omega = \{a\} \cup \Delta(a) \cup \Gamma(a)$ and hence

$$\Delta(b) = \Delta(b) \cap \Omega = [\Delta(b) \cap \{a\}] \cup [\Delta(b) \cap \Delta(a)] \cup [\Delta(b) \cap \Gamma(a)]$$
$$= [\Delta(b) \cap \Delta(a)] \cup [\Delta(b) \cap \Gamma(a)].$$

Hence $k = |\Delta(b)| = \mu + |\Delta(b) \cap \Gamma(a)|$.
On the other hand

$$\Gamma(a) = \Gamma(a) \cap \Omega = [\Gamma(a) \cap \{b\}] \cup [\Gamma(a) \cap \Delta(b)] \cup [\Gamma(a) \cap \Gamma(b)]$$
$$= \{b\} \cup [\Gamma(a) \cap \Delta(b)] \cup [\Gamma(a) \cap \Gamma(b)].$$

Therefore $l = |\Gamma(a)| = 1 + |\Gamma(a) \cap \Delta(b)| + \lambda_1$. This implies $k - l = \mu - 1 - \lambda_1$, i.e., $\lambda_1 = l - k + \mu - 1$.

Assume $b \in \Delta(a)$. Then $a \in \Delta(b)$ by 30.7. Therefore

$$\Delta(b) = [\Delta(b) \cap \{a\}] \cup [\Delta(b) \cap \Delta(a)] \cup [\Delta(b) \cap \Gamma(a)]$$
$$= \{a\} \cup [\Delta(b) \cap \Delta(a)] \cup [\Delta(b) \cap \Gamma(a)].$$

Thus $k = 1 + \lambda + |\Delta(b) \cap \Gamma(a)|$. Similarly $l = |\Gamma(a) \cap \Delta(b)| + \mu_1$, whence $k - l = 1 + \lambda - \mu_1$. $\qquad\square$

30.9 Lemma. *If* $|G|$ *is odd, then* $l = k$, $n = 2k + 1$ *and* $\lambda = \mu = \lambda_1 = \mu_1$.

PROOF. As $\Delta(a)^{\pi(a)} = \Gamma(a)$, we have $k = l$ by 30.3.

Let $a, b \in \Omega$ and $b \in \Delta(a)$. Then $a \in \Gamma(b)$ by 30.7 a). Therefore $\mu = |\Delta(b) \cap \Delta(a)| = |\Delta(a) \cap \Delta(b)| = \lambda$ by 30.5. Similarly $\lambda_1 = \mu_1$.

Moreover $\Delta(b) = [\Delta(b) \cap \Delta(a)] \cup [\Delta(b) \cap \Gamma(a)]$, as $a \notin \Delta(b)$, and hence $k = \lambda + |\Delta(b) \cap \Gamma(a)|$. Similarly $\Gamma(a) = [\Gamma(a) \cap \Delta(b)] \cup [\Gamma(a) \cap \Gamma(b)]$, whence $l = |\Gamma(a) \cap \Delta(b)| + \lambda_1$. As $k = l$, we get $\lambda = \lambda_1$. $\qquad\square$

30.10 Lemma. *The following statements are equivalent:*

(j) G *is imprimitive and* $k \leqslant l$.
(ij) $G_a \neq G_{\Gamma(a)}$.
(iij) *There exist* $a, b \in \Omega$ *with* $a \neq b$ *and* $\Gamma(a) = \Gamma(b)$.

PROOF. (j) implies (ij): Let Φ be a set of imprimitivity with $a \in \Phi$. As $\Phi \backslash \{a\} \neq \emptyset$, $\Omega \backslash \{a\}$, we have either $\Phi = \{a\} \cup \Delta(a)$ or $\Phi = \{a\} \cup \Gamma(a)$. From $\Phi = \{a\} \cup \Gamma(a)$, we deduce that $l + 1$ divides $1 + k + l$ and hence k. But then $l + 1 \leqslant k \leqslant l$. Hence $\Phi = \{a\} \cup \Delta(a)$. Now $G_a \subseteq G_\Phi = G_{\Gamma(a)}$. Moreover, G_Φ acts transitively on Φ. Thus $G_a \neq G_{\Gamma(a)}$.

(ij) implies (iij): We infer from $G_{\Gamma(a)} = G_{\{a\} \cup \Delta(a)}$ and $G_a \neq G_{\Gamma(a)}$ that there exists a $\delta \in G_{\Gamma(a)}$ with $b = a^\delta \neq a$. Obviously $\Gamma(a) = \Gamma(b)$.

(iij) implies (j): As G acts transitively on Ω, the set $\Gamma(a)$ cannot be invariant under G. Hence $\langle G_a, G_b \rangle \neq G$. Therefore G_a is not maximal. This yields that G acts imprimitively. Let Φ be a set of imprimitivity with $a \in \Phi$. As above, either $\Phi = \{a\} \cup \Delta(a)$ or $\Phi = \{a\} \cup \Gamma(a)$. Assume that $\Phi = \{a\} \cup \Gamma(a)$. Then $G_{\Gamma(a)} = G_a$. There exists $\gamma \in G$ with $a^\gamma = b$. Hence $\Phi^\gamma = \{b\} \cup \Gamma(b) = \{b\} \cup \Gamma(a)$. This yields $\Phi^\gamma = \Phi$ whence $a = b$, a contradiction. Therefore $\Phi = \{a\} \cup \Delta(a)$. This yields that $k + 1$ divides $k + l + 1$ and hence l. Thus $k \leqslant l$. $\qquad\square$

30.11 Corollary. *If* G *is imprimitive and* $k \leqslant l$, *then* $k + 1$ *divides* l. *Furthermore* G *has exactly one system of imprimitivity, namely the system consisting of all the sets* $\{a\} \cup \Delta(a)$. *Finally* $G_a = G_{\Delta(a)}$ *for all* $a \in \Omega$, *and* $\Delta(a) = \Delta(b)$ *implies* $a = b$.

30.12 Corollary. *If G is imprimitive, then G acts doubly transitively on its system of imprimitivity.*

30.13 Corollary. *Every rank-3-group of odd order acts primitively.*

We consider now the incidence structure (Ω, Ω, I) where $a \, I \, b$, if and only if $a \in \Delta(b)$. Obviously, G induces a group of automorphisms on (Ω, Ω, I). As G acts transitively on Ω, we see that (Ω, Ω, I) is a tactical configuration.

Let $c \in \Omega$. We count the pairs $(a, b) \in \Omega \times \Omega$ with $a \, I \, b, c$ and $b \neq c$. As (Ω, Ω, I) is a tactical configuration with k points per block, there also are k blocks through a given point. Let $a \in \Delta(c)$. Then there are $k - 1$ blocks b with $a \in \Delta(b)$ and $b \neq c$. Hence there are $k(k - 1)$ such pairs (a, b). On the other hand if $b \neq c$, then there are λ points a with $a \in \Delta(b) \cap \Delta(c)$ if $b \in \Delta(c)$ and μ such a's, if $b \in \Gamma(c)$. Hence the number of (a, b)'s is also $\lambda k + \mu l$. Thus $\lambda k + \mu l = k(k - 1)$. This yields

30.14 Lemma. $\mu l = k(k - \lambda - 1)$.

From 30.14 and 30.9 we get:

30.15 Corollary. *If $|G|$ is odd, then $\lambda = \mu = \frac{1}{2}(k - 1)$.*

30.16 Corollary. *The following statements are equivalent:*

(a) $|G| = 3$ or G is imprimitive and $k \leqslant l$.
(b) $\mu = 0$.
(c) $\lambda = k - 1$.

PROOF. (a) implies (b): Let G be imprimitive and $k \leqslant l$. Pick $a, b \in \Omega$. Then $\{a\} \cup \Delta(a)$ and $\{b\} \cup \Delta(b)$ are sets of imprimitivity by 30.11. If $b \in \Gamma(a)$, then $b \notin \{a\} \cup \Delta(a)$ whence $[\{a\} \cup \Delta(a)] \cap [\{b\} \cup \Delta(b)] = \emptyset$. Therefore $\Delta(a) \cap \Delta(b) = \emptyset$, if $b \in \Gamma(a)$. This proves $\mu = 0$.

If $|G| = 3$, then $3 = n = k + l + 1$. Hence $k = 1$. Thus $\mu = 0$ by 30.15.

(b) implies (c): This follows immediately from 30.14.

(c) implies (a): If $\lambda = 0$, then $k = 1$, as $\lambda = k - 1$ by assumption. Therefore $\Delta(a) = \{b\}$ for some $b \neq a$. In this case $G_a = G_{a,b} = G_b$. If $l = 1$, then $|G| = 3$. If $l \geqslant 2$, then $k = 1 < 2 \leqslant l$. Moreover a and b are the only fixed elements of $G_a = G_b$. Hence $\mathfrak{N}_G(G_{a,b}) \neq G_a$. Therefore G is imprimitive. We may thus assume that $\lambda > 0$. Then it follows from $\lambda = k - 1$ and 30.15 that $|G|$ is even. Let $b \in \Delta(a)$. Then $|\Delta(a) \cap \Delta(b)| = \lambda = k - 1$. As $|G|$ is even, $a \in \Delta(b)$ by 30.7 b). Since $a \notin \Delta(a)$, we therefore have that $\{a\} \cup (\Delta(a) \cap \Delta(b)) = \Delta(b)$. Therefore $\{b\} \cup \Delta(b) \subseteq \{a\} \cup \Delta(a)$, whence $\{a\} \cup \Delta(a) = \{b\} \cup \Delta(b)$. This implies $\Gamma(a) = \Gamma(b)$. Therefore, by 30.10, the group G is imprimitive and $k \leqslant l$. $\square$

30.17 Corollary. *If* $|G| > 3$, *then the following statements are equivalent*:

(a′) *G is imprimitive and* $l \leqslant k$.
(b′) $\lambda = k - l - 1$.
(c′) $\mu = k$.

PROOF. (a′) implies (b′): As G is imprimitive, it has even order. Furthermore $\mu_1 = 0$ by 30.15. By 30.8, we then have $0 = l - k + \lambda + 1$. Thus $\lambda = k - l - 1$.

(b′) implies (c′): By 30.14, we have

$$\mu l = k(k - \lambda - 1) = k(k - k + l + 1 - 1) = kl$$

whence $\mu = k$.

(c′) implies (a′): If $\mu = \lambda$, then $kl = k(k - k - 1)$ and thus $l = -1$, a contradiction. Thus $\mu \neq \lambda$. Therefore $|G|$ is even by 30.15. Hence $\lambda_1 = l - k + \mu - 1 = l - 1$ by 30.8. Thus G is imprimitive with $l \leqslant k$ by 30.16.

30.18 Corollary. *If* $|G| > 3$, *then G is primitive, if and only if* $\mu \neq 0, k$.

31. A Characterization of the Lüneburg Planes

The following theorem is due to R. Liebler 1972.

31.1 Theorem. *Let* $q = 2^{2\alpha+1} \geqslant 8$ *and let* $\mathfrak{A}$ *be a translation plane of order* q^2. *If* $\mathfrak{A}$ *admits a collineation group G isomorphic to* S(q), *then* $\mathfrak{A}$ *is a Lüneburg plane*.

PROOF. Let T be the translation group of $\mathfrak{A}$ and put $H = TG$. Then $T \cap G = \{1\}$, since G is simple. As T acts transitively on the set of points, we have $H_O \cong G$ where O is some point of $\mathfrak{A}$. Thus we may assume that $H_O = G$. Hence G fixes a non-incident point-line pair in the projective closure $\mathfrak{P}$ of $\mathfrak{A}$, namely (O, l_∞). By 28.11, G therefore has a point orbit $\mathfrak{p}_1$ with $|\mathfrak{p}_1| = (q^2 + 1)(q - 1)$ and a point orbit $\mathfrak{p}_2$ of length $(q^2 + 1)q(q - 1)$. In particular, H acts as a rank-3-group on the affine plane $\mathfrak{A}$. Moreover, H has two line orbits $\mathfrak{g}_1$ and $\mathfrak{g}_2$ with $|\mathfrak{g}_1| = (q^2 + 1)(q - 1)$ and $|\mathfrak{g}_2| = (q^2 + 1)q(q - 1)$. If l is a line through O, then H_l acts as a rank-3-group on the set Ω of points on l by 16.2 a). The lengths of the orbits of $H_{l,O}$ are 1, $q - 1$ and $q(q - 1)$ respectively. Let $|\Delta_l(O)| = q - 1$ and $|\Gamma_l(O)| = q(q - 1)$; also let λ and μ be the intersection numbers defined in 30.5. Then $0 \leqslant \lambda \leqslant q - 2$. By 30.14, we have $\mu q(q - 1) = (q - 1)(q - 2 - \lambda)$ and hence $\mu q = q - 2 - \lambda$ whence $\lambda + 2 \equiv 0 \bmod q$. Furthermore $2 \leqslant \lambda + 2 \leqslant q$ and therefore $\lambda + 2 = q$. This yields $\mu = 0$. It follows from 30.16 that H_l acts imprimitively on Ω. Thus by 30.11, $\Phi_l = \{O\} \cup \Delta_l(O)$ is a set of imprimitivity of H_l.

Put $T_{l,l} = \{\tau \mid \tau \in T, \ \Phi_l^\tau = \Phi_l\}$. Then $T_{l,l}$ is a subgroup of order q contained in $T(l \cap l_\infty)$.

As G acts transitively on l_∞, we infer from $|\mathfrak{g}_1| = (q^2 + 1)(q - 1)$ that each point of l_∞ is on exactly $q - 1$ lines of $\mathfrak{g}_1$ and on $q(q - 1)$ lines of $\mathfrak{g}_2$. It follows that H_P acts as a rank-3-group on the set of affine lines through P for all $P \mathrel{I} l_\infty$. As above, we see that H_P acts imprimitively and that $\Psi_{O,P} = \{OP\} \cup \{h \mid h \in \mathfrak{g}_1, P \mathrel{I} h\}$ is a set of imprimitivity. Put $\Sigma_P = \{\tau \mid \tau \in T, \ \Psi_{O,P}^\tau = \Psi_{O,P}\}$. Then $T(P) \subseteq \Sigma_P$ and hence $|\Sigma_P| = q^3$.

Let l and m be two distinct lines through O and put $P = l \cap l_\infty$ and $T_{l,m} = \Sigma_P \cap T(m \cap l_\infty)$. Then $|T_{l,m}| = q$.

Let π be the set of all $T_{l,l}$ and $T_{l,m}$. We shall show that π is a partition of T.

Let $l_1, \ldots, l_{q-1}$ be the lines of $\mathfrak{g}_1$ parallel to l and denote by $X_1, \ldots, X_{q-1}$ the points of $\mathfrak{p}_1$ which are on m. Then

$$T_{m,m} = \{1\} \cup \{\tau \mid \tau \in T \text{ and } O^\tau \in \{X_1, \ldots, X_{q-1}\}\}$$

and

$$T_{l,m} = \{1\} \cup \{\tau \mid \tau \in T \text{ and } O^\tau \in \{l_1 \cap m, \ldots, l_{q-1} \cap m\}\}.$$

By 28.10 a) we obtain $T_{m,m} \cap T_{l,m} = \{1\}$. Moreover, for $\gamma \in G$ we have $\gamma^{-1} T_{l,m} \gamma = T_{l^\gamma, m^\gamma}$ for all l and m distinct or not. As G has two point orbits other than $\{O\}$, we therefore obtain that $T = \bigcup_{X \in \pi} X$.

Furthermore, if $l \neq m$ then $\gamma^{-1} T_{l,m} \gamma = T_{l,m}$, if and only if

$$\gamma \in \mathfrak{Z}(\Pi) G_{l,m}$$

where Π is the Sylow 2-subgroup of G_m. This is seen as follows: We have $\gamma^{-1} T_{l,m} \gamma = T_{l,m}$ if and only if

$$T_{l,m} = \{1\} \cup \{\tau^\gamma \mid \tau \in T \text{ and } O^{\tau^\gamma} \in \{l_1 \cap m, \ldots, l_{q-1} \cap m\}\}$$

and hence if and only if

$$\{l_1 \cap m, \ldots, l_{q-1} \cap m\}^\gamma = \{l_1 \cap m, \ldots, l_{q-1} \cap m\}.$$

From this and $\{l_1 \cap m, \ldots, l_{q-1} \cap m\} \subseteq \mathfrak{p}_2$ we obtain the above assertion. Hence there are $(q^2 + 1)q$ groups of type $T_{l,m}$ with $l \neq m$. As there are $q^2 + 1$ groups of type $T_{l,l}$, we obtain $|\pi| = (q^2 + 1)q + q^2 + 1 = (q^2 + 1)(q + 1)$. Hence $|T| \leqslant 1 + (q^2 + 1)(q + 1)(q - 1) = q^4 = |T|$. Thus π is indeed a partition of T.

Next we prove that π is geometric. If $l \neq m$, then $T_{l,m} T_{m,m} = T(m \cap l_\infty)$. Hence $T_{l,m} T_{m,m}$ is π-admissible.

Again let $l \neq m$. Then $|G_{(l,m)}| = 2(q - 1)$. Hence there exist $q - 1$ elations in G which switch l and m. The axes of these elations are mutually distinct. Let σ be one such elation and denote by a its axis. Then

$$T_{l,a} = \{1\} \cup \{\tau \mid \tau \in T \text{ and } O^\tau \in \{l_1 \cap a, \ldots, l_{q-1} \cap a\}\}$$

$$= \{1\} \cup \{\tau \mid \tau \in T \text{ and } O^\tau \in \{l_1^\sigma \cap a, \ldots, l_{q-1}^\sigma \cap a\}\} = T_{m,a}.$$

This implies $T_{l,a} \subseteq T_{l,m}T_{m,l}$. From this we infer

$$T_{l,m}T_{m,l} = T_{l,m} \cup T_{m,l} \cup \bigcup_a T_{l,a}$$

where the union is taken over all those a's which are axes of elations in $G_{\{l,m\}}$. Hence $T_{l,m}T_{m,l}$ is π-admissible.

Next we remark: $T_{l,m}T_{m,l} = T_{u,v}T_{v,u}$ if and only if $\{l,m\} = \{u,v\}$. Assume first $\{u,v\} \cap \{l,m\} = \varnothing$. As $T_{u,v} \subseteq T_{l,m}T_{m,l}$, we have $T_{u,v} = T_{l,v}$ by what we have just proved. Let $\gamma \in G$ with $u^\gamma = l$ and $v^\gamma = v$. Then γ is in the normalizer of $T_{u,v}$ as well as in the normalizer of $T_{l,v}$. Therefore $\gamma \in \mathfrak{Z}(\Pi)G_{u,v} \cap \mathfrak{Z}(\Pi)G_{l,v} = \mathfrak{Z}(\Pi)$ where Π is the Sylow 2-subgroup of G_v. Hence $v^\gamma = v$, $u^\gamma = l$ and $l^\gamma = u$. Moreover, as $T_{m,l} \subseteq T_{u,v}T_{v,u}$, we have $T_{m,l} = T_{u,l}$. There also exists an involution $\delta \in G$ with $u^\delta = v$ and $l^\delta = l$. Hence $v^{\delta\gamma} = u^\gamma = l$, $l^{\delta\gamma} = l^\gamma = u$, $u^{\delta\gamma} = v^\gamma = v$. Thus 3 divides the order of $\delta\gamma$ and hence of G contradicting 21.10. Hence $\{u,v\} \cap \{l,m\} \neq \varnothing$. We may assume wlog that $v = m$. Then $T_{l,m} = T_{u,m}$. Assume $u \neq l$. From $T_{m,u} \subseteq T_{l,m}T_{m,l}$ we deduce that there exists an involution $\gamma \in G$ with $u^\gamma = u$ and $m^\gamma = l$, and from $T_{m,l} \subseteq T_{m,u}T_{u,m}$ we deduce that there also exists an involution δ with $l^\delta = l$ and $m^\delta = u$. Hence $u^{\gamma\delta} = u^\delta = m$, $m^{\gamma\delta} = l^\delta = l$, $l^{\gamma\delta} = m^\delta = u$. We thus reach again the contradiction that 3 divides the order of G. Hence $\{u,v\} = \{m,l\}$.

We have thus shown that there exist $\frac{1}{2}q^2(q^2+1)$ subgroups of the form $T_{l,m}T_{m,l}$ all of which are π-admissible. Furthermore, no subgroup $T_{l,m}T_{m,l}$ contains a group of the form $T_{a,a}$.

Let l and m be two distinct lines through O. Put $S = T_{l,m}T(l \cap l_\infty)$. If x is a line through O distinct from l then $S \cap T(x \cap l_\infty) = T_{l,x}$. Hence S is π-admissible. Furthermore G_l acts transitively on the set of all $T_{l,x}$ as well as on the set of all $T_{l,x}T_{x,l}$. Therefore the incidence structure whose points are the $T_{l,x}$ and whose blocks are the $T_{l,x}T_{x,l}$ is a tactical configuration with the parameters $v = b = q^2$ and $r = k = q$.

Let $Y_1, \ldots, Y_q$ be those $T_{l,y}T_{y,l}$ which contain $T_{l,x}$. Then

$$\left| S \setminus \bigcup_{i=1}^q (Y_i \setminus T_{l,x}) \right| = q^3 - q(q^2 - q) = q^2.$$

Moreover $S \setminus \bigcup_{i=1}^q (Y_i \setminus T_{l,x})$ is a union of components of π. Finally

$$(T_{l,x}T_{l,l}) \cap Y_i = T_{l,x}(T_{l,l} \cap Y_i) = T_{l,x}.$$

As $T_{l,l} \subseteq S$, we thus obtain

$$(T_{l,x}T_{l,l}) \cap \left[S \setminus \bigcup_{i=1}^q (Y_i \setminus T_{l,x}) \right] = T_{l,x}T_{l,l}.$$

Hence $T_{l,x}T_{l,l}$ is also π-admissible.

If $T_{l,x}T_{l,l} = T_{m,y}T_{m,m}$, then $l = m$, as there are q groups $T_{l,y}$ contained in $T_{l,x}T_{l,l}$. Using once again the fact that $T_{l,x}T_{l,l}$ contains exactly q

subgroups of the form $T_{l,\,y}$ we see that $q(q^2 + 1)$ is the number of groups of the form $T_{l,\,x}T_{l,\,l}$.

Consider the group $R_{m,\,l} = T_{m,\,m}T(l \cap l_\infty)$. Let Σ be a Sylow 2-subgroup of G_l. As $\mathfrak{Z}(\Sigma)$ fixes all the lines through $l \cap l_\infty$ and as $\mathfrak{Z}(\Sigma)$ centralizes $T(l \cap l_\infty)$, we have $T_{m^\sigma,\,m^\sigma} \subseteq R_{m,\,l}$ for all $\sigma \in \mathfrak{Z}(\Sigma)$. Now $T_{m^\sigma,\,m^\sigma}T_{m^\sigma,\,l}$ $= T_{m^\sigma,\,m^\sigma}T_{m,\,l}$ is π-admissible for all $\sigma \in \mathfrak{Z}(\Sigma)$. This implies that $R_{m,\,l}\setminus T(l \cap l_\infty)$ is the union of the sets $T_{m^\sigma,\,m^\sigma}T_{m,\,l}\setminus T_{m,\,l}$ where σ ranges over $\mathfrak{Z}(\Sigma)$. Hence $R_{m,\,l}$ is π-admissible. From $R_{l,\,m}R_{m,\,l} \supseteq T(l \cap l_\infty)T(m \cap l_\infty) = T$ and

$$q^6 = |R_{l,\,m}||R_{m,\,l}| = |T||R_{l,\,m} \cap R_{m,\,l}| = q^4|R_{l,\,m} \cap R_{m,\,l}|$$

we conclude that $R_{l,\,m} \cap R_{m,\,l} = T_{l,\,l}T_{m,\,m}$. Hence $T_{l,\,l}T_{m,\,m}$ is also π-admissible.

Let l, m and x be three distinct lines through O. Then $T_{x,\,x} \cap (T_{l,\,l}T_{m,\,m}) = \{1\}$: For assume $T_{x,\,x} \cap (T_{l,\,l}T_{m,\,m}) \neq \{1\}$. Then there exist lines l_1, m_1 with l_1 parallel to l and m_1 parallel to m such that $l_1 \cap m_1 \in \mathfrak{p}_1$. By 28.10, we have therefore $l_1, m_1 \in \mathfrak{g}_2$ and $|l_1 \cap \mathfrak{p}_1| = q = |m_1 \cap \mathfrak{p}_1|$. Hence there exist involutions ρ and σ in G with $l^\rho = l$, $m^\rho = x$ and $m^\sigma = m$, $l^\sigma = x$. This implies that 3 divides $|G|$, a contradiction. Thus $T_{x,\,x} \cap (T_{l,\,l}T_{m,\,m}) = \{1\}$. Therefore $T_{l,\,l}T_{m,\,m} = T_{u,\,u}T_{v,\,v}$, if and only if $\{u,v\} = \{l,m\}$.

Now there are

$(q^2 + 1)q(q + 1)$ pairs $X, Y \in \pi$ with $XY \in \kappa$, where $\kappa = \{T(P) | P \text{ I } l_\infty\}$,

$\frac{1}{2}q^2(q^2 + 1)q(q + 1)$ pairs $X, Y \in \pi$ with $XY = T_{l,\,m}T_{m,\,l}$,

$\frac{1}{2}q^2(q^2 + 1)q(q + 1)$ pairs $X, Y \in \pi$ with $XY = T_{l,\,l}T_{m,\,m}$,

$q(q^2 + 1)q(q + 1)$ pairs $X, Y \in \pi$ with $XY = T_{l,\,m}T_{l,\,l}$.

Hence there are $q(q + 1)(q^2 + 1)(q^2 + q + 1)$ pairs $X, Y \in \pi$ with $X \neq Y$ such that XY is π-admissible. The total number of pairs $X, Y \in \pi$ with $X \neq Y$ is also $q(q + 1)(q^2 + 1)(q^2 + q + 1)$ as $|\pi| = (q^2 + 1)(q + 1)$. Hence π is geometric.

It follows from 29.2 that T is a vector space of rank 4 over $GF(q)$. Hence $GF(q)$ is the kernel of $\mathfrak{A}$. Furthermore, it follows from 1.10 and the simplicity of G that G acts as a group of linear transformations on the $GF(q)$-vector space T. Therefore $\mathfrak{A}$ is a Lüneburg plane by 27.3. $\qquad\square$

CHAPTER V

Planes Admitting Many Shears

Next we collect results about unitary groups, we prove a characterization of A_5 which I extracted from J. Assion's Diplomarbeit, and we give a characterization of Galoisfields of odd characteristic due to A. A. Albert. All this is done in order to prove Hering's & Ostrom's theorem on collineation groups of translation planes generated by shears.

I would like to thank J. Assion for allowing me to make use of his unpublished Diplomarbeit.

32. Unitary Polarities of Finite Desarguesian Projective Planes and Their Centralizers

We start with a remarkable theorem of R. Baer 1946 b.

32.1 Theorem. *Let π be a polarity of a finite projective plane. Then π has at least one absolute point.*

PROOF (Ryser). Let $P_1, \ldots, P_v$ be the points of the plane $\mathfrak{P}$ and lable the lines of $\mathfrak{P}$ by $l_i = P_i^\pi$. Furthermore, define the incidence matrix $A = (a_{ij})$ of $\mathfrak{P}$ by $a_{ij} = 1$, if and only if $P_i \text{ I } l_j$, and $a_{ij} = 0$ otherwise. Then we have $a_{ij} = 1$, if and only if $P_i \text{ I } P_j^\pi$ which is equivalent to $P_j \text{ I } P_i^\pi$, as $\pi^2 = 1$. Therefore $a_{ij} = 1$ if and only if $a_{ji} = 1$. Hence A is symmetric.

If n is the order of the projective plane $\mathfrak{P}$, then $A^2 = AA^t = nE + J$ where E is the $(v \times v)$-identity matrix and J is the matrix all of whose entries are 1. It is easily seen that $(n + 1)^2$ is a simple eigenvalue and that n is an eigenvalue of multiplicity $n^2 + n$ of A^2.

The row sums of A are all equal to $n + 1$. Hence $n + 1$ is an eigenvalue of A. Let $e_1, \ldots, e_{v-1}$ be the remaining eigenvalues of A. As A is symmetric, A is diagonalizable. Hence e_i^2 are eigenvalues of A^2. Therefore $e_i^2 = n$ whence $e_i \in \{\sqrt{n}, -\sqrt{n}\}$ for all i. Let a be the multiplicity of $\sqrt{n}$ as an eigenvalue of A and let b be the multiplicity of $-\sqrt{n}$. Then $\text{trace}(A) = n + 1 + (a - b)\sqrt{n}$. Assume $\text{trace}(A) = 0$. Then $a - b \neq 0$, as $n + 1 \geqslant 3$. Hence $\sqrt{n}$ is rational. This yields $\sqrt{n} \in \mathbb{Z}$ whence $n + 1$ is divisible by $\sqrt{n}$ which is possible only for $n = 1$. Thus $\text{trace}(A) \neq 0$. Now $\text{trace}(A) = \sum_{i=1}^{v} a_{ii}$ and $a_{ii} = 1$ if and only if $P_i \text{ I } P_i^\pi$, i.e., if and only if P_i is absolute. As $\text{trace}(A) \neq 0$, we have that the number of absolute points is not nought. $\square$

32.2 Corollary (Baer 1946 b). *If π is a polarity of a finite projective plane of order n and if n is not a square, then π has exactly $n + 1$ absolute points.*

PROOF. As n is not a square, $\sqrt{n}$ is not rational. Therefore $\text{trace}(A) = n + 1 + (a - b)\sqrt{n}$ yields $a = b$, i.e., $\text{trace}(A) = n + 1$. $\square$

32.3 Lemma. *Let π be a polarity of a projective plane. If P is an absolute point of π, then P is on precisely one absolute line of π.*

PROOF. As P is absolute, $P \text{ I } P^\pi = l$. Then $l^\pi = P^{\pi^2} = P \text{ I } l$, whence l is an absolute line passing through P. Let $P \text{ I } m$ and $m^\pi \text{ I } m$. Assume that $l \neq m$. Then $Q = m^\pi \neq l^\pi = P$. Furthermore, $P = l \cap m$ and hence $l = P^\pi = l^\pi m^\pi = PQ$. On the other hand $Q = m^\pi \text{ I } m$. Therefore $m = PQ = l$, a contradiction. $\square$

Let V be a vector space of rank 3 over the skewfield K and let $\mathfrak{P}$ be the projective plane whose points are the subspaces of rank 1 of V and whose lines are the subspaces of rank 2. If π is a polarity of $\mathfrak{P}$, then it follows from 32.3 that not all points of $\mathfrak{P}$ are absolute points of π. Hence, by Baer [1952, Chapt. IV. In particular Prop. 1, p. 110], there exists an involutory antiautomorphism α of K and a mapping $f: V \times V \to K$ with the properties:

(1) $f(x + y, z) = f(x, z) + f(y, z)$ for all $x, y, z \in V$,
(2) $f(x, yk) = f(x, y)k$ for all $x, y \in V$ and all $k \in K$,
(3) $f(x, y) = f(y, x)^\alpha$ for all $x, y \in V$,

such that $X^\pi = \{y \mid y \in V, f(X, y) = 0\}$ for all subspaces X of V.
From $\{0\} = V^\pi$ we infer

(4) $f(V, y) = 0$ implies $y = 0$.

Let α be an involutory antiautomorphism of K and let f be a mapping from $V \times V$ in K such that the pair (α, f) satisfies (1) through (4). Then we say that (α, f) is a non-degenerate symmetric semibilinear form. Each non-degenerate symmetric semibilinear form (α, f) induces a polarity π on $\mathfrak{P}$ defined by $X^\pi = \{y \mid y \in V, f(X, y) = 0\}$ for all subspaces X of V. We shall say that π is represented by the pair (α, f).

32.4 Lemma. *Let the polarity π be represented by (α, f) and let $l = uK \oplus vK$ be a line such that $V = l \oplus l^\pi$. Then there exist vectors $u', v' \in l$ such that $l = u'K \oplus v'K$, $v'K = L \cap (vK)^\pi$ and $(u' + v'x^\alpha)K = l \cap [(u + vx)K]^\pi$ for all $x \in K$.*

PROOF. For a point P on l we define P^σ by $P^\sigma = l \cap P^\pi$. As l is not absolute, $P^\pi \neq l$. Thus P^σ is a point. Furthermore

$$P^{\sigma^2} = l \cap (l \cap P^\pi)^\pi = l \cap (l^\pi + P) = (l \cap l^\pi) + P = P,$$

since $P \subseteq l$ and $l \cap l^\pi = \{0\}$. This shows that σ is an involutory permutation of the set of points which are on l. The fixed points of σ are the absolute points of π which lie on l.

uK, vK, and $(u + v)K$ are three distinct points on l. Hence $(uK)^\sigma$, $(vK)^\sigma$, and $((u + v)K)^\sigma$ are three distinct points on l. Therefore there exist vectors $u'', v'' \in l$ with $(uK)^\sigma = u''K$, $(vK)^\sigma = v''K$, and $((u + v)K)^\sigma = (u'' + v'')K$. Put $t = f(v, u'')$ and assume $t = 0$. Then $u''K \subseteq (vK)^\pi \cap l = v''K$, a contradiction. Thus $t \neq 0$. Set $u' = u''t^{-1}$ and $v' = v''t^{-1}$. Then we have $(uK)^\sigma = u'K$, $(vK)^\sigma = v'K$ and $((u + v)K)^\sigma = (u' + v')K$. Moreover $f(u, u') = f(v, v') = f(u + v, u' + v') = 0$ and $f(v, u') = 1$. Hence

$$0 = f(u + v, u' + v') = f(u, u') + f(u, v') + f(v, u') + f(v, v') = f(u, v') + 1,$$

i.e., $f(u, v') = -1$. Therefore

$$f(u + vx, u' + v'x^\alpha) = f(u, u') + f(u, v')x^\alpha + x^\alpha f(v, u') + x^\alpha f(v, v')x^\alpha = 0.$$

□

32.5 Lemma. *Let π be a polarity represented by (α, f). If l is a line all of whose points are absolute, then $\alpha = 1$.*

PROOF. l is not absolute by the dual of 32.3. Hence $V = l \oplus l^\pi$. Let $l = uK \oplus vK$ and choose u', v' according to 32.4. Since $(uK)^\sigma = uK$, $(vK)^\sigma = vK$ and $((u + v)K)^\sigma = (u + v)K$, there exist $\lambda, \mu, \nu \in K^*$ with $(u + v)\lambda = u' + v'$, $u' = u\mu$, and $v' = v\nu$. As u and v are linearly independent, we obtain $\lambda = \mu = \nu$. Moreover, for each $x \in K$ there exists $\rho \in K^*$ with $(u + vx)\rho = u' + v'x^\alpha$, as $\sigma = 1$. This implies $u\rho + vx\rho = u\lambda + v\lambda x^\alpha$, whence $\rho = \lambda$ and $x^\alpha = \lambda^{-1}x\lambda$ for all $x \in K$. Therefore α is an automorphism as well. From this we deduce that K is commutative. Hence $x^\alpha = \lambda^{-1}x\lambda = x$. □

The polarity π is called *unitary*, if $\alpha \neq 1$. If K is finite and $\alpha \neq 1$, then $K = \mathrm{GF}(q^2)$. Conversely, if $K = \mathrm{GF}(q^2)$, then K admits an (anti-) automorphism α with $\alpha^2 = 1 \neq \alpha$, namely the one defined by $x^\alpha = x^q$. It is then easily seen that the desarguesian plane over $GF(q^2)$ admits a unitary polarity. Therefore, a finite desarguesian plane admits a unitary polarity, if and only if its order is a square.

Let π be a unitary polarity of the desarguesian plane over $GF(q^2)$. Then π has absolute points by 32.1. Let P be an absolute point and let l be a line through P other than P^π. Then $V = l \oplus l^\pi$ by 32.3. Let π be represented by (α, f). As $\alpha \neq 1$, there exists a point Q on l with $Q \neq Q^\sigma$. Since $P = P^\sigma$, the points P, Q, Q^σ are mutually distinct. Hence there exist vectors u and v with $Q = uK$, $Q^\sigma = vK$, and $P = (u + v)K$. Pick u' and v' according to 32.4. Then $u'K = (uK)^\sigma = vK$, $v'K = (vK)^\sigma = uK$, and $(u' + v')K = ((u + v)K)^\sigma = (u + v)K$. Hence $u' = v\lambda$, $v' = u\mu$, and $u' + v' = (u + v)\nu$ with $\lambda, \mu, \nu \in K^*$. It follows that $\lambda = \mu = \nu$.

The point $(u + vx)K$ is absolute, if and only if there exists $\rho \in K^*$ with $(u + vx)\rho = u' + v'x^\alpha = v\lambda + u\lambda x^\alpha$; hence if and only if $x\rho = \lambda$ and $\rho = \lambda x^\alpha$. This implies $1 = x^{\alpha+1}$, as $\lambda \neq 0$. Conversely, if $1 = x^{\alpha+1}$, then there exists $\rho \in K$ with $x\rho = \lambda$ whence $\rho = x^{\alpha+1}\rho = x^\alpha\lambda = \lambda x^\alpha$. As $x^{1+\alpha} = x^{1+q}$, there are exactly $q + 1$ elements $x \in K$ satisfying $x^{1+\alpha} = 1$. Therefore l carries exactly $q + 1$ absolute points of π.

Let $\mathfrak{U}$ be the incidence structure consisting of the absolute points of π and those lines of $\mathfrak{P}$ which carry at least two absolute points. Each line belonging to $\mathfrak{U}$ carries precisely $q + 1$ points of $\mathfrak{U}$, as we have just seen. Therefore $\mathfrak{U}$ is a block design with parameters v, b, $k = q + 1$, r, $\lambda = 1$. By 32.3 we see that $r = q^2$. Hence $v - 1 = \lambda(v - 1) = r(k - 1) = q^3$ whence $v = q^3 + 1$. As $vr = bk$, we finally get $b = q^2(q^2 - q + 1)$. The design $\mathfrak{U}$ is called the *unital* belonging to π. Thus we have proved:

32.6 Theorem. *If $\mathfrak{U}$ is the unital belonging to the unitary polarity π, then $\mathfrak{U}$ is a block design with parameters $v = q^3 + 1$, $b = q^2(q^2 - q + 1)$, $r = q^2$, $k = q + 1$, and $\lambda = 1$.*

32.7 Theorem. *If $\mathfrak{P}$ is a desarguesian projective plane of order q^2, then all unitary polarities of $\mathfrak{P}$ are conjugate.*

PROOF. Let π be a unitary polarity of $\mathfrak{P}$. By 32.3 there exists a point $P = b_1K$ which is not absolute. Therefore $f(b_1, b_1) \neq 0$. From $f(b_1, b_1)^\alpha = f(b_1, b_1)$ we infer $f(b_1, b_1) \in GF(q)$. As each element of $GF(q)$ is the norm of an element in $K = GF(q^2)$, there exists $\lambda \in GF(q^2)$ with $\lambda^{1+\alpha} = f(b_1, b_1)^{-1}$. Therefore $f(b_1\lambda, b_1\lambda) = \lambda^{1+\alpha}f(b_1, b_1) = 1$. We may therefore assume that $f(b_1, b_1) = 1$.

Let $\mathfrak{U}$ be the unital belonging to π. Then $(b_1K)^\pi$ is a line of $\mathfrak{U}$. This follows from 32.6, as $q^3 + 1 + q^2(q^2 - q + 1) = q^4 + q^2 + 1$. Therefore $(b_1K)^\pi$ carries $q + 1$ absolute points. Hence there exists $b_2 \in (b_1K)^\pi$ with $f(b_2, b_2) \neq 0$. As the above argument shows, we even find such a b_2 with $f(b_2, b_2) = 1$. Obviously $f(b_1, b_2) = 0 = f(b_2, b_1)$.

There are exactly $q + 1$ elements $x \in K$ with $x^{1+\alpha} = -1$. Let x be one such element. Then $f(b_1x + b_2, b_1x + b_2) = x^{1+\alpha} + 1 = 0$. Hence the line

$b_1K + b_2K$ is not absolute by the dual of 32.3. It follows that $(b_1K)^\pi \cap (b_2K)^\pi$ is not absolute. Hence there exists $b_3 \in (b_1K)^\pi \cap (b_2K)^\pi$ with $f(b_3, b_3) = 1$. Now $\{b_1, b_2, b_3\}$ is a basis of V and we have

$$f\left(\sum_{i=1}^{3} b_i x_i, \sum_{i=1}^{3} b_i y_i\right) = \sum_{i=1}^{3} x_i^\alpha y_i. \qquad \square$$

Let aK be a non-absolute point of π. Then $f(a, a) \neq 0$. Hence we may assume $f(a, a) = 1$. Let σ be a homology with centre aK and $\pi\sigma = \sigma\pi$. Then $(aK)^\pi$ is the axis of σ. Therefore σ is induced by the mapping (which we also call σ) defined by $x^\sigma = x + a\lambda f(a, x)$ where λ is a suitable element in K. As σ is linear and as π is centralized by σ, there exists $\mu \in K$ with $\mu f(x, y) = f(x^\sigma, y^\sigma)$ for all $x, y \in V$. This yields

$$\mu f(x, y) = f(x, y) + \lambda f(x, a) f(a, y) + \lambda^\alpha f(a, x)^\alpha f(a, y)$$
$$+ \lambda^{1+\alpha} f(a, x)^\alpha f(a, y).$$

There exist $x, y \in V$ such that $f(x, y) \neq 0 = f(x, a)$. Therefore $f(x, y) = \mu f(x, y)$ whence $\mu = 1$. Thus we obtain

$$0 = (\lambda + \lambda^\alpha + \lambda^{1+\alpha}) f(x, a) f(a, y)$$

for all $x, y \in V$. Putting $x = y = a$ yields $0 = \lambda + \lambda^\alpha + \lambda^{1+\alpha}$. This is equivalent to $(\lambda + 1)^{1+\alpha} = 1$. Conversely, if $(\lambda + 1)^{1+\alpha} = 1$, then $x^\sigma = x + a\lambda f(a, x)$ defines a homology with centre aK and $\pi\sigma = \sigma\pi$. As there exist exactly $q + 1$ such λ's, the group $\Delta(aK)$ of all homologies with centre aK which centralize π is cyclic of order $q + 1$. (The cyclicity of $\Delta(aK)$ follows from 3.1.) We note this as

32.8 Lemma. *If P is a non-absolute point of π, then $\Delta(P)$ is cyclic of order $q + 1$.*

If the characteristic of K is 2, then $f(x, y) = -f(y, x)^\alpha$. Assume that the characteristic of K is not 2. Then there exists $t \in K$ with $t^\alpha = -t \neq 0$. Put $g = tf$. Then π is also represented by the form (α, g) (which is, of course, not symmetric). Moreover

$$g(x, y) = tf(x, y) = -t^\alpha f(y, x)^\alpha = -(tf(y, x))^\alpha = -g(y, x)^\alpha.$$

Hence in either case there exists a form (α, g) representing π with $g(x, y) = -g(y, x)^\alpha$.

Denote by $PGU(3, q^2)$ the centralizer of π in $PG(3, q^2)$ and let τ be a non-trivial elation in $PGU(3, q^2)$ with centre aK. Then $(aK)^\tau$ is the axis of τ whence $aK \subseteq (aK)^\tau$. Thus aK is absolute. This implies $g(a, a) = 0$. As $(aK)^\tau$ is the axis of τ, we have $x^\tau = x + a\lambda g(a, x)$ with some $\lambda \in K$. Moreover, there exists $\mu \in K$ with $\mu g(x, y) = g(x^\tau, y^\tau)$ for all $x, y \in V$. A trivial computation shows that

$$\mu g(x, y) = g(x, y) + (\lambda - \lambda^\alpha) g(x, a) g(a, y)$$

for all $x, y \in V$. This implies $\mu = 1$ and $\lambda = \lambda^\alpha$, whence $\lambda \in \mathrm{GF}(q)$. Conversely, if $\lambda \in \mathrm{GF}(q)$, then $\lambda = \lambda^\alpha$ and the mapping defined by $x^\tau = x + a\lambda g(a, x)$ is an elation with centre aK belonging to $\mathrm{PGU}(3, q^2)$. Thus we have proved the first part of

32.9 Lemma. a) *If P is an absolute point of π and if $\mathrm{E}(P)$ denotes the group of all elations with centre P in $\mathrm{PGU}(3, q^2)$, then $|\mathrm{E}(P)| = q$.*
b) *If P and Q are two distinct absolute points of π, then $\langle \mathrm{E}(P), \mathrm{E}(Q) \rangle \cong \mathrm{SL}(2, q)$.*

PROOF. Let $P = aK$ and $Q = bK$. As $P + Q$ is not absolute, we may choose a and b such that $g(a, b) = 1$. Then $g(b, a) = -1$. Let $P^\pi \cap Q^\pi = cK$. Then $\{a, b, c\}$ is a basis of V. Pick $\sigma \in \mathrm{E}(P)$. Then $x^\sigma = x + a\lambda g(a, x)$ for some $\lambda \in \mathrm{GF}(q)$. Therefore σ is represented with respect to $\{a, b, c\}$ by the matrix

$$\begin{bmatrix} 1 & 0 & 0 \\ \lambda & 1 & 0 \\ 0 & 0 & 1 \end{bmatrix}.$$

If $\tau \in \mathrm{E}(Q)$, then there exists $\mu \in \mathrm{GF}(q)$ with $x^\tau = x - b\mu g(b, x)$. Hence τ is represented by the matrix

$$\begin{bmatrix} 1 & \mu & 0 \\ 0 & 1 & 0 \\ 0 & 0 & 1 \end{bmatrix}.$$

Therefore

$$\langle \mathrm{E}(P), \mathrm{E}(Q) \rangle = \left\langle \begin{bmatrix} 1 & 0 & 0 \\ \lambda & 1 & 0 \\ 0 & 0 & 1 \end{bmatrix}, \begin{bmatrix} 1 & \mu & 0 \\ 0 & 1 & 0 \\ 0 & 0 & 1 \end{bmatrix} \middle| \lambda, \mu \in \mathrm{GF}(q) \right\rangle$$

$$\cong \mathrm{SL}(2, q). \qquad \square$$

32.9 b) is actually all we shall need later on. But for the sake of completeness we also prove:

32.10 Theorem. *Let π be a unitary polarity of the projective plane over $\mathrm{GF}(q^2)$ and $\mathfrak{U}$ the unital belonging to π. If $G = \mathrm{PGU}(3, q^2)$ and $H = \mathrm{PSU}(3, q^2) = G \cap \mathrm{PSL}(3, q^2)$, then we have:*

(a) *G acts doubly transitively on the set of points of $\mathfrak{U}$.*
(b) *If P and Q are distinct points of $\mathfrak{U}$, then $G_{P,Q}$ is cyclic of order $q^2 - 1$.*
(c) *$|G| = (q^3 + 1)q^3(q^2 - 1)$.*
(d) *$|H| = (3, q + 1)^{-1}(q^3 + 1)q^3(q^2 - 1)$.*
(e) *If P is a point of $\mathfrak{U}$, then G_P contains a normal subgroup of order q^3 which acts transitively on the set of points of $\mathfrak{U}$ distinct from P.*
(f) *If Π and Π_1 are two distinct Sylow p-subgroups of G, where p is the characteristic of $\mathrm{GF}(q^2)$, then $\Pi \cap \Pi_1 = \{1\}$.*

(g) *If P is a point of $\mathfrak{U}$ and Π the Sylow p-subgroup of G containing $E(P)$, then $E(P) = \mathfrak{Z}(\Pi)$.*

(h) *Let Π be a Sylow p-subgroup of G. If $p = 2$, then $\mathfrak{Z}(\Pi) = \{\tau \mid \tau \in \Pi, \tau^2 = 1\}$. If $p \geqslant 3$, then $\tau^p = 1$ for all $\tau \in \Pi$.*

(i) *All elations of G are contained in H and conjugate under H.*

(j) *If $q \neq 2$, then H is simple.*

PROOF. (a) Let P be a point of $\mathfrak{U}$. Then $E(P)$ is a p-group, where p is the characteristic of $GF(q^2)$. As P is the only fixed point of $E(P)$ in $\mathfrak{U}$, the group G acts transitively on the set of points of $\mathfrak{U}$ by Gleason's lemma 15.1. If l is a line of $\mathfrak{U}$, then 32.9 b) implies that G_l acts doubly transitively on $l \cap \mathfrak{U}$. Hence we need only show that G_P acts transitively on the set of lines of $\mathfrak{U}$ passing through P for some point P of $\mathfrak{U}$.

Let $l_1, \ldots, l_{q^2}$ be all the lines of $\mathfrak{U}$ which pass through the point P of $\mathfrak{U}$. Put $q_i = l_i^\pi$. Then $\Delta(Q_i)$ has order $q + 1$ by 32.8. From $P \mathbin{I} l_i$ we infer $Q_i \mathbin{I} P^\pi$. The group $\Delta(Q_i)$ consists of homologies with axis l_i. Therefore $\Delta(Q_i)$ acts regularly on $\{l_1, \ldots, l_{q^2}\} \setminus \{l_i\}$. Applying Gleason's lemma once again using a prime divisor of $q + 1$, we obtain that G_P acts transitively on $\{l_1, \ldots, l_{q^2}\}$.

(b) If $\gamma \in G_{P,Q}$ fixes a third point of $P + Q$, then γ induces the identity on the line $P + Q$. Hence $G_{P,Q}/\Delta((P + Q)^\pi)$ is isomorphic to a group which acts regularly on the set of $q - 1$ points of $P + Q$ other than P and Q belonging to $\mathfrak{U}$. Hence $|G_{P,Q}| = (q + 1)s$, where s is a divisor of $q - 1$.

Let $P = b_1 K$, $Q = b_2 K$ and $P^\pi \cap Q^\pi = b_3 K$. Then $\{b_1, b_2, b_3\}$ is a basis of V. Define ρ by $b_1^\rho = b_1 \lambda^q$, $b_2^\rho = b_2 \lambda^{-1}$, $b_3^\rho = b_3$ where $\lambda \in K^*$. Then

$$f(b_1^\rho, b_1^\rho) = f(b_1\lambda^q, b_1\lambda^q) = 0 = f(b_1, b_1),$$

$$f(b_1^\rho, b_2^\rho) = f(b_1\lambda^q, b_2\lambda^{-1}) = f(b_1, b_2)\lambda^{q^2-1} = f(b_1, b_2),$$

$$f(b_1^\rho, b_3^\rho) = f(b_1\lambda^q, b_3) = 0 = f(b_1, b_3),$$

$$f(b_2^\rho, b_2^\rho) = f(b_2\lambda^{-1}, b_2\lambda^{-1}) = 0 = f(b_2, b_2),$$

$$f(b_2^\rho, b_3^\rho) = f(b_2\lambda^{-1}, b_3) = 0 = f(b_2, b_3),$$

$$f(b_3^\rho, b_3^\rho) = f(b_3, b_3).$$

This implies $\rho \in G$. Hence $G_{P,Q}$ contains a cyclic subgroup of order $q^2 - 1$, whence $|G_{P,Q}| = q^2 - 1$.

(c) is an immediate consequence of (a) and (b).

(d) From $|PGL(3, q^2) : PSL(3, q^2)| = (3, q^2 - 1)$ we infer $|G : H| \in \{1, 3\}$. Therefore H acts also doubly transitively on $\mathfrak{U}$. This implies $|G : H| = |G_{P,Q} : H_{P,Q}|$. Furthermore $\langle E(P), E(Q) \rangle \cong SL(2, q)$ by 32.9 b). Hence $\frac{1}{2}(q - 1)$ divides $H_{P,Q}$ as, obviously, all elations of G are contained in H. If 3 does not divide $q + 1$, then $\frac{1}{2}(q^2 - 1)$ divides $|H_{P,Q}|$, whence $|H_{P,Q}| = q^2 - 1$.

Assume that 3 divides $q + 1$ and let $P = aK$ with $f(a, a) = 1$. If $\sigma \in \Delta(P)$, then $x^\sigma = x + a\lambda f(a, x)$ with $(\lambda + 1)^{q+1} = 1$. As $a^\sigma = a[1 + \lambda f(a,$

$a)] = a(1 + \lambda)$, we see that $\lambda + 1$ is an eigenvalue of σ. Moreover $x^\sigma = x$ for all $x \in P^\pi$. Hence 1 is an eigenvalue of multiplicity 2. Therefore $\det(\sigma) = \lambda + 1$. We infer that σ belongs to H if and only if $\lambda + 1$ is a third power in $\mathrm{GF}(q^2)$. Therefore $H \neq G$ as 3 divides $q + 1$. Hence $|G : H| = 3$.

(e) Let P and Q be two distinct absolute points and let $R = P^\pi \cap Q^\pi$. There exist $b_1 \in P$ and $b_2 \in Q$ such that $f(b_1, b_2) = 1$. Pick $b_3 \in R$ such that $f(b_3, b_3) = 1$. Then $f(b_1, b_1) = f(b_2, b_2) = f(b_1, b_3) = f(b_2, b_3) = 0$. Let $\lambda, \mu \in K$ be such that $\mu^q + \mu = -\lambda^{1+q}$ and define $\tau(\lambda, \mu)$ by

$$b_1^{\tau(\lambda, \mu)} = b_1, \qquad b_2^{\tau(\lambda, \mu)} = b_1\mu + b_2 - b_3\lambda^q, \qquad b_3^{\tau(\lambda, \mu)} = b_1\lambda + b_3.$$

Putting $\tau = \tau(\lambda, \mu)$ we then obtain

$$f(b_1^\tau, b_1^\tau) = f(b_1, b_1),$$
$$f(b_1^\tau, b_2^\tau) = f(b_1, b_1\mu + b_2 - b_3\lambda^q) = f(b_1, b_2),$$
$$f(b_1^\tau, b_3^\tau) = f(b_1, b_1\lambda + b_3) = f(b_1, b_3),$$
$$f(b_2^\tau, b_2^\tau) = f(b_1\mu + b_2 - b_3\lambda^q, b_1\mu + b_2 - b_3\lambda^q)$$
$$\qquad = \mu^q + \mu + f(b_2, b_2) + \lambda^{1+q} = f(b_2, b_2),$$
$$f(b_2^\tau, b_3^\tau) = f(b_1\mu + b_2 - b_3\lambda^q, b_1\lambda + b_3) = \lambda - \lambda = 0 = f(b_2, b_3),$$
$$f(b_3^\tau, b_3^\tau) = f(b_1\lambda + b_3, b_1\lambda + b_3) = f(b_3, b_3).$$

Therefore $\tau \in G_P$.

The mapping $\mu \to \mu^q + \mu$ is a homomorphism of the additive group of $\mathrm{GF}(q^2)$ onto the additive group of $\mathrm{GF}(q)$. Hence for each $\lambda \in \mathrm{GF}(q^2)$ there exist precisely q elements μ such $\mu^q + \mu = -\lambda^{q+1}$. Therefore there exist exactly q^3 elements $\tau(\lambda, \mu)$.

Now

$$b_1^{\tau(\lambda, \mu)\tau(\lambda', \mu')} = b_1,$$
$$b_3^{\tau(\lambda, \mu)\tau(\lambda', \mu')} = (b_1\lambda + b_3)^{\tau(\lambda', \mu')} = b_1(\lambda + \lambda') + b_3,$$
$$b_2^{\tau(\lambda, \mu)\tau(\lambda', \mu')} = (b_1\mu + b_2 - b_3\lambda^q)^{\tau(\lambda', \mu')}$$
$$\qquad = b_1(\mu + \mu' - \lambda'\lambda^q) + b_2 - b_3(\lambda + \lambda')^q.$$

Furthermore

$$(\mu + \mu' - \lambda'\lambda^q)^q + \mu + \mu' - \lambda'\lambda^q = \mu^q + \mu + \mu'^q + \mu' - \lambda'\lambda^q - \lambda'^q\lambda$$
$$= -(\lambda^{q+1} + \lambda'^{q+1} + \lambda'\lambda^q + \lambda'^q\lambda)$$
$$= -(\lambda^q + \lambda'^q)(\lambda + \lambda')$$
$$= -(\lambda + \lambda')^{q+1}.$$

Therefore $\tau(\lambda, \mu)\tau(\lambda', \mu') = \tau(\lambda + \lambda', \mu + \mu' - \lambda'\lambda^q)$. This proves that the $\tau(\lambda, \mu)$'s form a group Π of order q^3. Moreover the orbit of Q under Π has length q^3.

As we have seen under b), the group $G_{P,Q}$ consists of all the mappings ρ defined by $b_1^\rho = b_1 \kappa^q$, $b_2^\rho = b_2 \kappa^{-1}$, $b_3^\rho = b_3$, where $\kappa \in K^*$. This yields

$$b_1^{\rho^{-1}\tau(\lambda,\ \mu)\rho} = b_1,$$

$$b_2^{\rho^{-1}\tau(\lambda,\ \mu)\rho} = b_1 \mu \kappa^{1+q} + b_2 - b_3 \lambda^q \kappa,$$

$$b_3^{\rho^{-1}\tau(\lambda,\ \mu)\rho} = b_1 \lambda \kappa^q + b_3.$$

Furthermore

$$\left(\mu\kappa^{1+q}\right)^q + \mu\kappa^{1+q} = (\mu^q + \mu)\kappa^{1+q} = -(\lambda\kappa)^{1+q} = -(\lambda\kappa^q)^{1+q},$$

as $\kappa^{q(1+q)} = \kappa^{1+q}$. Hence $\rho^{-1}\tau(\lambda,\ \mu)\rho = \tau(\lambda\kappa^q,\ \mu\kappa^{1+q})$. Hence Π is normal in G_P, since $G_P = \Pi G_{P,Q}$.

(f) is a consequence of (e).

(g) Obviously $E(P) = \{\tau(0,\ \mu)\mid \mu \in GF(q^2),\ \mu^q + \mu = 0\}$. Let $\tau(\lambda,\ \mu) \in 3(\Pi)$. Then $\lambda'\lambda^q = \lambda'^q\lambda$ for all $\lambda' \in GF(q^2)$. Hence with $\lambda' = 1$ we get $\lambda = \lambda^q$. If $\lambda \neq 0$, then $\lambda' = \lambda'^q$ for all λ', a contradiction. Hence $\lambda = 0$. Conversely, if $\lambda = 0$ then $\tau(0,\ \mu)\tau(\lambda',\ \mu') = \tau(\lambda',\ \mu + \mu') = \tau(\lambda',\ \mu')\tau(0,\ \mu)$.

(h) Induction shows that $\tau(\lambda,\ \mu)^i = \tau(i\lambda, i\mu - \frac{1}{2}i(i-1)\lambda^{1+q})$ for all $i \in \mathbb{N}$. Thus if $p = 2$, we have $1 = \tau(\lambda,\ \mu)^2 = \tau(0, \lambda^{1+q})$ if and only if $\lambda = 0$. If $p > 2$, then 2 divides $p - 1$, whence $\tau(\lambda,\ \mu)^p = \tau(p\lambda, p\mu - \frac{1}{2}p(p-1)\lambda^{1+q}) = 1$.

(i) The groups $E(X)$ are contained in H, as we remarked earlier. Moreover they are all conjugate, as H acts transitively on the set of points of $\mathfrak{U}$. Therefore, it suffices to show that all non-trivial elations of $E(P)$ are conjugate under H_P.

Now we know already (see the end of the proof of (e)) that $\rho^{-1}\tau(0,\ \mu)\rho = \tau(0,\ \mu\kappa^{1+q})$. Furthermore, the mapping $\kappa \to \kappa^{1+q}$ maps the group of all third powers of $GF(q^2)^*$ onto $GF(q)^*$. Since the set of all μ such that $\mu^q + \mu = 0$ is a subspace of the $GF(q)$-vector space $GF(q^2)$, the result follows.

(j) As $q > 2$, there exists $\kappa \in GF(q^2)$ with $\kappa^{3(q+1)} - 1 \neq 0$. Consider now the mapping ρ with κ replaced by κ^3. Then $\rho \in H$. Furthermore

$$\tau(0,\ \mu)^{-1}\rho^{-1}\tau(0,\ \mu)\rho = \tau\left(0, (\kappa^{3(q+1)} - 1)\mu\right) \in H'_P.$$

Hence $E(P) = 3(\Pi) \subseteq H'_P$. Moreover $\tau(\lambda,\ \mu)^{-1} = \tau(-\lambda,\ -\lambda^{1+q} - \mu)$ and hence $\tau(\lambda,\ \mu)^{-1}\rho^{-1}\tau(\lambda,\ \mu)\rho = \tau((\kappa^{3q} - 1)\lambda, z)$ with a suitable z. Since there exists κ such that $\kappa^{3q} - 1 \neq 0$, since λ is arbitrary, and since $\tau(\lambda,\ \mu)\tau(0,\ \mu') = \tau(\lambda,\ \mu + \mu')$, we find that $\Pi \subseteq H'_P$. On the other hand H_P/Π is cyclic. Therefore $\Pi = H'_P$.

Let γ be the mapping defined by $b_1^\gamma = b_2$, $b_2^\gamma = b_1$, and $b_3^\gamma = -b_3$. Then γ is in H. Furthermore $\gamma^{-1}H_{P,Q}\gamma = H_{P,Q}$. As H acts doubly transitively and $\gamma \notin H_P$, we therefore have

$$H = H_P \cup H_P \gamma H_P = H_P \cup \Pi H_{P,Q}\gamma\Pi.$$

Let $\tau(\lambda, \mu) \neq 1$. Then

$$b_1^{\gamma\tau(\lambda, \mu)\gamma} = b_2^{\tau(\lambda, \mu)\gamma} = (b_1\mu + b_2 - b_3\lambda^q)^\gamma = b_2\mu + b_1 + b_3\lambda^q \notin P.$$

Therefore $\gamma\tau(\lambda, \mu)\gamma \in \Pi H_{P, Q}\gamma\Pi$. Since we have to interpret this projectively, we have

(∗)   There exist $\lambda', \mu', \lambda'', \mu'', \alpha \in GF(q^2)$ and $\rho \in H_{P,Q}$ such that $v^{\gamma\tau(\lambda, \mu)\gamma} = v^{\tau(\lambda', \mu')\rho\gamma\tau(\lambda'', \mu'')}\alpha$ for all $v \in V$.

From (∗) we get

$$\rho\gamma = \tau(\lambda', \mu')^{-1}\tau(\lambda, \mu)\tau(\lambda, \mu)^{-1}\gamma\tau(\lambda, \mu)\gamma\tau(\lambda'', \mu'')^{-1}.$$

Therefore $\rho\gamma \in H'$ for all ρ that occur in (∗), as $\Pi \subseteq H'$ and as $\gamma = \gamma^{-1}$.

Next we show that each $\rho \in H_{P, Q}$ occurs in (∗). In order to do this we compute (∗) for $v = b_1$ and $v = b_3$. Recall that there is a $\kappa \in K^*$ such that $b_1^\rho = b_1\kappa^q$, $b_2^\rho = b_2\kappa^{-1}$ and $b_3^\rho = b_3$.

$$b_1^{\gamma\tau(\lambda, \mu)\gamma} = b_1 + b_2\mu + b_3\lambda^q,$$

$$b_1^{\tau(\lambda', \mu')\rho\gamma\tau(\lambda'', \mu'')}\alpha = (b_1\mu'' + b_2 - b_3\lambda''^q)\kappa^q\alpha,$$

$$b_3^{\gamma\tau(\lambda, \mu)\gamma} = -b_2\lambda + b_3,$$

$$b_3^{\tau(\lambda', \mu')\rho\gamma\tau(\lambda'', \mu'')}\alpha = [b_1(\mu''\kappa^q\lambda' - \lambda'') + b_2\kappa^q\lambda' - b_3(1 + \lambda''\kappa^q\kappa^q\lambda')]\alpha.$$

Thus we get in particular $\mu = \kappa^q\alpha$, $\lambda^q = -\lambda''^q\kappa^q\alpha$, $-\lambda = \kappa^q\lambda'\alpha$, $1 = -\alpha - \lambda''^q\kappa^q\lambda'\alpha$. The last two equations yield $\alpha = \lambda\lambda''^q - 1$. From the first and second equation we obtain $\lambda^q = -\lambda''^q\mu$. Hence $\alpha = -\lambda^{1+q}\mu^{-1} - 1$. Now λ and μ are linked by $\mu + \mu^q = -\lambda^{1+q}$. Therefore $\alpha = (\mu + \mu^q)\mu^{-1} - 1 = \mu^{q-1}$. Using the first equation once again, we find $\alpha = \mu^{q-2}\kappa^q\alpha$ whence $1 = \mu^{q-2}\kappa^q$. As μ can be any element in $GF(q^2)^*$ and as $(q^2 - 1, q - 2) = (q^2 - 4 + 3, q - 2) = (3, q - 2) = (3, q + 1)$, we see that $\rho\gamma \in H'$ for all $\rho \in H_{P, Q}$. This yields $\gamma = 1\gamma \in H'$ whence $H_{P, Q} \subseteq H'$. Thus $H_P \subseteq H'$. As this is true for all P, we have $H = H'$.

Since H_P is soluble, the simplicity of H now follows from Iwasawa's theorem (22.5).

33. A Characterization of A_5

Before we state the theorem characterizing A_5, we shall make a few observations about A_4 and A_5.

A) Let $\langle x, y \rangle = A_4$ and $o(x) = o(y) = 3$. Then $o(xy) = 3$, if and only if $o(x^{-1}y) = 2$.

PROOF. We may assume $x = (123)$. Then $x^{-1} = (132)$. The remaining elements of order 3 are $(124), (142), (134), (143), (234), (243)$. Furthermore

$$(123)(124) = (14)(23) \qquad (132)(124) = (134)$$
$$(123)(142) = (234) \qquad (132)(142) = (13)(24)$$
$$(123)(134) = (124) \qquad (132)(134) = (14)(23)$$
$$(123)(143) = (12)(34) \qquad (132)(143) = (243)$$
$$(123)(234) = (13)(24) \qquad (132)(234) = (142)$$
$$(123)(243) = (143) \qquad (132)(243) = (12)(34). \qquad \square$$

B) *If $A_4 = \langle x, y \rangle$ with $o(x) = o(y) = o(xy) = 3$, then $o(xyx) = 2$.*

PROOF. If $o(xyx) \neq 2$, then $o(xyx) = 3$. But then, by A), we obtain the contradiction $2 = o(y^{-1}x^{-1}x) = o(y^{-1}) = 3$. $\qquad \square$

C) *If $A_5 = \langle x, y \rangle$ with $o(x) = o(y) = 3$, then $\langle x, y^{xy^{-1}} \rangle \cong A_4$ and $y^{xy^{-1}} = x^{-1}x^y$. Also $o(xy^{-1}y^x) = 2$.*

PROOF. We may assume $x = (123)$ and $y = (345)$. Then $y^{xy^{-1}} = (345)^{(123)(354)} = (145)^{(354)} = (134)$ proving $\langle x, y^{xy^{-1}} \rangle \cong A_4$. Moreover $x^{-1}x^y = (132)(123)^{(345)} = (132)(124) = (134)$. Finally $(123)(354)(145) = (12)(34)$. $\qquad \square$

D) *Let $A_5 = \langle x, y \rangle$ with $o(x) = o(y) = 3$. Then $o(xy) = 5$ and $o(yy^x) = 2$.*

PROOF. We may assume $x = (123)$ and $y = (345)$. Then $xy = (12453)$ and $yy^x = (14)(35)$. $\qquad \square$

E) *Let $A_5 = \langle x, y \rangle$ with $o(x) = o(y) = 3$. If $z \in \langle x, y \rangle$ with $o(z) = 3$, $o(xz) = 2$ and $o(zy) = 5$, then $z \in \langle x^y, x^{y^{-1}} \rangle$.*

PROOF. Let $x = (123)$ and $y = (345)$. Moreover put $z = (abc)$. Then $|\{1,2,3\} \cap \{a,b,c\}| = 2$ and $|\{3,4,5\} \cap \{a,b,c\}| = 1$. If $3 \in \{a,b,c\}$, say $a = 3$, then $\{b,c\} \cap \{3,4,5\} = \emptyset$ whence $\{b,c\} = \{1,2\}$. Thus $z \in \langle x \rangle$ and hence $o(xz) = 1$ or 3. Thus $3 \notin \{a,b,c\}$. We may therefore assume that $\{a,b\} = \{1,2\}$ and $c \in \{4,5\}$. Hence $z = (124), (125), (214)$ or (215). Using $o(xz) = 2$ one gets $z = (124)$ or (125). Now $x^y = (124)$ and $x^{y^{-1}} = (125)$. $\qquad \square$

F) *Let $z \in A_5 = \langle x, y \rangle$ and $o(x) = o(y) = o(z) = 3$. If $(xz)^2 = 1 = (zy)^2$, then $z = y^{xy}$ or $z = y^{x^{-1}y^{-1}}$. In the first case $\langle z, y \rangle \cong A_4$, and in the second $\langle z, x \rangle \cong A_4$.*

PROOF. Let $x = (123)$, $y = (345)$, and $z = (abc)$. Then $|\{1,2,3\} \cap \{a,b,c\}| = 2 = |\{3,4,5\} \cap \{a,b,c\}|$. Hence $3 \in \{a,b,c\}$, say $a = 3$. Then $|\{1,2\} \cap \{b,c\}| = 1 = |\{4,5\} \cap \{b,c\}|$, whence

$$z \in \{(314), (315), (324), (325), (341), (351), (342), (352)\}.$$

Computing xz for all these cases and using $(xz)^2 = 1$ yields $z \in \{(314), (315), (342), (352)\}$. Finally $(zy)^2 = 1$ yields $z \in \{(315), (342)\}$.

On the other hand $y^{xy} = (315)$ and $y^{x^{-1}y^{-1}} = (342)$. ☐

G) *If* $A_5 = \langle x, y \rangle$ *and* $o(x) = o(y) = 3$, *then* $\langle x^y, y^x \rangle \cong A_4 \cong \langle x^y, y^{x^{-1}} \rangle$.

PROOF. We may assume $x = (123)$ and $y = (345)$. Then $x^y = (124)$ and $y^x = (145)$. Finally $y^{x^{-1}} = (245)$. This proves G). ☐

H) *If* $A_4 = \langle x, y \rangle$ *with* $o(x) = o(y) = o(xy) = 3$, *then* $\langle xy^{-1}, y^{-1}x \rangle$ *is the elementary abelian 2-group of order* 4 *of* A_4.

PROOF. Let $x = (123)$ and $y = (142)$. Then $xy^{-1} = (14)(23)$ and $y^{-1}x = (13)(24)$. ☐

33.1 Theorem (Assion). *Let* G *be a finite group and let* D *be a conjugacy class of elements of order* 3 *of* G. *Assume furthermore*:

(1) $G = \langle D \rangle$.
(2) *If* $a, b \in D$ *and* $b \notin \langle a \rangle$, *then* $\langle a, b \rangle \cong A_4$ *or* A_5.
(3) *There exist* $a, b \in D$ *with* $\langle a, b \rangle \cong A_5$.

Then $G \cong A_5$.

PROOF. As all elements of order 3 in A_5 are conjugate in A_5, it follows from (3) that $a \in D$ implies $a^{-1} \in D$. From this and (2) we infer that for all $a, b \in D$ the set $\langle a, b \rangle \cap D$ consists of all elements of order 3 in $\langle a, b \rangle$. We shall make continuous use of these remarks without mentioning them.

(a) Let $x, y, z \in D$ be such that $(x^{-1}y)^2 = (x^{-1}z)^2 = (y^{-1}z)^2 = 1$ and $z \notin \langle x, y \rangle$. Then $\langle x, y, z \rangle$ is a Frobenius group of order $2^4 \cdot 3$ with Frobenius kernel $\langle xy^{-1}, y^{-1}x \rangle \langle xz^{-1}, z^{-1}x \rangle$ and Frobenius complement $\langle x \rangle$.

Put $u = x^{-1}yz$. As $(x^{-1}z)^2 = (y^{-1}z)^2 = (x^{-1}y)^2 = 1$, we have $zx^{-1} = xz^{-1}$, $zy^{-1} = yz^{-1}$, and $x^{-1}y = y^{-1}x$. Therefore

$$u^2 = x^{-1}yzx^{-1}yz = x^{-1}yxz^{-1}yz = x^{-1}yxz^{-1}yz^{-1}z^{-1}$$

$$= x^{-1}yxz^{-1}zy^{-1}z^{-1}$$

$$= y^{-1}xxy^{-1}z^{-1} = y^{-1}x^{-1}y^{-1}z^{-1} = (zyxy)^{-1}.$$

This implies $u^2 \neq 1$, as $z \notin \langle x, y \rangle$.

Assume $o(u) = 5$. Then $u^3 = u^{-2} = zyxy$. Furthermore $y^{-1} = x^{-1}yx^{-1}$, as $(x^{-1}y)^2 = 1$. Therefore $y^{-1}x^{-1}y^{-1} = x^{-1}yx^{-3}yx^{-1} = x^{-1}y^{-1}x^{-1}$. This and B) imply $y^{-1}x^{-1}y^{-1} = xyx$, whence $u^2 = xyxz^{-1}$. Thus

$$zyxy = u^3 = xyxz^{-1}x^{-1}yz = x(z^{-1})^{x^{-1}y^{-1}}y^{-1}z.$$

This yields

$$(z^{-1})^{x^{-1}y^{-1}} = x^{-1}zyxyz^{-1}y = z^{-1}xyxzy^{-1}y = (xyx)^z,$$

a contradiction to B). Hence $o(u) \neq 5$. As $(y^{-1}z)^2 = 1$, it follows from A) that $yz \in D$. This together with $o(u) \neq 5$ implies $\langle x^{-1}, yz \rangle \cong A_4$. Therefore $o(u) = 3$. From this and A) we obtain $o(xyz) = 2$. Hence $1 = xyzy^{-1}x^{-1}y^{-1}x^{-1}z$, as $o(xy) = 3$. But $x^{-1}y^{-1}x^{-1} = xyx$ and $zy^{-1} = yz^{-1}$. Thus $1 = xy^2z^{-1}xyxz$ and hence $xy^{-1} = yx^{-1} = (xyx)^z$.

Now $(xy^{-1})^2 = xy^{-1}xy^{-1} = xx^{-1}yy^{-1} = 1$. Similarly $(xz^{-1})^2 = 1$ and $(yz^{-1})^2 = 1$. Therefore, replacing x, y, z by x^{-1}, y^{-1}, z^{-1} in the above argument, we obtain $y^{-1}x = (x^{-1}y^{-1}x^{-1})^{z^{-1}}$, i.e., $y^{-1}x = (xyx)^{z^{-1}}$. From this and H), we deduce that $M = \langle xy^{-1}, y^{-1}x \rangle$ is a normal subgroup of $\langle x, y, z \rangle = H$.

As $(x^{-1}z)^2 = (x^{-1}y)^2 = 1$ and $z^{-1}yz^{-1}y = z^{-1}zy^{-1}y = 1$, we may interchange the rôles of y and z in the above argument. Therefore $N = \langle xz^{-1}, z^{-1}x \rangle$ is a normal subgroup of H. Obviously $H = MN\langle x \rangle$, whence $|H| = 2^{2+s} \cdot 3$ with $1 \leqslant s \leqslant 2$, as $z \notin \langle x, y \rangle$. The group $\langle x, y \rangle$ contains 4 Sylow 3-subgroups of H. As $\langle z \rangle \not\subseteq \langle x, y \rangle$, the group H contains more than 4 Sylow 3-subgroups. Since the number of Sylow 3-subgroups is congruent to 1 modulo 3, the number of Sylow 3-subgroups of H is 16. Hence $|H| = 2^4 \cdot 3$ and H is a Frobenius group with Frobenius kernel MN and Frobenius complement $\langle x \rangle$.

(b) Let $a, b \in D$ and assume $\langle a, b \rangle \cong A_5$. If $c \in D$ is such that $\langle a, c \rangle \cong \langle b, c \rangle \cong A_4$, then $c \in \langle a, b \rangle$.

Replacing a by a^{-1} if necessary and b by b^{-1}, we may assume by A) that $(ac)^2 = 1 = (bc)^2$. Assume $c \notin \langle a, b \rangle$. Then

$$(ab^{-1}c)^2 = ab^{-1}cab^{-1}c = ab^{-1}a^{-1}acacc^{-1}b^{-1}c = ab^{-1}a^{-1}c^{-1}b^{-1}c$$

$$= ab^{-1}a^{-1}c^{-1}b^{-1}c^{-1}b^{-1}bc^2 = ab^{-1}a^{-1}bc^2 \neq 1.$$

It follows from A) that $b^{-1}c \in D$. Hence $\langle a, b^{-1}c \rangle \cong A_4$ or A_5.

Assume $\langle a, b^{-1}c \rangle \cong A_4$. Then $(ab^{-1}c)^3 = 1$, since $(ab^{-1}c)^2 \neq 1$. Therefore we have $ab^{-1}cab^{-1}c = c^{-1}ba^{-1}$. From $(ac)^2 = 1 = (bc)^2$ we obtain $ca = a^{-1}c^{-1}$ and $bc = c^{-1}b^{-1}$. Therefore $c^{-1}ba^{-1} = ab^{-1}a^{-1}c^{-1}b^{-1}c = ab^{-1}a^{-1}bc^2 = ab^{-1}a^{-1}bc^{-1}$, whence $(ba^{-1})^c = (b^{-1})^{a^{-1}}b$. But $o(ba^{-1}) = 5$ by D) and $o((b^{-1})^{a^{-1}}b) \leqslant 3$. This contradiction proves $\langle a, b^{-1}c \rangle \cong A_5$.

As $\langle a, b^{-1}c \rangle \cong A_5$, we obtain from D) that $o(b^{-1}c(b^{-1}c)^a) = 2$. Since $(b^{-1}c)^a = a^{-1}b^{-1}ca = a^{-1}b^{-1}a^{-1}c^{-1}$, we therefore have

$$1 = b^{-1}c(b^{-1}c)^a b^{-1}c(b^{-1}c)^a = b^{-1}ca^{-1}b^{-1}a^{-1}c^{-1}b^{-1}ca^{-1}b^{-1}a^{-1}c^{-1}$$

$$= b^{-1}ca^{-1}b^{-1}a^{-1}bc^2a^{-1}b^{-1}a^{-1}c^{-1} = b^{-1}ca^{-1}b^{-1}a^{-1}bacb^{-1}a^{-1}c^{-1}$$

$$= b^{-1}c(a^{-1})^{ba}cb^{-1}a^{-1}c^{-1} = b^{-1}c(a^{-1})^{ba}cb^{-1}a^{-1}bb^{-1}c^{-1}$$

$$= b^{-1}c(a^{-1})^{ba}c(a^{-1})^b cb = b^{-1}c(a^{-1})^{ba}(a^{-1})^{bc^{-1}}c^2b$$

$$= b^{-1}c(a^{-1})^{ba}(a^{-1})^{bc^{-1}}(b^{-1}c)^{-1}.$$

Hence $a^{bc^{-1}} = (a^{-1})^{ba}$. Moreover $(b^{-1}a)^c = c^{-1}b^{-1}ac = bcc^{-1}a^{-1} = ba^{-1}$. As $\langle a^{ba}, b^{-1}a \rangle$ is a subgroup of $\langle a, b \rangle$ generated by an element of order 5 (by D)) and an element of order 3, we have $\langle a^{ba}, b^{-1}a \rangle = \langle a, b \rangle$. Thus $\langle a, b \rangle$ is a normal subgroup of $\langle a, b, c \rangle$. We deduce that $\langle a, b \rangle \cap \langle a, b^{-1}c \rangle$ is normal in $\langle a, b^{-1}c \rangle \cong A_5$. This proves $\langle a, b \rangle = \langle a, b^{-1}c \rangle$ and hence $c \in \langle a, b \rangle$. This contradiction establishes (b).

(c) Let $a, b, c \in D$ be such that $\langle a, b \rangle \cong \langle a, c \rangle \cong A_5$ and $\langle b, c \rangle \cong A_4$. Then $c \in \langle a, b \rangle$.

We may assume by A) that $(bc)^2 = 1$. Assume furthermore $c \notin \langle a, b \rangle$. Then $b^{-1}c \in D$ by A). Therefore $\langle b^{-1}c, a \rangle \cong A_4$ or A_5, as $b^{-1}c \notin \langle a \rangle$, since otherwise $c \in \langle a, b \rangle$. Assume that $\langle b^{-1}c, a \rangle \cong A_4$. Then we infer from (b) and $\langle b^{-1}c, b \rangle = \langle c, b \rangle \cong A_4$ that $b^{-1}c \in \langle a, b \rangle$, a contradiction. Thus $\langle b^{-1}c, a \rangle \cong A_5$. Similarly $\langle bc^{-1}, a \rangle \cong A_5$.

If $c^{ac^{-1}} \in \langle b \rangle$, then $\langle b^c \rangle = \langle c^a \rangle \subseteq \langle a, c \rangle$, whence $\langle a, b \rangle = \langle a, c \rangle$ and $c \in \langle a, b \rangle$. Thus $c^{ac^{-1}} \notin \langle b \rangle$ and hence $\langle b, c^{ac^{-1}} \rangle \cong A_4$ or A_5.

Assume $\langle b, c^{ac^{-1}} \rangle \cong A_4$. By C), we have $\langle a, c^{ac^{-1}} \rangle \cong A_4$. Thus (b) yields $c^{ac^{-1}} \in \langle a, b \rangle$. Moreover $a^{-1}a^c = c^{ac^{-1}}$ by C). Hence $a^{-1}a^c \in \langle a, b \rangle$ and $a^c \in \langle a, b \rangle$. Furthermore

$$\langle a^c, b \rangle^{c^{-1}} = \langle a, b^{c^{-1}} \rangle = \langle a, cbc^{-1} \rangle = \langle a, b^{-1}c^{-2} \rangle = \langle c, b^{-1}c \rangle \cong A_5.$$

Therefore $\langle a^c, b \rangle \cong A_5$ whence $o(a^c b) = 5$ by D). Also $o(aa^c) = 2$. Thus, by E), $a^c \in \{a^b, a^{b^{-1}}\}$. If $a^c = a^b$, then $a = a^{bc^{-1}}$ which is impossible since $\langle a, bc^{-1} \rangle \cong A_5$. Hence $a^c = a^{b^{-1}}$. Now

$$(b^{-1}a^{-1}a^{b^{-1}})^2 = b^{-1}a^{-1}bab^{-1}b^{-1}a^{-1}bab^{-1} = b^{-1}b^a bb^a b^{-1}$$

$$\cdot \, b^{-2}(bb^a)^2 b^{-1} = b^3 = 1.$$

As $a^{-1}a^c = c^{ac^{-1}}$ and $a^c = a^{b^{-1}}$, we thus have $(b^{-1}c^{ac^{-1}})^2 = 1$, i.e., $((b^{-1})^c c^a)^2 = 1$. Moreover $(b^{-1})^c = (c^{-1}b^{-1})^2 bc^2 = bc^{-1}$ whence $\langle cb^{-1}, c^a \rangle \cong A_4$ and $((cb^{-1})^{-1}c^a)^2 = 1$. Hence $H = \langle c^{-1}, cb^{-1}, c^a \rangle$ is a Frobenius group of order $2^4 \cdot 3$ by (a). Moreover $b \in H$ whence $\langle b, c^a \rangle \cong A_4$. As $\langle c^a, a^c \rangle \cong A_4$ by G), we have $c^a \in \langle a^c, b \rangle$ by (b). As $a^c = a^{b^{-1}}$, we have $\langle a^c, b \rangle = \langle a^{b^{-1}}, b \rangle = \langle a, b \rangle$. Hence $c \in \langle a^c, b \rangle^{a^{-1}} = \langle a, b \rangle$. This proves that $\langle b, c^{ac^{-1}} \rangle \cong A_5$.

From $\langle c, cb^{-1} \rangle = \langle c, b \rangle \cong A_4$ and $\langle c, c^a \rangle \cong A_4$ and also $\langle cb^{-1}, c^a \rangle = \langle b, c^{ac^{-1}} \rangle^c \cong A_5$, we deduce using (b) that $c \in \langle cb^{-1}, c^a \rangle$.

As cc^a has order 2 by D) and $(cb^{-1}c)^2 = cb^{-1}c^{-1}b^{-1}c = c^{-1}(c^{-1}b^{-1})^2 c = 1$, we deduce from F) that $c \in \{c^{acb^{-1}c^a}, c^{abc^{-1}(c^{-1})^a}\}$. Now $(cc^a)^2 = 1$ implies $c^{(c^{-1})^a} = c^{-1}c^a$. Hence $c^{-1}c^a \in \{c^{acb^{-1}}, c^{abc^{-1}(c^{-2})^a}\} = \{c^{acb^{-1}}, c^{abc^{-1}c^a}\}$. Therefore $c^{-1}c^a \in \{c^{acb^{-1}}, c^{ab}\}$.

Assume $c^{-1}c^a = c^{ab}$. Then $(b^{-1}c)c^a = b^{-1}c^{-1}c^{-1}c^a b = b^{-1}c^{-1}b^{-1}c^a b = cbb^{-1}c^a b = cc^a b$. Furthermore $((b^{-1}c)c^a)^2 = cc^a bcc^a b = cc^a c^{-1}b^{-1}c^a b = cc^a c^{-1}c^{ab} = cc^a c^{-2}c^a = (cc^a)^2 = 1$. Therefore $\langle b^{-1}c, c^a \rangle \cong A_4$. Moreover $\langle a^c, c^a \rangle \cong A_4$ by G) and $\langle a, bc^{-1} \rangle \cong A_5$. This yields $A_5 \cong \langle a, bc^{-1} \rangle^c = \langle a^c, c^{-1}b \rangle = \langle a^c, b^{-1}c \rangle$ and hence $c^a \in \langle a^c, b^{-1}c \rangle$ by (b).

Therefore $a^{-1}a^c = c^{ac^{-1}} \in \langle a, cb^{-1} \rangle$, i.e., $a^c \in \langle a, cb^{-1} \rangle$. Moreover $(aa^c)^2 = 1$ and $a^c cb^{-1} = c^{-1}ac^{-1}b^{-1} = (ab)^c$ has order 5. Therefore, by E), $a^c \in \{a^{cb^{-1}}, a^{bc^{-1}}\}$. If $a^c = a^{cb^{-1}}$, then $b \in \langle a^c \rangle \subseteq \langle a, c \rangle$, a contradiction. Hence $a^{c^{-1}} = a^b$. From C) we get $o(ba^{-1}a^b) = 2$. Furthermore $a^{-1}a^b = a^{-1}a^{c^{-1}} = ((a^{-1})^c a)^{c^{-1}} = ((a^{-1}a^c)^{-1})^{c^{-1}} = ((c^{ac^{-1}})^{-1})^{c^{-1}} = (c^{-1})^{ac}$. Therefore $\langle b^{-1}, (c^{-1})^{ac} \rangle \cong A_4$. Moreover $\langle c, b^{-1} \rangle \cong A_4$ and $\langle c, (c^{-1})^{ac} \rangle = \langle c, (c^{-1})^a \rangle \cong A_4$ by D). Therefore $\langle c, b^{-1}, (c^{-1})^{ac} \rangle$ is soluble by (a). On the other hand $A_5 = \langle cb^{-1}, c^a \rangle \subseteq \langle c, b^{-1}, (c^{-1})^{ac} \rangle$, a contradiction. Thus we have $c^{-1}c^a = c^{acb^{-1}}$.

It follows from D) that $c^{-1}c^a = (c^{-1}c^ac)^{b^{-1}} = (c^{-2}(c^{-1})^a)^{b^{-1}} = (c(c^{-1})^a)^{b^{-1}} = bcb^{-1}(c^{-1})^{ab^{-1}} = c^{-1}b(c^{-1})^{ab^{-1}}$. Hence $b^{-1}c^a = (c^{-1})^{ab^{-1}}$. This implies $\langle b, c^a \rangle \cong A_4$. Moreover $cbc = b^{-1}$ and hence $A_5 \cong \langle a, bc^{-1} \rangle^{c^{-1}} = \langle a^{c^{-1}}, cbc \rangle = \langle a^{c^{-1}}, b \rangle$. Also $\langle c^a, a^{c^{-1}} \rangle \cong A_4$ by G). Hence $c^a \in \langle a^{c^{-1}}, b \rangle$ by (b). This implies $a^{-1}a^c = c^{ac^{-1}} \in \langle a^c, cbc^{-1} \rangle = \langle a^c, b^{-1}c \rangle$ and hence $a \in \langle a^c, b^{-1}c \rangle$. It follows therefore from D), E), and $\langle a, b^{-1}c \rangle \cong A_5$ that $a^c \in \{a^{b^{-1}c}, a^{c^{-1}b}\}$. As b does not centralize a, we get $a^c = a^{c^{-1}b}$.

As $\langle a^c, b \rangle = \langle a^{c^{-1}b}, b \rangle \cong A_5 \cong \langle a^c, c \rangle$ and $\langle c, b \rangle \cong A_4$, we may replace a by a^c in the above argument. Thus $c \in \langle a^c, b \rangle$ or $a^{c^{-1}} = (a^c)^c = (a^c)^{c^{-1}b} = a^b$. If $c \in \langle a^c, b \rangle$, then $a \in \langle a^c, b \rangle$, whence $\langle a^c, b \rangle = \langle a, b \rangle$ proving $c \in \langle a, b \rangle$. In the other case we have $c^a \in \langle a^{c^{-1}}, b \rangle = \langle a^b, b \rangle = \langle a, b \rangle$, whence $c \in \langle a, b \rangle$. This proves (c).

Pick $a, b \in D$ such that $\langle a, b \rangle \cong A_5$. We want to show that $D \subseteq \langle a, b \rangle$. Pick $w \in D$ and assume $w \notin \langle a, b \rangle$. Then $\langle a, w \rangle \cong \langle b, w \rangle \cong A_5$ by (b) and (c). Moreover $W = \langle b, w \rangle \cap \langle a, b \rangle$ is isomorphic to a subgroup of A_4, since 3 divides $|W|$ and $W \neq \langle a, b \rangle$. Hence W contains at most 8 elements of order 3. As there are 12 elements of order 3 in $\langle b, w \rangle$ which together with b generate an A_4, there exists $c \in \langle b, w \rangle$, $c \notin W$ such that $\langle b, c \rangle = A_4$. We infer from (b) and (c) that $c \in \langle a, b \rangle$, a contradiction. $\square$

34. A Characterization of Galois Fields of Odd Characteristic

All algebras to be considered in this section are algebras with 1 over fields of characteristic $\neq 2$. Algebras are not assumed to be associative unless otherwise stated. Moreover, only those homomorphisms are considered which map 1 onto 1.

Let K be a field and let J be a K-algebra. We shall call J a *Jordan algebra* if $a \circ b = b \circ a$ and $((a \circ a) \circ b) \circ a = (a \circ a) \circ (b \circ a)$ for all $a, b \in J$.

Examples of Jordan algebras are easily obtained as follows: Let A be an associative algebra. Then we define a new operation $\circ$ on A by $a \circ b = \frac{1}{2}(ab + ba)$ for all $a, b \in A$. It is easily checked that $A(+, \circ)$ is a Jordan algebra. This Jordan algebra is usually denoted by A^+.

The Jordan algebra J is called *special*, if there exists an associative algebra A and a monomorphism of J into A^+.

Let J be a Jordan algebra over K and let A be an associative K-algebra. If σ is a homomorphism of J into A^+, then the mapping σ is called an *associative specialization* of J into A.

Let J be a Jordan algebra and let s be an associative specialization of J into U. The pair (U,s) is called a *special universal envelope* of J, if for all pairs (A,σ) where σ is an associative specialization of J into A there exists a unique homomorphism η of U into A with $\sigma = s\eta$.

34.1 Theorem. *Let J be a Jordan algebra and let (U,s) be a special universal envelope of J. Then we have*:

a) *If (U',s') is a special universal envelope of J, then there exists exactly one isomorphism t from U onto U' with $s' = st$.*
b) *U is generated by J^s.*
c) *There exists exactly one involutory antiautomorphism α of U with $x^{s\alpha} = x^s$ for all $x \in J$.*
d) *If $\mathrm{rank}_K(J) = n < \infty$, then $\mathrm{rank}_K(U) \leqslant 2^{n-1}$.*

PROOF. a) As (U,s) and (U',s') are special universal envelopes of J, there exists exactly one homomorphism t of U into U' and exactly one homomorphism t' of U' into U with $s' = st$ and $s = s't'$. Hence $s = stt'$ and $s' = s't't$. This yields $tt' = \mathrm{id}_U$ and $t't = \mathrm{id}_{U'}$ by the universality of U repectively U'. Hence t is an isomorphism. This proves a).

b) Put $V = \langle J^s \rangle$. Then s is an associative specialization of J into V. Hence there exists a homomorphism η of U into V with $s = s\eta$. As $V \subseteq U$, the homomorphism η is in fact an endomorphism of U. Hence, by universality, $\eta = 1$, i.e., $U = V$.

c) Let $\overline{U}$ be the opposite structure of U. Then $\mu = 1$ is an antiisomorphism from U onto $\overline{U}$. This yields that μ is an isomorphism of U^+ onto $\overline{U}^+$. The mapping $\beta = s\mu$ is a homomorphism from J into $\overline{U}^+$. Hence there exists a homomorphism γ of U into $\overline{U}$ with $s\mu = \beta = s\gamma$. Put $\alpha = \gamma\mu^{-1}$. Then α is an antiendomorphism of U with $s = s\alpha$. This yields $s = s\alpha = s\alpha^2$ and hence $\alpha^2 = 1$ by universality. This proves the existence part of c). The uniqueness part follows from b).

d) Let $\{1, b_1, \ldots, b_{n-1}\}$ be a K-basis of J and put

$$V = \left\{ \sum x_i k_i \,\middle|\, x_i = y_{i1}^s \cdots y_{i,r_i}^s, y_{ij} \in J, k_i \in K \right\}.$$

Then V is a subalgebra of U and hence $V = U$ by b). As every y_{ij} is a K-linear combination of 1 and the b_i's, it suffices to show that every product of the form $b_{i(1)}^s b_{i(2)}^s \cdots b_{i(t)}^s$ is a K-linear combination of 1^s and the elements $b_{a(1)}^s \cdots b_{a(r)}^s$ with $1 \leqslant a(1) < a(2) < \cdots < a(r) \leqslant n - 1$.

Let N be the number of pairs (k,l) with $1 \leqslant k < l \leqslant t$ and $i(k) \geqslant i(l)$. Assume that t is minimal with respect to the property that $b_{i(1)}^s \cdots b_{i(t)}^s$ is not a K-linear combination of the required form. Moreover choose the product for the given t in such a way that N is minimal. Then $N \geqslant 1$. Hence there exists $k \in \{1, 2, \ldots, t\}$ with $i(k) \geqslant i(k + 1)$.

Assume $i(k) = i(k + 1)$. Then $b_{i(k)} = b_{i(k+1)}$. Hence $(b_{i(k)} \circ b_{i(k+1)})^s = b_{i(k)}^s b_{i(k+1)}^s$. Now $b_{i(k)} \circ b_{i(k+1)} = \sum_{j=0}^{n-1} b_j k_j$, where $b_0 = 1$. This yields

$$b_{i(1)}^s \cdots b_{i(t)}^s = \sum_{j=0}^{n-1} \left(b_{i(1)}^s \cdots b_{i(k-1)}^s b_j^s b_{i(k+2)}^s \cdots b_{i(t)}^s \right) k_j.$$

Using the minimality of t, we obtain that each summand of the right hand side and hence the left hand side of the above equation is a K-linear combination of the required form. This contradiction proves $i(k) > i(k + 1)$. Therefore

$$\left(b_{i(k)} \circ b_{i(k+1)} \right)^s = \tfrac{1}{2} \left(b_{i(k)}^s b_{i(k+1)}^s + b_{i(k+1)}^s b_{i(k)}^s \right)$$

and hence

$$b_{i(k)}^s b_{i(k+1)}^s = - b_{i(k+1)}^s b_{i(k)}^s + 2 \left(b_{i(k)} \circ b_{i(k+1)} \right)^s.$$

As $b_{i(k)} \circ b_{i(k+1)}$ is a K-linear combination of 1 and the b_i's we have

$$b_{i(1)}^s \cdots b_{i(t)}^s = - b_{i(1)}^s \cdots b_{i(k-1)}^s b_{i(k+1)}^s b_{i(k)}^s b_{i(k+2)}^s \cdots b_{i(t)}^s + L,$$

where L is a linear combination of the required form by the minimality of t. From the minimality of N we deduce that

$$b_{i(1)}^s \cdots b_{i(k-1)}^s b_{i(k+1)}^s b_{i(k)}^s b_{i(k+2)}^s \cdots b_{i(t)}^s$$

is also of the required form thus reaching again a contradiction. $\qquad\square$

34.2 Theorem. *Let J be a Jordan algebra. Then J possesses a special universal envelope.*

PROOF. Let T be the tensor algebra of J considered as a vector space over K and let I be the ideal of T generated by the elements $a \circ b - \tfrac{1}{2} (a \otimes b + b \otimes a)$ with $a, b \in J$ and $1_J - 1$ where 1_J is the unit element of J and 1 is the unit element of T. Put $U = T/I$ and let s be the mapping defined by $a^s = a + I$. We show that (U, s) is a special universal envelope of J.

First of all we note that s is linear. Moreover

$$(a \circ b)^s = a \circ b + I = \tfrac{1}{2} (a \otimes b + b \otimes a) + I \qquad \text{and} \qquad 1_J^s = 1 + I$$

by the definition of I. Hence s is an associative specialization of J into U. Let σ be an associative specialization of J into A. In particular, σ is a K-linear mapping of J into A. Therefore there exists a homomorphism δ of T into A with $x^\delta = x^\sigma$ for all $x \in J$. For $a, b \in J$ we then have

$$\left(a \circ b - \tfrac{1}{2} (a \otimes b + b \otimes a) \right)^\delta = (a \circ b)^\delta - \tfrac{1}{2} (a^\delta b^\delta + b^\delta a^\delta)$$

$$= (a \circ b)^\sigma - \tfrac{1}{2} (a^\sigma b^\sigma + b^\sigma a^\sigma) = 0,$$

as σ is an associative specialization of J. Hence $I \subseteq \text{Kern}(\delta)$. Therefore there exists a homomorphism η of U into A with $(a + I)^\eta = a^\delta$. This yields $\sigma = s\eta$.

Finally, η is unique, as T is generated by 1 and J. This proves 34.2. $\square$

If the algebra A admits an involutory antiautomorphism α, then we shall say that A is an algebra with involution α.

For an associative algebra A with involution α we put $H(A,\alpha) = \{x \mid x \in A, x^\alpha = x\}$. Obviously, $H(A,\alpha)$ is a subalgebra of A^+.

Let $F_r(K)$ be the free associative K-algebra in the free generators $x_1, \ldots, x_r$. Then $F_r(K)$ admits an involution ρ with $(x_{i_1} \cdots x_{i_k})^\rho = x_{i_k} \cdots x_{i_1}$. This involution ρ is called the *reversal operator* of $F_r(K)$.

Let $J_r(K)$ be the Jordan subalgebra of $F_r(K)^+$ generated by $x_1, \ldots, x_r$. Then $J_r(K) \subseteq H(F_r(K),\rho)$. P. M. Cohn proved that $J_r(K) \neq H(F_r(K),\rho)$ if $r \geqslant 4$. He also proved:

34.3 Theorem. *If $r \leqslant 3$, then $J_r(K) = H(F_r(K),\rho)$.*

PROOF. Denote by W the set of words in the $x_1, \ldots, x_r$. Then W is a K-basis for $F_r(K)$. From this we infer that $W_0 = \{\frac{1}{2}(w + w^\rho) \mid w \in W\}$ is a K-basis for $H(F_r(K),\rho)$. Hence it suffices to show $W_0 \subseteq J_r(K)$. For $w \in W$, we put $w^* = \frac{1}{2}(w + w^\rho)$. If the length of w is less than or equal to 2, then $w^* \in J_r(K)$. Let $w = x_{i_1} \cdots x_{i_n}$ with $n \geqslant 3$ and $1 \leqslant m < n$. Put $w_1 = x_{i_1} \cdots x_{i_m}$ and $w_2 = x_{i_{m+1}} \cdots x_{i_n}$. Then $w_1 w_2 = w$. Furthermore

$$4w_1^* \circ w_2^* = \frac{1}{2}((w_1 + w_1^\rho)(w_2 + w_2^\rho) + (w_2 + w_2^\rho)(w_1 + w_1^\rho))$$

$$= \frac{1}{2}(w + w_1 w_2^\rho + w_1^\rho w_2 + w_1^\rho w_2^\rho + w_2 w_1 + w_2 w_1^\rho + w_2^\rho w_1 + w^\rho)$$

$$= w^* + (w_1 w_2^\rho)^* + (w_1^\rho w_2)^* + (w_2 w_1)^*.$$

By the induction hypothesis, $4w_1^* \circ w_2^* \in J_r(K)$, i.e.,

(A) $w^* + (w_1 w_2^\rho)^* + (w_1^\rho w_2)^* + (w_2 w_1)^* \in J_r(K)$.

For $m = 1$ we obtain $w_1 = w_1^\rho$ and hence

$$(w_1 w_2^\rho)^* + (w_1^\rho w_2)^* = \frac{1}{2}(w_1 w_2^\rho + w_2 w_1 + w_1 w_2 + w_2^\rho w_1)$$

$$= \frac{1}{2}(w_1(w_2 + w_2^\rho) + (w_2 + w_2^\rho)w_1)$$

$$= 2w_1 \circ w_2^*.$$

As this is in $J_r(K)$, we have $(x_{i_1} \cdots x_{i_n})^* + (x_{i_2} \cdots x_{i_n} x_{i_1})^* \in J_r(K)$. By induction we get

(B) $(x_{i_1} \cdots x_{i_n})^* + (-1)^{k-1}(x_{i_{k+1}} \cdots x_{i_n} x_{i_1} \cdots x_{i_k})^* \in J_r(K)$.

This yields for n odd $2w^* \in J_r(K)$ and hence $w^* \in J_r(K)$. Thus we may assume that n is even. Take $m = 2$. Then (A) yields

$$(x_{i_1} \cdots x_{i_n})^* + (x_{i_1} x_{i_2} x_{i_n} \cdots x_{i_3})^* + (x_{i_2} x_{i_1} x_{i_3} \cdots x_{i_n})^*$$

$$+ (x_{i_3} \cdots x_{i_n} x_{i_1} x_{i_2})^* \in J_r(K).$$

By (B)

$$(x_{i_1} \cdots x_{i_n})^* - (x_{i_3} \cdots x_{i_n} x_{i_1} x_{i_2})^* \in J_r(K)$$

and

$$(x_{i_1} x_{i_2} x_{i_n} \cdots x_{i_3})^* - (x_{i_n} \cdots x_{i_3} x_{i_1} x_{i_2})^* \in J_r(K).$$

Trivially $(x_{i_n} \cdots x_{i_3} x_{i_1} x_{i_2})^* = (x_{i_2} x_{i_1} x_{i_3} \cdots x_{i_n})^*$. Therefore

$$2((x_{i_1} \cdots x_{i_n})^* + (x_{i_2} x_{i_1} x_{i_3} \cdots x_{i_n})^*) \in J_r(K).$$

Thus

$$(x_{i_1} \cdots x_{i_n})^* \equiv \mathrm{sgn}(1,2)(x_{i_2} x_{i_1} x_{i_3} \cdots x_{i_n})^* \bmod J_r(K).$$

As n is even, $\mathrm{sgn}(1,2,\ldots,n) = -1$. Hence by (b)

$$(x_{i_1} \cdots x_{i_n})^* \equiv \mathrm{sgn}(1,2,\ldots,n)(x_{i_2} \cdots x_{i_n} x_{i_1})^* \bmod J_r(K).$$

As $(1,2)$ and $(1,2,\ldots,n)$ generate S_n, we obtain

(C) $$(x_{i_1} \cdots x_{i_n})^* \equiv \mathrm{sgn}(\pi)(x_{i_{\pi(1)}} x_{i_{\pi(2)}} \cdots x_{i_{\pi(n)}})^* \bmod J_r(K)$$

for all $\pi \in S_n$.

Since $n > 2$ and n is even, $n \geqslant 4 > r$. Thus there exist $a, b \in \{1, 2, \ldots, n\}$ with $a \neq b$ and $x_{i_a} = x_{i_b}$. Let $\pi = (a,b)$. Then, by (C),

$$(x_{i_1} \cdots x_{i_n})^* \equiv -(x_{i_1} \cdots x_{i_n})^* \bmod J_r(K).$$

Therefore $(x_{i_1} \cdots x_{i_n})^* \in J_r(K)$. □

34.4 Theorem (Jacobson). *Let J be a Jordan algebra generated by at most 3 elements and let (U, h) be a special universal envelope of J. Then h is an epimorphism of J onto $H(U, \alpha)$, where α is the involution of J having the property that $h\alpha = h$.*

PROOF. As $h\alpha = h$, we have $J^h \subseteq H(U, \alpha)$. Let $y_1, \ldots, y_r$ generate J. If $F_r(K)$ is the free K-algebra with the free generators $x_1, \ldots, x_r$, then there exists a homomorphism σ from $F_r(K)$ into U with $x_i^\sigma = y_i^h$ for all i. Obviously, σ is also a homomorphism of $F_r(K)^+$ into U^+ and the restriction of σ to $J_r(K)$ maps $J_r(K)$ onto J^h. As J^h generates U, the homomorphism σ is surjective. Moreover, if $x \in J_r(K)$ then $x^{\sigma\alpha} = x^\sigma = x^{\rho\sigma}$. Hence $\sigma\alpha = \rho\sigma$, as $J_r(K)$ generates $F_r(K)$. Pick $b \in H(U, \alpha)$. Then there exists $a \in F_r(K)$ with $a^\sigma = b$. Therefore

$$b = \tfrac{1}{2}(b + b^\alpha) = \tfrac{1}{2}(a^\sigma + a^{\sigma\alpha}) = \tfrac{1}{2}(a + a^\rho)^\sigma.$$

By 34.3, $J_r(K) = H(F_r(K), \rho)$. Thus $H(U, \alpha) \subseteq J_r(K)^\sigma = J^h$. □

If J is a Jordan algebra, if $a \in J$ and if $n \in \mathbb{N}_0$, then we define $a^{\circ n}$ inductively by $a^{\circ 0} = 1$ and $a^{\circ(n+1)} = a^{\circ n} \circ a$.

34.5 Lemma. *If J is a special Jordan algebra, then J is power associative.*

PROOF. We may assume $J \subseteq A^+$ for an appropriate associative algebra A. Then $a^{\circ 0} = 1 = a^0$. Assume $a^{\circ n} = a^n$. Then

$$a^{\circ(n+1)} = a^n \circ a = \tfrac{1}{2}(a^n a + a a^n) = a^{n+1}.$$

Hence $a^{\circ n} = a^n$ for all n. From this we infer

$$a^{\circ m} \circ a^{\circ n} = a^m \circ a^n = \tfrac{1}{2}(a^m a^n + a^n a^m) = a^{m+n}.$$

Thus $(a^{\circ i} \circ a^{\circ k}) \circ a^{\circ l} = a^{i+k+l} = a^{\circ i} \circ (a^{\circ k} \circ a^{\circ l})$.   □

34.6 Lemma (McCrimmon). *Let A be an associative ring with involution α. Let A be generated by $H(A, \alpha)$ and assume that $H(A, \alpha) \backslash \{0\}$ consists only of units. If $\{0\}$ and A are the only α-invariant ideals of A, then A is a skewfield or $A = D \oplus D^\alpha$ where D is a skewfield.*

PROOF. Let Z_r be the set of all elements of A which do not have a right inverse and define Z_l similarly. Then $Z_r A \subseteq Z_r$ and $A Z_l \subseteq Z_l$. Let $z \in Z_r$ and put $H = H(A, \alpha)$. Then $z z^\alpha \in Z_r \cap H = \{0\}$. Hence $z z^\alpha = 0$. This yields $z \in Z_l$, i.e., $Z_r \subseteq Z_l$. Similarly $Z_l \subseteq Z_r$, i.e., $Z_r = Z_l = Z$.

If $Z = \{0\}$, then A is a skewfield. Assume $Z \neq \{0\}$. As $Z^\alpha = Z$ and $AZ \subseteq Z$ as well as $ZA \subseteq Z$, there are $x, y \in Z$ with $x + y \notin Z$. Otherwise, $Z = A$ and $1 \in Z$, a contradiction. Therefore, $x + y$ is a unit. Put $e = xu^{-1}$ and $f = yu^{-1}$, where $u = x + y$. Then $e, f \in Z$ and $1 = e + f$. Furthermore $e^\alpha e \in Z \cap H = \{0\}$. This yields $e^\alpha = e^\alpha(e + f) = e^\alpha f$. Similarly $f^\alpha = f^\alpha e$. Hence $f = e^\alpha f = e^\alpha$. This proves $1 = e + e^\alpha$. Moreover $e H e^\alpha, e^\alpha H e \subseteq Z \cap H = \{0\}$. Hence

$$H = 1H1 = (e + e^\alpha)H(e + e^\alpha) \subseteq eHe + e^\alpha He^\alpha \subseteq eAe + e^\alpha Ae^\alpha.$$

This implies $A = \langle H \rangle \subseteq eAe + e^\alpha Ae^\alpha$. Therefore $A = eAe + e^\alpha Ae^\alpha$. Pick $x \in eAe \cap e^\alpha Ae^\alpha$. Then $x = ex = 0$. Thus $A = eAe \oplus e^\alpha Ae^\alpha$. Put $D = eAe$. Then $D^\alpha = e^\alpha Ae^\alpha$. Finally, if $0 \neq x \in D$ then $0 \neq x + x^\alpha \in H$. Hence $x + x^\alpha$ is a unit. Therefore there exist $y \in D$ and $z \in D^\alpha$ with $1 = (x + x^\alpha)(y + z)$. As a consequence $e + f = 1 = xy + x^\alpha z$, whence $e = xy$. Similarly $e = yx$.   □

34.7 Lemma. *Let A be an associative algebra over a field of characteristic not 2. If $a, b \in A$ are such that $1 = \tfrac{1}{2}(ab + ba)$ and $a = \tfrac{1}{2}(a^2 b + ba^2)$, then a is a unit in A and $a^{-1} = b$.*

PROOF. We infer from $1 = \tfrac{1}{2}(ab + ba)$ that $\tfrac{1}{2}(a^2 b + aba) = a = \tfrac{1}{2}(aba + ba^2)$. Hence $a^2 b = ba^2$. Therefore $a = \tfrac{1}{2}(a^2 b + ba^2) = a^2 b = ba^2$. This yields $ab = ba^2 b = ba$. Thus $1 = ab = ba$.   □

Let J be a Jordan algebra and let $x \in J$. The element $y \in J$ is called a *Jordan inverse* of x, if $x \circ y = 1$ and $(x \circ x) \circ y = x$. The Jordan algebra J is called a *Jordan division algebra*, if every non-zero element of J has a Jordan inverse. Jordan division algebras may have zero-divisors, however. Every Jordan division algebra J is special by 34.2, as $\{0\}$ and J are the only ideals of J.

34.8 Theorem (Albert). *If J is a finite Jordan division algebra (of characteristic not 2), then J is a Galois field.*

PROOF (McCrimmon). As J is special, J is power associative by 34.5. Hence the subalgebra $\langle a \rangle$ generated by the element a is associative. Moreover, by 34.7, a^{-1} is a polynomial in a, as J is finite. Hence $a^{-1} \in \langle a \rangle$.

Let $a, b, c \in J$ and put $J_0 = \langle a, b, c \rangle$. As $x^{-1} \in J_0$ for all $x \in J_0$ by the remark just made, J_0 is a Jordan division algebra. It follows from 34.1, 34.2 and 34.4 that there exists a finite associative algebra A with involution α such that $J_0 = H(A, \alpha)$ and that $H(A, \alpha)$ generates A.

Let X be a proper α-invariant ideal of A and let σ be the canonical epimorphism from A onto $\overline{A} = A/X$. Then σ maps $H(A, \alpha)$ onto $H(\overline{A}, \overline{\alpha})$ where $\overline{\alpha}$ is defined by $(a + X)^{\overline{\alpha}} = a^{\alpha} + X$. Thus there exists an epimorphism from J_0 onto $H(\overline{A}, \overline{\alpha})$. As J_0 is simple, $J_0 \cong H(\overline{A}, \overline{\alpha})$. Therefore, we may assume, that $\{0\}$ and A are the only α-invariant ideals of A. Moreover, by 34.7, each non-zero element of $H(A, \alpha)$ is a unit of A. Thus, by 34.7, A is a skewfield or the direct sum of two skewfields. As A is finite, A is commutative. This yields $A^+ = A$. Therefore J_0 and hence J is associative. $\qquad\square$

34.9 Corollary (Glauberman). *Let V be a finite vector space over $\mathrm{GF}(p^s)$ where p is an odd prime. If $B \subseteq \mathrm{End}_{\mathrm{GF}(p^s)}(V)$ satisfies:*

a) $1 \in B$,
b) *If $0 \neq x \in B$, then $x \in \mathrm{GL}(V)$ and $x^{-1} \in B$;*
c) $B + B \subseteq B$,

then $BB \subseteq B$ and $B \cong \mathrm{GF}(p^r)$ for some r.

PROOF. We first show that $xyx \in B$ for all $x, y \in B$. For this we may assume $x, y \neq 0$ and $x \neq y^{-1}$. Put $z = y^{-1} - x$ and $w = x^{-1} + z^{-1}$. Then $xwz = x(x^{-1} + z^{-1})z = z + x = y^{-1}$. Hence $w = x^{-1}y^{-1}z^{-1}$. This yields $x - w^{-1} = x - zyx = x - (y^{-1} - x)yx = xyx$. As $x - w^{-1} \in B$ by b) and c), we have $xyx \in B$ for all $x, y \in B$.

Next we show $xy + yx \in B$ for all $x, y \in B$. This follows from a), c) and the remark just made, since $xy + yx = (1 + x)y(1 + x) - y - xyx$.

As $B(+)$ is an elementary abelian p-group, this yields that B is a subalgebra of $\mathrm{End}_{\mathrm{GF}(p)}(V)^+$. Moreover, B is a Jordan division algebra, as $x \circ x^{-1} = 1$ and $(x \circ x) \circ x^{-1} = x$. Hence $B(+, \circ)$ is a Galois field by 34.8. In particular $(B \setminus \{0\})(\circ)$ is cyclic. Let a be a primitive element of $(B \setminus \{0\})(\circ)$. Then $a^{\circ i} = a^i$. From this we infer that $\circ = \cdot$. $\qquad\square$

35. Groups Generated by Shears

In this section, we shall follow Hering 1972 b and Ostrom 1970 a, 1974.

Let V be a vector space of rank $2r > 0$ over a skewfield D and let ν be a partial spread of V. Furthermore, let G be a subgroup of $GL(V)$ leaving invariant ν. For $\emptyset \neq X \subseteq G$ we define V_X by $V_X = \{v \mid v \in V, v^\sigma = v$ for all $\sigma \in X\}$. If $X = \{\sigma\}$ then we shall write V_σ instead of $V_{\{\sigma\}}$. We shall call $\sigma \in G$ a ν-shear, if $V_\sigma \in \nu$ and if $(x - 1)^2$ is the minimal polynomial of σ. Moreover, V_σ is called the axis of σ. It is convenient to also call the identity a ν-shear, every $W \in \nu$ being an axis of it.

35.1 Lemma. *If $1 \neq \sigma \in G$ and if σ is a ν-shear, then the following is true:*

a) $V^{\sigma-1} = V_\sigma$.
b) $o(\sigma) = p$, *where p is the characteristic of D.*
c) *If $W \in \nu$ and $W^\sigma = W$, then $W = V_\sigma$.*

PROOF. a) As $V^{(\sigma-1)^2} = \{0\}$, we have $V^{\sigma-1} \subseteq \operatorname{Kern}(\sigma - 1)$. Moreover

$$2r = \operatorname{rk}(V) = \operatorname{rk}(V^{\sigma-1}) + \operatorname{rk}(\operatorname{Kern}(\sigma - 1))$$
$$= \operatorname{rk}(V^{\sigma-1}) + \operatorname{rk}(V_\sigma) = \operatorname{rk}(V^{\sigma-1}) + r.$$

Hence $\operatorname{rk}(V^{\sigma-1}) = r = \operatorname{rk}(V_\sigma)$ proving $V^{\sigma-1} = \operatorname{Kern}(\sigma - 1) = V_\sigma$.

b) If $p > 0$, then $\sigma^p - 1 = (\sigma - 1)^p = (\sigma - 1)^{p-2}(\sigma - 1)^2 = 0$. Hence $\sigma^p = 1$. As $\sigma \neq 1$, we have $o(\sigma) = p$.

Assume $p = 0$ and let $o(\sigma) = n > 0$. Then $o(\sigma) \geqslant 2$. Therefore

$$0 = (\sigma - 1)^n = \sum_{i=0}^{n} (-1)^i \binom{n}{i} \sigma^{n-i} = 1 + (-1)^n + \sum_{i=1}^{n-1} (-1)^i \binom{n}{i} \sigma^{n-i}.$$

Therefore, $(x - 1)^2$ divides $1 + (-1)^n + \sum_{i=1}^{n-1}(-1)^i\binom{n}{i}x^{n-i}$ and hence

$$(x - 1)^n - 1 - (-1)^n - \sum_{i=1}^{n-1} (-1)^i \binom{n}{i} x^{n-i} = x^n + (-1)^n - 1 - (-1)^n$$
$$= x^n - 1.$$

As $x^n - 1 = (x - 1)[\sum_{i=0}^{n-1}(x^i - 1) + n]$, we obtain the contradiction $n \equiv 0$ mod $(x - 1)$. This proves b).

c) Let ρ be the restriction of σ to W. Then $(\rho - 1)^2 = 0$. Hence $W^{\rho-1} \subseteq \operatorname{Kern}(\rho - 1)$. As $\operatorname{rk}(W^{\rho-1}) + \operatorname{rk}(\operatorname{Kern}(\rho - 1)) = \operatorname{rk}(W) = r > 0$, we therefore obtain $\operatorname{Kern}(\rho - 1) \neq \{0\}$. From this we infer $W \cap V_\sigma \neq \{0\}$, as $\operatorname{Kern}(\rho - 1) \subseteq \operatorname{Kern}(\sigma - 1) = V_\sigma$. Therefore, $W = V_\sigma$, as ν is a partial spread. $\square$

35.2 Lemma. *Let $W \in \nu$ and put $S = \{1\} \cup \{\sigma \mid \sigma \in G, V_\sigma = W, (\sigma - 1)^2 = 0\}$. Then S is an abelian group provided $|\nu| \geqslant 2$. If the characteristic p of D is not 0, then S is an elementary abelian p-group.*

PROOF. By 35.1 b), we need only show that S is an abelian group. Let $\sigma, \tau \in S \setminus \{1\}$. Then $W \subseteq V_{\sigma\tau}$. Moreover, $V^{\sigma-1} = V_\sigma = W = V_\tau = V^{\tau-1}$ by 35.1. Therefore, $V^{(\sigma-1)(\tau-1)} = V^{(\tau-1)^2} = \{0\} = V^{(\tau-1)(\sigma-1)}$. This yields $(\sigma - 1)(\tau - 1) = 0 = (\tau - 1)(\sigma - 1)$. Hence $\sigma\tau = \sigma + \tau - 1 = \tau\sigma$. Moreover, $\sigma\tau - 1 = \sigma - 1 + \tau - 1$, whence

$$V^{\sigma\tau - 1} \subseteq V^{\sigma-1} + V^{\tau-1} = W \subseteq V_{\sigma\tau}.$$

From this we obtain $(\sigma\tau - 1)^2 = 0$. Hence, if $W = V_{\sigma\tau}$ then $\sigma\tau \in S$.

Assume $V_{\sigma\tau} \neq W$. As $|v| \geq 2$, there exists $U \in v$ with $U \neq W$. Then $V = U \oplus W$ and hence $V_{\sigma\tau} \cap U \neq \{0\}$. This yields $U^{\sigma\tau} = U$. Therefore, $U^{\sigma\tau - 1} \subseteq U$. On the other hand, $U^{\sigma\tau - 1} \subseteq V^{\sigma\tau - 1} \subseteq W$. Hence $U^{\sigma\tau - 1} \subseteq U \cap W = \{0\}$. This proves $\sigma\tau = 1$. Thus $SS \subseteq S$. Finally, if $\sigma \in S$ then obviously $\sigma^{-1} \in S$. $\square$

35.3 Lemma. *Let π be a spread of V. If σ is a collineation of $\pi(V)$, then the following statements are equivalent:*

a) *σ is a shear of $\pi(V)$ fixing 0.*
b) *σ is a π-shear.*

PROOF. Assume a). Then $\sigma \in \Gamma L(V, K)$, where K is the kernel of $\pi(V)$ by 1.10. As σ fixes a component X of π pointwise, σ is linear. This yields that σ is also D-linear, as $D \subseteq K$. Moreover, $(X + v)^\sigma = X + v$, since σ is a shear with axis X. Therefore, $V^{\sigma-1} \subseteq X$. Furthermore, $X^{\sigma-1} = \{0\}$, whence $(\sigma - 1)^2 = 0$. As $V_\sigma = X \in \pi$, it follows that σ is a π-shear. Hence a) implies b). The converse is trivial. $\square$

From now on we shall assume that V is finite and that $|v| \geq 2$. Let p be the characteristic of D. Moreover, denote by $\mathfrak{S}$ the set of all nontrivial v-shears in G and assume $\mathfrak{S} \neq \emptyset$. For $\sigma \in \mathfrak{S}$ we put

$$S(\sigma) = \{1\} \cup \{\tau \mid \tau \in \mathfrak{S}, V_\tau = V_\sigma\}$$

and

$$\Sigma = \{S(\sigma) \mid \sigma \in \mathfrak{S}\}.$$

Finally, $L = \langle \mathfrak{S} \rangle$. By 35.2, the $S(\sigma)$ are elementary abelian p-groups.

35.4 Lemma. a) *If $S \in \Sigma$, then $\mathfrak{R}_L(S) \cap S^x = \{1\}$ for all $x \in L \setminus \mathfrak{R}_L(S)$.*
b) *The elements of Σ are conjugate in L.*

PROOF. a) If $|\Sigma| = 1$, then there is nothing to prove. Hence let $|\Sigma| > 1$ and let R and S be distinct elements of Σ. If $\sigma \in R \cap \mathfrak{R}_L(S)$, then σ fixes V_R and V_S by 35.1 c). Using 35.1 c) once more, we find $\sigma = 1$.
b) follows from 35.1 c) and 15.1. $\square$

Let R, S be two distinct elements of Σ and let $1 \neq \sigma \in R$ and $\tau \in S$. Then $V = V_R \oplus V_S$ and V_S^σ is a diagonal of $V_R \oplus V_S$. By 2.1, we may assume that $V = U \oplus U$ and $V(\infty) = V_R$, $V(0) = V_S$ and $V(1) = V_S^\sigma$. Then $(0,x)^\sigma = (0,x)$ and $(x,0)^\sigma = (x,0) + (x,0)^{\sigma-1}$. As $\sigma - 1$ is an isomorphism from V_S onto V_R, we see that $(x,0)^\sigma = (x,y)$ for some y. Moreover $(x,0)^\sigma \in V(1)$. Hence $(x,0)^\sigma = (x,x)$. Therefore $(x,y)^\sigma = (x,x+y)$ for all $x, y \in U$. Furthermore $V(\infty)^\tau \in \nu$. Thus there exists $\alpha \in GL(U)$ with $V(\infty)^\tau = \{(x^\alpha, x) | x \in U\}$. As $\tau - 1$ maps V_R onto V_S, we have $(0,x)^\tau = (x^\alpha, x)$ and hence $(x,y)^\tau = (x + y^\alpha, y)$ for all $x, y \in U$.

For $a, b, c, d \in \text{End}_K(U)$, we identify the mapping $\lambda : (x,y) \to (x^a + y^c, x^b + y^d)$ with the matrix $\begin{pmatrix} a & b \\ c & d \end{pmatrix}$. The mapping $\lambda \to \begin{pmatrix} a & b \\ c & d \end{pmatrix}$ is an isomorphism of all these mappings onto the ring of all (2×2)-matrices over $\text{End}_K(U)$.

We shall keep these notations fixed.

35.5 Lemma. *Let R, S be distinct elements of Σ such that $\langle R, S \rangle \cong SL(2, q)$, where q is a power of p. Then there exists a field $F \subseteq \text{End}_K(U)$ such that $|F| = q$ and*

$$\langle R, S \rangle = \left\{ \begin{pmatrix} a & b \\ c & d \end{pmatrix} \middle| a, b, c, d \in F, ad - cb = 1 \right\}.$$

Moreover

$$R = \left\{ \begin{pmatrix} 1 & b \\ 0 & 1 \end{pmatrix} \middle| b \in F \right\} \quad \text{and} \quad S = \left\{ \begin{pmatrix} 1 & 0 \\ b & 1 \end{pmatrix} \middle| b \in F \right\}.$$

Finally, there exists $\tau \in S$ with $(\sigma - 1)(\tau - 1) + (\tau - 1)(\sigma - 1) = 1$.

PROOF. Let R^* be a Sylow p-group of $\langle R, S \rangle$ containing R. As all subgroups of order p in $SL(2, q)$ are conjugate, R^* consists only of ν-shears. This yields $R^* = R$, i.e., R and S are Sylow p-subgroups of $\langle R, S \rangle$. This implies that there exists an isomorphism i of $SL(2, q)$ onto $\langle R, S \rangle$ mapping $\{\begin{pmatrix} 1 & \xi \\ 0 & 1 \end{pmatrix} | \xi \in GF(q)\}$ onto R and $\{\begin{pmatrix} 1 & 0 \\ \xi & 1 \end{pmatrix} | \xi \in GF(q)\}$ onto S and $\begin{pmatrix} 1 & 1 \\ 0 & 1 \end{pmatrix}$ onto σ.

Let $\rho \in R$. Then $V(0)^\rho \in \nu$. From this we infer the existence of $t \in \text{End}_K(U)$ with $(x,y)^\rho = (x, x^t + y)$ for all $x, y \in U$. Hence we have

$$\begin{pmatrix} 1 & \xi \\ 0 & 1 \end{pmatrix}^i = \begin{pmatrix} 1 & t'(\xi) \\ 0 & 1 \end{pmatrix}$$

with $t'(\xi) \in \text{End}_K(U)$ for all $\xi \in GF(q)$.

Similarly $\begin{pmatrix} 1 & 0 \\ \xi & 1 \end{pmatrix}^i = \begin{pmatrix} 1 & 0 \\ t(\xi) & 1 \end{pmatrix}$ with $t(\xi) \in \text{End}_K(U)$ for all $\xi \in GF(q)$.

As i is an isomorphism, we obtain that t and t' are additive. Moreover $t'(1) = 1$, as $\begin{pmatrix} 1 & 1 \\ 0 & 1 \end{pmatrix}^i = \sigma$.

Let $0 \neq \xi \in GF(q)$. Then

$$\begin{pmatrix} 1 & 0 \\ \xi & 1 \end{pmatrix} \begin{pmatrix} 1 & -\xi^{-1} \\ 0 & 1 \end{pmatrix} \begin{pmatrix} 1 & 0 \\ \xi & 1 \end{pmatrix} = \begin{pmatrix} 0 & \xi^{-1} \\ \xi & 0 \end{pmatrix}.$$

Therefore

(a) $$\begin{pmatrix} 0 & -\xi^{-1} \\ \xi & 0 \end{pmatrix}^i = \begin{pmatrix} 1 - t'(\xi^{-1})t(\xi) & -t'(\xi^{-1}) \\ 2t(\xi) - t(\xi)t'(\xi^{-1})t(\xi) & 1 - t'(\xi^{-1})t(\xi) \end{pmatrix}.$$

Furthermore

$$\begin{pmatrix} 0 & -\xi^{-1} \\ \xi & 0 \end{pmatrix}\begin{pmatrix} 1 & 1 \\ 0 & 1 \end{pmatrix} = \begin{pmatrix} 1 & 0 \\ -\xi^2 & 1 \end{pmatrix}\begin{pmatrix} 0 & -\xi^{-1} \\ \xi & 0 \end{pmatrix}.$$

An easy computation then shows

(b) $\qquad\qquad 1 - t'(\xi^{-1})t(\xi) = 0 \qquad$ for all $\xi \in \mathrm{GF}(q)^*$.

Hence by (a) and (b)

(c) $\qquad\qquad \begin{pmatrix} 0 & -\xi^{-1} \\ \xi & 0 \end{pmatrix}^i = \begin{pmatrix} 0 & -t(\xi)^{-1} \\ t(\xi) & 0 \end{pmatrix}.$

We infer from (b) and $t'(1) = 1$ that $t(1) = 1 = -t(-1)$. Moreover

$$\begin{pmatrix} 0 & 1 \\ -1 & 0 \end{pmatrix}\begin{pmatrix} 0 & -\xi^{-1} \\ \xi & 0 \end{pmatrix} = \begin{pmatrix} \xi & 0 \\ 0 & \xi^{-1} \end{pmatrix}.$$

Hence

$$\begin{pmatrix} \xi & 0 \\ 0 & \xi^{-1} \end{pmatrix}^i = \begin{pmatrix} 0 & 1 \\ -1 & 0 \end{pmatrix}\begin{pmatrix} 0 & -t(\xi)^{-1} \\ t(\xi) & 0 \end{pmatrix} = \begin{pmatrix} t(\xi) & 0 \\ 0 & t(\xi)^{-1} \end{pmatrix}.$$

From this we obtain $t(\xi\eta) = t(\xi)t(\eta)$ for all $\xi, \eta \in \mathrm{GF}(q)$. Hence t is an isomorphism from $\mathrm{GF}(q)$ onto $F = \{t(\xi)\,|\,\xi \in \mathrm{GF}(q)\}$. This proves the first assertion.

Finally, if $\tau = \begin{pmatrix} 1 & 0 \\ 1 & 1 \end{pmatrix}$, then $(\sigma - 1)(\tau - 1) + (\tau - 1)(\sigma - 1) = 1$, as is easily seen. $\qquad\square$

35.6 Lemma. *Assume* $|\Sigma| > 1$. *If there exists* $n \in \mathbb{N}$ *such that* $\langle X, Y \rangle \cong \mathrm{SL}(2, p^n)$ *for any two distinct* $X, Y \in \Sigma$, *then* $L \cong \mathrm{SL}(2, p^n)$.

PROOF. Let R, S, σ, U and F be as above. Pick $\varphi \in \mathfrak{S}$. Then $\varphi = \begin{pmatrix} 1+a & b \\ c & 1+d \end{pmatrix}$ with $a, b, c, d \in \mathrm{End}_K(U)$. Moreover, $b \in F$, as we shall prove now. If $\varphi \in S$, then $b = 0$. Hence we may assume $\varphi \notin S$. Then $S(\varphi) \neq S$. Applying 35.5 to $S(\varphi)$ and S, we see that there exists $\rho \in S$ with $(\rho - 1)(\varphi - 1) + (\varphi - 1)(\rho - 1) = 1$. As $\rho \in S$, we have $\rho = \begin{pmatrix} 1 & 0 \\ x & 1 \end{pmatrix}$ with $x \in F$. Thus

$$\begin{pmatrix} 1 & 0 \\ 0 & 1 \end{pmatrix} = \begin{pmatrix} a & b \\ c & d \end{pmatrix}\begin{pmatrix} 0 & 0 \\ x & 0 \end{pmatrix} + \begin{pmatrix} 0 & 0 \\ x & 0 \end{pmatrix}\begin{pmatrix} a & b \\ c & d \end{pmatrix} = \begin{pmatrix} bx & 0 \\ dx & 0 \end{pmatrix} + \begin{pmatrix} 0 & 0 \\ xa & xb \end{pmatrix},$$

whence $1 = bx = xb$ and hence $b = x^{-1} \in F$.

Let $Q \in \Sigma \backslash \{R\}$. We have to show $Q \subseteq \langle R, S \rangle$. By 35.5, there exists $\varphi \in Q$ with $(\sigma - 1)(\varphi - 1) + (\varphi - 1)(\sigma - 1) = 1$. Let $\varphi = \begin{pmatrix} 1+a & b \\ c & 1+d \end{pmatrix}$. Then $b \in F$, as we have seen. Moreover

$$\begin{pmatrix} 1 & 0 \\ 0 & 1 \end{pmatrix} = \begin{pmatrix} 0 & 1 \\ 0 & 0 \end{pmatrix}\begin{pmatrix} a & b \\ c & d \end{pmatrix} + \begin{pmatrix} a & b \\ c & d \end{pmatrix}\begin{pmatrix} 0 & 1 \\ 0 & 0 \end{pmatrix} = \begin{pmatrix} c & d \\ 0 & 0 \end{pmatrix} + \begin{pmatrix} 0 & a \\ 0 & c \end{pmatrix}.$$

Therefore, $c = 1$ and $a = -d$. As φ is a ν-shear, $(\varphi - 1)^2 = 0$. Thus

$$0 = \begin{pmatrix} a & b \\ 1 & -a \end{pmatrix}^2 = \begin{pmatrix} a^2 + b & ab - ba \\ 0 & b + a^2 \end{pmatrix}.$$

This yields $b = -a^2$. Hence

$$\varphi = \begin{pmatrix} 1 + a & -a^2 \\ 1 & 1 - a \end{pmatrix}.$$

If $x \in F$, then $\xi = \begin{pmatrix} 1 & x \\ 0 & 1 \end{pmatrix} \in \langle R, S \rangle$. Therefore $\varphi^\xi \in L$. Now

$$\varphi^\xi = \begin{pmatrix} 1 & -x \\ 0 & 1 \end{pmatrix}\begin{pmatrix} 1 + a & -a^2 \\ 1 & 1 - a \end{pmatrix}\begin{pmatrix} 1 & x \\ 0 & 1 \end{pmatrix} = \begin{pmatrix} 1 + a - x & -(a - x)^2 \\ 1 & 1 - a + x \end{pmatrix}.$$

By the remark made above $-(a - x)^2 \in F$ for all $x \in F$. In particular, $a^2, (a - 1)^2 \in F$. Hence $2a = a^2 - (a - 1)^2 + 1 \in F$. Consequently $a \in F$, if $p \neq 2$. Moreover, $(1 + a)(1 - a) + a^2 = 1$. Therefore, $\varphi \in \langle R, S \rangle$ by 35.5.

Assume finally that $p = 2$. If $a^2 = 0$, then φ leaves invariant V_S. Hence $V_\varphi = V_S$ by 35.1. This yields $Q = S$. Hence we may assume $a^2 \neq 0$. Let F_0 be the subfield of F generated by a^2. As the characteristic is 2, there exists $x \in F_0$ with $x^2 = a^2$. As x lies in the algebra generated by a^2, it lies in the algebra generated by a. Thus $ax = xa$, whence $(a - x)^2 = a^2 - x^2 = 0$. Therefore, $\varphi^\xi \in S$ by what we have already seen. As $\xi \in \langle R, S \rangle$, we get $\varphi \in \langle R, S \rangle$. This yields finally $Q \subseteq \langle R, S \rangle$. $\qquad \square$

35.7 Lemma. *Let $p \geqslant 3$ and assume $|\Sigma| > 1$. If for any $\sigma, \tau \in \mathfrak{S}$ with $\sigma\tau \neq \tau\sigma$, there exists an $r \in \mathbb{N}$ (depending on σ and τ) such that $\langle \sigma, \tau \rangle \cong \mathrm{SL}(2, p^r)$, then there exists $n \in \mathbb{N}$ with $L \cong \mathrm{SL}(2, p^n)$.*

PROOF. Pick $R, S \in \Sigma$ with $R \neq S$ and let $1 \neq \sigma \in R$. Let $1 \neq \tau \in S$. By assumption $H = \langle \sigma, \tau \rangle \cong \mathrm{SL}(2, p^r)$ for some r. Put $R^* = R \cap H$ and $S^* = S \cap H$. Let X be a Sylow p-subgroup of H containing R^*. As all subgroups of order p of H are conjugate within H, we deduce $X = R^*$. Likewise, S^* is a Sylow p-subgroup of H. Hence we may apply 35.5 to $H = \langle R^*, S^* \rangle$. Fix σ and denote by $F(\tau)$ the subfield of $\mathrm{End}_K(U)$ given by 35.5; also put $F = \bigcup_{\tau \in S \setminus \{1\}} F(\tau)$.

Let $\tau = \begin{pmatrix} 1 & 0 \\ x & 1 \end{pmatrix}$ and $\tau' = \begin{pmatrix} 1 & 0 \\ x' & 1 \end{pmatrix}$ with $x \in F(\tau)$ and $x' \in F(\tau')$. As $\tau\tau' = \begin{pmatrix} 1 & 0 \\ x + x' & 1 \end{pmatrix}$, we have $x + x' \in F(\tau\tau')$. Hence F is a group under addition. Moreover $1 \in F$ and for $0 \neq x \in F$, we have $x^{-1} \in F$. Thus F is a field by 34.9.

As σ and $\begin{pmatrix} 1 & 0 \\ -1 & 1 \end{pmatrix}$ are in $\langle R, S \rangle$, the group $\langle R, S \rangle$ contains

$$\begin{pmatrix} 1 & 1 \\ 0 & 1 \end{pmatrix}\begin{pmatrix} 1 & 0 \\ -1 & 1 \end{pmatrix}\begin{pmatrix} 1 & 1 \\ 0 & 1 \end{pmatrix} = \begin{pmatrix} 0 & 1 \\ -1 & 0 \end{pmatrix}$$

and hence all elements of the form

$$\begin{pmatrix} 0 & 1 \\ -1 & 0 \end{pmatrix}\begin{pmatrix} 1 & 0 \\ -x & 1 \end{pmatrix}\begin{pmatrix} 0 & -1 \\ 1 & 0 \end{pmatrix} = \begin{pmatrix} 1 & x \\ 0 & 1 \end{pmatrix}.$$

All these elements belong to R. Hence $\langle R, S \rangle \cong \mathrm{SL}(2, F)$. As $|F| = |R| = |X|$ for all $X \in \Sigma$ by 35.4, the lemma follows from 35.6. $\qquad \square$

35.8 Lemma. *Let ν be a partial spread of $V = U \oplus U$ with $V(0), V(1)$, $V(\infty), V(\alpha) \in \nu$, where $\alpha \in \mathrm{GL}(U)$. If the mapping ρ defined by $(x, y)^\rho = (x^{\alpha^{-1}}, y^\alpha)$ leaves ν invariant, then $V(\alpha^i) \in \nu$ for all $i \in \mathbb{N}$.*

PROOF. $V(\alpha^0)$ and $V(\alpha^1)$ are in ν. Let $n > 1$ and assume $V(\alpha^i) \in \nu$ for all $i < n$. Then

$$V(\alpha^{n-2})^\rho = \left\{ (x^{\alpha^{-1}}, x^{\alpha^{n-2}\alpha}) \mid x \in U \right\} = \left\{ (x, x^{\alpha^n}) \mid x \in U \right\} = V(\alpha^n). \qquad \square$$

35.9 Lemma. *Assume $p > 2$ and let σ and τ be two ν-shears with $\sigma\tau \neq \tau\sigma$. Then either $\langle \sigma, \tau \rangle \cong \mathrm{SL}(2, p^r)$ for some r or $p = 3$ and $\langle \sigma, \tau \rangle \cong \mathrm{SL}(2, 5)$.*

PROOF. We choose our notation as usual. Hence $\sigma = \left(\begin{smallmatrix} 1 & 1 \\ 0 & 1 \end{smallmatrix} \right)$ and $\tau = \left(\begin{smallmatrix} 1 & 0 \\ \alpha & 1 \end{smallmatrix} \right)$. Let A be the subalgebra of $\mathrm{End}_F(U)$ generated by α, where F is the prime field of K. Then we obtain by induction

$$(1) \qquad \langle \sigma, \tau \rangle \subseteq \left\{ \begin{pmatrix} a & b \\ c & d \end{pmatrix} \Big| a, b, c, d \in A,\ ad - bc = 1 \right\}.$$

Denote by μ the set of all components of ν which are axes of non-trivial ν-shears belonging to $\langle \sigma, \tau \rangle$ and let M be the set of all $a \in \mathrm{End}_K(U)$ with $\{(x^a, x) \mid x \in U\} \in \mu$. As μ is a partial spread, we have

(2) If $a, b \in M$ and $a \neq b$, then $a - b \in \mathrm{GL}(U)$. In particular $M \setminus \{0\} \subseteq \mathrm{GL}(U)$, as $0 \in M$.

V is also a vector space over $F = \mathrm{GF}(p)$. Let λ be an eigenvalue of α in some extension L of F. We may assume $L = F(\lambda)$.

Consider $V_L = V \otimes_F L$ and denote by $\mu \otimes L$ the set of all $X_L = X \otimes_F L$ for $X \in \mu$. Then $\mu \otimes L$ is a partial spread of V_L which is left invariant by $\sigma \otimes 1$ and $\tau \otimes 1$. Moreover $\langle \sigma, \tau \rangle \cong \langle \sigma \otimes 1, \tau \otimes 1 \rangle$. Therefore we may assume $K = L$.

As $\lambda \in K$, there exists $u \in U$ with $u^\alpha = u\lambda$ and $u \neq 0$. Hence $(0, u)^\tau = (u^\alpha, u) = (u\lambda, u)$. Moreover, $(u, 0)^\tau = (u, 0)$, $(0, u)^\sigma = (0, u)$, and $(u, 0)^\sigma = (u, u)$. Hence $W = (u, 0)K \oplus (0, u)K$ is a subspace of rank 2 of V which is left invariant by $\langle \sigma, \tau \rangle = H$. Moreover, $\left(\begin{smallmatrix} 1 & 1 \\ 0 & 1 \end{smallmatrix} \right)$ is the restriction of σ to W and $\left(\begin{smallmatrix} 1 & 0 \\ \lambda & 1 \end{smallmatrix} \right)$ is the restriction of τ to W. Let $\bar{H}$ be the restriction of H to W. By a theorem of Dickson (see e.g. Gorenstein [1968, Theorem 2.8.4, p. 44]), either $\bar{H} \cong \mathrm{SL}(2, K)$ or $K \cong \mathrm{GF}(9)$ and $\bar{H} \cong \mathrm{SL}(2, 5)$, since $K = F(\lambda)$. This yields in particular:

(3) If λ' is an eigenvalue of α, then $K = F(\lambda')$.

Next we prove:

(4) If $\bar{H} \cong \mathrm{SL}(2, K)$, then $S(\tau) = S = \{ \left(\begin{smallmatrix} 1 & 0 \\ a & 1 \end{smallmatrix} \right) \mid a \in M \}$. If $K \cong \mathrm{GF}(9)$ and $\bar{H} \cong \mathrm{SL}(2, 5)$, then $M = \{ \xi + \eta\alpha \mid \xi, \eta \in F \}$. In particular, $M(+)$ is a group.

Let $\rho \in S$. Then $\rho = \left(\begin{smallmatrix} 1 & 0 \\ t & 1 \end{smallmatrix} \right)$ with $t \in \mathrm{End}_K(U)$. Moreover, if $a \in M$, then $\{(x^a, x) \mid x \in U\}^\rho = \{(x^{a+t}, x) \mid x \in U\}$. Hence $M + t = M$. As $0 \in M$, we obtain $t \in M$.

Case 1: $\bar{H} \cong \mathrm{SL}(2, K)$. In this case, S acts transitively on $\mu \setminus \{V(0)\}$. By the remark made above, $S = \{ \left(\begin{smallmatrix} 1 & 0 \\ a & 1 \end{smallmatrix} \right) \mid a \in M \}$.

Case 2: $K \cong \mathrm{GF}(9)$ and $\overline{H} \cong \mathrm{SL}(2,5)$. As $\sigma^2 = \begin{pmatrix} 1 & -1 \\ 0 & -1 \end{pmatrix}$, we have $V(1), V(-1) \in \mu$. Moreover,

$$V(\infty)^{\langle \tau \rangle} = \big\{ V(\infty), \{(x^\alpha, x) \,|\, x \in U\}, \{(x^{2\alpha}, x) \,|\, x \in U\} \big\}.$$

$$V(1)^{\langle \tau \rangle} = \big\{ V(1), \{(x^{1+\alpha}, x) \,|\, x \in U\}, \{(x^{1+2\alpha}, x) \,|\, x \in U\} \big\},$$

$$V(-1)^{\langle \tau \rangle} = \big\{ V(-1), \{(x^{-1+\alpha}, x) \,|\, x \in U\}, \{(x^{-1+2\alpha}, x) \,|\, x \in U\} \big\}.$$

This implies $M \supseteq \{\xi + \alpha\eta \,|\, \xi, \eta \in F\}$. Let m be the minimal polynomial of α over $\mathrm{GF}(3)$. Then $m(\lambda) = 0$. Hence degree$(m) \geqslant 2$. Therefore, 1 and α are linearly independent over F. Hence $9 = |\{\xi + \alpha\eta \,|\, \xi, \eta \in F\}| \leqslant |M|$. As $\mathrm{SL}(2,5)$ contains exactly 10 Sylow 3-subgroups, $|M| \leqslant 9$. Hence $M = \{\xi + \eta\alpha \,|\, \xi, \eta \in F\}$. This proves (4).

(5) Let G be the subgroup of $\mathrm{GL}(V)$ which leaves invariant μ. Then $\begin{pmatrix} 1 & 0 \\ a & 1 \end{pmatrix} \in G$ for all $a \in M$.

This follows immediately from (4).

(6) $\begin{pmatrix} 0 & -1 \\ 1 & 0 \end{pmatrix} \in G$.

As $1 \in M$, we have $\rho = \begin{pmatrix} 1 & 0 \\ 1 & 1 \end{pmatrix} \in G$ by (5). Furthermore $\sigma^{-1} = \begin{pmatrix} 1 & -1 \\ 0 & 1 \end{pmatrix}$. Hence $\begin{pmatrix} 0 & -1 \\ 1 & 0 \end{pmatrix} = \begin{pmatrix} 1 & 0 \\ 1 & 1 \end{pmatrix}\begin{pmatrix} 1 & -1 \\ 0 & 1 \end{pmatrix}\begin{pmatrix} 1 & 0 \\ 1 & 1 \end{pmatrix} \in G$.

(7) $-\alpha^{-1} \in M$.

Put $\rho = \begin{pmatrix} 0 & -1 \\ 1 & 0 \end{pmatrix}$. Then $V(0)^{\tau\rho} \in \mu$ by (6). On the other hand $V(0)^{\tau\rho} = \{(-x^{\alpha^{-1}}, x) \,|\, x \in U\}$. Hence $-\alpha^{-1} \in M$.

(8) $\begin{pmatrix} 1 & a \\ 0 & 1 \end{pmatrix} \in G$ for all $a \in M$.

This follows from (6), (5), and $\begin{pmatrix} 0 & 1 \\ -1 & 0 \end{pmatrix}\begin{pmatrix} 1 & 0 \\ -a & 1 \end{pmatrix}\begin{pmatrix} 0 & -1 \\ 1 & 0 \end{pmatrix} = \begin{pmatrix} 1 & a \\ 0 & 1 \end{pmatrix}$.

(9)
$$\begin{pmatrix} \alpha & 0 \\ 0 & \alpha^{-1} \end{pmatrix} \in G.$$

This follows from (5), (6), (7), (8), and $\begin{pmatrix} 1 & \alpha \\ 0 & 1 \end{pmatrix}\begin{pmatrix} 1 & 0 \\ -\alpha^{-1} & 1 \end{pmatrix}\begin{pmatrix} 1 & \alpha \\ 0 & 1 \end{pmatrix}\begin{pmatrix} 0 & -1 \\ 1 & 0 \end{pmatrix} = \begin{pmatrix} \alpha & 0 \\ 0 & \alpha^{-1} \end{pmatrix}$.

(10) $M = A \cong K$.

By 35.8 and (9), $V(\alpha^{-i}) \in \mu$ for all $i \in \mathbb{N}$. On the other hand, $V(\alpha^{-i}) = \{(x, x^{\alpha^{-i}}) \,|\, x \in U\} = \{(x^{\alpha^i}, x) \,|\, x \in U\}$. Hence $\alpha^i \in M$ for all i. As $M(+)$ is a group, $A = \mathrm{GF}(p)[\alpha] \subseteq M$. Let m be the minimal polynomial of α over F. Then $m(\lambda) = 0$. Hence degree$(m) > [K:F]$. This yields $|A| \geqslant |K|$. On the other hand, $|M| = |K|$. Therefore, $M = A$. Thus M is a subalgebra of $\mathrm{End}_F(V)$. Hence, by (2), M is a field isomorphic to K.

Now we can finish the proof of 35.9. By (1), $H \subseteq I = \{\begin{pmatrix} a & b \\ c & d \end{pmatrix} \,|\, a,b,c,d \in A, \; ad - bc = 1\}$. By (10), $I = \mathrm{SL}(2,A)$. If $\overline{H} \cong \mathrm{SL}(2,K)$, then $H = I$ by (10). Assume $p = 3$ and $\overline{H} \cong \mathrm{SL}(2,5)$. Let $\xi \in H$ and assume that ξ induces the identity on W. Obviously $\xi = \begin{pmatrix} a & 0 \\ 0 & a \end{pmatrix}$ with $a \in A$. As $A = M$, we infer from (2) and $1 \in M$, that $a = 1$. Hence $\overline{H} = H$. $\square$

35.10 Theorem (Hering 1972b, Ostrom 1970a, 1974). *Let V be a vector space of rank $2r$ over $\mathrm{GF}(p)$, p being a prime, let ν be a partial spread of V*

with $|\nu| \geqslant 2$, *and let* $G \subseteq GL(V)$ *be a group leaving invariant* ν. *Moreover, let* $\mathfrak{S}$ *be the set of all non-trivial* ν-*shears in* G *and assume* $\mathfrak{S} \neq \emptyset$, *and denote by* μ *the set of all components of* ν *which are axes of shears in* $\mathfrak{S}$. *If* $L = \langle \mathfrak{S} \rangle$, *then one of the following holds*:

a) $|\mu| = 1$ *and* L *is an elementary abelian* p-*group.*

b) $L \cong SL(2, p^s)$ *for some* s. *Moreover* $|\mu| = p^s + 1$ *and* L *operates doubly transitively on* μ.

c) $p = 3$ *and* $L \cong SL(2,5)$. *Furthermore,* $|\mu| = 10$ *and* L *operates transitively on* μ.

d) $p = 2$ *and* $L \cong S(2^{2s+1})$ *for some* s. *In this case* $|\mu| = 2^{2(2s+1)} + 1$ *and* L *operates doubly transitively on* μ.

e) $p = 2$ *and* 4 *does not divide* $|L|$. *Here* $|\mu|$ *is odd and* L *acts transitively on* μ.

PROOF. If $|\mu| = 1$, then L is an elementary abelian p-group by 35.2. Therefore we may assume $|\mu| > 1$.

Case 1: $p = 2$. In this case, 35.4 determines the structure of L almost completely. In fact: By a theorem of Hering [1972a, Theorem 1], L is isomorphic to one of the groups $SL(2,q)$, $S(q)$, $PSU(3,q^2)$, or $SU(3,q^2)$, where q is a power of 2, or $L = \mathfrak{O}(L)S(\sigma)$ where $\mathfrak{O}(L)$ is the largest normal subgroup of odd order of L and $S(\sigma)$ is the group of all shears with axis V_σ and σ is any element of $\mathfrak{S}$. Moreover, $S(\sigma)$ is a Frobenius complement in the latter case. Hence $|S(\sigma)| = 2$, as $S(\sigma)$ is elementary abelian. This is case e).

If $L \cong SL(2,q)$ or $S(q)$, we obviously have case b) or d).

Assume $L \cong PSU(3,q^2)$ or $SU(3,q^2)$, then $\langle R, S \rangle \cong SL(2,q)$ as follows from 32.9, provided $R, S \in \Sigma$ and $R \neq S$. Hence $L \cong SL(2,q)$ by 35.6, a contradiction.

Case 2: $p > 2$. In this case $L \cong SL(2, p^s)$ by 35.9 and 35.7 unless $p = 3$ and $\langle \sigma, \tau \rangle \cong SL(2,5)$ for two elements $\sigma, \tau \in \mathfrak{S}$. Thus assume that $p = 3$ and that there exist $\sigma, \tau \in \mathfrak{S}$ with $\langle \sigma, \tau \rangle \cong SL(2,5)$. Let $X, Y \in \Sigma$ with $X \neq Y$. Pick $\xi \in X \setminus \{1\}$ and $\eta \in Y \setminus \{1\}$. By 35.9, either $\langle \xi, \eta \rangle \cong SL(2,3^s)$ for some s or $\langle \xi, \eta \rangle \cong SL(2,5)$. In either case, there exists a field $K(\eta)$ such that Y consists of elements of the form $\begin{pmatrix} 1 & 0 \\ \kappa & 1 \end{pmatrix}$ with $\kappa \in K(\eta)$. Put $K = \bigcup_{1 \neq \eta \in Y} K(\eta)$. Then K is a field by Glauberman's theorem. Moreover, $\langle \xi, Y \rangle \subseteq SL(2, K)$. It follows from 14.1 that either $\langle \xi, Y \rangle \cong SL(2,3^t)$ or $\langle \xi, Y \rangle \cong SL(2,5)$. In either case $\langle X, Y \rangle = \langle \xi, Y \rangle$. Moreover, $|Y| = 3^t$ in the first case. If $t > 1$, then this yields $\langle X, Y \rangle \cong SL(2,3^t)$ for each choice of X and Y. Hence we may assume $t = 1$ by 35.6.

Let L^* be the group induced by L on μ. Then L^* is generated by a conjugacy class D of elements of order 3 each of which has exactly one fixed element in μ. This implies that $a, b \in D$ and $ab = ba$ forces b to be an element of $\langle a \rangle$. Finally, for $a, b \in D$ with $b \notin \langle a \rangle$ we have $\langle a, b \rangle \cong A_4$ or A_5. Hence $L^* \cong A_5$ by 33.1 whence $L \cong SL(2,5)$. $\square$

There are actually examples of non-desarguesian translation planes of order 3^r where the group generated by all shears with axes through 0 is $SL(2,5)$ (Prohaska 1977). Moreover, in the Hall plane $H(q)$ of even order, the group generated by all those shears is dihedral of order $2(q + 1)$, as is easily seen. Also, if $\mathfrak{A}$ is a proper nearfield plane of even order, then the group generated by all shears of $\mathfrak{A}$ with axes through 0 is of type e).

CHAPTER VI

Flag Transitive Planes

In this chapter we give Huppert's description of all finite soluble 2-transitive permutation groups and Foulser's description of all soluble flag transitive collineation groups of finite affine planes. Using these and some characterizations of finite desarguesian projective planes involving the groups SL(2, q) and PSL(2, q), we are able to prove the theorem of Schulz and Czerwinski on finite translation planes admitting a collineation group acting 2-transitively on l_∞.

36. The Uniqueness of the Desarguesian Plane of Order 8

According to Hall, Swift and Walker 1956, there exists up to isomorphism only one plane of order 8. As this result was found by the aid of a computer, we shall give here a proof for the weaker result that there exists up to isomorphism only one translation plane of order 8. In order to do this, we have to show by 2.6 that there exists up to conjugacy exactly one $\Sigma \subseteq \mathrm{GL}(3,2)$ with the properties:

(1) $1 \in \Sigma$.
(2) If $A, B \in \Sigma$ and if $A \neq B$, then $A - B \in \mathrm{GL}(3,2)$.
(3) $|\Sigma| = 7$.

Now $|\mathrm{GL}(3,2)| = 2^3 \cdot 3 \cdot 7$. Hence, if $A \in \Sigma$, then $A^7 = 1$. Moreover, if $A \neq 1$, then A operates irreducibly on the vector space of rank 3 over GF(2). This yields that the minimal polynomial μ_A of A is irreducible of degree 3. Thus $\mu_A = x^3 + x^2 + 1$ or $\mu_A = x^3 + x + 1$. From this we infer that there are two conjugacy classes of elements of order 7 in $GL(3,2)$, namely

$$\Re_0 = \{A \mid o(A) = 7 \text{ and } \text{trace}(A) = 0\}$$

and

$$\Re_1 = \{A \mid o(A) = 7 \text{ and } \text{trace}(A) = 1\}.$$

There are exactly 8 Sylow 7-subgroups in $GL(3,2)$ and $GL(3,2)$ acts doubly transitively on the set of its Sylow 7-subgroups.

Let $A = \begin{bmatrix} 0 & 1 & 0 \\ 0 & 0 & 1 \\ 1 & 1 & 0 \end{bmatrix}$. Then A is the companion matrix of $x^3 + x + 1$.

Hence A is an element of order 7. Furthermore, $A^3 = \begin{bmatrix} 1 & 1 & 0 \\ 0 & 1 & 1 \\ 1 & 1 & 1 \end{bmatrix}$.

Therefore, $A \in \Re_0$ and $A^3 \in \Re_1$.

Let $B = \begin{bmatrix} 1 & 0 & 0 \\ 0 & 1 & 0 \\ 1 & 0 & 1 \end{bmatrix}$. Then $B = B^{-1}$ and $C = B^{-1}AB = \begin{bmatrix} 0 & 1 & 0 \\ 1 & 0 & 1 \\ 1 & 0 & 0 \end{bmatrix}$.

Moreover,

$$C^2 = \begin{bmatrix} 1 & 0 & 1 \\ 1 & 1 & 0 \\ 0 & 1 & 0 \end{bmatrix}, \qquad C^3 = \begin{bmatrix} 1 & 1 & 0 \\ 1 & 1 & 1 \\ 1 & 0 & 1 \end{bmatrix}, \qquad C^4 = \begin{bmatrix} 1 & 1 & 1 \\ 0 & 1 & 1 \\ 1 & 1 & 0 \end{bmatrix},$$

$$C^5 = \begin{bmatrix} 0 & 1 & 1 \\ 0 & 0 & 1 \\ 1 & 1 & 1 \end{bmatrix}, \qquad C^6 = \begin{bmatrix} 0 & 0 & 1 \\ 1 & 0 & 0 \\ 0 & 1 & 1 \end{bmatrix}.$$

Computing $A - C^i$ for $i = 1, \ldots, 6$ yields that only $A - C^3$ is an element of $GL(3,2)$. Also, among the elements $A^3 - C^i$, only $A^3 - C$ is in $GL(3,2)$. Thus we have proved

36.1 Lemma. *If Π_1 and Π_2 are distinct Sylow 7-subgroups of $GL(3,2)$ and if $1 \neq A \in \Pi_1$, then there exists exactly one $B \in \Pi_2$ with $B \neq 1$ and $A - B \in GL(3,2)$, Moreover,* $\text{trace}(A) = \text{trace}(B) + 1$.

Now we can establish:

36.2 Theorem. *If $\mathfrak{A}$ is a translation plane of order 8, then $\mathfrak{A}$ is desarguesian.*

PROOF. Let Σ be a subset of $GL(3,2)$ with the properties (1), (2) and (3). Let $A, B \in \Sigma \setminus \{1\}$ and assume that A and B belong to distinct Sylow 7-subgroups. Moreover, let $C \in \Sigma \setminus \{A, B, 1\}$. By 36.1, C belongs to a third Sylow 7-subgroup. Moreover, $\text{trace}(A) = \text{trace}(B) + 1$, $\text{trace}(B) =$

trace(C) + 1, and trace(A) = trace(C) + 1. Therefore, trace(A) = trace(B), a contradiction. Thus Σ is a Sylow 7-subgroup of GL(3, 2) and $\mathfrak{A}$ is unique up to isomorphism. $\qquad\square$

37. Soluble Flag Transitive Collineation Groups

In this section we shall determine all soluble flag transitive collineation groups of finite affine planes. At the same time, we shall determine all soluble doubly transitive permutation groups. All the results in this section are due to Huppert 1957 and Foulser 1964.

37.1 Lemma (Huppert). *Let V be an elementary abelian p-group of order p^r and let G be a subgroup of GL(r, p). If G contains a normal abelian subgroup A which acts irreducibly on V, then $L = \mathrm{GF}(p)[A]$ is a Galois field of order p^r and $G \subseteq \Gamma L(V, L) \cong \Gamma L(1, p^r)$.*

This follows immediately from 9.1.

37.2 Lemma. *Let V be an elementary abelian p-group of order p^r and assume that π is a p-primitive divisor of $p^r - 1$. If Σ is a non-trivial π-subgroup and N any subgroup of GL(r, p) which is normalized by Σ, then V is the direct sum of isomorphic irreducible N-submodules.*

PROOF. Σ, and hence ΣN, operate irreducibly on V by 10.5. Therefore, by Clifford's theorem (see e.g. Huppert [1967, V.17.3, p. 565]), V is a completely reducible N-module. Let $V_1, \ldots, V_s$ be the homogeneous components of the N-module V. Then ΣN, and hence Σ, act transitively on $\{V_1, \ldots, V_s\}$. Hence s is a power of π, as s divides $|\Sigma|$. Let $|V_1| = p^t$. Then $|V_i| = p^t$ for all i. Therefore $st = r$, as $V = \bigoplus_{i=1}^{s} V_i$. Since π is a p-primitive divisor of $p^r - 1$, we see that r is the order of p modulo π. This yields that $r = st$ divides $\pi - 1$. Hence $s = 1$. $\qquad\square$

37.3 Lemma. *Let V be an elementary abelian p-group of order p^r and let π be a p-primitive divisor of $p^r - 1$. Moreover, let G be a soluble subgroup of GL(r, p) with $\pi / |G|$. Assume that all abelian normal subgroups of G operate reducibly on V. If A is a maximal abelian normal subgroup of G, then $A = \mathfrak{Z}(G)$ and $|A|$ divides $p - 1$. Furthermore, if N is a minimal non-abelian normal subgroup of G, then N has the following properties:*

a) $N' \subseteq \mathfrak{Z}(N) \subseteq A$ *and* $|N'| = 2$.
b) $N/\mathfrak{Z}(N)$ *is an elementary abelian 2-group of order 2^{2m} and* $|\mathfrak{Z}(N)|$ *divides* 4.

Furthermore, we have:

c) $r = 2^m$ and $\pi = 2^m + 1 = |\Sigma|$, where Σ is a Sylow π-subgroup of G.

d) $G/\mathfrak{C}_G(N/\mathfrak{Z}(N))$ is isomorphic to a subgroup of the symplectic group $\mathrm{Sp}(2m, 2)$.

e) $\mathfrak{C}_G(N/\mathfrak{Z}(N))/A$ is isomorphic to a subgroup of the direct product of $2m$ copies of $\mathfrak{Z}(N)$.

PROOF. Let $\Sigma \in \mathrm{Syl}_\pi(G)$. Then $\Sigma \subseteq \mathfrak{C}_G(A)$ by 17.2. Moreover, $\Sigma \cap A = \{1\}$ by 10.5, as A acts reducibly on V. Therefore A is properly contained in $\mathfrak{C}_G(A)$. Since A is normal in G, we have that $\mathfrak{C}_G(A)$ is normal in G. Thus $\mathfrak{C}_G(A)$ is not abelian, as A is a maximal abelian normal subgroup. Hence $\mathfrak{C}_G(A)$ contains a minimal non-abelian normal subgroup N. As G is soluble N is, whence $N' \neq N$. Therefore N' is abelian by the minimality of N. From $N' \subseteq \mathfrak{C}_G(A)$ we infer that AN' is an abelian normal subgroup of G. This yields $N' \subseteq A$ by the maximality of A. Therefore $N' \subseteq \mathfrak{Z}(N)$, as N is centralized by A. Hence N is nilpotent of class 2. It follows that N contains a non-abelian Sylow q-group. As this group is characteristic in N, we see that N is a q-group by minimality. From 17.2 and the reducibility of A, we get that $|A|$ divides $p^t - 1$ for some $t < r$. Therefore, q divides $p^t - 1$, as $N' \neq \{1\}$. Hence $q \neq \pi$.

Let U be a normal subgroup of G which is properly contained in N. Then U is abelian and hence $U \subseteq A$, i.e., $U \subseteq A \cap N \subseteq \mathfrak{Z}(N)$. As $\mathfrak{Z}(N)$ is one such U, we obtain $A \cap N = \mathfrak{Z}(N)$. Moreover, $N/\mathfrak{Z}(N)$ is characteristically simple and hence an elementary abelian q-group.

N' and $\mathfrak{Z}(N)$ are cyclic since $N' \subseteq \mathfrak{Z}(N) \subseteq A$ and A is cyclic by 17.2. Let $\mu, \nu \in N$. Then $[\mu, \nu] = \mu^{-1}\nu^{-1}\mu\nu$ commutes with μ and ν. Therefore, by Huppert [1967, II.1.2, p. 253], $[\mu, \nu]^q = [\mu^q, \nu] = 1$ and $(\mu\nu)^n = \mu^n\nu^n[\nu, \mu]^{\binom{n}{2}}$ for all $n \in \mathbb{N}$. The first assertion yields that N' is an elementary abelian q-group. Therefore N' is cyclic of order q.

If $q = 2$, then $(\mu\nu)^4 = \mu^4\nu^4$, as $|N'| = 2$. Therefore $U = \{\mu \mid \mu \in N, \mu^4 = 1\}$ is a characteristic subgroup of N. Moreover, $\exp(U) = 4$, as $N' \neq \{1\}$. If $U \neq N$, then $U \subseteq A$ and U is cyclic of order 4. Also, U is the only subgroup of order 4 of N. Hence N is cyclic by Huppert [1967, III.8.4, p. 311]. This contradiction proves $U = N$ and $\exp(N) = 4$.

Let $N' = \langle z \rangle$. If $\mu, \nu \in N$, then $[\mu, \nu] = z^{f(\mu, \nu)}$ with $f(\mu, \nu) \in \mathrm{GF}(q)$. Obviously, $f(\mu, \mu) = 0$ and $f(\mu, \nu) = -f(\nu, \mu)$. Moreover if $\mu'\mu^{-1}, \nu'\nu^{-1} \in \mathfrak{Z}(N)$, then $f(\mu, \nu) = f(\mu', \nu')$. Hence g defined by $g(\mu\mathfrak{Z}(N), \nu\mathfrak{Z}(N)) = f(\mu, \nu)$ is a mapping from $N/\mathfrak{Z}(N)$ into $\mathrm{GF}(q)$. As $N' \subseteq \mathfrak{Z}(N)$, we have $[\mu\nu, \rho] = [\mu, \rho][\nu, \rho]$ by Huppert [1967, III.1.2.c), p. 253]. Hence $f(\mu\nu, \rho) = f(\mu, \rho) + f(\nu, \rho)$. Therefore g is a skew symmetric bilinear form on the $\mathrm{GF}(q)$-vector space $N/\mathfrak{Z}(N)$. Moreover, g is non-degenerate, as $f(\xi, \nu) = 0$ for all $\xi \in N$ implies $\nu \in \mathfrak{Z}(N)$. Therefore $|N/\mathfrak{Z}(N)| = q^{2m}$.

Let k be the degree of a faithful irreducible representation of N over $\mathbb{C}$. As N' centralizes N, each commutator is represented by a scalar matrix by Schur's lemma. Let χ be the character of the representation and let

$\mu \in N \setminus 3(N)$. Then there exists $\nu \in N$ with $[\nu, \mu] = cE \neq E$, where E is the $(k \times k)$-identity matrix. Thus

$$\chi(\mu^{-1}) = \chi(\nu^{-1}\mu^{-1}\nu) = \chi(\nu^{-1}\mu^{-1}\nu\mu\mu^{-1}) = \chi(c\mu^{-1}) = c\chi(\mu^{-1}).$$

As $c \neq 1$, we obtain $\chi(\mu^{-1}) = 0$. Therefore, by the orthogonality relation,

$$|N| = \sum_{\mu \in N} |\chi(\mu)|^2 = \sum_{\mu \in 3(N)} |\chi(\mu)|^2 = k^2 |3(N)|.$$

Hence $k = q^m$. Therefore every absolutely irreducible representation of N has degree q^m (see e.g. Huppert [1967, V.12.9, p. 530]).

By 37.2, V is the direct sum of isomorphic irreducible N-submodules $V_1, \ldots, V_t$. If N does not operate faithfully on V_i for some i, then $3(N)$ does not operate faithfully on V_i, since every non-trivial normal subgroup of N intersects $3(N)$ non-trivially. As the V_j are all isomorphic N-modules, $3(N)$ does not operate faithfully on V, a contradiction. Let K_i be the centralizer of N in $\operatorname{End}_{\mathrm{GF}(p)}(V_i)$. Then $3(N) \subseteq K_i^*$, i.e., each element of $3(N)$ acts as a scalar multiplication on the K_i-vector space V_i.

There exists a finite extension L_i of K_i such that $W_i = V_i \otimes_{K_i} L_i$ splits into a direct sum of absolutely irreducible N-submodules. Since each element of $3(N)$ acts as a scalar multiplication on V_i, every such element acts as the same scalar multiplication on W_i. Therefore, N operates faithfully on each N-submodule of W_i. By the remark made above, q^m divides $\operatorname{rk}_{L_i}(W_i) = \operatorname{rk}_{K_i}(V_i) = \operatorname{rk}_{\mathrm{GF}(p)}(V_i)[K_i : \mathrm{GF}(p)]^{-1}$. From this we infer that q^m divides r.

Σ centralizes $3(N)$ by 17.2. As the centralizer of Σ is cyclic, Σ does not centralize N, nor does any non-trivial subgroup of Σ. Hence Σ is isomorphic to a subgroup of $\operatorname{Aut}(N/3(N))$ by Gorenstein [1968, 5.3.2, p. 178]. Put $W = N/3(N)$. As $q \neq \pi$, we have $W = W_1 \oplus \cdots \oplus W_n$ with irreducible Σ-modules W_i by Maschke's theorem. As Σ is cyclic, Σ acts faithfully on at least one W_i, since Σ acts faithfully on W. We may assume $i = 1$. Put $|W_1| = q^b$. Then $|\Sigma|$ divides $q^b - 1$. Moreover, r divides $\pi - 1$. As q^m divides r, it divides $\pi - 1$. Therefore, assuming $b \leqslant m$, we obtain the contradiction $|\Sigma| \leqslant q^b - 1 < q^m \leqslant \pi - 1$. Thus $m < b \leqslant 2m$. Define h and g by $\pi = hq^m + 1$ and $q^b = g\pi + 1$. Then $h, g \geqslant 1$. Therefore, $q^b = g(hq^m + 1) + 1$, whence $g \equiv -1 \bmod q^m$. Put $g = kq^m - 1$. Then $k \geqslant 1$, as $g \geqslant 1$. Thus $q^b \geqslant (q^m - 1)(q^m + 1) + 1 = q^{2m}$. Hence $b = 2m$ and $W = W_1$. Moreover, $\pi = q^m + 1$. As π is a prime, $q = 2$. Furthermore, $|\Sigma|$ divides $2^b - 1 = 2^{2m} - 1 = (2^m - 1)\pi$, whence $|\Sigma| = \pi$. Finally, $2^m \leqslant r \leqslant \pi - 1 = 2^m$ and hence $r = 2^m$.

As $r = 2^m$, we have that Σ acts absolutely irreducibly on V. Therefore, by Huppert [1967, V.11.10, p. 525], the centralizer of N in $\operatorname{End}_{\mathrm{GF}(p)}(V)$ is the centre of $\operatorname{End}_{\mathrm{GF}(p)}(V)$. Therefore, $A \subseteq 3(G)$ and hence $A = 3(G)$ by the maximality of A. In particular, $|A|$ divides $p - 1$. Therefore, the assertions of the theorem up to c) are proved.

Let $\gamma \in G$ and $\mu, \nu \in N$. Then $[\mu^\gamma, \nu^\gamma] = [\mu, \nu]^\gamma = [\mu, \nu]$, as $N' \subseteq 3(N)$ $\subseteq A = 3(G)$. Therefore, G leaves invariant f and hence g. This proves d).

$3(N)$ is cyclic. Let $3(N) = \langle \zeta \rangle$ and let $\nu_1 3(N), \ldots, \nu_{2m} 3(N)$ be a basis of $N/3(N)$. If $\xi \in \mathfrak{C}_G(N/3(N))$, then $\xi^{-1} \nu_i \xi = \nu_i \xi^{a_i}$. The mapping $\xi \to (\zeta^{a_1}, \ldots, \zeta^{a_{2m}})$ is a homomorphism of G into the direct product of $2m$ copies of $3(N)$ the kernel of which is the centralizer of N in G. As V is an absolutely irreducible N-Module, $A = \mathfrak{C}_G(N)$, as we have remarked above. This proves e).   □

37.4 Corollary. *N acts absolutely irreducibly on V.*

37.5 Corollary. *Σ acts irreducibly and regularly on $N/3(N)$.*

37.6 Lemma. *Let Ω be a finite set and let G be a soluble permutation group acting primitively on Ω. Then G contains exactly one minimal normal subgroup. If V is this minimal normal subgroup, then V is an elementary abelian p-group and $|V| = |\Omega|$.*

The proof is obvious.

37.7 Theorem (Huppert 1957, Foulser 1964). *Let G be either a soluble doubly transitive permutation group on a finite set Ω or a soluble flag transitive collineation group of a finite affine plane $\mathfrak{A}$ of order n. Then G is the semidirect product of an elementary abelian p-group V of order $|\Omega|$ respectively n^2 and a group H, where $H \subseteq \Gamma L(1, |V|)$ unless $|V| \in \{3^2, 5^2, 7^2, 11^2, 23^2, 3^4\}$. In the latter cases, $H \subseteq \Gamma L(1, |V|)$ or H is one of 16 exceptional groups.*

PROOF. It follows from 37.6 and 15.16 that $G = VH$ where V is a normal elementary abelian p-group and $V \cap H = \{1\}$. By 37.1, we may assume that all normal abelian subgroups of H operate reducibly on V. Let $|V| = p^f$. Then $f = 2r$, if G is a flag transitive group. Assume that there exists a p-primitive divisor π of $p^f - 1$. Then π divides $|H|$, as either $p^f - 1$ or $p^r + 1$ divides $|H|$. Hence, applying 37.3, we obtain $f = 2^m$, $\pi = 2^m + 1$, $|3(N)| = 2^j$ with $j = 1$ or 2. Moreover, 2^j divides $p - 1$, as 2^j divides $|A|$ and $|A|$ divides $p - 1$. Also,

$$|H| = |H/\mathfrak{C}_H(N/3(N))| |\mathfrak{C}_H(N/3(N))/A| |A|.$$

Therefore $|H|$, and hence $p^{2^{m-1}} + 1$, divide $2^{m^2 + 2jm}(p - 1)\prod_{i=1}^m (2^{2i} - 1)$, since $|Sp(2m, 2)| = 2^{m^2} \prod_{i=1}^m (2^{2i} - 1)$ by Huppert [1967, II.9.13, p. 220].

If $m = 1$, then H permutes the subspaces of rank 1 of V transitively. As all these subspaces are fixed by A, we obtain that $p + 1$ divides $2^{1 + 2j} \cdot 3$. If $j = 2$, then 4 does not divide $p + 1$. Hence $p + 1 = 6$, i.e., $p = 5$. (Observe that $\pi = 3$ divides $p + 1$.) If $j = 1$ then $p + 1$ divides $2^3 \cdot 3$, whence $p = 5, 11$ or 23.

If $m \geqslant 2$, then $p^{2^{m-1}} + 1$ is not divisible by 4. Hence $p^{2^{m-1}} + 1$ divides $2\prod_{i=1}^m (2^{2i} - 1)$.

If $m = 2$, then $p^2 + 1$ divides $2 \cdot 3 \cdot 15$. As $p^2 \equiv 0 \bmod 3$ or $p^2 \equiv 1 \bmod 3$, we obtain that 3 does not divide $p^2 + 1$. Hence $p^2 + 1 = 10$, i.e. $p^2 = 9$.

Let $m \geqslant 3$. Choose v such that $3^v \leqslant p < 3^{v+1}$. Then $v \geqslant 1$. Moreover, $2(2^2 - 1)(2^4 - 1)(2^6 - 1) = 2 \cdot 3 \cdot 15 \cdot 63 < 3^8$. Furthermore, $2^3 < 3^2$ yields $2^i < 3^{i-1}$ for all $i \geqslant 3$. Therefore,

$$3^{v2^{m-1}} \leqslant p^{2^{m-1}} + 1 \leqslant 2 \prod_{i=1}^{m} (2^{2i} - 1) < 3^8 \prod_{i=4}^{m} 3^{2i-1}.$$

Hence $v2^{m-1} < 3 + 5 + \sum_{i=4}^{m}(2i - 1) = m^2 - 1$. From this inequality we obtain $m \leqslant 5$. As $\pi = 2^m + 1$ is a prime, $m = 4$. In particular, $\pi = 17$. By 37.3, π is the only p-primitive divisor of $p^f - 1$. Moreover, π^2 does not divide $|H|$. Hence $p^{2^3} + 1 = 2 \cdot 17$, a contradiction.

Hence we are left with the case where there is no p-primitive prime divisor of $p^f - 1$. In this case, either $p^f = 2^6$ or $f = 2$ and $p + 1 = 2^s$ by 6.2.

If $f = 2$ and $p + 1 = 2^s$, then H is a subgroup of $\mathrm{GL}(2, p)$ which induces a subgroup U of $\mathrm{PGL}(2, p)$ that acts transitively on the projective line over $\mathrm{GF}(p)$. It follows from 14.1 that U is isomorphic to a subgroup of S_4. Thus $p + 1$ divides 24, whence $p = 3$ or 7.

It remains to consider the case $p^f = 2^6$. Assume first that G is a flag transitive collineation group on a plane $\mathfrak{A}$ of order 8. As $\mathfrak{A}$ is desarguesian by 36.2, we have $H \subseteq \Gamma\mathrm{L}(2, 8)$. If 4 divides $|H|$ then $\mathrm{SL}(2, 8)$ is contained in H, as an inspection of Dickson's subgroup list will tell us. Hence 4 does not divide $|H|$, since H is soluble. Moreover, $|\Gamma\mathrm{L}(2, 8)| = 2^3 \cdot 3^3 \cdot 7^2$. Therefore $|H|$ divides $2 \cdot 3^3 \cdot 7^2$. Let $\overline{H}$ be the group induced by H in $\mathrm{P}\Gamma\mathrm{L}(2, 8)$. Then $|\overline{H}|$ divides $2 \cdot 3^3 \cdot 7^2$. Moreover, $2, 3, 9, 27, 6, 18, 54$ are all $\not\equiv 1 \bmod 7$. Thus $\overline{H}$ contains only one Sylow 7-subgroup Σ. If $\Sigma \neq \{1\}$, then $|\overline{H} : \Sigma| \leqslant 2$. On the other hand, 9 divides $|\overline{H}|$, as $\overline{H}$ acts transitively on l_∞. Thus $\Sigma = \{1\}$. This yields that there is only one Sylow 3-subgroup in $\overline{H}$. Therefore, if Π is a Sylow 3-subgroup of H, then $A = (H \cap 3\mathrm{GL}(2, 8))$ Π is a normal subgroup of H of order $7^i \cdot 3^{2+j}$ with $i = 0$ or 1 and $j = 0$ or 1. If $j = 0$ then A is abelian. Moreover, A acts irreducibly on V. Thus $H \subseteq \Gamma\mathrm{L}(1, 64)$. If $j = 1$, then $B = A \cap \mathrm{GL}(2, 8)$ is an abelian group of order $7^i \cdot 3^2$. Again, B acts irreducibly on V, whence $H \subseteq \Gamma\mathrm{L}(1, 64)$.

We may assume from now on that H acts transitively on $V \setminus \{1\}$. Let A be a maximal abelian normal subgroup H. We may assume that A acts reducibly on V. Let X be an irreducible A-submodule of V and put $\mathfrak{S} = \{X^\gamma \mid \gamma \in H\}$. As H acts transitively on $V \setminus \{1\}$, we obtain that $\mathfrak{S}$ is a partition of V. Moreover, A is contained in the multiplicative group of $\mathrm{K}(V, \mathfrak{S})$. Hence A is cyclic. Moreover, $\mathrm{rk}(X)$ divides $\mathrm{rk}(V) = 6$. Therefore $|A| = 3$ or $|A| = 7$. If $|A| = 7$, then $\mathfrak{S}(V)$ is the desarguesian plane of order 8 and A is not maximal, as we have seen above. Thus $|A| = 3$. As $|\mathrm{Aut}(A)| = 2$, we see that $|H : \mathfrak{C}_H(A)| \leqslant 2$. It follows that $\mathfrak{C}_H(A)$ acts also transitively on $V \setminus \{1\}$. Hence $A \neq \mathfrak{C}_H(A)$ and $\mathfrak{C}_H(A)$ is non-abelian. Let N be a minimal non-abelian normal subgroup of H which is contained in $\mathfrak{C}_H(A)$. It follows that $N' \subseteq A \cap N \subseteq \mathfrak{Z}(N)$, whence N is nilpotent of class

2. The minimality of N then implies that N is a 3-group. Moreover, $N/3(N)$ is an elementary abelian 3-group of order 3^2 or 3^3, as $\mathfrak{C}_H(A) \subseteq \mathrm{GL}(3,4)$ and $|\mathrm{GL}(3,4)| = 2^6 \cdot 3^4 \cdot 7 \cdot 5$. The arguments used in the proof of 37.3 show that $|N/3(N)| = 3^2$.

Let Σ be a Sylow 7-subgroup of $\mathfrak{C}_H(A)$. Then $|\Sigma| = 7$. Moreover Σ centralizes $N/3(N)$ and $3(N)$, as $|\mathrm{GL}(2,3)| = 3 \cdot 2^4$. Hence Σ centralizes N by Gorenstein [1968, 5.3.2, p. 178]. Moreover, $|A\Sigma| = 21$. Therefore, $A\Sigma$ operates irreducibly on V. By Schur's lemma, $\mathfrak{C}_H(A\Sigma)$ is cyclic. On the other hand, $N \subseteq \mathfrak{C}_H(A\Sigma)$, a contradiction. This proves that $3^2, 5^2, 7^2, 11^2, 23^2, 3^4$ are the only exceptional degrees.

We shall construct now all the exceptional groups. If the degree is p^2, then $H \subseteq \mathrm{GL}(2,p)$. Moreover, $|N| = 8$ unless $p = 5$; for in all other cases 4 does not divide $p - 1$. If $p = 5$, then $|N| = 8$ or 16.

Let $p = 3$. Then 4 divides $|H|$. As H is not contained in $\Gamma\mathrm{L}(1,9)$, we get that $H = \mathrm{SL}(2,3)$ or $H = \mathrm{GL}(2,3)$. This accounts for two exceptions.

Let $p \geqslant 5$. If p divides $|H|$, then H contains $\mathrm{SL}(2,p)$ which is impossible, since H is soluble.

As $3(\mathrm{GL}(2,p))$ is cyclic, it contains just one involution. Hence $3(N) \cap \mathrm{SL}(2,p) \neq \{1\}$. Therefore, $\nu^2 \in \mathrm{SL}(2,p)$ for all $\nu \in N$. From this we deduce $|N/(N \cap \mathrm{SL}(2,p))| \leqslant 2$, as $\mathrm{GL}(2,p)/\mathrm{SL}(2,p)$ is cyclic. Moreover $N \cap \mathrm{SL}(2,p)$ is normal in H. If $N \cap \mathrm{SL}(2,p) \neq N$, then $N \cap \mathrm{SL}(2,p) \subseteq 3(N)$ by 37.2 which is impossible, as N is not abelian. Hence $N \cap \mathrm{SL}(2,p) = N$, i.e., $N \subseteq \mathrm{SL}(2,p)$.

Let $p = 5$. Then 3 divides $|H \cap \mathrm{SL}(2,5)|$. Hence $|H \cap \mathrm{SL}(2,5)| = 2^3 \cdot 3$. In particular, $H \cap \mathrm{SL}(2,5) \cong \mathrm{SL}(2,3)$. This yields that $H \cap \mathrm{SL}(2,5)$ acts transitively on $V \setminus \{1\}$. Furthermore, $\mathfrak{N}_{\mathrm{GL}(2,5)}(H \cap \mathrm{SL}(2,5))/(H \cap \mathrm{SL}(2,5))$ is cyclic of order 4 by 14.5. Hence we have three possibilities for H; namely: $H/(H \cap \mathrm{SL}(2,5))$ is cyclic of order 1, 2, or 4. As all subgroups of $\mathrm{SL}(2,5)$ which are isomorphic to $\mathrm{SL}(2,3)$ are conjugate within $\mathrm{SL}(2,5)$ by 14.5, we have exactly three exceptions in this case.

Let $p = 7$. As $|\mathrm{GL}(2,7)| = 2^5 \cdot 3^2 \cdot 7$, we have $|H| = 2^{3+i} \cdot 3^j$ with $i, j \in \{0, 1, 2\}$. Let A be a maximal abelian normal subgroup of H. Then A acts reducibly on V. As H acts transitively on the set of subspaces of rank 1 of V, we see that A fixes all the subspaces of V. This yields that A is the centre of H. Let N be a minimal non-abelian normal subgroup of M. Then $N' \subseteq A$. This yields that N is nilpotent of class 2. In particular, N is either a 2-group or a 3-group. Since N is non-abelian and 3^3 does not divide $|H|$, we have that N is a 2-group. Using the ideas developed in the proof of 37.3, we get $|3(N)| = 2$, as 4 does not divide 6, and hence $N' = 3(N)$. Also $N \subseteq \mathrm{SL}(2,7)$, whence $N/3(N)$ is elementary abelian of order 4. In particular, N is the quaternion group of order 8.

Let C be a cyclic subgroup of order 4 of N and assume that C is normal in H. As 4 does not divide 6, we have that C operates irreducibly on V. Hence $H \subseteq \Gamma\mathrm{L}(1,7^2)$ by 37.1, a contradiction. Thus C is not normal in H.

Since N contains exactly three cyclic subgroups of order 4, we infer that $H/\mathfrak{C}_H(N)$ is not a 2-group. Therefore 3 divides $|H/\mathfrak{C}_H(N)|$.

N acts absolutely irreducibly on V. Hence $\mathfrak{C}_H(N)$ consists only of scalar matrices. Therefore $\mathfrak{C}_H(N)$ fixes all subspaces of rank 1 of V. As H acts transitively on the set of all these subspaces, we have that 8 divides $|H/\mathfrak{C}_H(N)|$. This proves $H/\mathfrak{C}_H(N) = \operatorname{Aut}(N)$. Therefore $|H| = 2^4 \cdot 3^j$ with $j \in \{1, 2\}$.

If $|H| = 2^4 \cdot 3^2$, then $H = \mathfrak{N}_{\mathrm{GL}(2,7)}(N)$. We infer from 14.4 that there is just one exception of this type. Thus we may assume $|H| = 2^4 \cdot 3$ from now on. As the number of subspaces of rank 1 of V is 8, each element σ of order 3 of H fixes at least two subspaces of rank 1. Let $1 \neq \nu \in \mathfrak{Z}(N)$, then $\langle \sigma \nu \rangle$ acts transitively on the set of non-zero vectors of at least one of these subspaces. Hence H acts transitively on $V \backslash \{0\}$. Thus H acts sharply transitively on $V \backslash \{0\}$, because $V \backslash \{0\} = 7^2 - 1 = |H|$. Hence $\sigma \in \mathfrak{Z}(\mathrm{GL}(2,7))$ or $\sigma \in \mathrm{SL}(2,7)$, as all other elements of order 3 fix non-zero vectors. We infer from $|H/\mathfrak{C}_H(N)| = 3 \cdot 8$ that the first case cannot occur. Furthermore, the Sylow 2-subgroups of H are generalized quaternion groups of order 16, as H is a Frobenius complement. Since all elements of order 8 of $\mathrm{GL}(2,7)$ are contained in $\mathrm{SL}(2,7)$, we therefore have $H \subseteq \mathrm{SL}(2,7)$. It follows from 14.4 that there is just one exception of this type.

Let $p = 11$. Then $|H|$ divides $2^4 \cdot 3 \cdot 5^2$, as $|\mathrm{GL}(2,11)| = 11 \cdot 2^4 \cdot 3 \cdot 5^2$. Since N is normal in H, each Sylow 5-subgroup of H centralizes N. Therefore each Sylow 5-subgroup of H is contained in $\mathfrak{Z}(H)$, whence it follows that 5^2 does not divide $|H|$. Moreover, $U = H \cap \mathrm{SL}(2,11) \cong \mathrm{SL}(2,3)$ and U is transitive on the set of subspaces of rank 1. Finally, $\mathfrak{N}_{\mathrm{GL}(2,11)}(U)/U$ is cyclic of order $2 \cdot 5$ by 14.5. Therefore we obtain four exceptional groups. H acts transitively on $V \backslash \{0\}$, if and only if 5 divides $|H|$.

Let $p = 23$. As $|\mathrm{GL}(2,23)| = 2^5 \cdot 3 \cdot 11^2 \cdot 23$, we have that $|H|$ divides $2^5 \cdot 3 \cdot 11^2$. Moreover, the Sylow 11-subgroup of H centralizes N which implies that it lies in the centre. Therefore 11^2 does not divide $|H|$. Moreover, $H \cap \mathrm{SL}(2,23)$ contains a subgroup U of index 1 or 2 which is isomorphic to $\mathrm{SL}(2,3)$. Since H acts transitively on the set of subspaces of rank 1 and since $\mathfrak{C}_H(N)$ fixes all these subspaces, 2^3 divides $|H/\mathfrak{C}_H(N)|$. Thus $|H/\mathfrak{C}_H(N)| = 2^3 \cdot 3 = |\operatorname{Aut}(N)|$. Therefore $|H| = 2^4 \cdot 3 \cdot 11$ or $|H| = 2^4 \cdot 3$. In the first case $H = Z(H \cap \mathrm{SL}(2,23))$, where Z is the Sylow 11-subgroup of $\mathfrak{Z}(\mathrm{GL}(2,23))$, and in the second case, $H \subseteq \mathrm{SL}(2,23)$. This yields two exceptions, the one where 11 divides $|H|$ being transitive on $V \backslash \{0\}$.

Finally, we have to handle the case 3^4. In this case $|\mathfrak{Z}(N)|$ divides $3 - 1 = 2$. Hence $\mathfrak{Z}(N)$ has order 2 and N has order 2^5. Let $\nu \in N$ be an element of order 4. Then $\nu^2 \in N' = \mathfrak{Z}(N)$. Therefore all cyclic subgroups of order 4 of N are normal in N. Let Z be one such subgroup. As N

operates irreducibly on V, we see that N is the direct product of
Z-irreducible subgroups $X_1, X_2, \ldots$. As Z is cyclic of order 4, we obtain
$|X_i| = 9$ and hence $V = X_1 \oplus X_2$. Assume that $\{X_1, X_2\}$ is left invariant by
N. Then X_1 is fixed by a subgroup U of N with $|U| = 2^4$. If U centralizes
Z, then the restriction of U to X_1 is cyclic by Schur's lemma. Hence
$U|_{X_1} = Z|_{X_1}$, as N is of exponent 4. Therefore U contains a subgroup W of
order 4 fixing X_1 pointwise. Moreover, W must operate faithfully and
hence irreducibly on X_2. Therefore W is cyclic of order 4, whence
$\mathfrak{Z}(N) \subseteq W$, a contradiction. This proves that U does not centralize Z. Put
$C = \mathfrak{C}_N(Z)$. Then C permutes X_1 and X_2, as $|C| = 2^4$ and $C \neq U$. (Recall
that $|\mathrm{Aut}(Z)| = 2$.) Therefore X_1 and X_2 are isomorphic Z-modules. If
$\{X_1, X_2\}$ is not left invariant by N, then there is an irreducible
Z-submodule X with $X \cap X_1 = X \cap X_2 = \{0\}$. We infer from this that X_1
and X_2 are also in this case isomorphic Z-modules. It now follows from 9.1
that N is isomorphic to a subgroup of $\Gamma L(2, 9)$ and that $|N : N \cap GL(2, 9)|$
$= 2$.

The group $GL(2, 9)/SL(2, 9)$ is cyclic of order 8. Therefore $(N \cap GL(2, 9))/(N \cap SL(2, 9)) \cong (N \cap GL(2, 9))SL(2, 9)/SL(2, 9)$ is cyclic of order 1 or
2, as the square of every element in N is contained in $\mathfrak{Z}(N)$. Since
$Z \subseteq \mathfrak{Z}(GL(2, 9))$, we see that $|N \cap SL(2, 9)| = 8$. In particular, $N \cap SL(2, 9)$
is the quaternion group of order 8. It follows from 14.4 that $N \cap SL(2, 9)$ is
uniquely determined up to conjugacy within $GL(2, 9)$. Hence $N \cap GL(2, 9)$
is uniquely determined, as $N \cap GL(2, 9) = Z(N \cap SL(2, 9))$.

Let

$$A = \begin{bmatrix} 0 & 1 & 0 & 0 \\ -1 & 0 & 0 & 0 \\ 0 & 0 & 0 & 1 \\ 0 & 0 & -1 & 0 \end{bmatrix}, \qquad B = \begin{bmatrix} 1 & 0 & 1 & 0 \\ 0 & 1 & 0 & 1 \\ 1 & 0 & -1 & 0 \\ 0 & 1 & 0 & -1 \end{bmatrix},$$

$$C = \begin{bmatrix} 0 & 0 & -1 & 0 \\ 0 & 0 & 0 & -1 \\ 1 & 0 & 0 & 0 \\ 0 & 1 & 0 & 0 \end{bmatrix}.$$

Then it is easily checked that $\langle B, C \rangle$ is a quaternion group which is
contained in the centralizer of A. Therefore we may assume $Z = \langle A \rangle$ and
$N \cap SL(2, 9) = \langle B, C \rangle$.

N is generated by its involutions, as N is a minimal non-abelian normal
subgroup of H. Hence there exists an involution $D \in N$ with $A^D = A^{-1}$.
As all cyclic subgroups of order 4 of N contain $\mathfrak{Z}(N) = N'$, they are
normal in N. Hence $C^D \in \langle C \rangle$ and $B^D \in \langle B \rangle$. The cosets of $N/\mathfrak{Z}(N)$
which are distinct from $\mathfrak{Z}(N)$ are:

$\{A, -A\}, \{B, -B\}, \{C, -C\}, \{D, -D\}, \{AB, -AB\}, \{AC, -AC\},$
$\{AD, -AD\}, \{BC, -BC\}, \{BD, -BD\}, \{CD, -CD\}, \{ABC, -ABC\},$
$\{ABD, -ABD\}, \{ACD, -ACD\}, \{BCD, -BCD\}, \{ABCD, -ABCD\}.$

If $BD = -DB$, then it is easily checked that six of these cosets consist of elements of order 4 and nine of elements of order 2. But this is impossible by 37.5. Hence $BD = DB$.

Assume that $C^D = -C$. Then CD is an involution with $A^{CD} = A^{-1}$ and $B^{CD} = -B$. This contradiction proves $C^D = C$. Using all these informations, an easy computation then shows

$$\begin{bmatrix} 1 & 0 & 0 & 0 \\ 0 & -1 & 0 & 0 \\ 0 & 0 & 1 & 0 \\ 0 & 0 & 0 & -1 \end{bmatrix} \in \{D, -D, AD, -AD\}.$$

Hence we may assume that

$$D = \begin{bmatrix} 1 & 0 & 0 & 0 \\ 0 & -1 & 0 & 0 \\ 0 & 0 & 1 & 0 \\ 0 & 0 & 0 & -1 \end{bmatrix}.$$

As $|GL(4,3)|$ is not divisible by 5^2, all subgroups of order 5 in $\mathfrak{N}_{GL(4,3)}(N)$ are conjugate. Therefore we may assume that

$$F = \begin{bmatrix} -1 & 1 & 1 & 1 \\ -1 & -1 & 1 & -1 \\ 0 & 1 & 1 & 0 \\ 0 & -1 & 1 & 0 \end{bmatrix}$$

is in H, as F normalizes N, i.e., H always contains the group $H_0 = \langle A, B, C, D, F \rangle$. Moreover, H_0 acts transitively on the set of non-zero vectors of V.

As $\langle F \rangle$ acts irreducibly on $N/\mathfrak{Z}(N)$, the group $\mathfrak{Z}(N)$ is a maximal abelian normal subgroup of H. Therefore by 37.3 e), $|\mathfrak{C}_H(N/\mathfrak{Z}(N))| \leqslant 2^5$. As $N \subseteq \mathfrak{C}_H(N/\mathfrak{Z}(N))$, we obtain $N = \mathfrak{C}_H(N/\mathfrak{Z}(N))$. Therefore H/N is isomorphic to a subgroup of $Sp(4,2)$. Moreover $Sp(4,2) \cong S_6$ (see e.g., Huppert [1967, II.9.22, p. 227]). This yields that H/N contains exactly one Sylow 5-subgroup since H/N is soluble. Thus H_0 is normal in H/N. Moreover, H/N is isomorphic to a subgroup of the group of all mappings $x \rightarrow ax + b$ with $a, b \in GF(5)$ and $a \neq 0$, as this group is isomorphic to the normalizer of a Sylow 5-subgroup in S_6. Hence H/H_0 is cyclic of order 1, 2 or 4.

Let $G = \begin{bmatrix} 1 & 0 & 1 & 0 \\ 0 & 0 & 0 & -1 \\ -1 & 0 & 1 & 0 \\ 0 & -1 & 0 & 0 \end{bmatrix}$. Then we have the relations:

$$GAG^{-1} = ABC, \qquad GBG^{-1} = B, \qquad GCG^{-1} = CDB,$$
$$GDG^{-1} = DB, \qquad GFG^{-1} = DCBF^2, \qquad G^4 = -B.$$

Hence $H_2 = \langle A, B, C, D, F, G \rangle$, $H_1 = \langle A, B, C, D, F, G^2 \rangle$ and H_0 are groups which occur as stabilizers of a point in an exceptional doubly transitive group of degree 3^4. (Recall that all exceptional groups of degree 3^4 are doubly transitive.) Let H be any such stabilizer. Then $H_0 \subseteq H$, as we have seen above. Let $|H:H_0| = 2$ and assume $H \neq H_1$. Then $\langle H, H_1 \rangle / H_0$ is elementary abelian of order 4. Therefore $\langle H, H_1 \rangle$ is soluble. But then $\langle H, H_1 \rangle / H_0$ is cyclic, a contradiction. If $|H:H_0| = 4$, then $H_1 \subseteq H$ by what we have just seen. Repeating the argument yields $H = H_2$. Hence there are exactly 3 exceptions of degree 3^4, all of them being doubly transitive. $\qquad\square$

37.8 Corollary. *The three exceptional doubly transitive groups of degree 3^4 operate as doubly transitive collineation groups on the nearfield plane of order 3^2.*

This follows from what we have seen above and 8.3.

37.9 Corollary. *Let G be a soluble flag transitive collineation group of the finite affine plane $\mathfrak{A}$ and let q be the order of $\mathfrak{A}$. If $q \neq 9$, then G also acts flag transitively on the desarguesian plane of order q. Moreover, the representation of G on the point set of $\mathfrak{A}$ is permutation isomorphic to the representation of G on the point set of the desarguesian plane.*

PROOF. This follows from 37.7 and the remark that translation planes of prime order are desarguesian by theorem 1.13. $\qquad\square$

37.10 Lemma. *Let p be a prime and let $\mathfrak{A}$ be an affine plane of order p^r. Assume that u is a p-primitive divisor of $p^r - 1$. If G is a soluble flag transitive collineation group of $\mathfrak{A}$ whose order is divisible by u, then $\mathfrak{A}$ is desarguesian.*

PROOF. $\mathfrak{A}$ is a translation plane by 15.16 and G contains the translation group T of $\mathfrak{A}$. If O is a point of $\mathfrak{A}$, then $G = TG_O$ and $T \cap G_O = \{1\}$. As $|T| = p^{2r}$, we have that u divides $|G_O|$. Moreover, $G_O \subseteq \Gamma L(1, p^{2r})$ by 37.7 or $r = 1$ and p is one of a finite number of primes or $p^r = 3^2$. In the second case, $\mathfrak{A}$ is desarguesian by 1.13. In the third case $p^r - 1 = 8$ so that there is no p-primitive divisor of $p^r - 1$. Hence we are left with the case that $G_O \subseteq \Gamma L(1, p^{2r})$. Moreover, we may assume $r \geq 2$ by 1.13. Now

$$|G_O| = |G_O : G_O \cap GL(1, p^{2r})| \, |G_O \cap GL(1, p^{2r})|$$

and $|G_O : G_O \cap GL(1, p^{2r})|$ is a divisor of $2r$. As $r > 1$, we have $u > 2$. Thus u does not divide $2r$, as u does not divide r. This implies that u divides $|G_O \cap GL(1, p^{2r})|$. Therefore $G_O \cap GL(1, p^{2r})$ contains a subgroup U of order u. As $G_O \cap GL(1, p^{2r})$ is a cyclic normal subgroup of G_O, the group U is normal in G_O. Moreover U fixes a point on l_∞, as u is a divisor of $p^r - 1$ which is greater than 2. We therefore infer from the transitivity of G_O on l_∞ that U fixes all the points on l_∞. Thus U is a group of (O, l_∞)-homologies. This yields that u divides the order of the multiplicative group of the kernel K of $\mathfrak{A}$. As u is a p-primitive prime divisor of $p^r - 1$, we finally see that $K \cong \mathrm{GF}(p^r)$. ◻

37.11 Theorem (Foulser). *Let $\mathfrak{A}$ be a finite affine plane. If $\mathfrak{A}$ admits a soluble collineation group acting doubly transitively on the set of points of $\mathfrak{A}$, then $\mathfrak{A}$ is desarguesian or $\mathfrak{A}$ is the nearfield plane of order* 9.

PROOF. Let G be a soluble doubly transitive collineation group of $\mathfrak{A}$. Then $\mathfrak{A}$ is a translation plane by 15.16 and G contains the translation group of $\mathfrak{A}$. If p^r is the order of $\mathfrak{A}$, then $p^{2r}(p^{2r} - 1)$ divides $|G|$. If $\mathfrak{A}$ is non-desarguesian, then there does not exist a p-primitive prime divisor of $p^r - 1$. Hence $p^r = 2^6$ or $r = 2$ and $p + 1 = 2^s$ by 6.2.

Assume $p^r = 2^6$. Then $G_O \subseteq \Gamma\mathrm{L}(1, 2^{12})$ by 37.7. As $|G_O / G_O \cap \mathrm{GL}(1, 2^{12})|$ divides 12, we see that 21 divides $|G_O \cap \mathrm{GL}(1, 2^{12})|$. Therefore G_O contains a normal subgroup U of order 21. As above, U consists entirely of (O, l_∞)-homologies whence it follows that $\mathfrak{A}$ is desarguesian.

Hence $r = 2$ and $p + 1 = 2^s$. Then

$$|G_O| = (p^4 - 1)k = (p - 1)2^s(p^2 + 1)k.$$

Therefore 2^{s+2} divides $|G_O|$. If $p > 3$, then $G_O \subseteq \Gamma\mathrm{L}(1, p^4)$ by 37.7. Thus $|G_O / G_O \cap \mathrm{GL}(1, p^4)|$ divides 4. This implies that $G_O \cap \mathrm{GL}(1, p^4)$ and hence G_O contain a cyclic normal subgroup V of order 2^s. Since 4 does not divide $p^2 + 1$, there exists a subgroup U of V of order 2^{s-1} fixing a point on l_∞. As V is cyclic, U is a characteristic subgroup of V and hence a normal subgroup of G_O. Therefore U fixes all the points of l_∞, i.e., U consists of (O, l_∞)-homologies only. From $p > 3$ we obtain $|U| \geqslant 4$. Therefore $|U|$ does not divide $p - 1$. This yields that the kernel of $\mathfrak{A}$ is isomorphic to $\mathrm{GF}(p^2)$. Hence $p = 3$ and $\mathfrak{A}$ is the nearfield plane of order 9 by 8.4. This proves 37.11. ◻

There are plenty of finite non-desarguesian affine planes admitting soluble flag transitive collineation groups. See e.g. M. L. N. Rao 1973.

38. Some Characterizations of Finite Desarguesian Planes

Let q be a power of a prime p and let $\mathfrak{P}$ be a projective plane of order q. Furthermore, let Δ be a collineation group of $\mathfrak{P}$ which is isomorphic to $\mathrm{PSL}(2, q)$.

38.1 Lemma. *If Δ does not fix a point or a line, then Δ has a point or a line orbit of length $q + 1$.*

PROOF. Let Σ be a Sylow p-subgroup of Δ. As $q^2 + q + 1 \equiv 1 \bmod p$, we see that Σ has a fixed point X. Put $N = \mathfrak{N}_\Delta(\Sigma)$ and assume $X^N = X$. If $\varphi \in \Delta \setminus N$, then $X^\varphi \neq X$, for otherwise X is fixed by $\langle \varphi^{-1}\Sigma\varphi, \Sigma \rangle = \Delta$. Therefore, $q + 1 = |\Delta : N| = |X^\Delta|$.

Assume now that there exists no point orbit of length $q + 1$. Then for every fixed point X of Σ there exists $\varphi \in N$ with $X^\varphi \neq X$. If all the fixed points of Σ are collinear, then this yields that N fixes a line and hence that Δ has a line orbit of length $q + 1$. Thus we may assume that not all the fixed points of Σ are collinear. Let $\mathfrak{F}$ be the substructure of $\mathfrak{P}$ consisting of all fixed points and all fixed lines of Σ. Then the lines joining two points of $\mathfrak{F}$ belong to $\mathfrak{F}$ as well as the intersections of two distinct lines of $\mathfrak{F}$. As the number of fixed points on a fixed line is congruent to $1 \bmod p$, it follows that each line of $\mathfrak{F}$ carries at least $1 + p \geq 3$ points of $\mathfrak{F}$. Thus $\mathfrak{F}$ is a subplane of $\mathfrak{P}$.

Let k be the order of $\mathfrak{F}$. If $1 \neq \sigma \in \Sigma$, then σ fixes a subplane $\mathfrak{F}_\sigma \supseteq \mathfrak{F}$ elementwise. By 4.4, the order of $\mathfrak{F}_\sigma$ is less than or equal to m, where $m = \max\{n \mid n^2 \leq q\}$. Let A be a point of $\mathfrak{F}$ and let l be a line through A which does not belong to $\mathfrak{F}$. Then $|l^\Sigma| \leq q - k < q$. Therefore, there exists $\sigma \in \Sigma \setminus \{1\}$ with $l^\sigma = l$. As Σ is abelian, σ fixes all the lines in l^Σ. On the other hand, each element in $\Sigma \setminus \{1\}$ fixes at most $m - k$ lines through A which are not in $\mathfrak{F}$. Therefore $|l^\Sigma| \leq m - k$.

Let B be a point fixed by a Sylow p-subgroup T other than Σ. If $\Sigma_B \neq \{1\}$, then $B^\Delta = B$, as $\langle \Sigma_B, T \rangle = \Delta$. Thus $\Sigma_B = \{1\}$, i.e., $|B^\Sigma| = q$. This yields that AB does not belong to $\mathfrak{F}$. Put $|B^\Sigma \cap AB| = h$. Then $|(AB)^\Sigma| = qh^{-1}$. Let $B^\Sigma \cap AB = \{P_1, \ldots, P_h\}$ and let A' be a point of $\mathfrak{F}$ other than A. Then $A'P_1, \ldots, A'P_h$ are h distinct lines, as $A' \notin AB$. Furthermore, $A'P_i \in (A'B)^\Sigma$. Hence $|(A'B)^\Sigma| \geq h$. If $qh^{-1} \leq m$, then $m^2 h^{-1} \leq qh^{-1} \leq m$, whence $m \leq h$. Therefore we have $|(AB)^\Sigma| \geq m$ or $|(A'B)^\Sigma| \geq m$. On the other hand $m - k \geq |(AB)^\Sigma|, |(A'B)^\Sigma|$, whence $m - k \geq m$, a contradiction. $\qquad\square$

38.2 Lemma. *If Δ fixes a point A, then the lines through A form an orbit of Δ.*

PROOF. Assume that Δ fixes all the lines through A. Then Δ consists only of perspectivities with centre A. But this implies that $|\Delta| = q(q^2 - 1)d$ where $d = (2, q - 1)^{-1}$ divides $q^2(q - 1)$ which is impossible. Thus there exists a line l through A with $|l^\Delta| \geq 2$. Therefore, $2 \leq |\Delta : \Delta_l| \leq q + 1$. It follows from 14.1 that either $|\Delta : \Delta_l| = q + 1$ or $q = 2, 3, 5, 7, 9$ or 11 and $|\Delta : \Delta_l| = 2, 3, 5, 7, 6$, or 10 respectively. We have to show that these seven exceptions cannot occur.

If $q = 2$ or 3 and $|\Delta : \Delta_l| = q$, then Δ_l is normal in Δ. From this we infer that Δ_l fixes q and hence all lines through A. Therefore, $q + 1 = |\Delta_l|$ divides $q^2(q - 1)$, a contradiction.

Let $q = 5, 7$ or 11. As q is not a square, the involutions in Δ are homologies. Moreover, Δ fixes a line l through A. As Δ is simple, Δ

operates faithfully on l as well as on the set of the remaining lines through A. Therefore, if σ is an involution in Δ, the axis of σ passes through A and is distinct from l and the centre of σ is on l and is distinct from A. Let m be a line through A other than l. Then $\Delta_m \cong A_4$, S_4, respectively A_5 for $q = 5$, 7, respectively 11. Since S_4 and A_5 are generated by their involutions, Δ_m is a group of homologies, if $q = 7$ or 11. But 24 does not divide $7 - 1$ and 60 does not divide $11 - 1$. If $q = 5$, then Δ_m contains an elementary abelian 2-group Λ of order 4. As Λ is abelian and since all elements of Λ have axis m, they must have the same centre. This implies that Λ is cyclic as $\mathfrak{P}$ is desarguesian, a contradiction.

Finally, we have to rule out the exceptional case $q = 9$. Here Δ fixes four lines through A. Moreover, Δ fixes at least four points on each of these lines. Therefore Δ fixes a subplane $\mathfrak{F}$ of $\mathfrak{P}$ elementwise. As $\Delta \neq \{1\}$ and $o(\mathfrak{F}) \geqslant 3$, we see that $\mathfrak{F}$ is a Baer subplane of $\mathfrak{P}$. From this we infer that Δ acts regularly on $\mathfrak{P} \backslash \mathfrak{F}$, whence $|\Delta| = 360$ divides $3^4 + 3^2 + 1 - 3^2 - 3 - 1 = 78$, a contradiction. Thus we have proved 38.2. $\square$

38.3 Lemma. *Let $\mathfrak{o}$ be a point orbit of length $q + 1$ of Δ. Then $\mathfrak{o}$ is an oval or $p = 2$ and $\mathfrak{o}$ consists of the points of a line.*

PROOF. Assume that $\mathfrak{o}$ consists of the points of a line. If $\sigma \in \Delta$ is an involution, then σ has at most 2 fixed points on l. Therefore, σ is a perspectivity. If $p > 2$, then σ has exactly 2 fixed points P and Q on l; one of them, say P, is the centre. The axis of σ passes through Q. There exists $\delta \in \Delta$ with $P^\delta = Q$ and $Q^\delta = P$. It follows from 4.8 that $\sigma\delta^{-1}\sigma\delta$ is an involutory homology with axis l. This contradicts the fact that Δ operates faithfully on $\mathfrak{o}$. Hence $p = 2$.

Let $P, Q, R \in \mathfrak{o}$ be non collinear. If $p = 2$, then Δ acts threefold transitively on $\mathfrak{o}$. Therefore, $\mathfrak{o}$ is an oval in this case. Thus we may assume $p > 2$. Then $|\Delta_{P,Q}| = \frac{1}{2}(q - 1)$. Assume that there exists $S \in (PQ \backslash \{P, Q\}) \cap \mathfrak{o}$. Then it follows that $|(PQ \backslash \{P, Q\}) \cap \mathfrak{o}| = \frac{1}{2}(q - 1)$ or $q - 1$; note that the latter case is impossible, since $R \nmid PQ$. Let Σ be a Sylow p-subgroup of Δ_p. Then Σ is transitive on $\mathfrak{o} \backslash \{P\}$. This implies that Σ_{PQ} is transitive on $(PQ \backslash \{P\}) \cap \mathfrak{o}$. Thus p divides $\frac{1}{2}(q + 1)$, a contradiction. This proves $PQ \cap \mathfrak{o} = \{P, Q\}$. As Δ acts doubly transitively on $\mathfrak{o}$, we finally obtain that $\mathfrak{o}$ is an oval. $\square$

38.4 Lemma. *If $\sigma \in \Delta$ is an involution, then σ is a perspectivity.*

PROOF. Dualizing if necessary, we may assume by 38.1 and 38.2 that Δ has a point orbit $\mathfrak{o}$ of length $q + 1$. If $\mathfrak{o}$ consists of the points of a line, then $p = 2$ by 38.3 and σ has exactly one fixed point on $\mathfrak{o}$. Hence σ is a perspectivity. We may thus assume that $\mathfrak{o}$ is an oval. If q is even, then σ has exactly one fixed point in $\mathfrak{o}$. Moreover, the knot K of $\mathfrak{o}$, i.e., the point K of $\mathfrak{P}$ which is on all the tangents of $\mathfrak{o}$, is also fixed by σ. If σ is not a perspectivity, then σ is a Baer involution and hence fixes $\sqrt{q} + 1$ lines through K and therefore $\sqrt{q} + 1$ points on $\mathfrak{o}$, a contradiction. Thus we may also assume that q is odd. If $q \equiv 3 \bmod 4$, then q is not a square, whence σ is a homology. If $q \equiv 1 \bmod 4$, then σ has exactly 2 fixed points on $\mathfrak{o}$. Let P and Q be these fixed points and denote by t_X the tangent of $\mathfrak{o}$ which carries $X \in \mathfrak{o}$. Then $X \to t_X \cap t_P$ is a bijection of $\mathfrak{o} \backslash \{P\}$ onto $t_P \backslash \{P\}$. Moreover, $(t_X \cap t_P)^\sigma = t_{X^\sigma} \cap t_P$. Therefore, $(t_X \cap t_P)^\sigma = t_X \cap t_P$ if and only if $X^\sigma = X$. Thus σ has exactly two fixed points on t_P, namely P and $t_Q \cap t_P$. This proves that σ is not a Baer involution. Hence σ is a homology. $\square$

38.5 Lemma. *Assume $p = 2$. If Δ fixes a point P, then Δ has exactly one point orbit $\mathfrak{o}$ of length $q + 1$. If no two of the involutions of Δ have the same centre, then $\mathfrak{o}$ is an oval and P is its knot. If two distinct involutions have the same centre, then $\mathfrak{o}$ consists of the points of a line. In either case, the points not in $\mathfrak{o}$ and other than P form an orbit of Δ.*

PROOF. By 38.2, the lines through P form an orbit of Δ. Hence Δ acts doubly transitively on the set of lines through P. Let l and m be two distinct lines through P. Then $\Delta_{l,m}$ is cyclic or order $q - 1$. Let $\delta \in \Delta_{l,m}$ be an element of order s where s is a prime. Then δ fixes a point Q on l other than P. Assume that δ fixes a third point R on l and let σ be an element of Δ switching l and m. Then $\Delta_{l,m}^\sigma = \Delta_{l,m}$. Therefore, as $\Delta_{l,m}$ is cyclic, $\langle \delta \rangle^\sigma = \langle \delta \rangle$. This shows that δ fixes the quadrangle Q, R, Q^δ, R^δ. Thus δ fixes a subplane of $\mathfrak{P}$. As $P^\delta = P$, we see that δ fixes at least three lines through P, a contradiction. This shows that δ fixes P and Q, but no other point on l. From this we infer that $\Delta_{l,m}$ fixes P and Q and operates regularly on $l \backslash \{P, Q\}$. Putting $\mathfrak{o} = \{Q^\eta \mid \eta \in \Delta\}$, either $|\mathfrak{o}| = q + 1$ or $|\mathfrak{o}| = q(q + 1)$. Let Σ be a Sylow 2-subgroup fixing l. Then Σ acts transitively on the set of the remaining lines through P. As the number of these lines is $q = |\Sigma|$, it follows that P is not the centre of an involution in Δ. Therefore, l is fixed pointwise by Σ. This yields $|\mathfrak{o}| = q + 1$. Obviously, $\mathfrak{o}$ is unique. Moreover, the points not in $\mathfrak{o}$ and other than P form an orbit of Δ. As all involutions of Δ are conjugate, either all or none of them have their centres in $\mathfrak{o}$. Since the number of involutions in Δ is $q^2 - 1$, there exist distinct involutions in Δ having the same centre if and only if all involutions have their centres in $\mathfrak{o}$. If all involutions have their centres in $\mathfrak{o}$ then, obviously, $\mathfrak{o}$ consists of the points of a line. On the other hand, if $\mathfrak{o}$ consists of the points of a line x, then $x^\Delta = x$, whence x carries all the centres. Now 38.3 yields the desired result. $\square$

38.6 Lemma. *Let $p = 2$ and assume that Δ has a point orbit $\mathfrak{o}$ which is an oval. Then Δ splits the set of points of $\mathfrak{P}$ into three orbits: The set $\mathfrak{P}_1 = \{K\}$, where K is the knot of $\mathfrak{o}$, the set $\mathfrak{P}_2 = \mathfrak{o}$, and the set $\mathfrak{P}_3$ of the remaining points. Moreover, Δ splits the sets of all lines into three orbits: The set $\mathfrak{L}_1$ of all the tangents of $\mathfrak{o}$, the set $\mathfrak{L}_2$ of all the secants, and the set $\mathfrak{L}_3$ of the remaining lines.*

PROOF. $\mathfrak{P}_1$ and $\mathfrak{P}_2$ are of course orbits. Applying 38.5 yields that $\mathfrak{P}_3$ is also an orbit. $\mathfrak{L}_1$ is an orbit, as $\mathfrak{P}_2$ is. Moreover, $\mathfrak{L}_2$ is an orbit, because Δ acts doubly transitively on $\mathfrak{o} = \mathfrak{P}_2$. Since Δ has as many point orbits as line orbits by the Dembowski-Hughes-Parker theorem, $\mathfrak{L}_3$ is also an orbit. $\square$

38.7 Lemma. *Let $p = 2$ and assume that Δ has a point orbit which is an oval. Then $\mathfrak{P}$ is desarguesian and $\mathfrak{o}$ is a conic section.*

PROOF. Using the notation of 38.6, we have $|\mathfrak{P}_1| = 1$, $|\mathfrak{P}_2| = q + 1$, $|\mathfrak{P}_3| = q^2 - 1$, $|\mathfrak{L}_1| = q + 1$, $|\mathfrak{L}_2| = \frac{1}{2} q(q + 1)$, and $|\mathfrak{L}_3| = \frac{1}{2} q(q - 1)$. We represent $\mathfrak{P}$ within Δ as follows:

a) To K we assign the set $\Pi = \{\mathfrak{N}_\Delta(\Sigma) \,|\, \Sigma \in \mathrm{Syl}_2(\Delta)\}$.
b) To $X \in \mathfrak{P}_2$ we assign the group Δ_X.
c) To $X \in \mathfrak{P}_3$ we assign the involution σ_X whose centre is X.
d) To each line l of $\mathfrak{P}$ we assign the group Δ_l.

The mapping defined in b) is a bijection from $\mathfrak{P}_2$ to Π. The one defined in c) is a bijection from $\mathfrak{P}_3$ onto the set of all involutions of Δ. If we denote the restriction to $\mathfrak{L}_i$ of the mapping defined in d) by σ_i, then σ_1 is a bijection from $\mathfrak{L}_1$ onto Π, and σ_2 is a bijection from $\mathfrak{L}_2$ onto the set of all dihedral subgroups of order $2(q - 1)$ of Δ, and σ_3 is a bijection from $\mathfrak{L}_3$ onto the set of all dihedral subgroups of order $2(q + 1)$ of Δ.

Incidence is described as follows: $K \mathrel{I} l$, if and only if $\Delta_l \in \Pi$. If $X \in \mathfrak{P}_2$, then $X \mathrel{I} l$ if and only if $\Delta_X = \Delta_l$ or $|\Delta_X \cap \Delta_l| = q - 1$. If $X \in \mathfrak{P}_3$, then $X \mathrel{I} l$ if and only if $\sigma_X \in \Delta_l$.

In order to show that $\mathfrak{P}$ is desarguesian, we have only to prove that the desarguesian plane of order q admits a collineation group $\Delta \cong \mathrm{PSL}(2, q)$ which fixes an oval $\mathfrak{c}$.

Let $\mathfrak{D}$ be the desarguesian plane over $F = \mathrm{GF}(q)$. Put $\mathfrak{c} = \{F(0, 1, 0), F(f, f^2, 1) \,|\, f \in F\}$. It is easily checked that $\mathfrak{c}$ is an oval. Let Δ be the group generated by the matrices

$$\begin{bmatrix} 1 & 0 & 0 \\ 1 & 1 & 0 \\ 0 & 0 & 1 \end{bmatrix} \quad \text{and} \quad \begin{bmatrix} a & 0 & 0 \\ 0 & a^2 & 0 \\ b & b^2 & 1 \end{bmatrix} \quad \text{with} \quad a, b \in F, \quad a \neq 0.$$

It is a trivial exercise to see that $\Delta \cong \mathrm{PSL}(2, q)$ and that $\mathfrak{c}$ is an orbit of Δ. Hence $\mathfrak{P} \cong \mathfrak{D}$. Finally, $\mathfrak{c} = \{F(x, y, z) \,|\, x^2 + yz = 0\}$. Thus $\mathfrak{c}$ and hence $\mathfrak{o}$ are conic sections. $\square$

38.8 Lemma. *Let q be odd. Then Δ has a point orbit $\mathfrak{o}$ of length $q + 1$ and a line orbit $\mathfrak{t}$ of length $q + 1$. The point orbit $\mathfrak{o}$ is an oval and the line orbit $\mathfrak{t}$ consists of the tangents of $\mathfrak{o}$. If $q \equiv 1 \bmod 4$, then the point P is the centre of an involution in Δ if and only if P is an exterior point of $\mathfrak{o}$. If $q \equiv 3 \bmod 4$, then the point P is the centre of an involution of Δ if and only if P is an interior point of $\mathfrak{o}$. In particular, each point of $\mathfrak{P}$ is the centre of at most one involution in Δ.*

PROOF. By 38.1 and 38.2, or the dual of 38.2, we know that Δ has an orbit of length $q + 1$. Dualizing if necessary we may assume that Δ has a point orbit $\mathfrak{o}$ of length $q + 1$. As q is odd, $\mathfrak{o}$ is an oval by 38.3. Since each point of $\mathfrak{o}$ is on exactly one tangent, the set $\mathfrak{t}$ of tangents is a line orbit of length $q + 1$. As $\mathfrak{t}$ is an oval in the dual of $\mathfrak{P}$, the first assertion of the lemma is proved.

Let σ and τ be involutions in Δ having the same centre. Then $\sigma\tau$ fixes $\mathfrak{o}$ pointwise. Consequently $\sigma\tau = 1$, i.e., $\sigma = \tau$, proving the last assertion of the lemma.

Assume $q \equiv 1 \bmod 4$. Then an involution $\sigma \in \Delta$ fixes two points, P and Q say, on $\mathfrak{o}$. Moreover, $t_P \cap t_Q = R$ is also fixed by σ. As σ is a homology, one of these points is the centre of σ. Obviously, P and Q are not centres of involutions in Δ. Hence R is the centre of σ. Therefore, centres of involutions are always exterior points in this case. As $\frac{1}{2} q(q + 1)$ is the number of exterior points as well as the number of involutions in Δ, the third assertion is proved.

Let $q \equiv 3 \bmod 4$. In this case the involutions of Δ act fixed-point-free on $\mathfrak{o}$. Therefore, their centres are interior points. As the number of interior points is the same as the number of involutions, namely $\frac{1}{2}(q - 1)q$, the fourth assertion is also proved. $\qquad\Box$

38.9 Lemma. *Assume $q \equiv 1 \bmod 4$ and let $\mathfrak{o}$ be the oval fixed by Δ. Then Δ acts transitively on the set of flags (P, l), where P is an interior point and l is an exterior line of $\mathfrak{o}$.*

PROOF. Let σ and τ be distinct involutions of Δ and assume $\sigma\tau = \tau\sigma$. Let l and m be the axes of σ respectively τ. By the dual of 38.8, we have $l \neq m$. Hence, by 4.8, $\sigma\tau$ is an involution with centre $l \cap m$. This yields by 38.8 that $l \cap m$ is an exterior point of $\mathfrak{o}$.

Let P be an interior point of $\mathfrak{o}$. The number of secants through P is $\frac{1}{2}(q + 1)$. Therefore, $\frac{1}{2}(q + 1)$ is also the number of exterior lines through P. Each of the secants is the axis of exactly one involution. As P is not a centre, the above argument shows that the only secant through P fixed by an involution of Δ_P is the axis of that involution. Therefore, by 15.1, the group Δ_P acts transitively on the set of secants through P. Thus $\frac{1}{2}(q + 1)$ divides $|\Delta_P|$. As $|\Delta_P|$ is even and $\frac{1}{2}(q + 1)$ is odd, $q + 1$ divides $|\Delta_P|$. It follows from 14.1 that Δ_P is a dihedral group of order $q + 1$. As $|\Delta : \Delta_P| = \frac{1}{2} q(q - 1)$ is the number of interior points, Δ permutes the interior points transitively.

By the dual argument, Δ acts transitively on the set of exterior lines. If l is such a line, then Δ_l is a dihedral group of order $q + 1$.

The number $\frac{1}{2}(q + 1)$ of exterior lines through P is odd, as $q \equiv 1 \bmod 4$. Therefore, each involution in Δ_P fixes at least one exterior line through P. Assume that there are distinct involutions in $\Delta_P \cap \Delta_l$, where l is an exterior line through P. Then $|\Delta_P \cap \Delta_l| \geqslant 3$. It then follows from 14.2 that $\Delta_P = \Delta_l$. Let Q be an interior point on l which is distinct from P. Then, using 14.2 again, $|\Delta_P \cap \Delta_Q| \leqslant 2$. If $|\Delta_P \cap \Delta_Q| = 2$, then there exists an involution τ fixing P and Q. As P and Q are interior points, $PQ = l$ is the axis of τ and hence l is a secant. This contradiction proves $\Delta_P \cap \Delta_Q = \{1\}$. From this and $\Delta_P = \Delta_l$ we infer that Δ_P acts regularly on the set of interior points on l which are distinct from P. Thus $q + 1$ divides $\frac{1}{2}(q + 1) - 1$, a contradiction. This proves $|\Delta_P \cap \Delta_l| \leqslant 2$ for all exterior lines l through P. Hence $|\Delta_P \cap \Delta_l| = 2$ for all such lines, as each involution in Δ_P fixes at least one exterior line through P. Invoking 15.1, we see that Δ_P acts transitively on the set of exterior lines through P. $\qquad\square$

38.10 Lemma. *Assume $q \equiv 3$ mod 4 and let $\mathfrak{o}$ be the oval fixed by Δ. Then Δ acts transitively on the set of flags (P, l) where P is an exterior point and l is a secant.*

PROOF. Let P be an exterior point of $\mathfrak{o}$. Then there are exactly two tangents through P. Let Q and R be their points of contact. Then $\Delta_P = \Delta_{\{Q, R\}}$, whence Δ_P is a dihedral group of order $q - 1$. Let s be a secant through P. Then $|\Delta_{P,s}| \leqslant 2$. Thus $q - 1 = |\Delta_P| \leqslant \frac{1}{2}(q - 1)|\Delta_{P,s}|$ and hence $|\Delta_{P,s}| = 2$. This shows that Δ_P acts transitively on the set of secants through P. Moreover, $|\Delta : \Delta_P| = \frac{1}{2}q(q + 1)$ is the number of exterior points. Therefore, Δ acts transitively on the set of exterior points. $\qquad\square$

In the following theorem we consider only the case $p \geqslant 3$. The case $p = 2$ will be dealt with later on.

38.11 Theorem (Lüneburg 1964, Yaqub 1966). *Let $p \geqslant 3$ be a prime and let q be a power of p. If $\mathfrak{P}$ is a projective plane of order q admitting a collineation group $\Delta \cong \mathrm{PSL}(2, q)$, then $\mathfrak{P}$ is desarguesian. If Γ is a second collineation group of $\mathfrak{P}$ with $\Gamma \cong \mathrm{PSL}(2, q)$, then Δ and Γ are conjugate in the collineation group of $\mathfrak{P}$.*

PROOF. As we know, Δ fixes an oval $\mathfrak{o}$.

Case 1: $q \equiv 1$ mod 4. We represent $\mathfrak{P}$ within Δ in the following way:

a) If P is an exterior point of $\mathfrak{o}$, then we assign to P the involution σ_P whose centre is P.

b) If $P \in \mathfrak{o}$, then we assign to P its stabilizer Δ_P.

c) If P is interior, then we assign to P the group Δ_P as well as the coset $\Pi(P) = \Pi\eta$, where Π is the stabilizer of a fixed interior point P_0 and η is such that $P_0^\eta = P$.

The mapping under a) is a bijection of the set of exterior points onto the set of involutions of Δ. The mapping under b) is a bijection of $\mathfrak{o}$ onto the set of all normalizers of Sylow p-subgroups. The first mapping under c) is a bijection of the set of interior points onto the set of all dihedral subgroups of order $q + 1$, whereas the second one is a bijection onto the set of right cosets of Π.

a') If l is a secant, then we assign to l the involution σ_l whose axis is l.

b') If l is a tangent, then we assign to l the group Δ_l.

c') If l is an exterior line, then we assign to l the group Δ_l as well as the coset $\Lambda(l) = \Lambda\eta$, where Λ is the stabilizer of an exterior line l_0 through P_0 and η is such that $l_0^\eta = l$.

What we have said about the mappings under a), b), and c) carries over mutatis mutandis to the mappings under a'), b'), and c').

Incidence is described as follows: Let P be an exterior point and l a line. If l is exterior or tangent, then P I l if and only if $\sigma_P \in \Delta_l$. If l is a secant, then P I l if and only if $\sigma_P \neq \sigma_l$ and $\sigma_P\sigma_l = \sigma_l\sigma_P$. Let $P \in \mathfrak{o}$ and l a line. If l is tangent, then P I l if and only if $\Delta_P = \Delta_l$. If l is a secant, then P I l if and only if $\sigma_l \in \Delta_P$. Let P be interior. If l is exterior, then P I l if and only if $\Pi(P) \cap \Lambda(l) \neq \emptyset$ by 38.9. If l is secant, then P I l if and only if $\sigma_l \in \Delta_P$.

Comparison with the desarguesian plane of order q now yields the desired results.

Case 2: $q \equiv 3 \bmod 4$. We represent $\mathfrak{P}$ within Δ in the following way:

a) If P is an interior point of $\mathfrak{o}$, then we assign to P the involution σ_P whose centre is P.

b) If $P \in \mathfrak{o}$, then we assign to P its stabilizer Δ_P.

c) If P is an exterior point, then we assign to P the group Δ_P as well as the coset $\Pi(P) = \Pi\eta$, where Π is the stabilizer of a fixed exterior point P_0 and η is such that $P_0^\eta = P$.

The mapping under a) is a bijection of the set of interior points onto the set of involutions of Δ. The mapping under b) is a bijection of $\mathfrak{o}$ onto the set of all normalizers of Sylow p-subgroups. The first mapping under c) is a bijection of the set of exterior points onto the set of all dihedral subgroups of order $q - 1$, whereas the second one is a bijection onto the set of right cosets of Π.

a') If l is an exterior line of $\mathfrak{o}$, then we assign to l the involution σ_l whose axis is l.

b') If l is a tangent, then we assign to l the group Δ_l.

c') If l is a secant, then we assign to l the group Δ_l as well as the coset $\Lambda(l) = \Lambda\eta$, where Λ is the stabilizer of a fixed secant l_0 passing through P_0 and η such that $l_0^\eta = l$.

What we have said about the mappings under a), b), and c) carries over mutatis mutandis to the mappings under a'), b'), and c').

Incidence is described as follows: Let P be an interior point and l a line. If l is a secant or a tangent, then P I l if and only if $\sigma_P \in \Delta_l$. If l is exterior, then P I l if and only if $\sigma_P \neq \sigma_l$ and $\sigma_P\sigma_l = \sigma_l\sigma_P$. Let $P \in \mathfrak{o}$ and let l be a line. If l is tangent, then P I l if and only if $\Delta_P = \Delta_l$. If l is a secant, then P I l if and only if $|\Delta_P \cap \Delta_l| = \frac{1}{2}(q-1)$. Let P be exterior. If l is a secant, then P I l if and only if $\Pi(P) \cap \Lambda(l) \neq \emptyset$ by 38.10. If l is a tangent, then P I l if and only if $|\Delta_P \cap \Delta_l| = \frac{1}{2}(q-1)$. If l is an exterior line, then P I l if and only if $\sigma_l \in \Delta_P$.

Comparison with the desarguesian plane of order q now yields that $\mathfrak{P}$ is desarguesian. This proves 38.11. □

38.12 Theorem (Lüneburg 1964, Yaqub 1966). *Let p be a prime and let q be a power of p. Assume furthermore that $\mathfrak{P}$ is a projective plane of order q. If $\mathfrak{P}$ admits a collineation group $\Delta \cong \mathrm{SL}(2,q)$, then $\mathfrak{P}$ is desarguesian. If $p = 2$, then the full collineation group of $\mathfrak{P}$ contains exactly three conjugacy classes of groups isomorphic to $\mathrm{SL}(2,q)$. If $p \geqslant 3$, the collineation group of $\mathfrak{P}$ contains just one conjugacy class of groups isomorphic to $\mathrm{SL}(2,q)$.*

PROOF. Assume first that $p = 2$. Then $\mathrm{SL}(2,q) \cong \mathrm{PSL}(2,q)$. Dualizing if necessary, we may assume by 38.1 and 38.2 that Δ has a point orbit $\mathfrak{o}$ of length $q + 1$. If $\mathfrak{o}$ is an oval, then the plane $\mathfrak{P}$ is desarguesian by 38.7. If $\mathfrak{o}$ is not an oval, then $\mathfrak{o}$ consists of the points of a line l by 38.3. By the dual of 38.5, we see that Δ has a line orbit $\mathfrak{L}$ of length $q + 1$. If $\mathfrak{L}$ is the dual of an oval, then $\mathfrak{P}$ is desarguesian by the dual of 38.7. Therefore we may assume that $\mathfrak{L}$ is not the dual of an oval. But then the dual of 38.3 implies that there exists a point P with P I x for all $x \in \mathfrak{L}$. Obviously, $P \not{\mathrm{I}} l$. By 38.5, the points of l are the centres of the involutions in Δ and the lines through P are their axes. Moreover, the points not on l and distinct from P form an orbit of Δ.

Assume now that $p \geqslant 3$. We shall show in this case that Δ fixes a non-incident point line pair (P, l), that the p-elements of Δ are elations with centres on l, and that Δ acts transitively on the set of points not on l and distinct from P.

As p is odd, $|\mathfrak{Z}(\Delta)| = 2$. Let $1 \neq \sigma \in \mathfrak{Z}(\Delta)$. If σ is not a homology, then σ fixes a Baer subplane $\mathfrak{Q}$ of $\mathfrak{P}$. Let s be the order of $\mathfrak{Q}$. Then $s^2 = q$. As σ is in the centre of Δ, the group Δ induces a group Δ^* of collineations of $\mathfrak{Q}$. Assume $\Delta^* = \{1\}$ and let l be a line of $\mathfrak{Q}$. Then Δ acts regularly on the $q - s$ points of l which do not belong to $\mathfrak{Q}$. Hence $|\Delta| = q(q^2 - 1)$ divides $q - s$, a contradiction. As $q = s^2 \geqslant 9$, the group $\Delta/\mathfrak{Z}(\Delta)$ is simple. Therefore $\Delta^* \cong \mathrm{PSL}(2,q)$. The number of points respectively of lines of $\mathfrak{Q}$ is $s^2 + s + 1 = q + 1 + s$. Note that $q = s^2$ is distinct from 5, 7, and 11. If $q \neq 9$, it therefore follows from 14.1 that each non trivial orbit of Δ has length at least $q + 1$. Moreover, $s^2 + s + 1$ does not divide $|\Delta| = \frac{1}{2}s^2(s^4 - 1)$. Hence Δ fixes a line l of $\mathfrak{Q}$, as the number of lines is $q + 1 + s < 2q + 1$. From $s + 1 < q + 1$, we infer that Δ fixes all the points of l which belong to $\mathfrak{Q}$. Hence each point orbit of Δ^* which is contained in $\mathfrak{Q}$ has length $\leqslant q + s + 1 - s - 1 = q < q + 1$, a contradiction. Thus

$q = 9$. Hence $|\Delta^*| = \frac{1}{2}9(9^2 - 1) = 5 \cdot 9 \cdot 8$ divides the order of the collineation group of $\mathfrak{Q}$. As $s = 3$, the plane $\mathfrak{Q}$ is desarguesian, whence $5 \cdot 9 \cdot 8$ divides $|\mathrm{PGL}(3,3)| = 27 \cdot 16 \cdot 13$, again a contradiction. Thus σ is a homology. If P is the centre and l the axis of σ, then $P^\Delta = P$ and $l^\Delta = l$. Moreover $P \nmid l$.

Δ cannot consist of (P,l)-homologies, as otherwise $|\Delta| = q(q^2 - 1)$ would divide $q - 1$. Hence Δ has a non trivial orbit t on l. If $Q \in t$, then $q(q^2 - 1) = |\Delta| = |t||\Delta_Q|$, whence $q(q^2 - 1) > |\Delta_Q| \geqslant q(q - 1)$. Assume $q \neq 5, 7, 9, 11$. Then it follows from 14.1 that $|\Delta_Q| = q(q - 1)$, i.e., $|t| = q + 1$. This yields that Δ acts doubly transitively on l.

Assume $q = 5, 7, 9$ or 11 and $|t| \neq q + 1$. Then it follows from 14.1 that $q = 5, 7, 11$ and $|t| = q$ or $q = 9$ and $|t| = 6$. In either case Δ fixes a point Q on l and hence the line PQ. As there are only $q - 1$ points other than P and Q on PQ, we infer that Δ fixes at least one more point on PQ which is impossible, as σ is a (P,l)-homology. Hence Δ acts in its natural 2-transitive representation on l.

Let Q and R be two distinct points on l. Then $|\Delta_{Q,R}| = q - 1$ and $\Delta_{Q,R}/\mathfrak{Z}(\Delta)$ is cyclic of order $\frac{1}{2}(q - 1)$. Hence $\Delta_{Q,R}$ is abelian. As the Sylow 2-subgroups of $\mathrm{SL}(2,q)$ are generalized quaternion groups, it follows that the Sylow 2-subgroups of $\Delta_{Q,R}$ are cyclic. Hence $\Delta_{Q,R}$ itself is cyclic.

Let S be a point on PQ other than P and Q and let $\delta \in \Delta_{Q,R}$ be such that $S^\delta = S$. There exists $\rho \in \Delta$ with $Q^\rho = R$ and $R^\rho = Q$. As $\Delta_{Q,R}^\rho = \Delta_{Q,R}$ and $\Delta_{Q,R}$ is cyclic, we have $\langle \delta \rangle^\rho = \langle \delta \rangle$. This yields that S^ρ is fixed by δ. Therefore δ fixes a quadrangle and hence a subplane. This implies that δ fixes more than two points on l whence $\delta \in \mathfrak{Z}(\Delta)$. This and $S^\delta = S$ imply $\delta = 1$. Therefore, $\Delta_{Q,R}$ acts transitively on the set of points on PQ which are distinct from P and Q. Hence Δ acts transitively on the set of points not on l and other than P.

Let Π be a Sylow p-subgroup of Δ_Q. Then Π fixes a point S on PQ which is distinct from P and Q. As $\Delta_{Q,R}$ normalizes Π and acts transitively on $PQ \setminus \{P,Q\}$, we see that Π fixes PQ pointwise. Therefore Π consists entirely of (Q,PQ)-elations.

Let p again be arbitrary, i.e., 2 or distinct from 2, and denote by $\mathfrak{P}^*$ the incidence structure we obtain by removing P and l and all the elements incident with P or l from $\mathfrak{P}$. Then Δ acts flag transitively on $\mathfrak{P}^*$: As we have seen above, Δ acts transitively on the set of points of $\mathfrak{P}^*$. If X is a point of $\mathfrak{P}^*$, then Δ_X is a Sylow p-subgroup of Δ. Since Δ_X acts transitively on the set of points on l which are distinct from $l \cap PX$, it follows that Δ_X acts transitively on the set of lines of $\mathfrak{P}^*$ which pass through X. Put $\Pi = \Delta_X$ for a fixed point X of $\mathfrak{P}^*$ and $\Lambda = \Delta_m$ for a fixed line m of $\mathfrak{P}^*$ which passes through X. If Y is a point of $\mathfrak{P}^*$, then we assign to Y the coset $\Pi\eta$ where $X^\eta = Y$. Similarly we assign to the line n the coset $\Lambda\delta$ where $m^\delta = n$. Then $Y \,\mathrm{I}\, n$ if and only if $\Pi\eta \cap \Lambda\delta \neq \varnothing$. Comparison with the desarguesian plane $\mathfrak{D}$ of order q now yields that $\mathfrak{P}$ and $\mathfrak{D}$ are isomorphic,

The statement about the conjugacy classes is now trivial. $\square$

38.13 Corollary. *Let p be an odd prime and let $\mathfrak{P}$ be a projective plane of order $q = p^r$. If Δ is a collineation group of $\mathfrak{P}$ isomorphic to $\mathrm{SL}(2,q)$, then Δ fixes a non-incident point line pair (P,l). Moreover, if Π is a Sylow p-subgroup of Δ, then Π consists entirely of (Q,PQ)-elations, where Q is a suitable point on l.*

39. Translation Planes Whose Collineation Group Acts Doubly Transitively on l_∞

We start with the following theorem.

39.1 Theorem (Lüneburg 1964). *Let $G = \mathrm{PSL}(2,q)$ and $K = \mathrm{P\Gamma L}(2,q)$ and let H be a subgroup of K containing G. If H admits a faithful representation as a doubly transitive permutation group of degree $n + 1$, then either $n = q$ and H is any group between G and K or one of the following is true:*

(1) $q = 4$, $n = 5$ and $H = \mathrm{PSL}(2,4)$ or $H = \mathrm{P\Gamma L}(2,4)$.
(2) $q = 5$, $n = 4$ and $H = \mathrm{PSL}(2,5)$ or $H = \mathrm{PGL}(2,5)$.
(3) $q = 7$, $n = 6$ and $H = \mathrm{PSL}(2,7)$.
(4) $q = 8$, $n = 27$ and $H = \mathrm{P\Gamma L}(2,8)$.
(5) $q = 9$, $n = 5$ and $H = \mathrm{PSL}(2,9)$ or $H = \mathrm{PSL}(2,9)\langle\sigma\rangle$, where σ is defined by $x^\sigma = x^3$.
(6) $q = 11$, $n = 10$ and $H = \mathrm{PSL}(2,11)$.

PROOF. First we show that these six exceptions really occur and that there are no other possibilities for H if q and n are as listed.

(1). It is easily seen and well known that $\mathrm{PSL}(2,4) \cong \mathrm{PSL}(2,5) \cong A_5$. Hence $\mathrm{PSL}(2,4)$ has a representation as a doubly transitive group of degree 6 coming from the isomorphism $\mathrm{PSL}(2,4) \cong \mathrm{PSL}(2,5)$. As every such representation is equivalent to the action of $\mathrm{PSL}(2,4)$ via inner automorphisms on the set of Sylow 5-subgroups of $\mathrm{PSL}(2,4)$, we see that H can be any group between $\mathrm{PSL}(2,4)$ and $\mathrm{P\Gamma L}(2,4)$. We infer from $|\mathrm{P\Gamma L}(2,4):\mathrm{PSL}(2,4)| = 2$ that we have for H just the two possibilities listed.

(2). We infer from $\mathrm{PSL}(2,5) \cong \mathrm{PSL}(2,4)$ that $\mathrm{PSL}(2,5)$ has a doubly transitive representation of degree 5. Every such representation is equivalent to the action of $\mathrm{PSL}(2,5)$ via inner automorphisms on the set of Sylow 2-subgroups, whence $H = \mathrm{PSL}(2,5)$ or $H = \mathrm{PGL}(2,5)$, as $\mathrm{PGL}(2,5) = \mathrm{P\Gamma L}(2,5)$.

(3). The groups $\mathrm{PSL}(2,7)$ and $\mathrm{PGL}(3,2)$ are isomorphic (see e.g. Huppert [1967, II.6.14 (4), p. 183]). From this we infer that $\mathrm{PSL}(2,7)$ has a doubly transitive representation of degree 7. Every such representation is equivalent to the action of $\mathrm{PSL}(2,7)$ via inner automorphisms on a conjugacy class of A_4's. As there are two such classes which fuse under $\mathrm{PGL}(2,7)$ into one, we see that H must be distinct from $\mathrm{PGL}(2,7)$. There are no further possibilities for H, since $\mathrm{PGL}(2,7) = \mathrm{P\Gamma L}(2,7)$.

(4). Let $G = \mathrm{PSL}(2,8)$ and let S be a Sylow 3-subgroup of G. Then $|S| = 9$, as $|G| = 8(8^2 - 1)$. It follows from 14.1 that S is cyclic. Moreover, its normalizer is a dihedral group of order 18. Hence there are $4 \cdot 7 = 28$ Sylow 3-subgroups in G. Therefore G and hence $K = \mathrm{P\Gamma L}(2,8)$ have a transitive representation of degree 28. Let T be a Sylow 3-subgroup of G distinct from S. Then $S \cap T = \{1\}$ by 14.3. Thus $S \cap \mathfrak{N}_G(T) = \{1\}$. Therefore, S splits the set $\mathfrak{T}$ of Sylow 3-subgroups of G which are distinct from S into orbits $\mathfrak{T}_1$, $\mathfrak{T}_2$, and $\mathfrak{T}_3$ each of length 9. Let S^* be a Sylow 3-subgroup of K which contains S. Then $|S^*| = 27$. Moreover S^* either leaves invariant all the $\mathfrak{T}_i$'s or acts transitively on $\mathfrak{T}$.

The polynomial $f = x^3 + x + 1$ is irreducible over $\mathrm{GF}(2)$, but reducible over $\mathrm{GF}(8)$. Hence there exists $a \in \mathrm{GF}(8) \backslash \mathrm{GF}(2)$ such that $f(a) = 0$. Moreover, a generates $\mathrm{GF}(8)^*$. Therefore

$$A_i = \begin{pmatrix} 0 & a^i \\ a^{-i} & 1 \end{pmatrix}, \qquad i = 0, 1, \ldots, 6$$

are 7 distinct matrices. Each of these matrices has order 3. Also

$$A_i^{-1} = \begin{pmatrix} 1 & a^i \\ a^{-i} & 0 \end{pmatrix}.$$

This proves that the A_i's lie in 7 distinct Sylow 3-subgroups of G. Put

$$C = \begin{pmatrix} a & a^4 \\ a^4 & a^2 \end{pmatrix}.$$

Then $C^3 = A_0$ and hence $o(C) = 9$. We may therefore assume that $S = \langle C \rangle$. Put $B_j = C^{-j} A_1 C^j$ for $j = 0, 1, \ldots, 8$. Then

$$B_0 = \begin{pmatrix} 0 & a \\ a^6 & 1 \end{pmatrix}, \qquad B_1 = \begin{pmatrix} a^6 & a^3 \\ 1 & a^2 \end{pmatrix}, \qquad B_2 = \begin{pmatrix} a^2 & a^4 \\ a^6 & a^6 \end{pmatrix},$$

$$B_3 = \begin{pmatrix} a^3 & a^4 \\ a & a \end{pmatrix}, \qquad B_4 = \begin{pmatrix} a^5 & a^3 \\ a^3 & a^4 \end{pmatrix}, \qquad B_5 = \begin{pmatrix} a^3 & a \\ a^4 & a \end{pmatrix},$$

$$B_6 = \begin{pmatrix} a^2 & a^6 \\ a^4 & a^6 \end{pmatrix}, \qquad B_7 = \begin{pmatrix} a^6 & 1 \\ a^3 & a^2 \end{pmatrix}, \qquad B_8 = \begin{pmatrix} 0 & a^6 \\ a & 1 \end{pmatrix}.$$

Hence A_1 and A_6 are conjugate under S, whereas there is no element in S which transforms A_1 into A_i or A_i^{-1} for $i = 2, \ldots, 5$. Hence the Sylow 3-subgroups containing A_1 and A_6 belong to $\mathfrak{T}_1$ say, whereas the Sylow 3-subgroups containing $A_2, \ldots, A_5$ are in $\mathfrak{T}_2 \cup \mathfrak{T}_3$. Now, if A is the mapping defined by $(x, y)A = (x^2, y^2)$, then $A^{-1} A_1 A = A_2$. Hence S^* is transitive on $\mathfrak{T}$. This proves that K acts doubly transitively on $\mathfrak{T} \cup \{S\}$. Moreover, $H = K$ is the only possibility for H, as $28 \cdot 27$ divides $|H|$.

(5). The groups $\mathrm{PSL}(2,9)$ and A_6 are isomorphic (see e.g. Huppert [1967, II.6.14 (6), p. 183]). Hence $\mathrm{PSL}(2,9)$ admits a doubly transitive representation of degree 6. Each such representation is equivalent to the action of $\mathrm{PSL}(2,9)$ via inner automorphisms on a conjugacy class of A_5's.

As $PSL(2, 9)$ contains exactly 12 subgroups isomorphic to A_5 which are all conjugate under $PGL(2, 9)$, we see that H cannot be equal to $PGL(2, 9)$. Let $\sigma \in P\Gamma L(2, 9)$ be defined by $x^\sigma = x^3$. Then σ fixes both conjugacy classes of A_5 by the Frattini argument. Hence $H = PSL(2, 9)\langle\sigma\rangle$ also admits a doubly transitive representation of degree 6.

(6). $G = PSL(2, 11)$ has a maximal subgroup M isomorphic to A_5. As $|PSL(2, 11)| = \frac{1}{2}11(11^2 - 1) = 11 \cdot 60$, we see that $PSL(2, 11)$ has a representation of degree 11. As 11 is a prime and as G is not soluble, G acts doubly transitively in this representation by a theorem of Burnside (see e.g. Huppert [1967, V.21.3, p. 609]). As $PSL(2, 11)$ has 22 subgroups isomorphic to A_5 which are all conjugate under $PGL(2, 11)$, we see that there is only one possibility for H.

Now we prove that these are the only exceptions. Let $G = PSL(2, 9) \subseteq H \subseteq K = P\Gamma L(2, 9)$ and assume that H admits a faithful representation as a doubly transitive permutation group of degree $n + 1$. If $q \leqslant 3$, then G has an abelian characteristic subgroup N of order $q + 1$. As N is also normal in H, we have that N acts transitively and hence regularly, whence $n + 1 = q + 1$, i.e., $n = q$. We may therefore assume $q > 3$. Then G is simple. Since G is normal in K, it is also normal in H. This yields that G is a minimal normal subgroup of H. Therefore, by Burnside [1955, §154] the group G acts primitively in the given representation of H. In order to determine n, we therefore have to compute the indices $n + 1$ of maximal subgroups of G and to check whether or not $n(n + 1)$ divides $|H|$. As $|H|$ is a divisor of $|K|$, we have

(j) $$n(n + 1) \text{ divides } sq(q^2 - 1),$$

where $q = p^s$ with p a prime.

Using 14.1, we see that the only candidates for maximal subgroups of G are:

1. Normalizers of Sylow p-subgroups.
2. Dihedral groups of order $q + 1$ or $2(q + 1)$ according to whether q is odd or even.
3. Dihedral groups of order $q - 1$ or $2(q - 1)$ according to whether q is odd or even.
4. Groups isomorphic to $PSL(2, p')$ with $sr^{-1} \geqslant 3$.
5. Groups isomorphic to $PGL(2, p')$ with $p > 2$ and $sr^{-1} = 2k \geqslant 4$.
6. Groups isomorphic to A_4, if $q \equiv \pm 3 \bmod 8$.
7. Groups isomorphic to S_4, if $q \equiv \pm 1 \bmod 8$.
8. Groups isomorphic to A_5, if $q \equiv \pm 1 \bmod 10$.

Case 1. Here we get $n = q$.

Case 2. Let D be a dihedral group of order $q + 1$ respectively $2(q + 1)$. Then $n + 1 = |G : D| = \frac{1}{2}q(q - 1)$ in either case. Therefore $\frac{1}{2}(p^s - 1)p^s - 1 = \frac{1}{2}(p^{2s} - p^s - 2)$ divides $sp^s(p^{2s} - 1)$ by (j). This yields that $\frac{1}{2}(p^{2s} - p^s - 2)$ divides $2s(p^s + 1)$, as $(\frac{1}{2}p^s(p^s - 1), \frac{1}{2}p^s(p^s - 1) - 1) = 1$. Hence

$p^{2s} - p^s - 2 \leqslant 4sp^s + 4s$. This implies $p^{2s} - (4s + 1)p^s \leqslant 4s + 2$, whence

$$\left(p^s - \frac{4s + 1}{2}\right)^2 \leqslant 4s + 2 + \frac{(4s + 1)^2}{4}.$$

From this we obtain the inequality

$$p^s - \frac{4s + 1}{2} < 4s + 1 + \frac{4s + 1}{2}$$

which yields $p^s < 8s + 2$. Assume $p^t \geqslant 8t + 2$. Then $p^{t+1} \geqslant 8tp + 2p$ and hence $p^{t+1} \geqslant 8(t + 1) + 2$. If $p \geqslant 11$, then $p > 10$ and hence $p^s \geqslant 8s + 2$ for all s. Thus $p^s < 8s + 2$ implies $p = 2, 3, 5,$ or 7. This together with the above argument yields $p^s = 2, 4, 8, 16, 32, 3, 9, 5,$ or 7. Using (j), we finally obtain $q = 4$ or 8. But then $n = 5$ respectively 27; i.e., we have case (1) or case (4).

Case 3. In this case we have $n + 1 = \frac{1}{2} q(q + 1)$. It follows from (j) that $\frac{1}{2} q(q + 1) - 1$ divides $sq(q^2 - 1)$ and hence $2s(q - 1)$. From this we deduce $p^{2s} + (1 - 4s)p^s \leqslant 2 - 4s < 0$ whence $p^s < 4s - 1$. Assume $p \geqslant 3$ and $p^s \geqslant 4s - 1$. Then $p^{s+1} \geqslant p(4s - 1) > 2(4s - 1) \geqslant 4(s + 1) - 2$. Therefore $p^{s+1} \geqslant 4(s + 1) - 1$. Thus $p^s \leqslant 4s - 1$ yields $p = 2$. Now $2^4 = 16 > 4 \cdot 4 - 1$. Assume $2^s > 4s - 1$. Then $s \geqslant 4$ and $2^{s+1} > 2(4s - 1) = 4 \cdot 2s - 2 = 4(s + 1) + 4(s - 1) - 2 > 4(s + 1) - 1$. Therefore $p^s \leqslant 4s - 1$ implies $p^s = 2, 4,$ or 8. Using (j) we finally obtain $p^s = 2$ and $n = 2 = q$.

Cases 4 and 5. Here we have $n + 1 = a^{-1}p^{s-r}(p^{2s} - 1)(p^{2r} - 1)^{-1}$ with $a = 1$ or 2. By (j) there exists $k \in \mathbb{N}$ with

$$a^{-1}p^{s-r}(p^{2s} - 1)(p^{2r} - 1)^{-1}\left[a^{-1}p^{s-r}(p^{2s} - 1)(p^{2r} - 1)^{-1} - 1\right]k$$
$$= sp^s(p^{2s} - 1).$$

Therefore

$$\left[a^{-1}p^{s-r}(p^{2s} - 1)(p^{2r} - 1)^{-1} - 1\right]k = sap^r(p^{2r} - 1).$$

As $s - r \geqslant 1$, we see that p does not divide n. Thus $a^{-1}p^{s-r}(p^{2s} - 1) \cdot (p^{2r} - 1)^{-1} - 1$ divides $as(p^{2r} - 1)$. Therefore

$$p^{s-r}\left[a^{-1}(p^{2s} - 1)(p^{2r} - 1)^{-1} - 1\right] < as(p^{2r} - 1).$$

Furthermore $s > s - r \geqslant 2r$ whence

$$a^{-1}(p^{2s} - 1)(p^{2r} - 1)^{-1} - 1 < as(p^{2r} - 1)p^{r-s} < as(p^{2r} - 1)(p^{2r} - 1)^{-1}$$
$$= as$$

and hence $p^{2s} - 1 \leqslant a^2s(p^{2r} - 1) < 4s(p^s - 1)$, since $a \leqslant 2$ and $2r < s$. Thus $p^s + 1 < 4s$ which yields the contradiction $p^s = 2$.

Case 6, 7, and 8. Let D_i ($i = 1, 2, 3$) be the stabilizer of a point and let the D_i be numbered in such a way that $D_1 = A_4$, $D_2 = S_4$ and $D_3 = A_5$. Then

$$n + 1 = |G : D_i| = \frac{1}{2} p^s(p^{2s} - 1)|D_i|^{-1}.$$

From this and (j) we infer that $|G : D_i| - 1$ divides $2s|D_i|$. Hence

$$p^s(p^{2s} - 1) \leqslant 4s|D_i|^2 + 2|D_i| < 8s|D_i|^2.$$

Assume $p^s(p^{2s} - 1) > 8s|D_i|^2$. Then

$$p^{s+1}(p^{2(s+1)} - 1) > p^{s+1}(p^{2s+2} - p^2) > 8p^3s|D_i|^2 > 8(s + 1)|D_i|^2.$$

$i = 1$: Then $p^s(p^{2s} - 1) < 8 \cdot 12^2s$. If $p \geqslant 11$, then $p(p^2 - 1) > 8 \cdot 12^2$. Hence we are left with $p = 3, 5,$ or 7. It is easily seen that $q = 3, 9, 5$ or 7. As $q \equiv \pm 3 \bmod 8$, we have $q = 5$, since we have assumed $q > 3$. But $q = 5$ yields $n = 4$. This is case (2).

$i = 2$: Then $p^s(p^{2s} - 1) < 8 \cdot 24^2s$. In this case we obtain $p^s = 3, 9, 5, 7,$ 11 or 13. As $q \equiv \pm 1 \bmod 8$, we are left with $q = 7$ or 9. For $q = 7$ we obtain $n = 6$, i.e., case (3), and $q = 9$ yields $n = 14$ which is impossible, as 14 does not divide $2 \cdot 9 \cdot 80$.

$i = 3$: Using the above inequality and $q \equiv \pm 1 \bmod 10$, we obtain $q = 9,$ 11, 19, or 29. By (j), the numbers 19 and 29 cannot occur. If $q = 9$ then $n = 5$, i.e., we have case (5); and if $q = 11$ then $n = 10$, i.e., we obtain case (6).   □

The next theorem has been known to several people for quite some time, but it has never appeared in print as far as I know. So I do not know to whom I shall ascribe it. The only thing I can trace is that Piper proved in 1963 that a plane satisfying the assumptions of the next theorem is a translation plane.

39.2 Theorem. *Let $\mathfrak{A}$ be a finite affine plane of order q, let P be an affine point of $\mathfrak{A}$ and let G be a group of collineations of $\mathfrak{A}$ generated by shears. If every affine line of $\mathfrak{A}$ is the axis of a non-trivial shear in G then one of the following is true:*

a) $\mathfrak{A}$ *is desarguesian,* $G = TG_P$, *and* $G_P \cong SL(2, q)$.
b) $\mathfrak{A}$ *is desarguesian of order 9 and* $G = TG_P$ *with* $G_P \cong SL(2, 5)$.
c) $\mathfrak{A}$ *is desarguesian of order* $q = 2^r$ *and* $G = TG_P$, *where* G_P *is dihedral of order* $2(q + 1)$.
d) $\mathfrak{A}$ *is a Lüneburg plane of order* $q = 2^{2(2r+1)}$ *and* $G = TG_P$ *with* $G_P \cong S(2^{2r+1})$.

PROOF. First we show that $\mathfrak{A}$ is a translation plane and that G contains the translation group T of $\mathfrak{A}$. Let X be a point on l_∞ and let $l_1, \ldots, l_q$ be the affine lines of $\mathfrak{A}$ carrying X. Furthermore, denote by S_i the group of all shears with axis l_i in G and by E the group of all elations in G whose centre is X. By 1.1 and 1.6, E is an elementary abelian p-group. Moreover $|S_i| > 1$ and

$$|E| = |G(X)| + \sum_{i=1}^{q} (|S_i| - 1) = |G(X)| - q + \sum_{i=1}^{q} |S_i|.$$

(Recall that $G(X) = G \cap T(X)$.) As $|G(X)| \geqslant 1$ and $|E| \equiv |S_i| \equiv q \equiv 0 \bmod p$, it follows that $|G(X)| > 1$ for all X. As E is a p-group for all X on l_∞ which induces a group of permutations on l_∞ possessing only one fixed point on l_∞, we obtain by Gleason's lemma (15.1) that G acts transitively on l_∞. Therefore there exists an integer $h \geqslant 2$ with $|G(X)| = h$ for all X I l_∞. Thus $\mathfrak{A}$ is a translation plane and G contains the translation group of $\mathfrak{A}$ by 15.2. This yields $G = TG_P$ and $T \cap G_P = \{1\}$.

Let η be a shear in G. Then there exists $\tau \in T$ such that the axis of $\tau^{-1}\eta\tau = \epsilon$ passes through P. Hence $\eta = \tau\epsilon\tau^{-1} = \tau(\epsilon\tau^{-1}\epsilon^{-1})\epsilon$ is the product of a translation and a shear the axis of which passes through P.

Let $\gamma \in G$. As G is generated by shears, we have by the remark just made that $\gamma = \tau_1\epsilon_1\tau_2\epsilon_2 \cdots \tau_n\epsilon_n$, where the τ_i's are translations and the ϵ_i's are shears with axes through P. Using induction, we obtain $\gamma = \tau\epsilon_1\epsilon_2 \cdots \epsilon_n$ where τ is a translation. This yields in particular that G_P is generated by shears whose axes pass through P. Therefore 35.10 applies to G_P; we have:

A) $G_P \cong \mathrm{SL}(2,q)$.
B) $q = 9$ and $G_P \cong \mathrm{SL}(2,5)$.
C) $q = 2^r$ and $G_P = C\langle\epsilon\rangle$ where C is a normal subgroup of odd order of G_P and ϵ is an elation with axis through P.
D) $q = 2^{2(2r+1)}$ and $G_P \cong \mathrm{S}(2^{2r+1})$.

In case A), the plane is desarguesian by 38.12. In case B), the plane $\mathfrak{A}$ is desarguesian by 8.4 and 8.3. In case D), the plane is a Lüneburg plane by 31.1.

It remains to consider the case C). In this case G_P is soluble by the Feit-Thompson theorem. As G_P acts transitively on l_∞, the group G acts flag transitively on $\mathfrak{A}$. Hence $G_P \subseteq \Gamma\mathrm{L}(1,q^2)$ by 37.7. Moreover, by 37.9 the group G_P acts on the set of points of the desarguesian plane $\mathfrak{D}$ of order q in the same way it acts on the set of points of the plane $\mathfrak{A}$. As q is even, all involutions in $\Gamma\mathrm{L}(1,q^2)$ are conjugate. This shows that the shears of $\mathfrak{A}$ which are contained in G_P are also shears of $\mathfrak{D}$. This yields that the lines of $\mathfrak{A}$ which pass through P are also lines of $\mathfrak{D}$. Therefore $\mathfrak{A}$ is desarguesian. This yields finally that G_P is dihedral of order $2(q + 1)$. $\square$

39.3 Theorem (Schulz 1971, Czerwinski 1972). *Let $\mathfrak{A}$ be a finite translation plane admitting a collineation group G acting doubly transitively on l_∞. Assume furthermore that G does not contain a Baer involution if the order of $\mathfrak{A}$ is odd. Then $\mathfrak{A}$ is either desarguesian or a Lüneburg plane.*

PROOF. As $\mathfrak{A}$ is a translation plane, we may assume $G = G_P$ for some affine point P of $\mathfrak{A}$. Let q be the order of $\mathfrak{A}$ and assume that q is even. If l is a line through P, then $G_{P,l}$ acts transitively on the set of points on l_∞ which are distinct from $l \cap l_\infty$. Therefore, it follows from 15.5 that $G_{P,l}$ contains an elation $\sigma \neq 1$. As $P^\sigma = P$ and $l^\sigma = l$, it follows that l is the axis of σ. This yields that $\mathfrak{A}$ is either desarguesian or a Lüneburg plane by 39.2.

Assume that q is odd. If G contains an involutory homology whose axis passes through P, then by 3.17 and the double transitivity of G we see that G contains non-trivial shears, whence $\mathfrak{A}$ is desarguesian by 39.2. We may thus assume that G does not contain an involutory homology with affine axis. It then follows by our assumption on the involutions in G that all involutions in G are (P, l_∞)-homologies. Hence, $|G|$ being even, G contains exactly one involution. This implies that the Sylow 2-subgroups of G are either cyclic or generalized quaternion groups.

Let $\overline{G}$ be the group induced on l_∞ by G and assume that the Sylow 2-subgroups of $\overline{G}$ are cyclic. Then $\overline{G}$ contains a normal subgroup N of odd order such that $\overline{G}/N$ is a 2-group (see e.g. Gorenstein [1968, 7.6.1, p. 257]). As the degree of $\overline{G}$ is $q + 1$, the group N having odd order cannot act transitively on l_∞. Therefore $N = \{1\}$ by the double transitivity of $\overline{G}$. But then $\overline{G}$ is a 2-group, a contradiction. Therefore the Sylow 2-subgroups of G are generalized quaternion groups and hence the Sylow 2-subgroups of $\overline{G}$ are dihedral, since the only involution in G fixes l_∞ pointwise.

Suppose first that $\overline{G}$ is soluble. Then $\overline{G}$ contains an elementary abelian normal subgroup E. As $\overline{G}$ acts doubly transitively on l_∞, we have that $|E| = q + 1$. Since q is odd, E is an elementary abelian 2-group. This yields $|E| = 4$, since the Sylow 2-subgroups of $\overline{G}$ are dihedral. Therefore $q = 3$ and $\mathfrak{A}$ is desarguesian. We may thus assume that $\overline{G}$ is not soluble. Moreover, $\overline{G}$ does not contain a non-trivial normal subgroup of odd order, since $q + 1$ divides the order of any non-trivial normal subgroup of $\overline{G}$. Therefore, by Gorenstein & Walter 1965, $\overline{G}$ is either isomorphic to A_7 or $\overline{G}$ is isomorphic to a group H with $\mathrm{PSL}(2, q^*) \subseteq H \subseteq \mathrm{P\Gamma L}(2, q^*)$ for some odd prime power q^*.

Assume first $\overline{G} \cong A_7$, then $q(q + 1)$ divides $\frac{1}{2} 7! = 2^3 \cdot 3^2 \cdot 5 \cdot 7$. As q is odd and not a prime (otherwise $\mathfrak{A}$ would be desarguesian), $q = 3^2$ and $\mathfrak{A}$ is the nearfield plane of order 9 by 8.4. But 7 does not divide the order of the collineation group of this plane by 8.3. Thus $\overline{G}$ is isomorphic to a group H with $\mathrm{PSL}(2, q^*) \subseteq H \subseteq \mathrm{P\Gamma L}(2, q^*)$ with q^* odd. Using this and the fact that q is odd, then 39.1 tells us that either $q^* = q$ or $q^* = 9$ and $q = 5$. In the latter case, $\mathfrak{A}$ is desarguesian of order 5. But the group $\mathrm{SL}(2, 9)$ does not operate on that plane. Hence $q^* = q$. This shows that G contains a group S which induces a group S^* on l_∞ isomorphic to $\mathrm{PSL}(2, q)$ in its natural doubly transitive representation. Let S be minimal among all such subgroups of G and let K be the kernel of the homomorphism $S \to S^*$. The group $\mathrm{PSL}(2, q)$ is simple, as $q > 3$. Hence $KS' = S$. Moreover, $(S')^* = S^*$. Therefore $S' = S$ by the minimality of S. Since K consists only of (P, l_∞)-homologies, K is cyclic. Thus $\mathrm{Aut}(K)$ is abelian. We deduce from this and $S' = S$ that $K \subseteq \mathfrak{Z}(S)$. Moreover 2 divides $|K|$, as the Sylow 2-subgroups of G are quaternion. Hence by Schur [1907, Satz III and Satz IX], we have either $S \cong \mathrm{SL}(2, q)$ or $q = 9$ and $|K| = 6$. But the latter case cannot occur, since $|K|$ divides $q - 1$. Hence $S \cong \mathrm{SL}(2, q)$ and $\mathfrak{A}$ is desarguesian by 38.12. $\square$

40. A Theorem of Burmester and Hughes

40.1 Theorem *Let $\mathfrak{A}$ be a finite translation plane of order q and let G be a group of collineations of $\mathfrak{A}$ fixing two distinct points P and Q on l_∞ and an affine point O. If q is even, assume that G does not contain a Baer involution, and if q is odd, assume that the rank of the translation group over its kernel is twice an odd number. Then G is soluble.*

PROOF. Using standard notations, we may assume that $V(0) = OP$ and $V(\infty) = OQ$. If K is the kernel of $\mathfrak{A}$, then G consists of K-semilinear mappings. If $\alpha \in G$, denote by comp(α) the companion automorphism of α. Then comp is a homomorphism of G into Aut(K). If G_1 is the kernel of comp, then G/G_1 is cyclic, as Aut(K) is cyclic. Let $\alpha \in G_1$. Then there exist $\alpha_1, \alpha_2 \in GL(X)$ with $(x, y)^\alpha = (x^{\alpha_1}, y^{\alpha_2})$, where X is such that $V(0) = \{(x, 0) \mid x \in X\}$ and $V(\infty) = \{(0, x) \mid x \in X\}$. The mapping $\alpha \rightarrow (\det(\alpha_1), \det(\alpha_2))$ is a homomorphism of G_1 into $K^* \times K^*$. Hence G_1/G_2 is abelian, where G_2 is the kernel of the mapping $\alpha \rightarrow (\det(\alpha_1), \det(\alpha_2))$. Therefore, G is soluble if G_2 is.

Next we show that G does not contain a Baer involution. By way of contradiction, assume that α is a Baer involution in G. Then the order of $\mathfrak{A}$ is odd by assumption. Moreover α centralizes a subspace U of $V(0)$ with $2\mathrm{rk}(U) = \mathrm{rk}(V(0))$. From this we infer that 4 divides the K-rank of the translation group of $\mathfrak{A}$, a contradiction.

Let α be an involution in G_2. Then α is a central collineation by the remark just made. As α fixes O, P and Q, we see that α is a homology. Therefore the order of $\mathfrak{A}$ is odd. Let α_1 and α_2 be as above. Then at least one of the α_i maps every $x \in X$ onto $-x$. For this α_i we then have $1 = \det(\alpha_i) = (-1)^{\mathrm{rk}(X)}$. This shows that rk($X$) is even, as the characteristic is odd. Therefore 4 divides the K-rank of the translation group, a contradiction. Thus $|G_2|$ is odd, whence G_2 is soluble by the Feit-Thompson theorem.   □

40.2 Corollary. *Let $\mathfrak{A}$ be a finite translation plane of non-square order. If G is a group of collineations of $\mathfrak{A}$ fixing two points on l_∞, then G is soluble.*

PROOF. Let O be an affine point of $\mathfrak{A}$ and put $H = TG$. Then $H = TH_O$ and H_O is soluble by 40.1. Therefore H is soluble which implies that G is soluble.   □

41. Bol Planes

An affine plane $\mathfrak{A}$ is called a *Bol plane*, if there exist two distinct points P and Q on l_∞ such that each line of $\mathfrak{A}$ which does not pass through either of the points P and Q is the axis of an involutory perspectivity which interchanges P and Q. By 4.9, a Bol plane is always a translation plane. Thus, by the remarks after 3.10, the plane $\mathfrak{A}$ may be represented by a set Σ of fixed point free linear mappings of a vector space X such that $\rho\Sigma^{-1}\rho = \Sigma$ for all $\rho \in \Sigma$. This implies, as we have seen earlier, that $\Sigma^{-1} = \Sigma$ and that $\rho^i \in \Sigma$ for all $\rho \in \Sigma$ and all $i \in \mathbb{Z}$. Let K be the centralizer of Σ in the endomorphism ring of X and assume that X is finite. Then $K = \mathrm{GF}(q)$. Denote by d the rank of X over K. Assume furthermore that X may be identified with the additive group of $\mathrm{GF}(q^d)$ such that $\Sigma \subseteq \Gamma\mathrm{L}(1, q^d)$. If $\sigma \in \Sigma$, then there exist $\lambda(\sigma) \in \{0, 1, \ldots, d-1\}$ and $a \in \mathrm{GF}(q^d)^*$ with $x^\sigma = x^{q^{\lambda(\sigma)}}a$.

41.1 Lemma. *If $\rho, \sigma \in \Sigma$, then $\lambda(\rho\sigma\rho) = 2\lambda(\rho) + \lambda(\sigma)$.*

PROOF. By the remarks made above, $\rho\sigma\rho \in \Sigma$. Thus $\lambda(\rho\sigma\rho)$ is defined. Let $x^\rho = x^{q^{\lambda(\rho)}}a$ and $x^\sigma = x^{q^{\lambda(\sigma)}}b$ and $x^{\rho\sigma\rho} = x^{q^{\lambda(\rho\sigma\rho)}}c$. Then

$$x^{q^{\lambda(\rho\sigma\rho)}}c = x^{\rho\sigma\rho} = x^{q^{2\lambda(\rho)+\lambda(\sigma)}}a^{q^{\lambda(\rho)+\lambda(\sigma)}+1}b^{q^{\lambda(\rho)}},$$

whence $\lambda(\rho\sigma\rho) = 2\lambda(\rho) + \lambda(\sigma)$. (Note that equality in this context means congruence modulo d.) $\qquad\square$

41.2 Lemma. *If $\sigma \in \Sigma$ and $k \in \mathbb{N}$, then $\lambda(\sigma^k) = k\lambda(\sigma)$.*

PROOF. Let $x^\sigma = x^{q^{\lambda(\sigma)}}a$ and assume that $\lambda(\sigma^k) = k\lambda(\sigma)$. Then

$$x^{\sigma^{k+1}} = \left(x^{q^{k\lambda(\sigma)}}b\right)^\sigma = x^{q^{(k+1)\lambda(\sigma)}}b^{q^{\lambda(\sigma)}}a.$$

Thus $\lambda(\sigma^{k+1}) = (k+1)\lambda(\sigma)$. $\qquad\square$

41.3 Lemma. *λ is a mapping from Σ onto the set $\{0, 1, \ldots, d-1\}$.*

PROOF. As $K = \mathrm{GF}(q)$ is the centralizer of Σ, we have that $\Gamma = \langle (x \to x^{q^{\lambda(\sigma)}}) \mid \sigma \in \Sigma \rangle$ is the Galois group of $\mathrm{GF}(q^d):\mathrm{GF}(q)$. Therefore it suffices to prove that $\Delta = \{(x \to x^{q^{\lambda(\sigma)}}) \mid \sigma \in \Sigma\}$ is a subgroup of Γ. This is certainly true, if $d = 1$. Assume $d > 1$. Then there exists $\rho \in \Sigma$ with $0 < \lambda(\rho) < d$. Let ρ be such that $\lambda(\rho)$ is minimal. Let $\sigma \in \Sigma$ and put $f = (\lambda(\rho), \lambda(\sigma))$. Then either $f = (2\lambda(\rho), \lambda(\sigma))$ or $f = (\lambda(\rho), 2\lambda(\sigma))$. Let $f = (2\lambda(\rho), \lambda(\sigma))$. Then there exist integers r and s with $f = 2r\lambda(\rho) + s\lambda(\sigma)$. From this we obtain by 41.2 and 41.1 that $f = \lambda(\rho^r\sigma^s\rho^r)$. Similarly $f = \lambda(\sigma^s\rho^r\sigma^s)$, if $f = (\lambda(\rho), 2\lambda(\sigma))$. Thus $(x \to x^{q^{\lambda(f)}}) \in \Delta$. By the minimality of $\lambda(\rho)$, we have $f \geqslant \lambda(\rho) \geqslant (\lambda(\rho), \lambda(\sigma)) = f$. Thus $f = \lambda(\rho)$, whence $\lambda(\rho)$ divides $\lambda(\sigma)$. This together with 41.2 yields $\Delta = \{(x \to x^{q^{k\lambda(\rho)}}) \mid k \in \mathbb{N}\}$. $\qquad\square$

For $0 \leqslant i \leqslant d-1$ we put $\Sigma_i = \{\sigma \mid \lambda(\sigma) = i\}$.

41.4 Lemma. *If $\rho \in \Sigma$, then $\rho\Sigma_i\rho = \Sigma_{i+2\lambda(\rho)}$.*

This follows immediatly from 41.1.

41.5 Lemma. Σ_0 *is a cyclic group.*

PROOF. Let $\sigma \in \Sigma_0$. Then $x^\sigma = xw^i$, where w is a generator of $\mathrm{GF}(q^d)^*$. Let $1 \neq \sigma$ be such that i is minimal. Pick $\tau \in \Sigma_0$ and let $x^\tau = xw^j$. Put $f = (i, j)$. Then either $f = (2i, j)$ or $f = (i, 2j)$. Assume $f = (2i, j)$. Then there exist $s, t \in \mathbb{Z}$ such that $f = 2is + jt$. Put $\rho = \sigma^s \tau' \sigma^s$. Then $\lambda(\rho) = 2s\lambda(\sigma) + t\lambda(\tau) = 0$. Thus $\rho \in \Sigma_0$. Moreover, $x^\rho = xw^f$. Therefore $f \geqslant i$. On the other hand $f = (i, j) \leqslant i$. Hence $f = i$ and i divides j. This proves $\Sigma_0 \subseteq \langle \sigma \rangle$. Now 41.2 yields $\langle \sigma \rangle \subseteq \Sigma_0$, whence $\Sigma_0 = \langle \sigma \rangle$. □

41.6 Lemma. Σ_0 *is normalized by* Σ.

This follows from the fact that Σ_0 is a subgroup of $\mathrm{GL}(1, q^d)$ and that all subgroups of a cyclic group are characteristic subgroups of that group.

41.7 Lemma. *If* $\rho \in \Sigma_1$, *then* $\Sigma_0 \langle \rho^2 \rangle = \Sigma_0 \cup \Sigma_2 \cup \Sigma_4 \cup \cdots$. *Moreover,* $\rho^{2k}\Sigma_0 = \Sigma_{2k}$ *and* $\sigma^2 \in \Sigma_0 \langle \rho^2 \rangle$ *for all* $\sigma \in \Sigma$.

PROOF. By 41.6, 41.4 and 41.2, we have

$$\rho^{2k}\Sigma_0 = \rho^k \Sigma_0 \rho^k = \Sigma_{0 + 2k\lambda(\rho)} = \Sigma_{2k}.$$ □

41.8 Lemma. *If* d *is odd, then* Σ *is a group.*

PROOF. Since d is odd, there exists for each $i \in \{0, 1, \ldots, d - 1\}$ a $k \in \{0, 1, \ldots, d - 1\}$ such that $i \equiv 2k \bmod d$. By 41.3, there exists a $\sigma \in \Sigma$ with $\lambda(\sigma) = k$. Therefore,

$$\Sigma_i = \Sigma_{2\lambda(\sigma)} = \sigma \Sigma_0 \sigma = \Sigma_0 \sigma^2 \subseteq \Sigma_0 \langle \rho^2 \rangle$$

by 41.7. Hence $\Sigma = \Sigma_0 \langle \rho^2 \rangle$. □

41.9 Theorem (Kallaher & Ostrom 1971). *Let* $\mathfrak{A}$ *be a finite Bol plane. If the rank of the translation group of* $\mathfrak{A}$ *over its kernel is twice an odd number, then* $\mathfrak{A}$ *is a nearfield plane.*

PROOF. We may assume that the points of $\mathfrak{A}$ are the elements of a vector space $V = X \oplus X$ where X is a vector space of rank d over the kernel $K = \mathrm{GF}(q)$ of $\mathfrak{A}$. We may even further assume that $V(\sigma)$ is the axis of an involutory perspectivity interchanging $V(0)$ and $V(\infty)$ for all $\sigma \in \Sigma$. This perspectivity is represented by the mapping $(x, y) \to (y^{\sigma^{-1}}, x^\sigma)$, as we know. Combining this collineation with the collineation $(x, y) \to (y, x)$ yields that $(x, y) \to (x^\sigma, y^{\sigma^{-1}})$ is a collineation of $\mathfrak{A}$. Moreover, by 40.1 the group of all collineations fixing $V(0)$ and $V(\infty)$ is soluble. Therefore, $\langle \Sigma \rangle$ is soluble. As $\langle \Sigma \rangle$ acts transitively on $X \setminus \{0\}$, we obtain from 37.7 that $\langle \Sigma \rangle \subseteq \Gamma\mathrm{L}(1, q^d)$. Applying 41.8, we see that Σ is a group. Thus, by 3.2, $\mathfrak{A}$ is a nearfield plane. □

41.10 Theorem (Kallaher 1972). *If $\mathfrak{A}$ is a finite Bol plane of even order, then $\mathfrak{A}$ is a nearfield plane.*

PROOF. Let Λ be the group generated by all the shears interchanging $V(0)$ and $V(\infty)$. As each $V(\sigma)$ is the axis of exactly one shear in Λ it follows from 35.10 and the Feit-Thompson Theorem that Λ is soluble. This implies that $\langle \Sigma \rangle$ is soluble. Again, by 37.7 we have $\langle \Sigma \rangle \subseteq \Gamma L(1, q^d)$. By 41.3, there exists $\rho \in \Sigma$ with $\lambda(\rho) = 1$. If n is the order of ρ, then $0 = \lambda(\rho^n) \equiv n \bmod d$. As $\langle \rho \rangle \subseteq \Sigma$ and as each element of Σ acts fixed-point-free on $V \setminus \{0\}$, we see that n divides $q^d - 1$. Therefore d divides $q^d - 1$. As q is even, d is odd. Thus $\mathfrak{A}$ is a nearfield plane by 41.9.   □

For more information on Bol planes see a paper by Kallaher and Ostrom which will appear in J. Algebra.

CHAPTER VII

Translation Planes of Order q^2 Admitting SL(2,q) as a Collineation Group

This chapter gives the complete description of all translation planes of order q^2, having GF(q) contained in their kernels, and admitting SL(2,q) as a collineation group. Before we can give this description, we have to prove some results on ovals in finite desarguesian planes of odd order, among them Segre's famous result that any oval in such a plane is a conic. Moreover, we have to investigate twisted cubics in projective 3-space and similar configurations in projective 3-spaces of characteristic 2.

42. Ovals in Finite Desarguesian Planes

Most of the results in this section are due to B. Segre. See Segre 1961 for bibliographical details.

A set $\mathfrak{a}$ of k points in a finite projective plane is called a k-arc, if no three points of $\mathfrak{a}$ are collinear. Hence if l is a line of the plane $\mathfrak{P}$ and if $\mathfrak{a}$ is a k-arc, then $|l \cap \mathfrak{a}| \leqslant 2$. We shall call l a *secant*, a *tangent* or an *exterior line* of $\mathfrak{a}$, if $|l \cap \mathfrak{a}| = 2$, 1 or 0 respectively.

42.1 Lemma. *For the number k of points of a k-arc in a projective plane of order n we have:*

a) $k \leqslant n + 2$, *if n is even.*
b) $k \leqslant n + 1$, *if n is odd.*

PROOF. Let $\mathfrak{a}$ be a k-arc and pick $P \in \mathfrak{a}$. Then there are $k - 1$ secants through P, whence $k - 1 \leqslant n + 1$. Hence $k \leqslant n + 2$ in all cases. Pick

$Q \notin \mathfrak{a}$. Let r be the number of secants through Q and s be the number of tangents. Then $2r + s = k$. If $k = n + 2$, then there are no tangents of $\mathfrak{a}$, whence $s = 0$ and $n = 2(r - 1)$. This proves b). □

We have also proved:

42.2 Lemma. *Let $\mathfrak{a}$ be a k-arc and let Q be a point not in $\mathfrak{a}$. If s is the number of tangents of $\mathfrak{a}$ through Q, then $s \equiv k \bmod 2$.*

Next we prove:

42.3 Lemma. *Let $\mathfrak{P}$ be a projective plane of order $q > 3$, q being odd. If $\mathfrak{a}$ is a q-arc of $\mathfrak{P}$, then there exists a point Q of $\mathfrak{P}$ with $Q \notin \mathfrak{a}$ such that Q is on at least 5 tangents of $\mathfrak{a}$.*

PROOF. Assume that there is no such point and denote by s_Q the number of tangents through a point Q of $\mathfrak{P}$. Then $s_Q = 2$ for all $Q \in \mathfrak{a}$, as $\mathfrak{a}$ is a q-arc, and $s_Q \in \{1,3\}$ for all other points by 42.2. Let t be a tangent and put $P = \mathfrak{a} \cap t$. Then $\sum(s_Q - 1) = 2(q - 1)$ where the summation is over all points Q on t other than P. Let r be the number of those points Q for which $s_Q = 3$. Then $2r = 2(q - 1)$. Hence $r = q - 1$. Therefore there is exactly one point X on t such that $s_X = 1$.

Let $\mathfrak{J}$ be the incidence structure whose points are the points Q of $\mathfrak{P}$ with $s_Q = 3$ and whose lines are the tangents of $\mathfrak{a}$. Then $\mathfrak{J}$ is a tactical configuration with $v = q^2 + q + 1 - q - 2q = (q - 1)^2$, $b = 2q$, $k = q - 1$ and $r = 3$. Therefore $3(q - 1)^2 = 2q(q - 1)$. This yields $3(q - 1) = 2q$, i.e. $q = 3$, a contradiction. Therefore there exists Q with $s_Q > 3$. Thus $s_Q \geqslant 5$ by 42.2, as q is odd. □

42.4 Lemma (B. Segre). *Let $\mathfrak{P}$ be the desarguesian plane over $\mathrm{GF}(q) = K$ and let $\mathfrak{a}$ be a k-arc containing the points $P_1 = (1,0,0)K$, $P_2 = (0,1,0)K$, $P_3 = (0,0,1)K$. Then the tangent lines t_{ij} through P_i have equations:*

$$t_{1j} : \quad x_2 = k_{1j}x_3, \qquad t_{2j} : \quad x_3 = k_{2j}x_1, \qquad t_{3j} : \quad x_1 = k_{3j}x_2,$$

with $\prod_{i=1}^{3} \prod_{j=1}^{h} k_{ij} = -1$, where $h = q - k + 2$.

PROOF. Pick $P = (a_1,a_2,a_3)K \in \mathfrak{a} \setminus \{P_1,P_2,P_3\}$. Then $a_i \neq 0$ for all i, as no three points of $\mathfrak{a}$ are collinear. Put $b_1 = a_2 a_3^{-1}$, $b_2 = a_3 a_1^{-1}$ and $b_3 = a_1 a_2^{-1}$. Then the lines PP_i have equations $x_2 = b_1 x_3$, $x_3 = b_2 x_1$ and $x_1 = b_3 x_2$ respectively. As no line l with equation $x_2 = bx_3$ and $b \neq 0$ contains P_2 or P_3, we have $l = t_{1j}$ for some j or $l = PP_1$ for some $P \in \mathfrak{a} \setminus \{P_1,P_2,P_3\}$. Hence

$$\prod_{j=1}^{h} k_{1j} \prod_{\substack{P \in \mathfrak{a} \\ P \neq P_i}} b_1(P) = -1,$$

as the product over all non-zero elements of K is -1. Similarly

$$\prod_{j=1}^{h} k_{ij} \prod_{\substack{P \in \mathfrak{a} \\ P \neq P_u}} b_i(P) = -1 \quad \text{for} \quad i = 2,3.$$

Moreover $b_1(P)b_2(P)b_3(P) = 1$. Thus

$$-1 = \prod_{i=1}^{3} \prod_{j=1}^{h} k_{ij} \prod_{\substack{P \in \mathfrak{a} \\ P \neq P_u}} b_i(P) = \prod_{i=1}^{3} \prod_{j=1}^{h} k_{ij} . \qquad \square$$

42.5 Lemma. *Let $\mathfrak{P}$ be the desarguesian plane over* GF(q), *q being odd and assume that $\mathfrak{a}$ is a q-arc of $\mathfrak{P}$. Let $t_1, t_2, \ldots, t_r$ be tangents of $\mathfrak{a}$ passing through the point P. Put $Q_i = t_i \cap \mathfrak{a}$ and denote by s_i the second tangent of $\mathfrak{a}$ through Q_i. Then there is a conic $\mathfrak{c}$ of $\mathfrak{P}$ containing all the Q_i and such that s_i is a tangent of $\mathfrak{c}$ for all i.*

PROOF. We may assume $r \geqslant 3$. We may further assume that $Q_1 = (1,0,0)K$, $Q_2 = (0,1,0)K$, $Q_3 = (0,0,1)K$, where $K = $ GF(q), and that s_1 and s_2 have the equations $x_2 = -x_3$ and $x_3 = -x_1$ respectively. Then the conic $\mathfrak{c}$ determined by the equation $x_1 x_2 + x_2 x_3 + x_3 x_1 = 0$ passes through Q_1, Q_2, Q_3 and has tangents s_1, s_2. The equations of t_1, t_2, t_3 and s_3 are $x_2 = k_{11} x_3$, $x_3 = k_{21} x_1$, $x_1 = k_{31} x_2$, $x_1 = k_{32} x_2$. Hence by 42.3 we have $k_{11} k_{21} k_{31} k_{32} = -1$. As t_1, t_2, t_3 are confluent

$$0 = \det \begin{bmatrix} 0 & 1 & -k_{11} \\ -k_{21} & 0 & 1 \\ 1 & -k_{31} & 0 \end{bmatrix} = 1 - k_{11} k_{21} k_{31} .$$

Hence $k_{32} = -1$. From this we infer that s_3 is also a tangent of $\mathfrak{c}$.

The triangle Q_1, Q_2, Q_3 and the triangle whose sides are s_1, s_2, s_3 are perspective, their centre of perspectivity being the point $(1,1,1)K$ as is easily seen. As Q_1, Q_2, Q_3 can be replaced by any three of the points $Q_1, \ldots, Q_r$, it follows that the triangle Q_i, Q_j, Q_k and the triangle whose sides are s_i, s_j, s_k are perspective and hence axial, since $\mathfrak{P}$ is desarguesian.

Pick $Q \in \{Q_4, \ldots, Q_r\}$. If $Q = Q_i$, put $s = s_i$. Let $Q = (a_1, a_2, a_3)K$ and let s have the equation $b_1 x_1 + b_2 x_2 + b_3 x_3 = 0$. Then the triangles Q, Q_1, Q_2 and s, s_1, s_2 are axial. This means that the points $s \cap Q_1 Q_2$, $s_1 \cap QQ_2$ and $s_2 \cap QQ_1$ are collinear. Now $Q_1 Q_2$ has equation $x_3 = 0$. Hence $s \cap Q_1 Q_2 = (b_2, -b_1, 0)K$. The line QQ_2 has equation $a_3 x_1 - a_1 x_3 = 0$. Hence $s_1 \cap QQ_2 = (a_1, -a_3, a_3)K$. Finally QQ_1 has equation $a_3 x_2 - a_2 x_3 = 0$ and hence $s_2 \cap QQ_1 = (-a_3, a_2, a_3)K$. Thus

$$0 = \det \begin{bmatrix} b_2 & -b_1 & 0 \\ a_1 & -a_3 & a_3 \\ -a_3 & a_2 & a_3 \end{bmatrix} = -a_3 b_2(a_2 + a_3) + a_3 b_1(a_1 + a_3)$$

whence $a_3b_1(a_1 + a_3) = a_3b_2(a_2 + a_3)$. As $\mathfrak{a}$ is a k-arc, we have $a_3 \neq 0$. Hence $b_1(a_1 + a_3) = b_2(a_2 + a_3)$. Similarly $b_2(a_1 + a_2) = b_3(a_1 + a_3)$ and $b_3(a_2 + a_3) = b_1(a_2 + a_1)$. As s is a tangent through Q and $Q \neq Q_1, Q_2, Q_3$, we see that $b_i \neq 0$ for all i. Hence there exists $\lambda \in K^*$ with $\lambda(b_1, b_2, b_3) = (a_2 + a_3, a_3 + a_1, a_1 + a_2)$. Thus

$$0 = (a_2 + a_3)a_1 + (a_3 + a_1)a_2 + (a_1 + a_2)a_3 = 2(a_1a_2 + a_2a_3 + a_3a_1)$$

whence $Q \in \mathfrak{c}$, as q is odd. Moreover, if $(x_1, x_2, x_3)K$ is on s, then $0 = (a_2 + a_3)x_1 + (a_3 + a_1)x_2 + (a_1 + a_2)x_3$. Hence s is a tangent of $\mathfrak{c}$. This proves 42.5. $\qquad\square$

42.6 Theorem (B. Segre). *Let $\mathfrak{o}$ be an oval in the desarguesian plane $\mathfrak{P}$ over $GF(q)$, q being odd. Then $\mathfrak{o}$ is a conic.*

PROOF. This is certainly true if $q = 3$. Thus we may assume $q \geqslant 5$. Pick $P \in \mathfrak{o}$. Then all the secants of $\mathfrak{o}$ through P are tangents of the q-arc $\mathfrak{o} \backslash \{P\}$. Hence there is a conic $\mathfrak{c}$ with $\mathfrak{c} \supseteq \mathfrak{o} \backslash \{P\}$ such that all the tangents of $\mathfrak{o}$ which have their points of contact in $\mathfrak{o} \backslash \{P\}$ are also tangents of $\mathfrak{c}$ by 42.5. Pick $Q \in \mathfrak{o} \backslash \{P\}$. The same argument yields a conic $\mathfrak{d}$ with $\mathfrak{d} \supseteq \mathfrak{o} \backslash \{Q\}$ such that the tangents of $\mathfrak{d}$ through points of $\mathfrak{o} \backslash \{Q\}$ are also tangents of $\mathfrak{o}$. Hence $\mathfrak{c}$ and $\mathfrak{d}$ have at least $q - 1 \geqslant 4$ points and at least as many tangents in common. This is more than enough to make sure that $\mathfrak{c} = \mathfrak{d}$. $\qquad\square$

A collineation of a desarguesian projective plane will be called *projective*, if it is induced by a linear mapping of the underlying vector space.

42.7 Lemma. *Let P, P_1, P_2, P_3, Q and R be the points of a 6-arc in a projective plane $\mathfrak{P}$ over a commutative field. Let σ and τ be projective collineations of $\mathfrak{P}$ with the properties:*

a) $P_i^\sigma = P_i = P_i^\tau$ *for* $i = 1, 2, 3$.
b) $Q^\sigma = P$ *and* $R^\tau = P$.

Then there exists a conic $\mathfrak{c}$ with $\mathfrak{c} \supseteq \{P, P_1, P_2, P_3, Q, R\}$, if and only if P, P^σ and P^τ are collinear.

PROOF. Assume that P, P^σ and P^τ are collinear. As $P \neq Q, R$ and $Q^\sigma = P$, $R^\tau = P$, we have $P \neq P^\sigma, P^\tau$. Thus $PP^\sigma = PP^\tau = l$ is a line. As σ is projective, it induces a projectivity of the pencil of lines through Q onto the pencil of lines through P. Hence

$$\mathfrak{c} = \{x \cap x^\sigma \mid x \text{ is a line through } Q\}$$

is a conic (Steiner's construction of conics). Moreover $P, Q, P_1, P_2, P_3 \in \mathfrak{c}$, as $QP_i \cap (QP_i)^\sigma = QP_i \cap PP_i = P_i$. Furthermore $l = PP^\sigma = (QP)^\sigma$ and hence l is a tangent of $\mathfrak{c}$.

Similarly $\mathfrak{d} = \{y \cap y^\sigma \mid y \text{ is a line through } R\}$ is a conic with P, $R, P_1, P_2, P_3 \in \mathfrak{d}$ and l is a tangent of $\mathfrak{d}$. As $\mathfrak{c}$ and $\mathfrak{d}$ have the four points P,

P_1, P_2, P_3 and the tangent l in common, $\mathfrak{c} = \mathfrak{d}$. This proves one half of the lemma.

To prove the converse let $\mathfrak{c}$ be a conic containing P, Q, R, P_1, P_2, P_3. Then

$$\mathfrak{c} = \{ x \cap x^\sigma \,|\, x \text{ is a line through } Q \} = \{ y \cap y^\tau \,|\, y \text{ is a line through } R \}$$

by Steiner's theorem. Hence PP^σ and PP^τ are tangents of $\mathfrak{c}$ through P, whence $PP^\sigma = PP^\tau$. $\square$

42.8 Theorem (B. Segre). *Let $\mathfrak{P}$ be the projective plane over* GF(q), $q \geqslant 5$ *being odd. If $\mathfrak{o}$ is a q-arc in $\mathfrak{P}$, then there exists exactly one point P in $\mathfrak{P}$ such that $\mathfrak{o} \cup \{P\}$ is an oval. In particular, there exists exactly one conic $\mathfrak{c}$ with $\mathfrak{o} \subseteq \mathfrak{c}$.*

PROOF. To prove uniqueness assume that there are two such points P and Q. Then $\mathfrak{c} = \mathfrak{o} \cup \{P\}$ and $\mathfrak{d} = \mathfrak{o} \cup \{Q\}$ are conics by 42.6. Moreover $\mathfrak{c}$ and $\mathfrak{d}$ have at least q points in common. As $q \geqslant 5$, we see that $\mathfrak{c} = \mathfrak{d}$, whence $P = Q$.

To prove the existence of such a P we have only to show that there exists a conic which contains $\mathfrak{o}$. By 42.3 there exists a point Q in $\mathfrak{P}$ which is on $r \geqslant 5$ tangents of $\mathfrak{o}$. Let $t_1, \ldots, t_r$ be these tangents and put $Q_i = t_i \cap \mathfrak{o}$. By 42.5 there exists a conic $\mathfrak{c}$ with $Q_i \in \mathfrak{c}$ for all $i = 1, 2, \ldots, r$ and such that the tangents s_i of $\mathfrak{c}$ through Q_i are also tangents of $\mathfrak{o}$. Moreover $s_i \neq t_i$ for all i.

We may assume that $Q_1 = (1,0,0)K$, $Q_2 = (0,1,0)K$, $Q_3 = (0,0,1)K$ and $Q_4 = (1,1,1)K$. Let $Q = (b_1, b_2, b_3)K$. Then $\mathfrak{c}$ has the equation $a_1 x_2 x_3 + a_2 x_3 x_1 + a_3 x_1 x_2 = 0$ with $a_1 + a_2 + a_3 = 0$ and $a_i \neq 0$ for all i. Moreover t_1, t_2, t_3, t_4 have equations

$$b_2 x_3 - b_3 x_2 = 0, \qquad b_3 x_1 - b_1 x_3 = 0, \qquad b_1 x_2 - b_2 x_1 = 0,$$

$$(b_2 - b_3)x_1 + (b_3 - b_1)x_2 + (b_1 - b_2)x_3 = 0,$$

and s_1, s_2, s_3, s_4 have equations

$$a_2 x_3 + a_3 x_2 = 0, \qquad a_3 x_1 + a_1 x_3 = 0, \qquad a_1 x_2 + a_2 x_1 = 0,$$

$$(a_2 + a_3)x_1 + (a_3 + a_1)x_2 + (a_1 + a_2)x_3 = 0.$$

Pick $X \in \mathfrak{o} \backslash \mathfrak{c}$. Then X is on neither of the lines $Q_1 + Q_2, Q_2 + Q_3, Q_3 + Q_1$. Hence all the coordinates of X are distinct from 0. Thus $X = (p_1^{-1}, p_2^{-1}, p_3^{-1})K$. Any line through X has an equation of the form $u(p_1 x_1 - p_3 x_3) - v(p_2 x_2 - p_3 x_3) = 0$. As neither of the points Q_1, Q_2, Q_3 is on a tangent of $\mathfrak{o}$ through X, the tangents through X may be represented by the equations $(p_1 x_1 - p_3 x_3) - \alpha(p_2 x_2 - p_3 x_3) = 0$ and $(p_1 x_1 - p_3 x_3) - \beta(p_2 x_2 - p_3 x_3) = 0$ with $\alpha \neq 0 \neq \beta$.

Since the points Q_1, Q_2, X are not collinear, the vectors $(1,0,0) = e_1$, $(0,1,0) = e_2$ and $(p_1^{-1}, p_2^{-1}, p_3^{-1}) = f$ form a basis of the underlying vector space. Denote by (y_1, y_2, y_3) the coordinates with respect to this basis and let s and t be the tangents of $\mathfrak{o}$ through X. We may assume that s has the equation $(p_1x_1 - p_3x_3) - \alpha(p_2x_2 - p_3x_3) = 0$ in the original coordinates. In the new coordinates, s_1, s_2, s, t_1, t_2, t have equations

$$y_2 - \kappa_1 y_3 = 0, \qquad y_3 - \kappa_2 y_1 = 0, \qquad y_1 - \kappa_3 y_2 = 0,$$
$$y_2 - \lambda_1 y_3 = 0, \qquad y_3 - \lambda_2 y_1 = 0, \qquad y_1 - \lambda_3 y_2 = 0.$$

Put $(0,0,1) = e_3$. Then

$$y_1 e_1 + y_2 e_2 + y_3 f = (y_1 + y_3 p_1^{-1}) e_1 + (y_2 + y_3 p_2^{-1}) e_2 + y_3 p_3^{-1} e_3.$$

Hence $y_1 + y_3 p_1^{-1} = x_1$, $y_2 + y_3 p_2^{-1} = x_2$, $y_3 p_3^{-1} = x_3$. From this we get $y_3 = x_3 p_3$, $y_2 = x_2 - x_3 p_3 p_2^{-1}$, $y_1 = x_1 - x_3 p_3 p_1^{-1}$. If one puts these into the above equations, one finds that s_1, s_2, s, t_1, t_2, t have the equations

$$(\kappa_1 + p_2^{-1})x_3 - p_3^{-1}x_2 = 0, \qquad (1 + \kappa_2 p_1^{-1})x_1 - p_3^{-1}x_3 = 0,$$
$$p_1^{-1}(p_1 x_1 - p_3 x_3) - \kappa_3 p_2^{-1}(p_2 x_2 - p_3 x_3) = 0,$$
$$(\lambda_1 + p_2^{-1})x_3 - p_3^{-1}x_2 = 0, \qquad (1 + \lambda_2 p_1^{-1})x_1 - p_3^{-1}x_3 = 0,$$
$$p_1^{-1}(p_1 x_1 - p_3 x_3) - \lambda_3 p_2^{-1}(p_2 x_2 - p_3 x_3) = 0.$$

Comparison with the original equations for these lines yields

$$\lambda_1 = (b_2 p_3^{-1} - b_3 p_2^{-1})b_3^{-1}, \qquad \lambda_2 = b_3(b_1 p_3^{-1} - b_3 p_1^{-1})^{-1},$$
$$\lambda_3 = \alpha p_2 p_1^{-1}, \qquad \kappa_1 = -(a_2 p_3^{-1} + a_3 p_2^{-1})a_3^{-1},$$
$$\kappa_2 = -a_3(a_1 p_3^{-1} + a_3 p_1^{-1})^{-1}, \qquad \kappa_3 = \beta p_2 p_1^{-1}.$$

Put $c_1 = (a_2 p_2 + a_3 p_3)(b_2 p_2 - b_3 p_3)$, $c_2 = (a_3 p_3 + a_1 p_1)(b_3 p_3 - b_1 p_1)$, $c_3 = (a_1 p_1 + a_2 p_2)(b_1 p_1 - b_2 p_2)$. Then $\alpha\beta = c_2 c_1^{-1}$, as $-1 = \kappa_1 \kappa_2 \kappa_3 \lambda_1 \lambda_2 \lambda_3$ by 42.4.

Next we use the basis e_1, e_3, f. Denote by (y_1, y_2, y_3) the coordinates with respect to this basis. Then s_1, s, s_3, t_1, t, t_3 have equations

$$y_2 - \kappa_1 y_3 = 0, \qquad y_3 - \kappa_2 y_1 = 0, \qquad y_1 - \kappa_3 y_2 = 0,$$
$$y_2 - \lambda_1 y_3 = 0, \qquad y_3 - \lambda_2 y_1 = 0, \qquad y_1 - \lambda_3 y_2 = 0.$$

(The κ's and λ's have, of course, a new meaning.) The coordinates (y_1, y_2, y_3) are connected with the coordinates (x_1, x_2, x_3) by the equations $y_1 = x_1 - p_1^{-1} p_2 x_2$, $y_2 = x_2 p_2$, $y_3 = x_3 - p_3^{-1} p_2 x_2$. Putting these into the above equations, and comparing with the original equations for the lines $s_1, \ldots, t_3$ yields

$$\kappa_1 = -a_2(a_2 p_3^{-1} + a_3 p_2^{-1})^{-1}, \qquad \kappa_2 = p_1(p_3(1 - \alpha))^{-1},$$
$$\kappa_3 = -(a_1 p_2^{-1} + a_2 p_1^{-1})a_2^{-1}, \qquad \lambda_1 = b_2(b_3 p_2^{-1} - b_2 p_3^{-1}),$$
$$\lambda_2 = p_1(p_3(1 - \beta))^{-1}, \qquad \lambda_3 = (b_1 p_2^{-1} - b_2 p_1^{-1})b_2^{-1}.$$

From $-1 = \kappa_1 \kappa_2 \kappa_3 \lambda_1 \lambda_2 \lambda_3$ we obtain after a little computation $(1 - \alpha)(1 - \beta) = c_3 c_1^{-1}$. Hence $\alpha + \beta = (c_1 + c_2 - c_3)c_1^{-1}$, as $\alpha\beta = c_2 c_1^{-1}$.

Finally we play the same game with the basis e_1, f, and $e_4 = (1,1,1)$. Then s_1, s, s_4, t_1, t, t_4 have equations

$$y_2 - \kappa_1 y_3 = 0, \qquad y_3 - \kappa_2 y_1 = 0, \qquad y_1 - \kappa_3 y_2 = 0,$$
$$y_2 - \lambda_1 y_3 = 0, \qquad y_3 - \lambda_2 y_1 = 0, \qquad y_1 - \lambda_3 y_2 = 0.$$

Moreover the coordinates (y_1, y_2, y_3) and (x_1, x_2, x_3) are connected by the equations $x_1 = y_1 + y_2 p_1^{-1} + y_3$, $x_2 = y_2 p_2^{-1} + y_3$, $x_3 = y_2 p_3^{-1} + y_3$. Put these values in the original equations; then comparison with the above equations yields

$$\kappa_1 = -(a_2 + a_3)(a_2 p_3^{-1} + a_3 p_2^{-1}), \qquad \kappa_2 = -p_1(p_1 - p_3 + \alpha(p_2 - p_3))^{-1},$$

$$\kappa_3 = -((a_2 + a_3)p_1^{-1} + (a_3 + a_1)p_2^{-1} + (a_1 + a_2)p_3^{-1})(a_2 + a_3)^{-1},$$

$$\lambda_1 = (b_3 - b_2)(b_2 p_3^{-1} - b_3 p_2^{-1}), \qquad \lambda_2 = -p_1(p_1 - p_3 + \beta(p_2 - p_3))^{-1},$$

$$\lambda_3 = -((b_2 - b_3)p_1^{-1} + (b_3 - b_1)p_2^{-1} + (b_1 - b_2)p_3^{-1})(b_2 - b_3)^{-1}.$$

As $-1 = \kappa_1 \kappa_2 \kappa_3 \lambda_1 \lambda_2 \lambda_3$, $\alpha\beta = c_2 c_1^{-1}$ and $\alpha + \beta = (c_1 + c_2 - c_3)c_1^{-1}$, we obtain after some computations that

$$2(a_1 p_1 + a_2 p_2 + a_3 p_3)(b_1 p_1^2(p_3 - p_2) + b_2 p_2^2(p_1 - p_3) + b_3 p_3^2(p_2 - p_1)) = 0.$$

Assume $a_1 p_1 + a_2 p_2 + a_3 p_3 = 0$. Then $a_1 p_2^{-1} p_3^{-1} + a_2 p_3^{-1} p_1^{-1} + a_3 p_1^{-1} p_2^{-1} = 0$. But this is not the case, as X is not on $\mathfrak{c}$. Hence $b_1 p_1^2(p_3 - p_2) + b_2 p_2^2(p_1 - p_3) + b_3 p_3^2(p_2 - p_1) = 0$, since $2 \neq 0$. This is equivalent to

$$\det \begin{pmatrix} b_1 & b_2 & b_3 \\ p_1^{-1} & p_2^{-1} & p_3^{-1} \\ p_1^{-2} & p_2^{-2} & p_3^{-2} \end{pmatrix} = 0.$$

Thus the points Q, X and $Y = (p_1^{-2}, p_2^{-2}, p_3^{-2})K$ are collinear. Let σ be the projective collineation induced by

$$\begin{pmatrix} p_1^{-1} & 0 & 0 \\ 0 & p_2^{-1} & 0 \\ 0 & 0 & p_3^{-1} \end{pmatrix}.$$

Then $Q_i^\sigma = Q_i$ for $i = 1, 2, 3$ and $Q_4^\sigma = X$ and $X^\sigma = Y$. Using Q_1, Q_2, Q_3 and Q_5 instead of Q_1, Q_2, Q_3, Q_4 we obtain a projective collineation τ with $Q_i^\tau = Q_i$ for $i = 1, 2, 3$ and $Q_5^\tau = X$ such that X^τ is on the line $X + Q$. Therefore Q_1, Q_2, Q_3, Q_4, Q_5 and X are on a conic $\mathfrak{d}$ by 42.7. Since $Q_1, \ldots, Q_5 \in \mathfrak{c} \cap \mathfrak{d}$, we have $\mathfrak{c} = \mathfrak{d}$ whence $X \in \mathfrak{c}$, a contradiction. This contradiction proves $\mathfrak{o} \subseteq \mathfrak{c}$. $\square$

43. Twisted Cubics

Let $\mathfrak{P}$ be a projective space of rank 4 over a commutative field K. A set of points of $\mathfrak{P}$ which can be mapped by a collineation of $\mathfrak{P}$ onto the set $\mathfrak{C} = \{(s^3, s^2 t, st^2, t^3) K \mid s, t \in K, \ (s,t) \neq (0,0)\}$ will be called a *twisted cubic*. The set $\mathfrak{C}$ is contained in the intersection of the two cones

$$\mathfrak{R}_1 = \left\{ (x_1, x_2, x_3, x_4) K \mid x_1 x_3 - x_2^2 = 0 \right\}$$

and

$$\mathfrak{R}_2 = \left\{ (x_1, x_2, x_3, x_4) K \mid x_2 x_4 - x_3^2 = 0 \right\}.$$

The centre of $\mathfrak{R}_1$ is the point $(0,0,0,1)K$ and the centre of $\mathfrak{R}_2$ is the point $(1,0,0,0)K$. Moreover the line $L = \{(x,0,0,y) \mid x,y \in K\}$ is contained in $\mathfrak{R}_1 \cap \mathfrak{R}_2$. The tangent plane T_1 at $\mathfrak{R}_1$ containing L has equation $x_3 = 0$ and the tangent plane T_2 at $\mathfrak{R}_2$ containing L has equation $x_2 = 0$. Conversely we have

43.1 Theorem. *Let $\mathfrak{R}_1$ and $\mathfrak{R}_2$ be two cones of $\mathfrak{P}$ with distinct centres P_1 and P_2. Assume that $P_1 + P_2$ is contained in $\mathfrak{R}_1 \cap \mathfrak{R}_2$ and denote by T_i the tangent plane of $\mathfrak{R}_i$ with $P_1 + P_2 \subseteq T_i$. If $T_1 \neq T_2$, then $\mathfrak{R}_1 \cap \mathfrak{R}_2$ is the union of $P_1 + P_2$ and a twisted cubic $\mathfrak{C}$. Moreover $P_1, P_2 \in \mathfrak{C}$.*

PROOF. We may assume $P_1 = (0,0,0,1)K$ and $P_2 = (1,0,0,0)K$. As the stabilizer of $L = P_1 + P_2$ acts doubly transitively on the set of planes through L, we may assume that T_1 has equation $x_3 = 0$ and that T_2 has equation $x_2 = 0$. Let $\mathfrak{R}_i$ be represented by the equation

$$a_{1;i} x_1^2 + a_{2;i} x_2^2 + a_{3;i} x_3^2 + a_{4;i} x_4^2 + a_{12;i} x_1 x_2 + \cdots = 0.$$

Then $a_{1;i} = a_{4;i} = 0$. As $L \subseteq \mathfrak{R}_1 \cap \mathfrak{R}_2$ we have $a_{14;i} x_1 x_4 = 0$ for all x_1, x_4. Hence $a_{14;i} = 0$. Furthermore $(x_1, 1, 0, x_4)K \notin \mathfrak{R}_1$, since T_1 is a tangent plane of $\mathfrak{R}_1$. Hence $a_{2;1} + a_{12;1} x_1 + a_{24;1} x_4 \neq 0$ for all x_1, x_4. Thus $a_{2;1} \neq 0$ and $a_{12;1} = a_{24;1} = 0$. We may assume $a_{2;1} = 1$. Similarly $(x_1, 0, 1, x_4)K \in \mathfrak{R}_2$. Hence $a_{3;2} + a_{13;2} x_1 + a_{34;2} x_4 \neq 0$ for all x_1, x_4. Therefore $a_{13;2} = a_{34;2} = 0$ and $a_{3;2} \neq 0$. Again we may assume $a_{3;2} = 1$. Thus $\mathfrak{R}_1$ is represented by

$$x_2^2 + a_{3;1} x_3^2 + a_{13;1} x_1 x_3 + a_{23;1} x_2 x_3 + a_{34;1} x_3 x_4 = 0.$$

and $\mathfrak{R}_2$ is represented by

$$a_{2;2} x_2^2 + x_3^2 + a_{12;2} x_1 x_2 + a_{23;2} x_2 x_3 + a_{24;2} x_2 x_4 = 0.$$

The plane T_2 contains two lines of $\mathfrak{R}_1$, namely the line L and the line represented by the two equations $x_2 = 0$ and $a_{3;1} x_3 + a_{13;1} x_1 + a_{34;1} x_4 = 0$. Since the group of all elations with axis T_1 and centres on L fixes T_2 and acts transitively on $T_2 \backslash L$, we may assume that $(0,0,1,0)K$ is on $\mathfrak{R}_1$. This yields $a_{3;1} = 0$. As $(0,0,1,0)K$ is on T_2, we may also have

$(0,1,0,0)K \in \Re_2$. This yields $a_{2;2} = 0$. As $\Re_1$ is a cone, $(0,0,0,1)K + (0,0,1,0)$ $K \subseteq \Re_1$. Therefore $a_{34;1} = 0$. Likewise $a_{12;2} = 0$. Thus $\Re_1$ is represented by $x_2^2 + (a_{13;1}x_1 + a_{23;1}x_2)x_3 = 0$ and $\Re_2$ is represented by $x_3^2 + (a_{23;2}x_3 + a_{24;2}x_4)x_2 = 0$. As a cone is not contained in the union of two planes, we have $a_{13;1} \neq 0$ and $a_{24;2} \neq 0$. Therefore the matrix

$$\begin{bmatrix} -a_{13;1} & -a_{23;1} & 0 & 0 \\ 0 & 1 & 0 & 0 \\ 0 & 0 & 1 & 0 \\ 0 & 0 & -a_{23;2} & -a_{24;2} \end{bmatrix}$$

is regular. Introducing new coordinates by means of this matrix, we see that we may assume that $\Re_1$ and $\Re_2$ are represented by the equations $x_2^2 - x_1x_3 = 0$ respectively $x_3^2 - x_2x_4 = 0$.

Let $P = (x_1,x_2,x_3,x_4)K$ be a point of $\Re_1 \cap \Re_2$. If $x_3 = 0$, then $x_2 = 0$, whence P is on L. Hence we may assume $x_3 = 1$. This yields $x_2 \neq 0$ and

$$P = \left(x_2^2, x_2, 1, x_2^{-1}\right)K = \left(x_2^3, x_2^2, x_2, 1\right)K.$$

Hence P is on $\mathfrak{C} = \{(s^3,s^2t,st^2,t^3)K \mid s,t \in K, (s,t) \neq (0,0)\}$. Finally

$$(0,0,0,1)K, (1,0,0,0)K \in \mathfrak{C}.$$

□

The proof of 43.1 shows that the following is also true.

43.2 Theorem. *Let $\Re_1,\Re_2,\Re_1',\Re_2'$ be two pairs of cones in $\mathfrak{P}$ with centres P_1,P_2,P_1',P_2' respectively. Assume $P_1 \neq P_2$ and $P_1' \neq P_2'$ as well as $P_1 + P_2 \subseteq \Re_1 \cap \Re_2$ and $P_1' + P_2' \subseteq \Re_1' \cap \Re_2'$. Denote by T_i respectively T_i' the tangent plane of $\Re_i$ resp. $\Re_i'$ with $P_1 + P_2 \subseteq T_i$ and $P_1' + P_2' \subseteq T_i'$. If $T_1 \neq T_2$ and $T_1' \neq T_2'$, then there exists a projective collineation σ of $\mathfrak{P}$ with $\Re_1^\sigma = \Re_1'$ and $\Re_2^\sigma = \Re_2'$.*

Next we prove:

43.3 Theorem. *Let $\mathfrak{C}$ be a twisted cubic and let G be its stabilizer in the projective group of $\mathfrak{P}$. Then G is isomorphic to PGL$(2,K)$ provided $|K| > 4$. Moreover, G acts sharply triply transitively on $\mathfrak{C}$.*

PROOF. We may assume that

$$\mathfrak{C} = \{(s^3,s^2t,st^2,t^3)K \mid s,t \in K, (s,t) \neq (0,0)\}.$$

Let $a,b,c,d \in K$ be such that $ad - bc \neq 0$ and consider the mapping φ from GL$(2,K)$ into GL$(4,K)$ defined by

$$\varphi : \begin{pmatrix} a & b \\ c & d \end{pmatrix} \rightarrow \begin{bmatrix} a^3 & 3a^2b & 3ab^2 & b^3 \\ a^2c & a^2d+2abc & b^2c+2abd & b^2d \\ ac^2 & bc^2+2acd & ad^2+2bcd & bd^2 \\ c^3 & 3c^2d & 3cd^2 & d^3 \end{bmatrix}.$$

Then φ is a homomorphism from $GL(2,K)$ into $GL(4,K)$, as is easily seen. Let ψ be the canonical homomorphism from $GL(4,K)$ into $PGL(4,K)$ and put $\pi = \psi\varphi$. If $\begin{pmatrix} a & b \\ c & d \end{pmatrix}$ is in the kernel of π, then $b^3 = c^3 = 0$, whence $b = c = 0$ and hence $a^3 = a^2d$. This yields $a = d$. Thus the kernel of π is equal to the set of all scalar matrices. Hence $GL(2,K)/\mathrm{Kern}(\pi)$ is isomorphic to $PGL(2,K)$. Put $s' = as + bt$ and $t' = cs + dt$. Then

$$\begin{pmatrix} a^3 & 3a^2b & 3ab^2 & b^3 \\ a^2c & a^2d + 2abc & b^2c + 2abd & b^2d \\ ac^2 & bc^2 + 2acd & ad^2 + 2bcd & bd^2 \\ c^3 & 3c^2d & 3cd^2 & d^3 \end{pmatrix} \begin{pmatrix} s^3 \\ s^2t \\ st^2 \\ t^3 \end{pmatrix} = \begin{pmatrix} s'^3 \\ s'^2t' \\ s't'^2 \\ t'^3 \end{pmatrix}$$

This shows that G contains a subgroup H isomorphic to $PGL(2,K)$ which acts sharply triply transitively on $\mathfrak{C}$.

Let $\gamma \in G$ fix three distinct points of $\mathfrak{C}$. We may assume that γ fixes $(1,0,0,0)K$, $(0,0,0,1)K$ and $(1,1,1,1)K$. Then γ fixes $\mathfrak{R}_1$ and $\mathfrak{R}_2$ and also T_1 and T_2. (Here we use $|K| > 4$.) This yields that γ is induced by a matrix of the form

$$\begin{pmatrix} a_{11} & a_{12} & a_{13} & 0 \\ 0 & 1 & 0 & 0 \\ 0 & 0 & 1 & 0 \\ 0 & a_{42} & a_{43} & a_{44} \end{pmatrix}$$

with $a_{11} + a_{12} + a_{13} = 1 = a_{42} + a_{43} + a_{44}$. Now the intersection of $\mathfrak{R}_1$ with the plane having equation $x_4 = 0$ consists of the points $(s^2,st,t^2,0)K$. The line $(s^2,st,t^2,0)K + (0,0,0,1)K$ has a point distinct from $(0,0,0,1)K$ in common with $\mathfrak{C}$, if and only if $s \neq 0$. This yields that the plane with equation $x_1 = 0$ which is tangent to $\mathfrak{R}_1$ is fixed by γ. Therefore $a_{12} = 0 = a_{13}$ and $a_{11} = 1$. Similarly $a_{42} = a_{43} = 0$ and $a_{44} = 1$. Hence $H = G$. $\qquad\square$

The plane $\mathfrak{D}$ with equation $x_1 = 0$ is said to be the *osculating plane* of $\mathfrak{C}$ in the point $(0,0,0,1)K$ and the line $\{(0,0,x,y)K \mid x,y \in K\}$ is called the *tangent* of $\mathfrak{C}$ in $(0,0,0,1)K$. Using the group G, one sees that the osculating plane in $(s^3,s^2t,st^2,t^3)K$ is the plane with the equation $t^3x_1 - 3t^2sx_2 + 3ts^2x_3 - s^3x_4 = 0$.

43.4 Theorem. *Let $\mathfrak{P}$ be the projective 3-space over the field K and assume that the characteristic of K is not 3 and that $|K| > 2$. If $\mathfrak{C}$ is a twisted cubic in $\mathfrak{P}$, then there exists exactly one symplectic polarity π of $\mathfrak{P}$ such P^π is the osculating plane of $\mathfrak{C}$ in P for all $P \in \mathfrak{C}$.*

PROOF. That there is at most one follows from the fact that any subset of four points of $\mathfrak{C}$ generate $\mathfrak{P}$. In order to prove that there is one, we may assume $\mathfrak{C} = \{(s^3,s^2t,st^2,t^3)K \mid s,t \in K, (s,t) \neq (0,0)\}$. Then the skew symmetric form f defined by $f(x,y) = x_1 y_4 - x_4 y_1 - 3(x_2 y_3 - x_3 y_2)$ defines such a polarity. $\qquad\qquad\qquad\qquad\qquad\qquad\qquad\qquad\qquad\qquad\qquad\qquad$ $\square$

The tangent at $\mathfrak{C}$ in $(1,0,0,0)K$ is the line $\{(x,y,0,0)\,|\,x,y \in K\}$. This line has no point in common with the tangent at $\mathfrak{C}$ in $(0,0,0,1)K$ which is the line $\{(0,0,x,y)\,|\,x,y \in K\}$. Also, G acts triply transitively on the set of tangents by 43.3. Thus we have:

43.5 Theorem. *The set of tangents of a twisted cubic of $\mathfrak{P}$ is a partial spread of $\mathfrak{P}$.*

Let φ again denote the homomorphism of $GL(2,K)$ into $GL(4,K)$ defined in the proof of 43.3. Then the restriction φ^* of φ to $SL(2,K)$ is a monomorphism. In fact, if $\begin{pmatrix} a & b \\ c & d \end{pmatrix}$ is in the kernel of φ^*, then $b^3 = c^3 = 0$ and $a^3 = a^2 d = d^3 = 1$. This yields $b = c = 0$ and $a = d$. As $\begin{pmatrix} 1 & 0 \\ 0 & 1 \end{pmatrix}$ and $\begin{pmatrix} -1 & 0 \\ 0 & -1 \end{pmatrix}$ are the only diagonal matrices in $SL(2,K)$, it follows from $a^3 = 1$ that $a = d = 1$. Denote the image of $SL(2,K)$ under φ^* by S.

43.6 Theorem. *If $|K| > 2$ and if the characteristic of K is not 3, then S acts irreducibly on $V = K \oplus K \oplus K \oplus K$. If the characteristic of K is 3, then S fixes the line $\{(0,x,y,0)\,|\,x,y \in K\}$. If $|K| = 2$, then S fixes the plane spanned by $\mathfrak{C}$.*

PROOF. The image of $\begin{pmatrix} 1 & 1 \\ 0 & 1 \end{pmatrix}$ under φ^* is

$$\tau = \begin{bmatrix} 1 & 3 & 3 & 1 \\ 0 & 1 & 2 & 1 \\ 0 & 0 & 1 & 1 \\ 0 & 0 & 0 & 1 \end{bmatrix}.$$

Hence

$$(\tau - 1)^2 = \begin{bmatrix} 0 & 0 & 6 & 6 \\ 0 & 0 & 0 & 2 \\ 0 & 0 & 0 & 0 \\ 0 & 0 & 0 & 0 \end{bmatrix}, \qquad (\tau - 1)^3 = \begin{bmatrix} 0 & 0 & 0 & 6 \\ 0 & 0 & 0 & 0 \\ 0 & 0 & 0 & 0 \\ 0 & 0 & 0 & 0 \end{bmatrix}$$

and $(\tau - 1)^4 = 0$. Thus if $\mathrm{Char}(K) \neq 2,3$, then $(x - 1)^4$ is the minimal polynomial of τ. Hence V is a cyclic primary τ-module. Therefore, $\{0\}$, $\{(x,0,0,0)\,|\,x \in K\}$, $\{(x,y,0,0)\,|\,x,y \in K\}$, $\{(x,y,z,0)\,|\,x,y,z \in K\}$ and V are the only τ-invariant subspaces of V. From this we infer that S acts irreducibly, if $\mathrm{Char}(K) \neq 2,3$.

If $\mathrm{Char}(K) = 3$, then S fixes $\{(0,x,y,0)\,|\,x,y \in K\}$, as is seen immediately. Thus assume $\mathrm{Char}(K) = 2$ and $|K| > 2$. Let π be the polarity defined in 43.4. If S fixes the subspace U, then it also fixes U^π, since S centralizes π.

Case 1: U is not a line. Then we may assume that U is a point. Let $U = (x_1, x_2, x_3, x_4)K$. Then (x_1, x_2, x_3, x_4) is fixed by

$$A = \begin{bmatrix} 1 & b & b^2 & b^3 \\ 0 & 1 & 0 & b^2 \\ 0 & 0 & 1 & b \\ 0 & 0 & 0 & 1 \end{bmatrix},$$

as all the eigenvalues of this matrix are 1. Hence $(x_1, x_2, x_3, x_4) = (x_1 + bx_2 + b^2 x_3 + b^3 x_4, x_2 + b^2 x_4, x_3 + bx_4, x_4)$ for all b. Putting $b = 1$ yields $x_4 = 0$ and $x_2 = x_3$. Hence $bx_2(1 + b) = 0$ for all b. This yields $x_2 = 0$, as $|K| > 2$. Hence $U = (1,0,0,0)K$ is on $\mathfrak{C}$, a contradiction.

Case 2: U is a line. Consider again the above matrix A. As U is fixed by S, it contains an eigenvector of A. Each such eigenvector has the form $(x_1, x_2, x_3, 0)$ with $x_2 + bx_3 = 0$. As U contains at most two points of $\mathfrak{C}$ and as S acts doubly transitively on $\mathfrak{C}$, we see that U does not contain a point of $\mathfrak{C}$. Hence $x_2 \neq 0 \neq x_3$. We may therefore assume that $x_3 = 1$, whence $(x_1, x_2, x_3, 0) = (x_1, b, 1, 0)$. If $b \neq 1$, then $(x_1, b, 1, 0)$ and $(x_1', 1, 1, 0)$ are linearly independent. This shows that U is contained in the plane with equation $x_4 = 0$. Using the transpose of A which is also in S one sees that $U = \{(0, x, y, 0) \mid x, y \in K\}$. Then $U^\pi = \{(x, 0, 0, y) \mid x, y \in K\}$; hence U^π contains the points $(1,0,0,0)K$ and $(0,0,0,1)K$, which is impossible. This contradiction proves our theorem. $\qquad \square$

We conclude this section with another beautiful theorem of B. Segre.

43.7 Theorem. *Let q be a power of an odd prime and assume $q > 3$. If $\mathfrak{P}$ is the projective 3-space over $\mathrm{GF}(q)$ and if $\mathfrak{C}$ is a set of $q + 1$ points of $\mathfrak{P}$ such that no four points of $\mathfrak{C}$ are coplanar, then $\mathfrak{C}$ is a twisted cubic.*

PROOF. Pick $P \in \mathfrak{C}$. Then $\mathfrak{P}/P$ is the projective plane over $\mathrm{GF}(q)$. The q chords through P are q points of $\mathfrak{P}/P$ no three of which are collinear. Hence they are contained in a unique conic of $\mathfrak{P}/P$ by 42.8. Thus there is a unique line through P such that this line together with the q chords through P form a cone $\mathfrak{K}_1$. Pick another point $Q \in \mathfrak{C}$. Then we obtain likewise a cone $\mathfrak{K}_2$ such that the q chords of $\mathfrak{C}$ through Q are on $\mathfrak{K}_2$. In order to prove the theorem, we have only to show by 43.1 that the tangent planes T_i of $\mathfrak{K}_i$ containing $P + Q$ are distinct.

Suppose $T_1 = T_2$. Then we may assume $P = (1,0,0,0)K$, $Q = (0,0,0,1)K$ and that T_1 is represented by the equation $x_2 + x_3 = 0$. We may further assume that $(0,1,0,0)K$ and $(0,0,1,0)K$ are contained in the intersection of $\mathfrak{K}_1$ and $\mathfrak{K}_2$. This implies $\mathfrak{K}_2$ is represented by $ax_1(x_2 + x_3) + x_2 x_3 = 0$ and $\mathfrak{K}_1$ is represented by $bx_4(x_2 + x_3) + x_2 x_3 = 0$. Therefore, if $X = (x_1, x_2, x_3, x_4)K$ is in the intersection then $(ax_1 - bx_4)(x_2 + x_3) = 0$. Thus if $X \neq P, Q$, we have $ax_1 - bx_4 = 0$. Hence at least

$q - 1$ of the points of $\mathfrak{C}$ are contained in the plane with equation $ax_1 - bx_4 = 0$ whence $q - 1 \leqslant 3$, i.e., $q = 3$, a contradiction. This proves 43.7.   □

44. Irreducible Representations of SL(2,2^r)

Theorem 43.7 is not true if $q = 2^r$, as we shall see now. In fact, if $r > 2$, then there are exactly $\frac{1}{2} \varphi(r)$ projectively inequivalent classes of point sets $\mathfrak{C}$ with the property that no four points of $\mathfrak{C}$ are coplanar and such that $\mathfrak{C}$ admits a group induced by collineations isomorphic to SL(2,q) operating in its natural representation on $\mathfrak{C}$.

Throughout this section let $q = 2^r \geqslant 4$, $K = \mathrm{GF}(q)$ and $V = K^{(4)}$. Furthermore, for α a generator of $\mathrm{Aut}(\mathrm{GF}(q))$, denote by $\mathfrak{C}(\alpha)$ the set of points $(ss^\alpha, ts^\alpha, st^\alpha, tt^\alpha)K$ with $s,t \in K$ and $(s,t) \neq (0,0)$. Define the mapping Φ from GL(2,q) into GL(4,q) by

$$\Phi\begin{pmatrix} a & b \\ c & d \end{pmatrix} = \begin{bmatrix} aa^\alpha & ba^\alpha & ab^\alpha & bb^\alpha \\ ca^\alpha & da^\alpha & cb^\alpha & db^\alpha \\ ac^\alpha & bc^\alpha & ad^\alpha & bd^\alpha \\ cc^\alpha & dc^\alpha & cd^\alpha & dd^\alpha \end{bmatrix}.$$

Then $\Phi\begin{pmatrix} a & b \\ c & d \end{pmatrix}$ is just the Kronecker product of the two matrices $\begin{pmatrix} a & b \\ c & d \end{pmatrix}$ and $\begin{pmatrix} a^\alpha & b^\alpha \\ c^\alpha & d^\alpha \end{pmatrix}$. This yields that Φ is a homomorphism, since α is an automorphism of K. Denote by $G(\alpha)$ the image of GL(2,q) under Φ. A simple computation shows that $G(\alpha)$ leaves invariant the set $\{(ss^\alpha, ts^\alpha, st^\alpha, tt^\alpha) \mid s,t \in K, (s,t) \neq (0,0)\}$. If $G^*(\alpha)$ is the collineation group induced by $G(\alpha)$, then $G^*(\alpha) \cong \mathrm{PSL}(2,q)$ and $G^*(\alpha)$ acts in its natural triply transitive representation on $\mathfrak{C}(\alpha)$.

44.1 Lemma. a) *No four points of $\mathfrak{C}(\alpha)$ are coplanar.*
b) *If $q > 4$, then $G^*(\alpha) = \mathrm{PGL}(V)_{\mathfrak{C}(\alpha)}$.*
c) *If $q = 4$, then $\mathrm{PGL}(V)_{\mathfrak{C}(\alpha)} \cong S_5$ and $|\mathrm{PGL}(V)_{\mathfrak{C}(\alpha)} : G^*(\alpha)| = 2$.*

PROOF. In order to prove the first assertion, we may assume that the four points are $(1,0,0,0)K$, $(0,0,0,1)K$, $(1,1,1,1)K$ and $(ss^\alpha, s^\alpha, s, 1)K$ with $s \neq 0,1$, as $G^*(\alpha)$ acts triply transitively on $\mathfrak{C}(\alpha)$. Then $s^\alpha + s \neq 0$, since α generates $\mathrm{Aut}(K)$. Thus

$$\det \begin{bmatrix} 1 & 0 & 0 & 0 \\ 0 & 0 & 0 & 1 \\ 1 & 1 & 1 & 1 \\ ss^\alpha & s^\alpha & s & 1 \end{bmatrix} = s + s^\alpha \neq 0.$$

This proves the first assertion.

Let $0 \neq b \in K$ and put $M(b) = \Phi(\begin{smallmatrix} 1 & b \\ 0 & 1 \end{smallmatrix})$. Then

$$M(b) = \begin{bmatrix} 1 & b & b^\alpha & bb^\alpha \\ 0 & 1 & 0 & b^\alpha \\ 0 & 0 & 1 & b \\ 0 & 0 & 0 & 1 \end{bmatrix}.$$

As all eigenvalues of $M(b)$ are equal to 1, the fixed points of $M(b)$ consist entirely of fixed vectors. Let (x_1, x_2, x_3, x_4) be such a fixed vector. Then $x_4 = 0$ and $bx_2 + b^\alpha x_3 = 0$. Therefore the Sylow 2-subgroup $\Sigma = \{M(b) \mid b \in K\}$ induces a group of elations in the plane $T(\Sigma)$ with equation $x_4 = 0$ such that each line through $(1,0,0,0)K$ contained in $T(\Sigma)$ is the axis of an elation except the lines $L_1(\Sigma) = \{(x,0,y,0) \mid x,y \in K\}$ and $L_2(\Sigma) = \{(x,y,0,0) \mid x,y \in K\}$.

Denote by $\Re$ the set of chords of $\mathbb{C}(\alpha)$ through $(1,0,0,0)K$. Then $\{L_1(\Sigma), L_2(\Sigma)\} \cup \Re$ is a $(q+2)$-arc in $V/(1,0,0,0)K$. As a $(q+2)$-arc is determined by any q of its points, it follows that $\mathrm{PGL}(V)_{\mathbb{C}(\alpha),(1,0,0,0)K}$ fixes the set $\{L_1(\Sigma), L_2(\Sigma)\}$ and hence $T(\Sigma)$.

Let $\sigma \in \mathrm{PGL}(V)_{\mathbb{C}(\alpha)}$ fix the points $(1,0,0,0)K$, $(0,0,0,1)K$ and $(1,1,1,1)K$. Then σ fixes $T(\Sigma)$ by the remark made above and, by the same token, also the plane with equation $x_1 = 0$. Hence σ is induced by a matrix of the form

$$\begin{bmatrix} a & 0 & 0 & 0 \\ 0 & c & d & 0 \\ 0 & e & f & 0 \\ 0 & 0 & 0 & g \end{bmatrix}.$$

As $(1,1,1,1)K$ is fixed by σ, we have $a = c + d = e + f = g$. We may assume $a = 1$. Assume $L_1(\Sigma)^\sigma = L_1(\Sigma)$. Then $L_2(\Sigma)^\sigma = L_2(\Sigma)$. Since

$$\begin{bmatrix} 1 & 0 & 0 & 0 \\ 0 & c & d & 0 \\ 0 & e & f & 0 \\ 0 & 0 & 0 & 1 \end{bmatrix} \begin{bmatrix} x \\ y \\ 0 \\ 0 \end{bmatrix} = \begin{bmatrix} x \\ cy \\ ey \\ 0 \end{bmatrix}$$

and

$$\begin{bmatrix} 1 & 0 & 0 & 0 \\ 0 & c & d & 0 \\ 0 & e & f & 0 \\ 0 & 0 & 0 & 1 \end{bmatrix} \begin{bmatrix} x \\ 0 \\ y \\ 0 \end{bmatrix} = \begin{bmatrix} x \\ dy \\ fy \\ 0 \end{bmatrix},$$

we have $e = d = 0$ and $c = f = 1$, whence $\sigma = 1$. Thus, if $\sigma \notin G^*(\alpha)$, then $L_1(\Sigma)^\sigma = L_2(\Sigma)$ and $L_2(\Sigma)^\sigma = L_1(\Sigma)$. This yields $c = f = 0$ and $e = d = 1$. Now

$$\begin{bmatrix} 1 & 0 & 0 & 0 \\ 0 & 0 & 1 & 0 \\ 0 & 1 & 0 & 0 \\ 0 & 0 & 0 & 1 \end{bmatrix} \begin{bmatrix} ss^\alpha \\ s^\alpha \\ s \\ 1 \end{bmatrix} = \begin{bmatrix} ss^\alpha \\ s \\ s^\alpha \\ 1 \end{bmatrix}.$$

Therefore there exist $\rho, \lambda, \mu \in K$ with $\rho s s^\alpha = \lambda \lambda^\alpha$, $\rho s = \mu \lambda^\alpha$, $\rho s^\alpha = \mu^\alpha \lambda$, $\rho = \mu \mu^\alpha$. Putting $v = \lambda \mu^{-1}$ yields $s = v^\alpha$, $s^\alpha = v$. Hence $s = s^{\alpha^2}$ for all $s \in K$. Thus $\alpha^2 = 1$ whence $r = 2$. This proves b) and $|\mathrm{PGL}(V)_{\mathfrak{C}(\alpha)} : G^*(\alpha)| \leqslant 2$, if $q = 4$. On the other hand, assume $q = 4$ and let σ be the collineation induced by

$$\begin{bmatrix} 1 & 0 & 0 & 0 \\ 0 & 0 & 1 & 0 \\ 0 & 1 & 0 & 0 \\ 0 & 0 & 0 & 1 \end{bmatrix}.$$

Then $\sigma \in G^*(\alpha)$ and $\mathfrak{C}(\alpha)^\sigma = \mathfrak{C}(\alpha)$. As σ restricted to $\mathfrak{C}(\alpha)$ is a transposition and since $G^*(\alpha)$ acts triply transitively on $\mathfrak{C}(\alpha)$, we finally see that $\mathrm{PGL}(V)_{\mathfrak{C}(\alpha)} \cong S_5$. $\qquad\square$

44.2 Lemma. *Let α and β each generate* $\mathrm{Aut}(\mathrm{GF}(q))$. *Then the following statements are equivalent*:

a) *There exists a collineation mapping* $\mathfrak{C}(\alpha)$ *onto* $\mathfrak{C}(\beta)$.
b) *There exists a projective collineation mapping* $\mathfrak{C}(\alpha)$ *onto* $\mathfrak{C}(\beta)$.
c) $\alpha \in \{ \beta, \beta^{-1} \}$.

PROOF. a) implies b): Let ρ be a collineation mapping $\mathfrak{C}(\alpha)$ onto $\mathfrak{C}(\beta)$. Then ρ is induced by a semilinear mapping σ with companion automorphism v. Let τ be the mapping defined by $(x_1, \ldots)^\tau = (x_1^{v^{-1}}, \ldots)$. Then τ leaves invariant $\mathfrak{C}(\beta)$, since $\mathrm{Aut}(\mathrm{GF}(q))$ is abelian. Moreover, $\sigma\tau$ is linear. Thus $\sigma\tau$ induces a projective collineation mapping $\mathfrak{C}(\alpha)$ onto $\mathfrak{C}(\beta)$.

b) implies a) trivially.

b) implies c): We may assume that the projective collineation which maps $\mathfrak{C}(\alpha)$ onto $\mathfrak{C}(\beta)$ fixes $(1,0,0,0)K$, $(0,0,0,1)K$ and $(1,1,1,1)K$. It follows that it also fixes the planes with equations $x_1 = 0$ and $x_4 = 0$. Hence it is induced by a matrix of the form

$$\begin{bmatrix} 1 & 0 & 0 & 0 \\ 0 & a & b & 0 \\ 0 & c & d & 0 \\ 0 & 0 & 0 & 1 \end{bmatrix}$$

with $a + b = c + d = 1$. Moreover it fixes the set

$$\{ \{(x,y,0,0) \,|\, x,y \in K\}, \{(x,0,y, 0) \,|\, x,y \in K\} \}.$$

If it fixes each of these lines individually, we get $c = b = 0$ and $a = d = 1$, whence $\alpha = \beta$. If it switches the two lines, then $a = d = 0$ and $c = b = 1$. Now

$$\begin{bmatrix} 1 & 0 & 0 & 0 \\ 0 & 0 & 1 & 0 \\ 0 & 1 & 0 & 0 \\ 0 & 0 & 0 & 1 \end{bmatrix} \begin{bmatrix} ss^\alpha \\ s^\alpha \\ s \\ 1 \end{bmatrix} = \begin{bmatrix} ss^\alpha \\ s \\ s^\alpha \\ 1 \end{bmatrix}.$$

Hence there exist $\rho, t \in K$ with $\rho s s^\alpha = t t^\beta$, $\rho s = t^\beta$, $\rho s^\alpha = t$, $\rho = 1$. Thus $s = s^{\alpha\beta}$ for all s, whence $\alpha = \beta^{-1}$.

c) implies b): We may restrict ourselves to the case $\alpha = \beta^{-1}$. Then the projective collineation induced by

$$\begin{bmatrix} 1 & 0 & 0 & 0 \\ 0 & 0 & 1 & 0 \\ 0 & 1 & 0 & 0 \\ 0 & 0 & 0 & 1 \end{bmatrix}$$

maps $\mathfrak{C}(\alpha)$ onto $\mathfrak{C}(\beta)$, as is easily seen. $\qquad\square$

Call the plane with equation $x_4 = 0$ the *osculating plane* of $\mathfrak{C}(\alpha)$ in $(1,0,0,0)K$. Using the group $G^*(\alpha)$, one sees that the *osculating plane* of $\mathfrak{C}(\alpha)$ in $(ss^\alpha, ts^\alpha, st^\alpha, tt^\alpha)K$ is the plane with equation $tt^\alpha x_1 + st^\alpha x_2 + ts^\alpha x_3 + ss^\alpha x_4 = 0$.

44.3 Lemma. *There exists precisely one symplectic polarity π such that P^π is the osculating plane of $\mathfrak{C}(\alpha)$ in P for all $P \in \mathfrak{C}(\alpha)$.*

PROOF. The uniqueness follows at once from $|\mathfrak{C}(\alpha)| = q + 1 \geqslant 5$ and 44.1 a). Define $f: V \times V \to K$ by $f(x,y) = x_1 y_4 + x_4 y_1 + x_2 y_3 + x_3 y_2$. Then f is a non-degenerate skew symmetric form which obviously defines a polarity of the kind required. $\qquad\square$

Denote by $S(\alpha)$ the image of the restriction of Φ to $SL(2,q)$. Then $S(\alpha) \cong SL(2,q)$. Also, we have

44.4 Theorem. $S(\alpha)$ *acts irreducibly on V.*

PROOF. If $S(\alpha)$ fixes U, then $S(\alpha)$ fixes U^π, as $G^*(\alpha)$ centralizes the polarity π defined in 44.3.

Case 1: U is not a line. Then we may assume that U is a point. Let $U = (x_1, x_2, x_3, x_4)K$. Then (x_1, x_2, x_3, x_4) is fixed by

$$M(b) = \begin{bmatrix} 1 & b & b^\alpha & bb^\alpha \\ 0 & 1 & 0 & b^\alpha \\ 0 & 0 & 1 & b \\ 0 & 0 & 0 & 1 \end{bmatrix}$$

for all $b \in K$, as all eigenvalues of $M(b)$ are 1. Hence $x_1 = x_1 + bx_2 + b^\alpha x_3 + bb^\alpha x_4$, $x_2 = x_2 + b^\alpha x_4$, $x_3 = x_3 + bx_4$. Putting $b = 1$ yields $x_4 = 0$ and $x_2 = x_3$. Therefore $(b + b^\alpha)x_2 = 0$ for all b. As α is a non trivial automorphism, there exists b with $b + b^\alpha \neq 0$. Hence $x_2 = x_3 = 0$. Thus $U = (1,0,0,0)K$, a contradiction.

Case 2: U is a line. Then U contains an eigenvector of $M(b)$. This eigenvector has the form $(x_1, x_2, x_3, 0)$ with $bx_2 + b^\alpha x_3 = 0$. Since $S(\alpha)$ has no fixed point on $\mathfrak{C}(\alpha)$ and also no subgroup of index 2, we see that no point of U is on $\mathfrak{C}(\alpha)$. Hence $x_3 \neq 0$. We may therefore assume $x_3 = b$, if $b \neq 0$. This yields $(x_1,x_2,x_3,0) = (x_1,b^\alpha,b,0)$. If $b \neq 1$, then $b^\alpha \neq b$. Hence $(x_1,b^\alpha,b,0)$ and $(x_1,1,1,0)$ are linearly independent. This shows that U is contained in the plane with equation $x_4 = 0$. Using the matrices

$$\begin{bmatrix} 1 & 0 & 0 & 0 \\ b & 1 & 0 & 0 \\ b^\alpha & 0 & 1 & 0 \\ bb^\alpha & b^\alpha & 0 & 1 \end{bmatrix},$$

one sees that U is also contained in the plane with equation $x_1 = 0$. Therefore $U = \{(0,x,y,0) \,|\, x,y \in K\}$. Now $U^\pi = \{(x,0,0,y) \,|\, x,y \in K\}$ which is impossible, as U^π carries the points $(1,0,0,0)K$ and $(0,0,0,1)K$. This contradiction proves our theorem.   □

44.5 Theorem. *Let $\mathfrak{C}$ be a set of $q + 1$ points of the vector space V such that no four points of $\mathfrak{C}$ are coplanar. If S is a subgroup of $\mathrm{GL}(V)$ isomorphic to $\mathrm{SL}(2,q)$ such that S leaves $\mathfrak{C}$ invariant and acts triply transitively on $\mathfrak{C}$, then there exists an automorphism α of $\mathrm{GF}(q)$ generating $\mathrm{Aut}(\mathrm{GF}(q))$ and a collineation κ such that $\mathfrak{C}^\kappa = \mathfrak{C}(\alpha)$.*

PROOF. Pick $P \in \mathfrak{C}$. Then there exists $\Pi \in \mathrm{Syl}_2(S)$ with $P^\Pi = P$. It follows that Π fixes a plane $T(P)$ through P, since the number of planes through P is $q^2 + q + 1$. As Π acts transitively on $\mathfrak{C}\backslash\{P\}$, it follows that P is the only point common to $\mathfrak{C}$ and $T(P)$. Assume that there is a $\sigma \in \Pi\backslash\{1\}$ fixing $T(P)$ pointwise. Pick $Q \in \mathfrak{C}\backslash\{P\}$. Then $Q^\sigma \neq Q$ and $(Q^\sigma + Q) \cap T(P)$ is the centre of the elation σ. Since $|\mathfrak{C}| = q + 1 \geqslant 5$, there exists $R \in \mathfrak{C}\backslash\{P,Q,Q^\sigma\}$. The plane $Q + Q^\sigma + R$ passes through the centre of σ. Therefore it is fixed by σ. Hence $R^\sigma = R$, as no four points of $\mathfrak{C}$ are coplanar. This yields $\sigma = 1$, since Π acts regularly on $\mathfrak{C}\backslash\{P\}$. This contradiction shows that each $\sigma \in \Pi\backslash\{1\}$ induces a non trivial elation in $T(P)$.

Again let Q be a point of $\mathfrak{C}$ other than P. Then it follows from our assumptions that $\{T(P) \cap (X + Q) \,|\, X \in \mathfrak{C}\backslash\{Q\}\}$ is a q-arc containing P. Pick $\sigma \in \Pi\backslash\{1\}$, then $T(P) \cap (Q^\sigma + Q)$ is a fixed point of σ in $T(P)$ other than P. Hence $P + [T(P) \cap (Q^\sigma + Q)]$ is the axis of the elation of $T(P)$ induced by σ. Since $\{T(P) \cap (X + Q) \,|\, X \in \mathfrak{C}\backslash\{Q\}\}$ is a q-arc, we see that exactly $q - 1$ of the lines of $T(P)$ which pass through P are axes of elations induced by elements of $\Pi\backslash\{1\}$. All these lines are fixed by Π, because Π is abelian. Since Π consists only of projective collineations and since $q - 1 \geqslant 3$, it follows that Π fixes all the lines of $T(P)$ which pass through P. Denote by $N_1(P)$ and $N_2(P)$ the two lines through P which are not axes. Then $\{N_1(P),N_2(P)\}$ is fixed by $M = \mathfrak{N}_S(\Pi)$. As $|M| = q(q - 1)$, it follows that $N_1(P)$ and $N_2(P)$ are fixed individually by M.

Denote by $\mathfrak{R}_i$ the orbit of $N_i(P)$ under S. Then $|\mathfrak{R}_i| = q + 1$. Moreover choose the notation so that $N_i(Q) \in \mathfrak{R}_i$ for all $Q \in \mathfrak{C}$. We shall show that $\mathfrak{R}_1$ is a regulus and that $\mathfrak{R}_2$ is the regulus opposite to $\mathfrak{R}_1$.

Let Q be a point of $\mathfrak{C}$ distinct from P. Then $Z = M_Q$ is cyclic of order $q - 1$. The group Z fixes the line $T(P) \cap T(Q)$ and has two fixed points on it, namely $N_i(P) \cap T(P) \cap T(Q)$. As all involutions in Π are conjugate under Z, the group Z acts transitively on the set of axes through P and hence transitively on the set of the $q - 1$ points on $T(P) \cap T(Q)$ which are distinct from $N_i(P) \cap T(P) \cap T(Q)$.

Assume $N_1(P) \cap N_1(Q) \neq \{0\}$. Then $N_1(P) \cap N_1(Q)$ is a point on $T(P) \cap T(Q)$ which is fixed by Z. It follows that $N_2(P) \cap N_2(Q)$ is the second fixed point of Z on $T(Q) \cap T(P)$. There exists an involution $\sigma \in S$ mapping P onto Q whence $N_1(P)^\sigma = N_1(Q)$ and $N_2(P)^\sigma = N_2(Q)$. It follows that σ fixes 2 and hence all points on $T(P) \cap T(Q)$. But σ cannot be an elation of the space. Hence all fixed points of σ are on $T(P) \cap T(Q)$. From this we infer that $T(P) \cap T(Q)$ meets $\mathfrak{C}$ in a point which is distinct from P, a contradiction. Hence $N_1(P) \cap N_1(Q) = \{0\}$ for all $Q \in \mathfrak{C} \setminus \{P\}$. Thus $\mathfrak{R}_1$ consists of pairwise skew lines. Likewise $\mathfrak{R}_2$ consists also of pairwise skew lines. On the other hand

$$\{N_i(P) \cap T(P) \cap T(Q) \mid i = 1,2\} = \{N_i(Q) \cap T(P) \cap T(Q) \mid i = 1,2\}.$$

Therefore $N_1(P) \cap N_2(Q) \neq \{0\}$ which is also true if $P = Q$. This proves that $\mathfrak{R}_1$ and $\mathfrak{R}_2$ are reguli which are opposite to each other.

Let $\mathfrak{Q}$ be the quadric carrying $\mathfrak{R}_1$ and $\mathfrak{R}_2$. As all ruled quadrics are projectively equivalent, we may assume that $\mathfrak{Q}$ is defined by the equation $x_1 x_4 - x_2 x_3 = 0$. As no two of the points $P = (1,0,0,0)K$, $Q = (0,0,0,1)K$ and $R = (1,1,1,1)K$ are on a line belonging to $\mathfrak{R}_1 \cup \mathfrak{R}_2$, we may assume $P,Q,R \in \mathfrak{C}$. It follows that $T(P)$ has equation $x_4 = 0$ and $T(Q)$ has equation $x_1 = 0$. Consider the groups S_1 and S_2 consisting of all the matrices

$$\begin{bmatrix} \alpha & \beta & 0 & 0 \\ \gamma & \delta & 0 & 0 \\ 0 & 0 & \alpha & \beta \\ 0 & 0 & \gamma & \delta \end{bmatrix}, \quad \text{respectively} \quad \begin{bmatrix} A & 0 & B & 0 \\ 0 & A & 0 & B \\ C & 0 & D & 0 \\ 0 & C & 0 & D \end{bmatrix},$$

with $\alpha\delta - \beta\gamma = 1 = AD - BC$. Then $S_1 \cong S_2 \cong SL(2,q)$. As S is simple, we have $S \subseteq S_1 \times S_2$. Furthermore, $S \cap S_i = 1$. Hence S is a diagonal of $S_1 \times S_2$; i.e., there exists an automorphism ρ from S_1 onto S_2 such that $S = \{\sigma\sigma^\rho \mid \sigma \in S_1\}$. Let τ be an element of order 2 of S fixing $(1,0,0,0)$. Then there are $\alpha,\beta,\gamma,\delta,A,B,C,D \in K$ with $\alpha\delta - \beta\gamma = 1$ and $AD - BC = 1$ and

$$\tau = \begin{bmatrix} \alpha & \beta & 0 & 0 \\ \gamma & \delta & 0 & 0 \\ 0 & 0 & \alpha & \beta \\ 0 & 0 & \gamma & \delta \end{bmatrix} \begin{bmatrix} A & 0 & B & 0 \\ 0 & A & 0 & B \\ C & 0 & D & 0 \\ 0 & C & 0 & D \end{bmatrix} = \begin{bmatrix} \alpha A & \beta A & \alpha B & \beta B \\ \gamma A & \delta A & \gamma B & \delta B \\ \alpha C & \beta C & \alpha D & \beta D \\ \gamma C & \delta C & \gamma D & \delta D \end{bmatrix}.$$

Hence $\alpha A = 1$, $\gamma A = \alpha C = \gamma C = 0$, whence $\gamma = C = 0$. As all eigenvalues of τ are 1, we have $\delta A = \alpha D = \delta D = 1$. Hence $\alpha = \delta$ and $A = D$. We infer from $1 = \alpha\delta = \alpha^2$ and $1 = AD = A^2$ that $\alpha = \delta = A = D = 1$. Therefore

$$\tau = \begin{bmatrix} 1 & \beta & 0 & 0 \\ 0 & 1 & 0 & 0 \\ 0 & 0 & 1 & \beta \\ 0 & 0 & 0 & 1 \end{bmatrix} \begin{bmatrix} 1 & 0 & B & 0 \\ 0 & 1 & 0 & B \\ 0 & 0 & 1 & 0 \\ 0 & 0 & 0 & 1 \end{bmatrix}.$$

Since ρ is an isomorphism, B is uniquely determined by β. Therefore $B = \beta^f$ where f is an automorphism of the additive group of K.

There exists an involution $\mu \in S$ with $P^\mu = Q$ and $R^\mu = R$. Again the mapping μ has the form

$$\mu = \begin{bmatrix} \alpha A & \beta A & \alpha B & \beta B \\ \gamma A & \delta A & \gamma B & \delta B \\ \alpha C & \beta C & \alpha D & \beta D \\ \gamma C & \delta C & \gamma D & \delta D \end{bmatrix}$$

It follows from $P^\mu = Q$ that $\alpha A = \gamma A = \alpha C = 0$ and $\gamma C \neq 0$. Hence $\alpha = A = 0$. Moreover $Q^\mu = P$ implies $\beta B \neq 0 = \delta B = \beta D = \delta D$ whence $\delta = D = 0$. Since $\mu^2 = 1$ and $R^\mu = R$, we see that $(1,1,1,1)$ is fixed by μ. This yields $\beta B = \gamma B = \beta C = \gamma C = 1$. Thus

$$\mu = \begin{bmatrix} 0 & 0 & 0 & 1 \\ 0 & 0 & 1 & 0 \\ 0 & 1 & 0 & 0 \\ 1 & 0 & 0 & 0 \end{bmatrix}.$$

From this we obtain that all the involutions of S fixing Q are of the form

$$\mu\tau\mu = \begin{bmatrix} 1 & 0 & 0 & 0 \\ \beta & 1 & 0 & 0 \\ \beta^f & 0 & 1 & 0 \\ \beta\beta^f & \beta^f & \beta & 1 \end{bmatrix}.$$

Write $\tau(\beta)$ instead of τ and put $\sigma(\beta) = \mu\tau(\beta)\mu$. Now for $\beta \neq 0$ we obtain

$$\lambda(\beta) = \sigma(\beta)\tau(\beta^{-1})\sigma(\beta) = \begin{bmatrix} 0 & 0 & 0 & \beta^{-1}\beta^{-f} \\ 0 & 0 & \beta\beta^{-f} & 0 \\ 0 & \beta^{-1}\beta^f & 0 & 0 \\ \beta\beta^f & 0 & 0 & 0 \end{bmatrix}.$$

There exists $\beta \neq 0$ such that $P^{\sigma(\beta)} = R$. Hence $(1,\beta,\beta^f,\beta\beta^f) \in (1,1,1,1)K$ which yields $1^f = 1$. Hence $\lambda(1) = \mu$. Therefore

$$\lambda(1)\lambda(\beta^{-1}) = \begin{bmatrix} \beta^{-1}\beta^{-f} & 0 & 0 & 0 \\ 0 & \beta\beta^{-f} & 0 & 0 \\ 0 & 0 & \beta^{-1}\beta^f & 0 \\ 0 & 0 & 0 & \beta\beta^f \end{bmatrix}.$$

This yields

$$
\begin{bmatrix}
\beta^{-1} & 0 & 0 & 0 \\
0 & \beta & 0 & 0 \\
0 & 0 & \beta^{-1} & 0 \\
0 & 0 & 0 & \beta
\end{bmatrix}^{\rho}
=
\begin{bmatrix}
\beta^{-f} & 0 & 0 & 0 \\
0 & \beta^{-f} & 0 & 0 \\
0 & 0 & \beta^{f} & 0 \\
0 & 0 & 0 & \beta^{f}
\end{bmatrix}.
$$

From this we infer that f is an automorphism of K.

As $\mathfrak{C} = \{P\} \cup \{Q^{\tau(\beta)} \mid \beta \in K\}$, we obtain

$$\mathfrak{C} = \{(ss^f, ts^f, st^f, tt^f)K \mid s,t \in K, (s,t) \neq (0,0)\}.$$

Moreover, if $s \neq 0, 1$, then P, Q, R and $(ss^f, s^f, s, 1)K$ are not collinear. Hence

$$
0 \neq \det
\begin{bmatrix}
1 & 0 & 0 & 0 \\
0 & 0 & 0 & 1 \\
1 & 1 & 1 & 1 \\
ss^f & s^f & s & 1
\end{bmatrix}
= s^f + s.
$$

This proves that f generates $\mathrm{Aut}(\mathrm{GF}(q))$. □

45. The Hering and the Schäffer Planes

In this section we shall describe the planes discovered by Hering 1970, 1971, Ott 1975, and Schäffer 1975 arising from irreducible representations of $\mathrm{SL}(2,q)$.

45.1 Theorem. *Let q be a power of a prime satisfying $q \equiv -1 \bmod 3$ and let V be the vector space of rank 4 over $\mathrm{GF}(q)$. Assume that $\mathfrak{C}$ is either a twisted cubic of V or a point set of the type $\mathfrak{C}(\alpha)$ described in section 44 and let S be the subgroup of $\mathrm{GL}(V)_{\mathfrak{C}}$ isomorphic to $\mathrm{SL}(2,q)$. Then we have:*

a) *If q is odd, then there exists exactly one spread of V which is left invariant by S.*

b) *If q is even, then there exist exactly two spreads of V which are left invariant by S.*

PROOF. Let $\rho \in S$ be of order 3 and pick a point $P \in \mathfrak{C}$. As $q \equiv -1 \bmod 3$, the mapping ρ acts regularly on $\mathfrak{C}$. Hence $E = P + P^\rho + P^{\rho^2}$ is a plane left fixed by ρ. By Maschke's theorem, E is a completely reducible ρ-module. We infer from $P \neq P^\rho$ that ρ does not act trivially on E. Hence $E = Q \oplus l$ with Q a trivial ρ-module and l a line on which ρ acts irreducibly. Again by Maschke's theorem V/l is a completely reducible ρ-module. As $(Q + l)/l$ is a trivial ρ-submodule of V/l, we deduce that V/l is a trivial ρ-module. Invoking Maschke's theorem a third time, we see that $V = l \oplus m$ with m a trivial ρ-module.

Let n be a third line fixed by ρ. Then $l \cap n = \{0\}$, since ρ has no fixed point on l. If $m \cap n \neq \{0\}$, then ρ fixes the plane $m + n$ and hence the point $(m + n) \cap l$. Thus $m \cap n = \{0\}$. Therefore l, m, and n are pairwise skew. But then any transversal of $\{l,m,n\}$ is fixed by ρ contrary to the fact that l is an irreducible ρ-module. Hence l and m are the only fixed lines of ρ. It follows from this fact that the planes fixed by ρ are exactly the planes containing l.

The normalizer N of $\langle \rho \rangle$ in S is a group of order $2(q + 1)$ which is at the same time a maximal subgroup of S. As S acts irreducibly on V by 43.7 and 44.4, we have $S_l = N = S_m$. Hence $|l^S| = |m^S| = \frac{1}{2} q(q - 1)$. Moreover $l^S \cap m^S = \emptyset$ by the same token.

Now let π be a spread of V left invariant by S.

Case 1: q is odd. Let $P \in \mathfrak{C}$ and let Σ be a Sylow p-subgroup of S fixing P. Furthermore, let $X \in \pi$ be such that $P \subseteq X$. Then $X^\Sigma = X$. As we have seen in the proof of 43.7, every $\tau \in \Sigma \setminus \{1\}$ has minimal polynomial $(x - 1)^4$; thus every element $\tau \in \Sigma \setminus \{1\}$ fixes exactly one subspace of rank 2 of V. Hence X is the tangent of $\mathfrak{C}$ in P. Thus π contains the set $\mathfrak{X}$ of all tangents of $\mathfrak{C}$. Moreover $\mathfrak{X} \cap l^S = \emptyset = \mathfrak{X} \cap m^S$. We infer from $|\pi \setminus \mathfrak{X}| = q(q - 1) \equiv 2 \bmod 3$ that ρ fixes at least two lines belonging to $\pi \setminus \mathfrak{X}$. Hence $l,m \in \pi \setminus \mathfrak{X}$ whence $\pi = \mathfrak{X} \cup l^S \cup m^S$. This proves the uniqueness part in this case.

Case 2: q is even. Let $\mathfrak{R}_1$ and $\mathfrak{R}_2$ be the two reguli belonging to $\mathfrak{C}$. (See section 44.) Pick $P \in \mathfrak{C}$ and $X \in \pi$ such that $P \subseteq X$. Then X is fixed by S_P. This yields $X \in \mathfrak{R}_1$ or $X \in \mathfrak{R}_2$. From this we infer as above that $\pi = \mathfrak{R}_1 \cup l^S \cup m^S$ or $\pi = \mathfrak{R}_2 \cup l^S \cup m^S$. Hence we have proved the uniqueness part in all cases.

By 43.4 and 44.3 there exists a symplectiv polarity σ such that P^σ is the osculating plane in P for all $P \in \mathfrak{C}$. Pick a point Q on m. Then Q^σ is a plane fixed by ρ. Hence $l \subseteq Q^\sigma$ by the remark made above. This yields $m^\sigma = l$ and $l^\sigma = m$. Let $t \in \mathfrak{X}$, respectively $t \in \mathfrak{R}_1 \cup \mathfrak{R}_2$. Then $l \cap t \neq \{0\}$ if and only if $m \cap t \neq \{0\}$, as $t^\sigma = t$. But if $m \cap t \neq \{0\}$, then $t \cap t^\rho \neq \{0\}$, whence $t = t^\rho$, a contradiction, or $\mathfrak{R}_1^\rho = \mathfrak{R}_2$ and $\mathfrak{R}_2^\rho = \mathfrak{R}_1$ which yields the contradiction that 2 divides $o(\rho) = 3$.

Let ρ' be an element of order 3 of S which is not in $\langle \rho \rangle$ and let l' and m' have the meaning for ρ' that the lines l and m have for ρ. We want to show that $\{l,m,l',m'\}$ is a partial spread of V. If $l \cap l' \neq \{0\}$, then $(l \cap l')^\sigma = m + m'$ yields $m \cap m' \neq \{0\}$. If $l \cap m' \neq \{0\}$, then $l \cap m' = (l \cap m')^{\rho'} = l^{\rho'} \cap m'$ yields $l \cap l^{\rho'} \neq \{0\}$. Therefore, replacing ρ' by $\rho'^{-1} \rho \rho'$ if necessary, we may assume $m \cap m' \neq \{0\}$. The group $H = \langle \rho, \rho' \rangle$ fixes $m \cap m'$ vectorwise. If H has odd order then we infer from 14.1 that $|H|$ divides $q(q - 1)$ or that $|H|$ divides $q + 1$ and H is cyclic. The first case cannot hold, as $q \equiv -1 \bmod 3$ and 3 divides $|H|$. The second case cannot hold either, as $\rho' \notin \langle \rho \rangle$. Hence $|H|$ is even. If q is odd, then S contains only one involution. If α is this involution, then $\alpha \in H$ and $x^\alpha = x$ for all $x \in m \cap m'$. On the other hand $v^\alpha = -v$ for all $v \in V$. This contradiction shows that q is even. As H is generated by two elements of odd order, H cannot contain a subgroup of index 2. Therefore 4 divides $|H|$. We infer from this fact that there exist two distinct commuting involutions α and β in H. Since two distinct commuting involutions of S have exactly one fixed point in common and since this fixed point is on $\mathfrak{C}$, we finally obtain the contradiction $m \cap m' \in \mathfrak{C}$. This contradiction shows that $\{l,m,l',m'\}$ is a partial spread of V. As $|l^S| = |m^S| = \frac{1}{2}q(q-1)$, we finally get that $\pi = \mathfrak{X} \cup l^S \cup m^S$, respectively $\pi = \mathfrak{R}_i \cup l^S \cup m^S$, is a spread.   □

The planes of the form $(\mathfrak{X} \cup l^S \cup m^S)(V)$ are called *Hering planes* and the planes of the form $(\mathfrak{R}_i \cup l^S \cup m^S)(V)$ are called *Schäffer planes*.

If $q = 2^r > 4$, then there are $\frac{1}{2}\varphi(r)$ inequivalent $\mathfrak{C}(\alpha)$'s, as we have seen above. We shall see in section 50 that the $\varphi(r)$ Schäffer planes determined by them are pairwise non-isomorphic.

46. Three Planes of Order 25

Let $K = GF(5)$ and denote by $\tau(s)$ the matrix

$$\begin{bmatrix} 1 & 0 & 0 & 0 \\ s & 1 & 0 & 0 \\ 3s^2 & s & 1 & 0 \\ s^3 & 3s^2 & s & 1 \end{bmatrix}$$

and by ρ the matrix

$$\begin{bmatrix} 0 & 1 & 0 & 0 \\ -1 & 0 & 0 & 0 \\ 0 & 0 & 0 & 1 \\ 0 & 0 & -1 & 0 \end{bmatrix}.$$

Then $\{\tau(s) \,|\, s \in K\}$ is a group of order 5, as is easily seen. Let S be the group generated by the $\tau(s)$'s and ρ. Then S fixes the subspace $L = \{(0,0,x,y) \,|\, x,y \in K\}$. Moreover S induces groups isomorphic to $SL(2,5)$ on L and V/L, where $V = K^{(4)}$. The kernels of the two homomorphisms coincide: it is the group N fixing L and V/L vectorwise. We want to show $N = \{1\}$. Put $U_0 = \{(x,y,0,0) \,|\, x,y \in K\}$. Then $U_0 = \{(x, -sx + y,0,0) \,|\, x,y \in K\}$. Since

$$\begin{bmatrix} 1 & 0 & 0 & 0 \\ s & 1 & 0 & 0 \\ 3s^2 & s & 1 & 0 \\ s^3 & 3s^2 & s & 1 \end{bmatrix} \begin{bmatrix} x \\ -sx + y \\ 0 \\ 0 \end{bmatrix} = \begin{bmatrix} x \\ y \\ 2s^2x + sy \\ -2s^3x + 3s^2y \end{bmatrix},$$

we see that U_0 is mapped by $\tau(s)$ onto

$$U_s = \{(x,y,2s^2x + sy, -2s^3x + 3s^2y)\,|\,x,y \in K\}.$$

An easy computation shows that ρ maps U_s onto U_{2s^3}. Therefore $\{U_s \,|\, s \in K\}$ is an orbit of S. Pick $\nu \in N$. Then ν is induced by a matrix of the form

$$\begin{bmatrix} 1 & 0 & 0 & 0 \\ 0 & 1 & 0 & 0 \\ a & b & 1 & 0 \\ c & d & 0 & 1 \end{bmatrix}.$$

Applying this to U_0 yields the subspace

$$U' = \{(x,y,ax + by, cx + dy)\,|\,x,y \in K\}.$$

As this is one of the U_s, we get $a = 2s^2$, $b = s$, $c = -2s^3$, and $d = 3s^2$. Now ν^2 is induced by the matrix

$$\begin{bmatrix} 1 & 0 & 0 & 0 \\ 0 & 1 & 0 & 0 \\ 2a & 2b & 1 & 0 \\ 2c & 2d & 0 & 1 \end{bmatrix}.$$

Hence there exists $t \in K$ with $4s^2 = 2a = 2t^2$ and $2s = 2b = t$. Hence $t^2 = 2t^2$, i.e., $t = 0$, whence $s = 0$. This proves $N = \{1\}$, i.e., $S \cong \mathrm{SL}(2,5)$.

S acts reducibly on V, as L is left invariant by S, but S does not act completely reducibly, since the minimal polynomial of $\tau(1)$ is $(x - 1)^4$.

Put $\sigma(s) = \rho\tau(-s)\rho$. Then

$$\sigma(s) = \begin{bmatrix} 1 & s & 0 & 0 \\ 0 & 1 & 0 & 0 \\ 3s^2 & s^3 & 1 & s \\ s & 3s^2 & 0 & 1 \end{bmatrix}.$$

Moreover, for $0 \neq s \in K$ we find

$$\tau(s)\sigma(-s^{-1})\tau(s) = \begin{bmatrix} 0 & -s^{-1} & 0 & 0 \\ s & 0 & 0 & 0 \\ 0 & 0 & 0 & -s^{-1} \\ 0 & 0 & s & 0 \end{bmatrix}.$$

Finally, if

$$\delta(s) = \begin{bmatrix} s & 0 & 0 & 0 \\ 0 & s^{-1} & 0 & 0 \\ 0 & 0 & s & 0 \\ 0 & 0 & 0 & s^{-1} \end{bmatrix},$$

then $\delta(s) = \rho\tau(s)\sigma(-s^{-1})\tau(s) \in S$ for all $s \in K^*$.

Next we want to determine all spreads left invariant by S. First of all we note:

46.1 Lemma. *If π is a spread of V left invariant by S, then $L \in \pi$.*

PROOF. The minimal polynomial of $\tau(1)$ is $(x - 1)^4$. Hence there exists exactly one subspace P of rank 1 and exactly one subspace X of rank 2 with $P^{\tau(1)} = P$ and $X^{\tau(1)} = X$. Hence $L = X$. Moreover there exists exactly one $Y \in \pi$ with $P \subseteq Y$. Therefore $P \subseteq Y \cap Y^{\tau(1)}$ whence $Y = Y^{\tau(1)}$. This yields $L = Y \in \pi$. $\qquad\square$

Put $\Delta = \{\delta(s) \mid s \in K^*\}$. Then Δ is a group of order 4 which is normalized by ρ, as $\rho^{-1}\delta(s)\rho = \delta(s^{-1})$. The group Δ fixes the two subspaces

$$X_1 = \{(x,0,y,0) \mid x,y \in K\} \text{ and } X_2 = \{(0,x,0,y) \mid x,y \in K\}$$

pointwise. Moreover $X_1^\rho = X_2$ and $X_2^\rho = X_1$.

46.2 Lemma. *Let π be a spread of V left invariant by S. Then for $X \in \pi$ the following statements are equivalent:*

a) $X^\Delta = X$.
b) $\mathrm{rk}(X \cap X_1) = 1$.
c) $\mathrm{rk}(X \cap X_2) = 1$.

PROOF. We infer from $\mathrm{rk}(X_i \cap L) = 1$ and 46.1 that $X_i \notin \pi$.

a) implies b) and c): It follows from Maschke's theorem that Δ has two fixed points on X. As $X_i \notin \pi$, we deduce that Δ has exactly two fixed points on X. Hence $\mathrm{rk}(X \cap X_i) = 1$.

b) as well as c) imply a): Pick $\delta \in \Delta$. Then $X \cap X_i = (X \cap X_i)^\delta \subseteq X \cap X^\delta$ whence $X = X^\delta$. $\qquad\square$

46.3 Corollary. *If S leaves invariant the spread π, then Δ fixes exactly six components of π including L.*

PROOF. This follows immediately from 46.2. $\qquad\square$

As ρ normalizes Δ and fixes L and since ρ^2 is the identity on the projective space belonging to V, it follows from 46.3 that ρ fixes either one, three or five of the components other than L of π which are fixed by Δ.

46.4 Theorem. *There exists exactly one spread π_1 of V left invariant by S such that ρ fixes all the components of π_1 fixed by Δ.*

PROOF. Let P be a point of X_1 which is not on L. Then $P = (1,0,x,0)K$. Put

$$C_x = \{(k,l,xk,xl) \mid k,l \in K\}.$$

Then $C_x^\rho = C_x$ and $C_x \cap C_y = \{0\}$ for $x \neq y$. Moreover

$$C_x^{\tau(s)} = \{(k,l,2s^2k + sl + xk, -2s^3k + 3s^2l + lx) \mid k,l \in K\}.$$

Assume $C_y \cap C_x^{\tau(s)} \neq \{0\}$. Then there exist $k,l \in K$ with $(k,l) \neq (0,0)$ and $2s^2k + sl + xk = yk, -2s^3k + 3s^2l + xl = yl$. Putting $z = y - x$ yields $(2s^2 - z)k + sl = 0, -2s^3k - (2s^2 + z)l = 0$. Therefore $0 = -(2s^2 + z) \cdot (2s^2 - z) + 2s^4 = -2s^4 + z^2$. As 2 is not a square, we get $s = 0$ and $x = y$. From this we infer that

$$\pi_1 = \bigcup_{x \in K} \{C_x^{\tau(s)} | s \in K\} \cup \{L\}$$

is a partial spread. Moreover $|\pi_1| = 5^2 + 1$. whence π_1 is a spread of V. This proves the uniqueness part of the theorem.

In order to prove the existence, we have to show that π_1 is left invariant by S. As π_1 is left invariant by all the $\pi(s)$'s, it suffices to prove that π_1 is left invariant by ρ. But $C_x^{\tau(s)\rho} = C_x^{\tau(2s^3)}$ as is easily seen. This proves 46.4. $\quad\square$

46.5 Lemma. *Denote by $\alpha(b)$ the matrix*

$$\begin{bmatrix} 1 & 0 & 0 & 0 \\ 0 & 1 & 0 & 0 \\ b & 0 & 1 & 0 \\ 0 & b & 0 & 1 \end{bmatrix}.$$

Then $\alpha(b)$ centralizes S and fixes X_1 and X_2. Moreover $C_x^{\alpha(b)} = C_{x+b}$.

PROOF. This follows from straightforward computations.

46.6 Corollary. *$\alpha(b)$ induces a shear with axis L on $\pi_1(V)$.*

PROOF. This follows at once from 46.5 and the fact that $C_x \in \pi_1$ for all $x \in K$.

46.7 Corollary. *Let G be the subgroup of $GL(V)$ fixing π_1. Then $L^G = L$ and G acts transitively on $\pi_1 \backslash \{L\}$.*

PROOF. It follows from 46.5, 46.6, and the proof of 46.4 that G_L acts transitively on $\pi_1 \backslash \{L\}$. If $G \neq G_L$, then G contains a subgroup isomorphic to $SL(2,5^2)$ by 35.10 and 46.6. Hence $\pi_1(V)$ is desarguesian by 38.12. But then all subgroups of G isomorphic to $SL(2,5)$ are conjugate within G, whence S cannot act in the above way. This contradiction proves $G = G_L$.
$\quad\square$

Put $P_u = (1,0,u,0)K$ and $Q_v = (0,1,0,v)K$.

46.8 Lemma. *$\mathfrak{S} = \{(P_u + Q_v)^{\tau(s)} | s \in K\}$ is a partial spread, if and only if $u - v \in \{0,1,-1\}$.*

PROOF. It is

$$P_u + Q_v = \{(k,l,uk,vl) | k,l \in K\}$$

$$= \{(k, -sk + l, uk, -vsk + vl) | k,l \in K\}.$$

Hence

$$(P_u + Q_v)^{\tau(s)} = \left\{ \left(k,l,(2s^2 + u)k + sl,(-2s^3 + (u - v)s)k \right. \right.$$
$$\left. \left. + (3s^2 + v)l \right) \mid k,l \in K \right\}.$$

As $\mathfrak{S}$ is a partial spread, if and only if $(P_u + Q_v) \cap (P_u + Q_v)^{\tau(s)} = \{0\}$ for all $s \in K^*$, we get that $\mathfrak{S}$ is a partial spread if and only if

$$0 \neq \det \begin{pmatrix} 2s^2 & s \\ -2s^3 + (u - v)s & 3s^2 \end{pmatrix} = s^2 (3s^2 - (u - v))$$

for all $s \in K^*$, i.e., if and only if $u - v \in \{0,1,-1\}$. $\square$

46.9 Lemma. $\{(P_r + Q_{r+1})^\sigma \mid \sigma \in S\}$ *is a partial spread consisting of* 10 *lines.*

PROOF. We may assume by 46.5 that $r = 0$. It follows from 46.8 that $\{(P_0 + Q_1)^{\tau(s)} \mid s \in K\}$ and $\{(P_1 + Q_0)^{\tau(s)} \mid s \in K\}$ are partial spreads. Now

$$(P_1 + Q_0)^{\tau(s)} = \left\{ \left(k,l,(2s^2 + 1)k + sl,(-2s^3 + s)k + 3s^2 l \right) \mid k,l \in K \right\}.$$

If $(k,l,0,l) \in (P_1 + Q_0)^{\tau(s)}$, then $(2s^2 + 1)k + sl = 0$ and $(-2s^3 + s)k + 3s^2 l = l$. From

$$\det \begin{pmatrix} 2s^2 + 1 & s \\ -2s^3 + s & 3s^2 - 1 \end{pmatrix} = (2s^2 + 1)(3s^2 - 1) + 2s^4 - s^2 = 3s^4 - 1 \neq 0,$$

we infer $k = l = 0$. This shows that

$$\left\{ (P_0 + Q_1)^{\tau(s)} \mid s \in K \right\} \cup \left\{ (P_1 + Q_0)^{\tau(s)} \mid s \in K \right\}$$

is a partial spread of 10 lines. Moreover, this partial spread is left invariant by ρ, as is easily checked, and hence by S. This proves the lemma. $\square$

46.10 Theorem. *There exist exactly five spreads* $\pi_{2,r}$ $(r \in K)$ *of* V *left invariant by* S *such that* ρ *fixes exactly four of the six components of* $\pi_{2,r}$ *fixed by* Δ. *The five translation planes* $\pi_{2,r}(V)$ *are all isomorphic.*

PROOF. Assume that there exists a spread π of this type. Then it contains a component of the form $P_r + Q_s$ with $r \neq s$. It follows from 46.8 that $r - s \in \{1,-1\}$. As $(P_r + Q_s)^\rho = P_s + Q_r$, we may assume $s = r + 1$. Thus

$$\pi = \{L\} \cup \left\{ (P_r + Q_{r+1})^\sigma \mid \sigma \in S \right\} \cup \bigcup_{i=2}^{4} \left\{ (C_{r+i}^\sigma) \mid \sigma \in S \right\} = \pi_{2,r}.$$

Hence, by 46.5, the theorem is proved, as soon as we have proved that $\pi_{2,0}$ is a spread.

We infer from $\{L\} \cup \bigcup_{i=2}^{4} \{C_i^\sigma \mid \sigma \in S\} \subseteq \pi_1$ and 46.9 that $\{L\} \cup \bigcup_{i=2}^{4} \{C_i^\sigma \mid \sigma \in S\}$ and $\{(P_0 + Q_1)^\sigma \mid \sigma \in S\}$ are partial spreads. Furthermore $L \cap (P_0 + Q_1)^\sigma = L \cap (P_0 + Q_1) = \{0\}$.

As $C_i^{\tau(s)\rho} = C_i^{\tau(2s^3)}$, we have $\{C_i^\sigma \mid \sigma \in S\} = \{C_i^{\tau(s)} \mid s \in K\}$. Assume $(0,0,0,0) \neq (x_1,x_2,x_3,x_4) \in C_i^{\tau(s)} \cap (P_0 + Q_1)$. Then we infer from $P_0 + Q_1 =$

$\{(k,l,0,l)\,|\,k,l,\in K\}$ and $C_i^{\tau(s)} = \{(k,l,2s^2 + sl + ik, -2s^3 + 3s^2l + il)\,|\,k,l \in K\}$ that $x_3 = 0$ and $x_2 = x_4$. Hence $2s^2k + sl + ik = 0$ and $l = -2s^3k + 3s^2l + il$. Therefore

$$(2s^2 + i)k + sl = 0$$

$$-2s^3k + (3s^2 + i - 1)l = 0.$$

Hence $0 = (2s^2 + i)(3s^2 + i - 1) + 2s^4 = 3s^4 + 3s^2 + 6i(i - 1)$. Thus $0 = s^4 + s^2 + 2i(i - 1) = s^4 + 2s^23 + 3^2 + 1 + 2i(i - 1) = (s^2 + 3)^2 + 2i(i - 1) + 1$. We infer that $2i(i - 1) + 1$ is a square, as -1 is a square. Now $i = 2$, 3 or 4 whence $2i(i - 1) + 1 = 0$, 3 or 0. This yields $s^2 = -3$, a contradiction. Hence $\pi_{2,0}$ is a spread. $\qquad\square$

46.11 Theorem. *There exist exactly five spreads $\pi_{3,r}$ $(r \in K)$ of V left invariant by S such that ρ fixes exactly two of the six components of $\pi_{3,r}$ fixed by Δ. The five planes $\pi_{3,r}(V)$ are all isomorphic.*

PROOF. Assume there exists a spread π of this type. Then it contains exactly one component of the type C_{r+4}. It then follows from 46.8 that

$$\pi = \{L\} \cup \{C_{r+4}^{\sigma}\,|\,\sigma \in S\} \cup \bigcup_{i=0}^{1} \{(P_{r+2i} + Q_{r+2i+1})^{\sigma}\,|\,\sigma \in S\} = \pi_{3,r}.$$

Hence, by 46.5, the theorem is proved as soon as we have proved that $\pi_{3,0}$ is a spread. Using the spreads $\pi_{2,r}$ constructed above, it suffices to show that $\{(P_0 + Q_1)^{\sigma}\,|\,\sigma \in S\} \cup \{(P_2 + Q_3)^{\sigma}\,|\,\sigma \in S\}$ is a partial spread.

Assume $(0,0,0,0) \neq (x_1,x_2,x_3,x_4) \in (P_0 + Q_1) \cap (P_2 + Q_3)^{\sigma}$. Then $x_3 = 0$ and $x_2 = x_4$. Moreover, there exists $s \in K$ such that $(P_2 + Q_3)^{\sigma} = (P_2 + Q_3)^{\tau(s)}$ or $(P_2 + Q_3)^{\sigma} = (P_3 + Q_2)^{\tau(s)}$. The first case yields

$$0 = \det\begin{pmatrix} 2s^2 + 2 & s \\ -2s^3 - s & 3s^2 + 1 \end{pmatrix} = -2(s^2 - 1)^2 - 1$$

which is impossible, as 2 is not a square, whereas the second case yields

$$0 = \det\begin{pmatrix} 2s^2 + 3 & s \\ -2s^3 + s & 3s^2 + 2 \end{pmatrix} = -2(s^2 - 3)^2 - 1$$

which is also impossible. This proves 46.11. $\qquad\square$

46.12 Theorem. *Let V be the vector space of rank 4 over GF(5) and let G be a subgroup of GL(V) isomorphic to SL(2,5). If G fixes a subspace of rank 2 of V and if all elements of order 5 of G have minimal polynomial $(x - 1)^4$, then there exists $\gamma \in GL(V)$ with $\gamma^{-1}G\gamma = S$, where S is the group defined at the beginning of this section.*

PROOF. Let L be a subspace of rank 2 of V fixed by G. If τ is an element of order 5 of G, then V is a cyclic τ-module since $(x - 1)^4$ is the minimal polynomial of τ. Therefore, τ does not induce the identity on L. Thus the restriction of G to L is a non-trivial subgroup of $GL(L) \cong GL(2,5)$. Hence G acts faithfully on L. Similarly, G acts faithfully on V/L.

As all automorphisms of $SL(2,5)$ are induced by inner automorphisms of $GL(2,5)$, we have that the elements of G may be represented by matrices of the form

$$
\begin{bmatrix}
a & b & 0 & 0 \\
c & d & 0 & 0 \\
e & f & a & b \\
g & h & c & d
\end{bmatrix}
$$

with $ad - bc = 1$. From

$$
\begin{bmatrix}
3 & 0 & 0 & 0 \\
0 & 2 & 0 & 0 \\
e & f & 3 & 0 \\
g & h & 0 & 2
\end{bmatrix}^2
=
\begin{bmatrix}
-1 & 0 & 0 & 0 \\
0 & -1 & 0 & 0 \\
e & 0 & -1 & 0 \\
0 & -h & 0 & -1
\end{bmatrix}
$$

we infer in this particular case that $e = 0 = h$. Replacing the basis $(1,0,0,0),(0,1,0,0),(0,0,1,0),(0,0,0,1)$ by the basis $(1,0,0,g),(0,1,-f,0),(0,0,1,0)$, $(0,0,0,1)$, we see that we may assume $g = f = 0$. Hence the element ρ_a represented by

$$
\begin{bmatrix}
a & 0 & 0 & 0 \\
0 & a^{-1} & 0 & 0 \\
0 & 0 & a & 0 \\
0 & 0 & 0 & a^{-1}
\end{bmatrix}
$$

belongs to G.

Consider the element $\tau(s)$ defined by

$$
\begin{bmatrix}
1 & 0 & 0 & 0 \\
s & 1 & 0 & 0 \\
e_s & f_s & 1 & 0 \\
g_s & h_s & s & 1
\end{bmatrix}.
$$

Then we have $e_{s+t} = e_s + tf_s + e_t$ whence $tf_s = sf_t$, as $\tau(s)\tau(t) = \tau(t)\tau(s)$. Hence $f_s = sf_1 = sk$. As $(x - 1)^4$ is the minimal polynomial of $\tau(s)$, we get $k \neq 0$. Conjugation by

$$
\begin{bmatrix}
1 & 0 & 0 & 0 \\
0 & 1 & 0 & 0 \\
0 & 0 & k^{-1} & 0 \\
0 & 0 & 0 & k^{-1}
\end{bmatrix}
$$

shows that we may assume $k = 1$. Hence $e_{s+t} = e_s + e_t + st$. Moreover $h_{s+t} = h_s + h_t + st$. Now

$$
\rho_a \tau(s) \rho_a^{-1} =
\begin{bmatrix}
1 & 0 & 0 & 0 \\
a^2 s & 1 & 0 & 0 \\
e_s & a^2 s & 1 & 0 \\
a^2 g_s & h_s & a^2 s & 1
\end{bmatrix}.
$$

(Observe $a^{-2} = a^2$.) Therefore $e_s = e_{a^2s}$, $h_s = h_{a^2s}$ and $g_{a^2s} = a^2 g_s$. In particular $e_1 = e_4$ and $e_2 = e_3$ as well as $h_1 = h_4$ and $h_2 = h_3$. As $0 = e_{1+4} = e_1 + e_4 + 4 = 2e_1 + 4$, we have $e_1 = 3$. Similarly $h_1 = 3$. Furthermore, $0 = e_{2+3} = e_2 + e_3 + 6 = 2e_2 + 6$, whence $e_2 = 2$. Likewise $h_2 = 2$. Therefore $e_s = h_s = 3s^2$ for all s.

As $\begin{pmatrix} 0 & 1 \\ -1 & 0 \end{pmatrix}\begin{pmatrix} a & 0 \\ 0 & a^{-1} \end{pmatrix} = \begin{pmatrix} a^{-1} & 0 \\ 0 & a \end{pmatrix}\begin{pmatrix} 0 & 1 \\ -1 & 0 \end{pmatrix}$, we have

$$\begin{bmatrix} 0 & 1 & 0 & 0 \\ -1 & 0 & 0 & 0 \\ u & v & 0 & 1 \\ x & y & -1 & 0 \end{bmatrix} \operatorname{diag}(a, a^{-1}, a, a^{-1})$$

$$= \operatorname{diag}(a^{-1}, a, a^{-1}, a) \begin{bmatrix} 0 & 1 & 0 & 0 \\ -1 & 0 & 0 & 0 \\ u & v & 0 & 1 \\ x & y & -1 & 0 \end{bmatrix}.$$

Hence $ua = ua^{-1}$ and $ya^{-1} = ya$ for all $a \neq 0$. This yields $u = y = 0$. Furthermore,

$$\begin{bmatrix} 0 & 1 & 0 & 0 \\ -1 & 0 & 0 & 0 \\ 0 & v & 0 & 1 \\ x & 0 & -1 & 0 \end{bmatrix}\begin{bmatrix} 0 & 1 & 0 & 0 \\ -1 & 0 & 0 & 0 \\ 0 & v & 0 & 1 \\ x & 0 & -1 & 0 \end{bmatrix} = \operatorname{diag}(-1, -1, -1, -1).$$

This yields $v = x$. Also,

$$\begin{bmatrix} 0 & 1 & 0 & 0 \\ -1 & 0 & 0 & 0 \\ 0 & v & 0 & 1 \\ v & 0 & -1 & 0 \end{bmatrix}^{-1} = \begin{bmatrix} 0 & -1 & 0 & 0 \\ 1 & 0 & 0 & 0 \\ 0 & -v & 0 & -1 \\ -v & 0 & 1 & 0 \end{bmatrix}.$$

Thus we obtain

$$\begin{bmatrix} 0 & 1 & 0 & 0 \\ -1 & 0 & 0 & 0 \\ 0 & v & 0 & 1 \\ v & 0 & -1 & 0 \end{bmatrix}\begin{bmatrix} 1 & 0 & 0 & 0 \\ s & 1 & 0 & 0 \\ 3s^2 & s & 1 & 0 \\ g_s & 3s^2 & s & 1 \end{bmatrix}\begin{bmatrix} 0 & -1 & 0 & 0 \\ 1 & 0 & 0 & 0 \\ 0 & -v & 0 & -1 \\ -v & 0 & 1 & 0 \end{bmatrix}$$

$$= \begin{bmatrix} 1 & -s & 0 & 0 \\ 0 & 1 & 0 & 0 \\ 3s^2 & 2vs - g_s & 1 & -s \\ -s & -2v + 3s^2 & 0 & 1 \end{bmatrix}.$$

From $\begin{pmatrix} 1 & 0 \\ 1 & 1 \end{pmatrix}\begin{pmatrix} 1 & -1 \\ 0 & 1 \end{pmatrix}\begin{pmatrix} 1 & 0 \\ 1 & 1 \end{pmatrix} = \begin{pmatrix} 0 & -1 \\ 1 & 0 \end{pmatrix}$ we deduce therefore

$$
\begin{bmatrix} 1 & 0 & 0 & 0 \\ 1 & 1 & 0 & 0 \\ 3 & 1 & 1 & 0 \\ g_1 & 3 & 1 & 1 \end{bmatrix}
\begin{bmatrix} 1 & -1 & 0 & 0 \\ 0 & 1 & 0 & 0 \\ 3 & -2v - g_1 & 1 & -1 \\ -1 & -2v + 3 & 0 & 1 \end{bmatrix}
\begin{bmatrix} 1 & 0 & 0 & 0 \\ 1 & 1 & 0 & 0 \\ 3 & 1 & 1 & 0 \\ g_1 & 3 & 1 & 1 \end{bmatrix}
$$

$$
= \begin{bmatrix} 0 & -1 & 0 & 0 \\ 1 & 0 & 0 & 0 \\ 0 & -v & 0 & -1 \\ -v & 0 & 1 & 0 \end{bmatrix}
$$

and hence $-3g_1 - 2v + 3 = 0 = -2g_1 - v - 3$. This yields $g_1 = 1$ and $v = 0$. This shows that G is generated by the matrices

$$
\begin{bmatrix} 1 & 0 & 0 & 0 \\ 1 & 1 & 0 & 0 \\ 3 & 1 & 1 & 0 \\ 1 & 3 & 1 & 1 \end{bmatrix} \quad \text{and} \quad
\begin{bmatrix} 0 & 1 & 0 & 0 \\ -1 & 0 & 0 & 0 \\ 0 & 0 & 0 & 1 \\ 0 & 0 & -1 & 0 \end{bmatrix}. \qquad \square
$$

46.13 Corollary. *The planes $\pi_1(V)$, $\pi_{2,0}(V)$ and $\pi_{3,0}(V)$ are pairwise non-isomorphic.*

PROOF. Let $\pi_x, \pi_y \in \{\pi_1, \pi_{2,0}, \pi_{3,0}\}$ and assume that $\pi_x(V)$ and $\pi_y(V)$ are isomorphic. Then, by 46.12, there exists $\gamma \in \mathfrak{N}_{GL(V)}(S)$ mapping π_x onto π_y. Hence, obviously $x = y$. $\qquad \square$

Finally we show:

46.14 Theorem. *The Hering plane of order 5^2 is isomorphic to $\pi_{3,r}(V)$.*

PROOF. Conjugating the group S of all the matrices

$$
\begin{bmatrix} a^3 & 3a^2b & 3ab^2 & b^3 \\ a^2c & a^2d + 2abc & b^2c + 2abd & b^2d \\ ac^2 & bc^2 + 2acd & ad^2 + 2cdb & bd^2 \\ c^3 & 3c^2d & 3cd^2 & d^3 \end{bmatrix}
$$

with the matrix $\operatorname{diag}(1,1,3,1)$ yields a group S^* with one of its Sylow 5-subgroups being the group of all matrices

$$
\begin{bmatrix} 1 & 0 & 0 & 0 \\ s & 1 & 0 & 0 \\ 3s^2 & s & 1 & 0 \\ s^3 & 3s^2 & s & 1 \end{bmatrix}.
$$

Moreover,

$$\delta = \begin{bmatrix} 0 & 0 & 0 & 1 \\ 0 & 0 & 3 & 0 \\ 0 & 3 & 0 & 0 \\ -1 & 0 & 0 & 0 \end{bmatrix} \in S^*.$$

The group S^* leaves invariant a twisted cubic whose set of tangents is the set $\mu = \{L\} \cup \{C_0^{7(s)} | s \in \mathrm{GF}(5)\}$. Furthermore, δ leaves invariant $\pi_{3,1} \backslash \mu$, as is easily checked. The theorem now follows from 45.1 and 46.11.   □

The planes $\pi_1(V)$ and $\pi_{2,0}(V)$ are called the *exceptional Walker planes*, since they were discovered by Walker 1973.

47. Quasitransvections

Let V be a vector space over K and let U be a subspace of V. If $\sigma \in \mathrm{End}_K(V)$ is such that it induces the identity on U and on V/U, then we call σ a *quasitransvection* of V. If σ is a quasitransvection of V, then we call $C(\sigma) = V^{\sigma-1}$ the *centre* and $A(\sigma) = \mathrm{kern}(\sigma - 1)$ the *axis* of σ.

47.1 Lemma. *Let σ be a quasitransvection of the vector space V and let U be a subspace of V such that σ induces the identity on U and on V/U. Then $C(\sigma) \subseteq U \subseteq A(\sigma)$. Furthermore $(\sigma - 1)^2 = 0$ and σ is a bijection.*

PROOF. Put $\alpha = \sigma - 1$. Then $v^\alpha \in U$ for all $v \in V$, as σ induces the identity on V/U. Hence $C(\sigma) \subseteq U$. Moreover, $u^\alpha = u^\sigma - u = u - u = 0$, as σ induces the identity on U. Hence $U \subseteq A(\sigma)$. As $v^\alpha \in C(\sigma) \subseteq A(\sigma)$, we have $v^{\alpha^2} = 0$, whence $\alpha^2 = 0$. Finally

$$(1 - \alpha)\sigma = (1 - \alpha)(1 + \alpha) = 1 - \alpha^2 = 1 = \sigma(1 - \alpha).$$

Hence σ is bijective.   □

47.2 Lemma. *Let $\sigma \in \mathrm{GL}(V)$ be such that $(\sigma - 1)^2 = 0$. Then σ is a quasitransvection.*

PROOF. Put $U = \mathrm{kern}(\sigma - 1)$. Then $u^\sigma = u$ for all $u \in U$. Moreover $v^{(\sigma-1)^2} = 0$ implies $v^{\sigma-1} \in U$ for all $v \in V$. Hence σ is a quasitransvection.   □

47.3 Lemma. *Let V be a vector space and let U and W be subspaces of V with $U \subseteq W$. If $T(U,W)$ is the set of all quasitransvections σ with $C(\sigma) \subseteq U$ and $W \subseteq A(\sigma)$, then $T(U,W)$ is an abelian group which is isomorphic to $\mathrm{Hom}(V/W, U)$. In particular, $T(U,W)$ is an elementary abelian p-group, if the characteristic of the underlying skew field is $p > 0$.*

PROOF. Put $H = \{\alpha \mid \alpha \in \text{Hom}(V, U), \ W \subseteq \text{kern}(\alpha)\}$. Then H is a subgroup of $\text{Hom}(V, U)$ which is isomorphic to $\text{Hom}(V/W, U)$. For $\alpha \in H$ we put $\tau(\alpha) = 1 + \alpha$. Then $(\tau(\alpha) - 1)^2 = \alpha^2 = 0$, as $U \subseteq W \subseteq \text{kern}(\alpha)$. Therefore $\tau(\alpha)$ is a quasitransvection by 47.2. Furthermore $C(\tau(\alpha)) = V^\alpha \subseteq U$, whence $\tau(\alpha) \in T(U, W)$.

For $\alpha, \beta \in H$ we have $\alpha\beta = 0$. Hence

$$\tau(\alpha)\tau(\beta) = (1 + \alpha)(1 + \beta) = 1 + \alpha + \beta + \alpha\beta = \tau(\alpha + \beta).$$

Moreover $\tau(\alpha) = 1$, if and only if $\alpha = 0$. Hence τ is a monomorphism. Finally, if $\sigma \in T(U, W)$, then $\alpha = \sigma - 1 \in H$ and $\tau(\alpha) = \sigma$. Thus τ is an isomorphism from H onto $T(U, W)$. □

47.4 Lemma. *Let σ be a quasitransvection of V. Then σ fixes no subspace of rank 1 outside of $A(\sigma)$.*

PROOF. This follows from the fact that σ has no eigenvalue other than 1, as $(\sigma - 1)^2 = 0$. □

47.5 Lemma. *If V has finite rank n and if σ is a quasitransvection of V, then $\text{rk}(C(\sigma)) \leqslant \frac{1}{2}n$ and $\text{rk}(A(\sigma)) \geqslant \frac{1}{2}n$.*

PROOF. This follows from $\text{rk}(C(\sigma)) + \text{rk}(A(\sigma)) = n$ and $C(\sigma) \subseteq A(\sigma)$. □

47.6 Lemma. *Let V be a vector space of rank $4n$ over K and let σ and τ be two commuting quasitransvections of V with $\text{rk}(A(\sigma)) = \text{rk}(A(\tau)) = 2n$ and $\text{rk}(A(\sigma) \cap A(\tau)) = n$. Then $\sigma\tau$ is a quasitransvection, if and only if $\text{char}(K) = 2$.*

PROOF. Assume $\text{char}(K) = 2$. Then $(\sigma\tau - 1)^2 = (\sigma\tau)^2 - 1 = \sigma^2\tau^2 - 1 = 0$, whence $\sigma\tau$ is a quasitransvection by 47.2.

Assume conversely that $\sigma\tau$ is a quasitransvection. Put $A(\sigma) \cap A(\tau) = X_0$. There exist subspaces X_1, X_2, X_3 with $\text{rk}(X_i) = n$ and $A(\sigma) = X_0 \oplus X_1$, $A(\tau) = X_0 \oplus X_2$ and $V = X_0 \oplus X_1 \oplus X_2 \oplus X_3$. Using this decomposition, σ and τ are represented by matrices

$$\sigma = \begin{bmatrix} I & 0 & A_1 & A_2 \\ 0 & I & A_3 & A_4 \\ 0 & 0 & I & 0 \\ 0 & 0 & 0 & I \end{bmatrix} \quad \text{and} \quad \tau = \begin{bmatrix} I & B_1 & 0 & B_2 \\ 0 & I & 0 & 0 \\ 0 & B_3 & I & B_4 \\ 0 & 0 & 0 & I \end{bmatrix}.$$

As σ and τ commute, we have $A(\sigma)^\tau = A(\sigma)$ and $A(\tau)^\sigma = A(\tau)$. Hence $A_3 = 0$ and $B_3 = 0$. Computing $\sigma\tau$ and $\tau\sigma$ and using $\sigma\tau = \tau\sigma$ yields $A_1 B_4 = B_1 A_4$. Since $\sigma\tau$ is a quasitransvection, $(\sigma\tau - 1)^2 = 0$. A simple computation then yields $2B_1 A_4 = 0$. As $\sigma - 1$ and $\tau - 1$ both have rank $2n$ by assumption, we conclude that B_1 and A_4 are regular matrices, whence $B_1 A_4 \neq 0$. Therefore $\text{char}(K) = 2$. □

47.7 Corollary. *Let p be an odd prime and let V be the vector space of rank 4 over* GF(p'). *Suppose that Σ is a group of quasitransvections with* rk($A(\sigma)$) = 2 *for all $\sigma \in \Sigma \backslash \{1\}$. Then Σ is an elementary abelian p-group and $A(\sigma) = A(\tau)$ for all $\sigma, \tau \in \Sigma \backslash \{1\}$.*

PROOF. As $\sigma^p - 1 = (\sigma - 1)^p = 0$ for all $\sigma \in \Sigma$, we see that Σ is a p-group of exponent p. Let $1 \neq \sigma \in \mathfrak{Z}(\Sigma)$ and $1 \neq \tau \in \Sigma$. Then $A(\sigma)^\tau = A(\sigma)$. Since $A(\sigma)$ is a p-group, we find that $A(\sigma) \cap A(\tau) \neq \{0\}$. Hence $A(\sigma) = A(\tau)$ by 47.6. This proves the last statement of the Corollary. Finally, $\Sigma \subseteq \mathrm{T}(A(\sigma), A(\sigma))$, whence Σ is abelian by 47.3. $\qquad\square$

48. Desarguesian Spreads in $V(4, q)$

In this section we shall study the geometry of reguli and desarguesian spreads in the vector space V of rank 4 over GF(q). We shall call a spread π of V *desarguesian*, if the plane $\pi(V)$ is desarguesian.

48.1 Lemma. *Let $\mathfrak{P}$ be the projective space of rank 4 over* GF(q). *If $\mathfrak{Q}$ is a ruled quadric in $\mathfrak{P}$, then there are exactly $\frac{1}{2} q^2 (q - 1)^2$ lines in $\mathfrak{P}$ which do not meet $\mathfrak{Q}$.*

PROOF. $\mathfrak{Q}$ carries exactly two reguli $\mathfrak{R}_1$ and $\mathfrak{R}_2$ and each line of $\mathfrak{P}$ which is completely contained in $\mathfrak{Q}$ belongs to $\mathfrak{R}_1 \cup \mathfrak{R}_2$. Hence there are $2(q + 1)$ lines which are completely contained in $\mathfrak{Q}$. Pick $P \in \mathfrak{Q}$. Then there is exactly one $l_i \in \mathfrak{R}_i$ with $P \subseteq l_i$. The lines through P contained in $l_1 + l_2$ and distinct from l_1, l_2 meet $\mathfrak{Q}$ only in P. The q^2 remaining lines through P carry exactly 2 points of $\mathfrak{Q}$. Hence there are exactly $(q + 1)^2 (q - 1)$ lines having exactly one point in common with $\mathfrak{Q}$, whereas the number b of secants satisfies $(q + 1)^2 q^2 = 2b$, whence $b = \frac{1}{2} q^2 (q + 1)^2$. As $(q^2 + 1)(q^2 + q + 1)$ is the total number of lines in $\mathfrak{P}$, the number of lines which do not meet $\mathfrak{Q}$ is

$$(q^2 + 1)(q^2 + q + 1) - 2(q + 1) - (q + 1)^2(q - 1) - \tfrac{1}{2} q^2 (q + 1)^2$$
$$= \tfrac{1}{2} q^2 (q - 1)^2, \qquad\qquad\qquad\qquad\qquad\qquad\qquad \square$$

48.2 Lemma. *Let V be the vector space of rank 4 over* GF(q). *Then the number of desarguesian spreads of V is $\frac{1}{2} q^4 (q^3 - 1)(q - 1)$.*

PROOF. Let π and π' be desarguesian spreads of V and denote by K resp. K' their kernels. Then $|K| = |K'| = q^2$, whence $K \cong$ GF(q^2) $\cong K'$. Moreover, GF(q) is contained in K as well as in K'. As K and K' are isomorphic and as rk$_K(V) = 2 = $ rk$_{K'}(V)$, there exists a bijective semilinear mapping σ from the K-vector space V onto the K'-vector space V. Since π consists of all subspaces of K-rank 1 of V and since π' consists of all subspaces of

K'-rank 1, we have $\pi^\sigma = \pi'$. Moreover, we deduce from $\mathrm{GF}(q) \subseteq K \cap K'$ that σ induces a semilinear mapping on the $\mathrm{GF}(q)$-vector space V, i.e., $\sigma \in \Gamma\mathrm{L}(V) = \Gamma\mathrm{L}(4,q)$.

Putting $\pi = \pi'$ in the above considerations, one sees that the group of all semilinear mappings of the $\mathrm{GF}(q)$-vector space V which leave invariant π is isomorphic to $\Gamma\mathrm{L}(2,q^2)$. Hence the number of desarguesian spreads is $N = |\Gamma\mathrm{L}(4,q)| \, |\Gamma\mathrm{L}(2,q^2)|^{-1}$. As

$$|\Gamma\mathrm{L}(4,q)| = |\mathrm{Aut}(\mathrm{GF}(q))| q^6 (q^4 - 1)(q^3 - 1)(q^2 - 1)(q - 1)$$

and

$$|\Gamma\mathrm{L}(2,q^2)| = |\mathrm{Aut}(\mathrm{GF}(q^2))| q^2 (q^4 - 1)(q^2 - 1)$$

and $|\mathrm{Aut}(\mathrm{GF}(q^2))| = 2|\mathrm{Aut}(\mathrm{GF}(q))|$, we finally obtain $N = \frac{1}{2} q^4 (q^3 - 1) \cdot (q - 1)$. $\qquad\square$

48.3 Lemma. *The number of reguli in a projective space of rank 4 over* $\mathrm{GF}(q)$ *is* $q^4 (q^3 - 1)(q^2 + 1)$.

PROOF. It is easily seen and, in fact, well known that all reguli are in one orbit of $\mathrm{PGL}(4,q)$. Moreover the stabilizer of a regulus in $\mathrm{PGL}(4,q)$ is isomorphic to $\mathrm{PGL}(2,q) \times \mathrm{PGL}(2,q)$. Hence the number of reguli is $|\mathrm{PGL}(4,q)| \, |\mathrm{PGL}(2,q)|^{-2} = q^4 (q^3 - 1)(q^2 + 1)$. $\qquad\square$

Let U be a vector space of rank 2 over $\mathrm{GF}(q)$. Then

$$V = U \otimes_{\mathrm{GF}(q)} \mathrm{GF}(q^2)$$

is a vector space of rank 4 over $\mathrm{GF}(q)$ and a vector space of rank 2 over $\mathrm{GF}(q^2)$. Pick $u \in U$ with $u \neq 0$. Then $l_u = \{ u \otimes x \mid x \in \mathrm{GF}(q^2) \}$ is a subspace of rank 2 of the $\mathrm{GF}(q)$-vector space V. Moreover $l_u \cap l_{u'} \neq \{0\}$, if and only if $u\mathrm{GF}(q) = u'\mathrm{GF}(q)$. In this case we have $l_u = l_{u'}$. Consider $\mathfrak{R} = \{ X \mid X = l_u \text{ for some } u \in U \}$. Then $\mathfrak{R}$ is a regulus the opposite regulus being the set of subspaces $\{ y \otimes k \mid y \in U \}$ for $0 \neq k \in \mathrm{GF}(q^2)$. Moreover each $X \in \mathfrak{R}$ is a subspace of rank 1 of the $\mathrm{GF}(q^2)$-vector space V. This shows that each desarguesian spread of V contains a regulus. More precisely:

48.4 Lemma. *Let* V *be a vector space of rank 4 over* $\mathrm{GF}(q)$. *Then each desarguesian spread of* V *contains exactly* $q(q^2 + 1)$ *reguli and each regulus is contained in* $\frac{1}{2} q(q - 1)$ *desarguesian spreads.*

PROOF. Let π be a desarguesian spread of V and let G be the subgroup of $\mathrm{GL}(V)$ leaving π invariant. Then G acts triply transitively on π. Hence each set of three distinct components of π is contained in a regulus which consists of lines of π by the remark made above. As three pairwise skew lines are in exactly one regulus, π contains $\binom{q^2 + 1}{3}\binom{q + 1}{3}^{-1} = q(q^2 + 1)$ reguli.

Consider the incidence structure whose points are the reguli of V, whose lines are the desarguesian spreads of V, and whose incidence relation is the inclusion. Then this incidence structure is a tactical configuration with parameters $v = q^4(q^3 - 1)(q^2 + 1)$, $b = \frac{1}{2}q^4(q^3 - 1)(q - 1)$, $k = q(q^2 + 1)$ and r. From $vr = bk$ we obtain $r = \frac{1}{2}q(q - 1)$.   □

48.5 Lemma. *Let $\Re$ be a regulus of the vector space V of rank 4 over $\mathrm{GF}(q)$ and let $G \cong \mathrm{GL}(2,q)$ be the subgroup of the stabilizer of $\Re$ in $\mathrm{GL}(V)$ which fixes the opposite regulus of $\Re$ linewise. Then G leaves invariant each desarguesian spread which contains $\Re$.*

PROOF. Let $\Re$ be represented by means of a tensor product as before. Then $\sigma \to \sigma \otimes 1$ gives an imbedding of $\mathrm{GL}(U)$ into $\mathrm{GL}(V)_\pi$ where π consists of the subspaces of rank 1 of the $\mathrm{GF}(q^2)$-vector space V. Moreover, this imbedding is such that it leaves the opposite regulus invariant linewise. This proves the lemma, since $\mathrm{GL}(V)$ acts transitively on the set of pairs $(\Re', \pi')$ where $\Re'$ is a regulus, π' a desarguesian spread and $\Re' \subseteq \pi'$.   □

48.6 Lemma. *Let $\Re$ be a regulus, $\Re'$ its opposite regulus, and assume that π is a desarguesian spread containing $\Re$. If S is a subgroup of $\mathrm{GL}(V)$ isomorphic to $\mathrm{SL}(2,q)$ fixing $\Re'$ linewise, then $\pi^S = \pi$ and S acts transitively on $\pi \backslash \Re$.*

PROOF. It follows from 48.5 that $\pi^S = \pi$. We represent $\Re$ and π by means of a tensor product as in the proof of 48.5. We also imbed $\mathrm{SL}(2,q)$ into $\mathrm{GL}(V)_\pi \cong \mathrm{GL}(2,q^2)$ by the mapping $\sigma \to \sigma \otimes 1$. Interpreting this in the projective space belonging to V, we obtain an imbedding of $S^* \cong \mathrm{PSL}(2,q)$ into $G = \mathrm{PGL}(2,q^2)$. Let Z be a cyclic subgroup of S^* of order $\frac{1}{2}(q + 1)$ or $q + 1$ accordingly to whether q is odd or even. Then Z is contained in a cyclic subgroup C of order $q^2 - 1$ of $\mathrm{PGL}(2,q^2)$. Hence Z fixes two lines X and Y of π, since $\mathrm{PGL}(2,q^2)$ acts in its natural representation on π. As Z operates regularly on $\Re$, we see that X and Y do not belong to $\Re$. Now $S_X^* \subseteq G_X$ and $|G_X| = q^2(q^2 - 1)$. Moreover the Sylow p-subgroups of S^* act regularly on $\pi \backslash \Re$. Therefore S_X^* is cyclic. As $Z \subseteq S_X^*$ and as Z is a maximal cyclic subgroup of S^*, we get $Z = S_X^*$. Thus $|X^{S^*}| = q(q - 1)$. This proves the last assertion of the lemma.   □

48.7 Lemma. *Let π and π' be two desarguesian spreads of V and assume that $\Re$ is a regulus contained in $\pi \cap \pi'$. If $|\pi \cap \pi'| > q + 1$, then $\pi = \pi'$.*

PROOF. Let $\Re'$ be the opposite regulus of $\Re$ and let S be a subgroup of $\mathrm{GL}(V)$ isomorphic to $\mathrm{SL}(2,q)$ which fixes $\Re'$ linewise. Pick $X \in (\pi \cap \pi') \backslash \Re$. Then $\pi' \backslash \Re = X^S = \pi \backslash \Re$ by 48.6. Hence $\pi = \pi'$.   □

48.8 Lemma. *Let σ be a spread of V containing a regulus $\Re$ and let S be a subgroup of $\mathrm{GL}(V)$ isomorphic to $\mathrm{SL}(2,q)$ leaving invariant $\Re$ and σ.*

a) *If S fixes $\mathfrak{R}$ linewise, then $\sigma(V)$ is a Hall plane.*

b) *If S fixes the opposite regulus of $\mathfrak{R}$ linewise, then $\sigma(V)$ is desarguesian.*

PROOF. Let $\mathfrak{X}' \in \{\mathfrak{R},\mathfrak{R}'\}$ be the regulus fixed linewise by S. Then the opposite regulus $\mathfrak{X}$ of $\mathfrak{X}'$ is contained in $\frac{1}{2}q(q-1)$ desarguesian spreads $\pi_1, \ldots, \pi_{\frac{1}{2}q(q-1)}$. Moreover $\pi_i \cap \pi_j = \mathfrak{X}$ by 48.7, if $i \neq j$. Hence $U = \bigcup_{i=1}^{(1/2)q(q-1)}(\pi_i \setminus \mathfrak{X})$ contains $\frac{1}{2}q^2(q-1)^2$ lines which do not meet the ruled quadric defined by $\mathfrak{X}$. As these are all the lines of V which do not meet this quadric by 48.1, there exists $\pi \in \{\pi_1, \ldots, \pi_{(1/2)q(q-1)}\}$ such that $(\sigma \setminus \mathfrak{X}) \cap (\pi \setminus \mathfrak{X})$ is not empty. Therefore $\sigma \setminus \mathfrak{X} = \pi \setminus \mathfrak{X}$. Hence $\sigma = \pi$, if $\mathfrak{X} = \mathfrak{R}$, and $\sigma = (\pi \setminus \mathfrak{R}') \cup \mathfrak{R}$, if $\mathfrak{X} = \mathfrak{R}'$. $\square$

49. Translation Planes of Order q^2 Admitting SL(2,q) as a Collineation Group

In this section we shall determine all translation planes of order q^2 whose kernels contain GF(q) and which admit a group of collineations isomorphic to SL(2,q). The results in this section are due to Walker and Schäffer.

We assume throughout this section that p is a prime and $q = p^r$. Moreover V will always denote the vector space of rank 4 over GF(q) = K.

49.1 Lemma. *Let S be a subgroup of GL(V) isomorphic to SL(2,q) with the following properties*:

1) *If $1 \neq \zeta \in \mathfrak{Z}(S)$, then $v^\zeta = -v$ for all $v \in V$.*
2) *If $p = 2$, then S fixes no subspaces of rank 1.*
3) *If $1 \neq \tau$ is a p-element of S, then τ is a quasitransvection with rk($A(\tau)$) = 2.*
4) *If $p = 2$ and $\Sigma \in \mathrm{Syl}_2(S)$, then $A(\sigma) = A(\tau)$ for all $\sigma,\tau \in \Sigma \setminus \{1\}$.*

Then the following is true:

a) *If $\Sigma \in \mathrm{Syl}_p(S)$, then $A(\sigma) = A(\tau)$ for all $\sigma,\tau \in \Sigma \setminus \{1\}$.*
b) *If $F(\Sigma)$ denotes the subspace of V left fixed vectorwise by $\Sigma \in \mathrm{Syl}_p(S)$, then $\mathfrak{R} = \{F(\Sigma) | \Sigma \in \mathrm{Syl}_p(S)\}$ is a partial spread. Moreover, there exist at least two distinct transversals of $\mathfrak{R}$ and each transversal of $\mathfrak{R}$ is fixed by S.*

PROOF. If $p = 2$, then a) is just the assumption 2). If $p > 2$, then a) follows from 47.7.

Assume that $F(\Sigma) \cap F(\mathrm{T}) \neq \{0\}$ for two distinct Sylow p-subgroups Σ and T. Then S fixes a point $P \subseteq F(\Sigma) \cap F(\mathrm{T})$ vectorwise, as $\langle \Sigma,\mathrm{T} \rangle = S$. By 2) we have therefore $p > 2$, whence $|\mathfrak{Z}(S)| = 2$. Let $1 \neq \zeta \in \mathfrak{Z}(S)$. Then $v = v^\zeta = -v$ for all $v \in P$ by 1). This contradiction proves $F(\Sigma) \cap F(\mathrm{T}) = \{0\}$. Thus $\mathfrak{R}$ is a partial spread.

Let U be a transversal of $\Re$ and pick $\Sigma \in \mathrm{Syl}_p(S)$. Then U^σ is also a transversal for all $\sigma \in \Sigma$. Moreover $U \cap A(\sigma) = U^\sigma \cap A(\sigma)$. This yields $U = U^\sigma$, as $|\Re| = q + 1 \geqslant 3$. As S is generated by its p-elements, we have $U^S = U$.

In order to prove that $\Re$ has at least two transversals we first consider the case $p = 2$. Pick $F, G \in \Re$ with $F \neq G$. Then $Z = S_{F,G}$ is cyclic of order $q - 1$ and acts transitively on $\Re \backslash \{F, G\}$, since S acts in its natural representation on $\Re$. It follows from Maschke's theorem that F is a completely reducible Z-module. Hence Z has at least two fixed points on F, as Z cannot act irreducibly on F. Let P be such a fixed point. There exists an involution $\sigma \in S$ with $F^\sigma = G$. Since σ normalizes Z, it follows that P^σ is a fixed point of Z on G. Moreover, $A(\sigma)$ meets $P + P^\sigma$ non-trivially, because $P + P^\sigma$ is fixed by σ and $P + P^\sigma$ is a 2-group. Thus $P + P^\sigma$ is a transversal of $\{F, G, A(\sigma)\}$. As Z acts transitively on $\Re \backslash \{F, G\}$ and fixes $P + P^\sigma$, we conclude that $P + P^\sigma$ is a transversal of $\Re$. Hence $\Re$ has at least two transversals in this case, since Z has at least two fixed points on F.

Now let $p > 2$ and pick $\Sigma, \mathrm{T} \in \mathrm{Syl}_p(S)$ with $\Sigma \neq \mathrm{T}$. Put $A = F(\Sigma)$, $B = F(\mathrm{T})$, and $Z = S_{A,B}$. Then Z is cyclic of order $q - 1$. It follows from Maschke's theorem that Z has at least two fixed points on A, as Z cannot act irreducibly on A.

Case 1: Z has at least three fixed points on A. Then Z fixes all the points on A, as Z induces projective collineations on the line A. Since there exists an element in S interchanging A and B, it follows that Z fixes also all the points on B.

Let X be a plane containing A. Then $X \cap B$ is a point, whence $X^Z = X$. Moreover Σ induces a group of transvections on X. Since all subgroups of order p of Σ are conjugate under Z and since Z fixes A pointwise, it follows that all the transvections induced by Σ on X have the same centre P. Hence $P + (X \cap B)$ is a line fixed by Σ. It then follows from the transitivity of Σ on $\Re \backslash \{A\}$ that $P + (X \cap B)$ is a transversal of $\Re$. As different planes yield different transversals, $\Re$ has $q + 1$ transversals in this case.

Case 2: Z has exactly two fixed points on A. There exist $\sigma \in \Sigma$ and $\tau \in \mathrm{T}$ such that $\langle \sigma, \tau \rangle$ acts transitively on $\Re$ (see e.g. Gorenstein [1968, p. 44]). Let α be a linear mapping such that α fixes A vectorwise, leaves B invariant, and maps B^σ onto A^τ; furthermore, let β be a second linear mapping such that β fixes B vectorwise, leaves A invariant, and maps A^τ onto B^σ. Then $\alpha\beta = \beta\alpha$ and $\alpha\beta$ leaves A, B, A^τ, and B^σ invariant. It follows that $(\alpha\beta)^{-1}\sigma\alpha\beta = \sigma'$ is a quasitransvection with $A(\sigma') = A$ which maps B onto B^σ. Pick $b \in B$. Then

$$b^{\sigma'} = b^{\beta^{-1}\alpha^{-1}\sigma\alpha\beta} = b^{\alpha^{-1}\sigma\alpha\beta} = \left(b^{\alpha^{-1}} + a\right)^{\alpha\beta} = (b + a)^\beta = b + a^\beta.$$

On the other hand, there exist $b' \in B$ and $a' \in A$ with $b^{\sigma'} = b'^\sigma = b' + a'$. Hence $b - b' = a' - a \in A \cap B = \{0\}$. Therefore $b^{\sigma'} = b^\sigma$. This shows that

σ and $\alpha\beta$ commute. Similarly $\tau\alpha\beta = \alpha\beta\tau$. Hence $\alpha\beta$ centralizes $\langle\sigma,\tau\rangle$. As $\langle\sigma,\tau\rangle$ acts transitively on $\Re$, it follows that $\alpha\beta$ fixes each line of $\Re$. Similar arguments then show that $\alpha\beta$ centralizes S.

Let P and Q be the two fixed points of Z on A. Then

$$\{P,Q\}^{\alpha\beta} = \{P,Q\}.$$

Assume that F is a fixed point of $\alpha\beta$ on A. If $F \in \{P,Q\}$, then $\alpha\beta$ has the two fixed points P and Q. If $F \notin \{P,Q\}$, then F is not a fixed point of Z, whence $\alpha\beta$ has two fixed points in this case, too. Let L be the unique transversal of A,B,A^τ through F. Then $L = F + (L \cap B)$. As $F^\beta = F^{\alpha\beta}$ $= F$ and $(L \cap B)^\beta = L \cap B$, we have $L = L^\beta$. Thus L is a transversal of $A,B,A^{\tau\beta} = B^\sigma$. Therefore L is a transversal of A,B,A^τ,B^σ. This implies that L is fixed by σ and by τ and hence by $\langle\sigma,\tau\rangle$. Since $\langle\sigma,\tau\rangle$ acts transitively on $\Re$, we see that L is a transversal of $\Re$. Therefore $\Re$ has at least two transversals, if $\alpha\beta$ has a fixed point on A.

If $\alpha\beta$ fixes a point X not on A, then $X + A$ is a plane which is met by B in exactly one point. As this point is fixed by $\alpha\beta$, we also find a fixed point of $\alpha\beta$ on A by the transitivity of S on $\Re$. Hence we may assume that $\alpha\beta$ has no fixed point. But P and Q are fixed by $(\alpha\beta)^2$. Hence $(\alpha\beta)^2$ has at least two fixed points on each line belonging to $\Re$.

Assume $(\alpha\beta)^2 \neq 1$. Then $(\alpha\beta)^2$ has exactly two fixed points on each line of $\Re$, as is easily seen using the fact that $(\alpha\beta)^2$ is linear. $P + B$ is a plane fixed by $(\alpha\beta)^2$. This plane is not fixed pointwise by $(\alpha\beta)^2$, as B is not fixed pointwise by $(\alpha\beta)^2$. Hence the set of fixed points of $(\alpha\beta)^2$ which are contained in $P + B$ does not contain a quadrangle. As $X \cap (P + B)$ is a fixed point of $(\alpha\beta)^2$ for all $X \in \Re\setminus\{B\}$, we see that $(\alpha\beta)^2$ has at least $q + 2$ fixed points on $P + B$. This yields that $q + 1$ of them are on a line and the remaining one is the second fixed point of $(\alpha\beta)^2$ on B. This yields a transversal of $\Re$ through P. Similarly we find a transversal through Q.

It remains to consider the case where $(\alpha\beta)^2 = 1$ and $\alpha\beta$ has no fixed point. Then the set $\mathfrak{S}$ of lines fixed by $\alpha\beta$ is a spread with $\Re \subseteq \mathfrak{S}$. Moreover the kernel of $\mathfrak{S}$ is $\mathrm{GF}(q^2)$ and S is contained in the group fixing $\mathfrak{S}$, as S centralizes $\alpha\beta$. This means that S acts as a subgroup of $\mathrm{GL}(2,q^2)$ on V considered as a vector space over $\mathrm{GF}(q^2)$. Since all subgroups of $\mathrm{GL}(2,q^2)$ which are isomorphic to $\mathrm{SL}(2,q)$ are conjugate in $\mathrm{GL}(2,q^2)$, we infer that $\Re$ is a regulus in this case. $\quad\square$

49.2 Lemma. *Let S be a subgroup of $\mathrm{GL}(V)$ isomorphic to $\mathrm{SL}(2,q)$ and let U and W be subspaces of rank 2 of V such that the following is true:*

a) $V = U \oplus W$.
b) $U^S = U$ and $W^S = W$.
c) *U and W are irreducible S-modules.*

Let $\rho \in S$ be of order t where t divides $q + 1$ but not $p^s + 1$ for all $s < r$. If ρ fixes a subspace X of rank 2 other than U and W, then $\Re =$

$\{F(\Sigma)\,|\,\Sigma \in \mathrm{Syl}_p(S)\}$ *is a regulus and S fixes the regulus opposite to $\Re$ linewise.*

PROOF. Obviously, $\Re$ is a set of $q + 1$ pairwise skew lines of V, and U and W are transversals of $\Re$, and all transversals of $\Re$ are fixed by S. Hence it suffices to prove that $\Re$ has a third transversal.

Let $1 \neq \tau \in \Sigma \in \mathrm{Syl}_p(S)$. Then τ induces a quasitransvection on U and on W. Hence τ is a quasitransvection of V. Thus τ fixes all the planes containing $F(\Sigma)$.

Let us consider the case $r = 1$ first. Then $|\Sigma| = p$. Hence $\Sigma = \langle \tau \rangle$ for $\tau \in \Sigma \backslash \{1\}$. Pick a plane E containing $F(\Sigma)$ other than $F(\Sigma) + U$ and $F(\Sigma) + W$. Then Σ induces a group of elations on E. Therefore there exists a point P on $F(\Sigma)$ such that τ fixes all the lines through P which are in E. Pick $\mathrm{T} \in \mathrm{Syl}_p(S)\backslash\{\Sigma\}$. Then $P + (E \cap F(\mathrm{T}))$ is a line in E. We infer from $\Sigma = \langle \tau \rangle$ that $\langle \tau \rangle$ acts transitively on $\Re \backslash \{F(\Sigma)\}$. Hence $P + (E \cap F(\mathrm{T}))$ is a transversal of $\Re$ which is distinct from U and W, as these two lines are not contained in E. This settles the case $r = 1$. Thus we may assume $r > 1$.

We have $V = U \oplus W = U \oplus X = W \oplus X$ as ρ-modules. Hence $U \cong V/X \cong W$ as ρ-modules. Moreover U, W, X are irreducible ρ-modules, as is easily seen. Therefore $K[\rho]$ is a field by 9.1. This implies that ρ fixes a spread π of V linewise. Let P be a point on $F(\Sigma)$ other than $F(\Sigma) \cap U$ and $F(\Sigma) \cap W$. There exists exactly on $Y \in \pi$ with $P \subseteq Y$. As ρ acts irreducibly on Y we see that Y intersects at least 3 lines of $\Re$ unless $t = 4$. As t divides $q + 1$, we obtain $q \equiv 3 \bmod 4$ in the latter case whence $p \equiv 3 \bmod 4$. This yields $p + 1 \equiv 0 \bmod 4$. Hence $r = 1$ contrary to the assumption $r > 1$. Thus Y intersects at least 3 lines of $\Re$. Let $F(\mathrm{T}_1)$ and $F(\mathrm{T}_2)$ be two distinct lines other than $F(\Sigma)$ which meet Y. Then there exists a $\tau \in \Sigma$ such that $F(\mathrm{T}_1)^\tau = F(\mathrm{T}_2)$. As τ induces an elation in $F(\Sigma) + Y = E$, the centre of this elation and the points $F(\mathrm{T}_1) \cap E$ and $F(\mathrm{T}_2) \cap E$ are collinear. Since $Y = (F(\mathrm{T}_1) \cap E) + (F(\mathrm{T}_2) \cap E)$, we see that $Y^\tau = Y$. Consider the group $H = \langle \tau, \rho \rangle$. Then this group acts irreducibly on Y. Moreover p divides $|H|$. It follows from 14.1 that either $H \cong \mathrm{SL}(2, p^s)$ or $H/\mathfrak{Z}(S) \cong A_4, S_4$, or A_5. In the first case, t divides $p^s + 1$ whence $s = r$, i.e., $H = S$ and Y is a transversal of $\Re$.

Assume $H/\mathfrak{Z}(S) \cong A_4$. From $t \neq 4$ we get $t = 3$ and $p = 2$ and hence $q = 2$, as 3 divides $2 + 1$.

Assume $H/\mathfrak{Z}(S) \cong S_4$. Thus $p = 3$ and $t = 8$ or $p = 2$ and $t = 3$. But $3^{2j+i} \equiv 3^i \bmod 8$. Hence $q + 1 \equiv 2$ or $4 \bmod 8$ in the former case. In the latter case we obtain again $q = 2$.

Assume finally $H/\mathfrak{Z}(S) \cong A_5$. Then $p = 2, 3$, or 5. If $p = 2$, then $t = 3$ or 5, as A_5 does not contain an element of order 15. As $r > 1$, we obtain $t = 5$. This yields $q = 4$ whence $H = S$ and we are done. If $p = 5$, then $t = 3$ or 6, as $t \neq 4$. In either case $q = 5$, a contradiction. Thus $p = 3$. Then $t = 5$ or 10 and hence $q = 9$. In this case, H acts transitively on the set of points of Y. Therefore Y is a transversal in this case, too. $\square$

Remark. *As the proof of* 49.2 *shows, we can drop the assumptions on* ρ *without affecting the conclusions of* 49.2, *if* $r = 1$.

49.3 Lemma. *Let* $\mathfrak{P}$ *be a finite projective space. If* π *is a spread of* $\mathfrak{P}$, *then* π *is also a spread of the dual of* $\mathfrak{P}$.

PROOF. This follows from the fact that the number of points in $\mathfrak{P}$ is equal to the number of hyperplanes in $\mathfrak{P}$. $\qquad\square$

The finiteness assumption in 49.3 is essential. See Bruen & Fisher 1969.

49.4 Lemma. *Let* τ *be a linear mapping of order* p *of* V *leaving invariant a spread. Then the minimal polynomial of* τ *is* $(x - 1)^2$ *or* $(x - 1)^4$. *If it is* $(x - 1)^4$, *then* $p \geqslant 5$.

PROOF. We have $0 = 1 - \tau^p = (1 - \tau)^p$. Hence the minimal polynomial μ of τ divides $(x - 1)^p$. Thus $\mu = (x - 1)^i$. As $\tau \neq 1$, it follows that $2 \leqslant i \leqslant \mathrm{rk}(V) = 4$. Moreover $i \leqslant p$, whence $p \geqslant 5$ if $\mu = (x - 1)^4$.
Assume $\mu = (x - 1)^3$. Then τ has the rational normal form

$$\begin{bmatrix} 0 & 1 & 0 & 0 \\ 0 & 0 & 1 & 0 \\ 1 & -3 & 3 & 0 \\ 0 & 0 & 0 & 1 \end{bmatrix}.$$

Hence V, considered as a ρ-module, is the direct sum of a cyclic ρ-module U with $\mathrm{rk}(U) = 3$ and a cyclic ρ-module W with $\mathrm{rk}(W) = 1$. Let σ be the spread left invariant by τ. By 49.3, there exists exactly one $X \in \sigma$ with $X \subseteq U$. Moreover, there exists precisely one $Y \in \sigma$ with $W \subseteq Y$. Therefore $X^\tau = X$ and $Y^\tau = Y$. This yields $U = X \oplus (Y \cap U)$ where X and $Y \cap U$ are τ-modules. This contradicts the fact that U is a cyclic τ-module isomorphic to the $K[x]$-module $K[x]/\mu K[X]$. $\qquad\square$

49.5 Lemma. *Let* τ *be a linear mapping of order* p *of* V *leaving invariant a spread* σ. *Then the following statements are equivalent*:

a) *The minimal polynomial of* τ *is* $(x - 1)^2$.
b) *The subspace* $A(\tau) = \{v \,|\, v \in V, \, v^\tau = v\}$ *has rank 2.*
c) τ *is a shear or a Baer collineation.*

PROOF. a) implies b): Obviously, τ is a quasitransvection. Hence $\mathrm{rk}(A(\tau)) \geqslant 2$ by 47.5. Assume $\mathrm{rk}(A(\tau)) = 3$. Then $A(\tau)$ contains exactly one $X \in \sigma$ by 49.3. The remaining q^2 components of σ meet $A(\tau)$ in a point. Hence $Y^\tau = Y$ for all $Y \in \sigma$. This yields that τ is in the kernel of σ whence $o(\tau) = p$ divides $q^2 - 1$, a contradiction. This proves $\mathrm{rk}(A(\tau)) = 2$.
b) implies c): If $A(\tau) \in \sigma$, then τ is a shear. If $A(\tau) \notin \sigma$, then $A(\tau)$ consists of the points of a Baer subplane of $\sigma(V)$, since $\mathrm{rk}(A(\tau)) = 2$. Hence τ is a Baer collineation in this case.

c) implies a): If $(x - 1)^2$ is not the minimal polynomial of τ, then τ has minimal polynomial $(x - 1)^4$ by 49.4. Hence V is a cyclic τ-module. As a result $\text{rk}(A(\tau)) = 1$, whence τ has exactly q affine fixed points in $\sigma(V)$, a contradiction.   □

The next theorem was proved by Walker for q odd and by Schäffer for q even.

49.6 Theorem (Walker 1973, Schäffer 1975). *Let $\mathfrak{T}$ be a translation plane of order q^2 such that $\text{GF}(q)$ is contained in the kernel of $\mathfrak{T}$. If $\mathfrak{T}$ admits a group S of collineations isomorphic to $\text{SL}(2,q)$, then one of the following holds:*

a) *S operates completely reducibly but not irreducibly on the translation group of $\mathfrak{T}$ and $\mathfrak{T}$ is either desarguesian or a Hall plane.*
b) *S operates irreducibly on the translation group of $\mathfrak{T}$, and $q \equiv -1 \bmod 3$, and $\mathfrak{T}$ is either a Hering plane or a Schäffer plane.*
c) *S operates reducibly but not completely reducibly on the translation group of $\mathfrak{T}$, the order of $\mathfrak{T}$ is 25, and $\mathfrak{T}$ is either the Hering plane of order 25 or one of the two exceptional Walker planes described in section 46.*

PROOF. It follows from our assumptions that $\mathfrak{T} \cong \pi(V)$ where V is a vector space of rank 4 over $\text{GF}(q)$ and π is a spread of V left invariant by S.

As all translation planes of order 4 and 9 are known, we may assume $q > 3$.

1) *Let $1 \neq \zeta \in \mathfrak{Z}(S)$. Then $v^\zeta = -v$ for all $v \in V$.*

We have $\zeta^2 = 1$ whence $V = V^+ \oplus V^-$ where $V^+ = \{v \mid v \in V,\ v^\zeta = v\}$ and $V^- = \{v \mid v \in V,\ v^\zeta = -v\}$. Assume $V^+ \neq \{0\}$. As ζ is either a homology or a Baer collineation of $\pi(V)$, we see that $|V^+| = q^2$. As S fixes V^+ we obtain a nontrivial homomorphism of S into $\text{GL}(V^+)$ $\cong \text{GL}(2,q)$. As $\text{GL}(2,q)$ does not contain a subgroup isomorphic to $\text{SL}(2,q)/\mathfrak{Z}(\text{SL}(2,q))$, we deduce that S acts trivially on V^+. (Remember $q > 3$.) If $V^+ \in \pi$, then S consists only of perspectives with axis V^+ and centres on l_∞. This yields that S is a Frobenius group which is absurd. Hence $V^+ \notin \pi$ whence V^+ is a Baer subplane. But then S acts regularly on the $q(q - 1)$ points of l_∞ which do not belong to V^+. As a consequence $q(q^2 - 1) = |S|$ divides $q(q - 1)$, a contradiction. Therefore $V^+ = \{0\}$ which proves 1).

2) *S fixes no point and no plane of V.*

Assume that 2) is false. By 49.3 we may assume that S fixes a point P. Then S acts trivially on P. There exists exactly one $X \in \pi$ with $P \subseteq X$. It follows that S fixes X. This yields that S acts trivially on X, since S operates trivially on P. Therefore S consists only of perspectivities with axis X which is absurd.

PROOF OF a). We infer from 2) that there is no irreducible S-submodule of rank 1 or 3 of V. It follows therefore that there exist two irreducible S-submodules U and W of V of rank 2 with $V = U \oplus W$. From this and 1) we infer that S satisfies the assumptions of 49.1. Hence $\Re = \{F(\Sigma)| \Sigma \in \mathrm{Syl}_p(S)\}$ is a set of $q + 1$ pairwise skew lines and U and W are transversals of $\Re$. Since S acts transitively on $\Re$, we have $\Re \subseteq \pi$ or $\Re \cap \pi = \varnothing$.

Case 1: $\Re \cap \pi = \varnothing$. Pick $\Sigma \in \mathrm{Syl}_p(S)$ and let P be a point on $F(\Sigma)$. Then there exists exactly one $X \in \pi$ with $P \subseteq X$. Hence $X^\Sigma = X$. The plane $E = F(\Sigma) + X$ is fixed by Σ. Moreover all elements of Σ induce elations in E with axis $F(\Sigma)$ and centre $P = F(\Sigma) \cap X$, as $X^\Sigma = X$. Also Σ acts transitively on the set of points of Y other than P for all lines Y with $P \subseteq Y \subseteq E$ and $Y \ne F(\Sigma)$. Pick $\mathrm{T} \in \mathrm{Syl}_p(S)$ with $\mathrm{T} \ne \Sigma$. Then $F(\mathrm{T}) \cap E \not\subseteq F(\Sigma)$. Hence $P + (F(\mathrm{T}) \cap E)$ is a transversal of $\Re$ by the above remarks. Thus $\Re$ is a regulus and the opposite regulus $\Re'$ of $\Re$ is fixed linewise by S. We deduce from this fact that $\Re_S(\Sigma)$ fixes $F(\Sigma)$ pointwise. Hence X is fixed by $\Re_S(\Sigma)$. Let T again be a Sylow p-subgroup of S distinct from Σ. Pick $\delta \in (\Re_S(\Sigma) \cap \Re_S(\mathrm{T})) \backslash \mathfrak{Z}(S)$. This is possible since $q > 3$. Then δ fixes $F(\mathrm{T})$ pointwise. As the collineation induced by δ is not the identity, all fixed points of δ are on $F(\Sigma)$ or on $F(\mathrm{T})$. As X is fixed by δ, Maschke's theorem implies that δ has a fixed point Q on X other than P. Hence $Q = X \cap F(\mathrm{T})$. Thus $X \in \Re'$, i.e., $\Re' \subseteq \pi$. It follows now from 48.8 that $\pi(V)$ is a Hall plane.

Case 2: $\Re \subseteq \pi$. If $r = 1$, then $\Re$ is a regulus by the remark made after 49.2. As the opposite regulus is fixed linewise by S, it follows from 48.8 that $\pi(V)$ is desarguesian in this case. Hence we may assume $r > 1$. Let u be a p-primitive prime divisor of $p^{2r} - 1$. Then u divides $p^r + 1$ but u does not divide $p^s + 1$ for $s < r$. If there exists no p-primitive prime divisor of $p^{2r} - 1$, then $p = 2$ and $r = 3$ by 6.2. In this case we put $u = 2^3 + 1 = 3^2$. Obviously u is odd in all cases. Let ρ be an element of order u in S. As ρ leaves invariant $\pi \backslash \Re$ and as $|\pi \backslash \Re| = q(q - 1)$, it follows that ρ fixes an $X \in \pi \backslash \Re$, since u is a power of a prime satisfying $(q(q - 1), u) = 1$. Hence $\Re$ is a regulus by 49.2. We infer from 49.1 that the opposite regulus of $\Re$ is fixed linewise by S. Therefore, $\pi(V)$ is desarguesian by 48.8.

PROOF OF b). As the p-elements behave quite differently in this case according to whether $p = 2$ or $p > 2$, we have to separate the discussion into two cases. First we discuss the case $p > 2$. In order to make this apparent, we shall label our statements by O1), O2), . . .

O1) *If $\tau \in S$ has order p, then the minimal polynomial of τ is $(x - 1)^4$.*

This follows from 49.4 and 49.1 by using 1).

If τ is an element of order p of S, then V is a cyclic primary τ-module by O1). Hence there exist exactly one point $P(\tau)$, exactly one line $L(\tau)$, and exactly one plane $U(\tau)$ left invariant by τ. Moreover $P(\tau) \subseteq L(\tau)$ and $L(\tau) \subseteq U(\tau)$.

O2) Let $\Sigma \in \mathrm{Syl}_p(S)$. Then $P(\tau) = P(\tau')$, $L(\tau) = L(\tau')$, and $U(\tau) = U(\tau')$ for all $\tau, \tau' \in \Sigma \backslash \{1\}$.

This follows from the fact that Σ is abelian.

As $P(\tau)$ etc. depend only on Σ we shall write $P(\Sigma)$ etc. instead of $P(\tau)$ etc.

Pick $\Sigma \in \mathrm{Syl}_p(S)$ and put $H = \mathfrak{N}_S(\Sigma)$. Then there exists a cyclic subgroup Δ of H such that $H = \Delta\Sigma$ and $\Delta \cap \Sigma = \{1\}$. It follows that $|\Delta| = q - 1$. Moreover we have $P(\Sigma)^\Delta = P(\Sigma)$, $L(\Sigma)^\Delta = L(\Sigma)$, and $U(\Sigma)^\Delta = U(\Sigma)$. The line $L(\Sigma)$ is a completely reducible Δ-module by Maschke's theorem. Hence there exists a point Q with $L(\Sigma) = P(\Sigma) + Q$ and $Q^\Delta = Q$. Moreover Σ acts sharply transitively on the set of points on $L(\Sigma)$ other than $P(\Sigma)$. Let R be a point on $L(\Sigma)$ with $R \neq P(\Sigma)$, Q and assume that $\delta \in \Delta$ fixes R. There exists exactly one $\tau \in \Sigma$ with $R^\tau = Q$. This yields $R^{\delta^{-1}\tau\delta} = R^{\tau\delta} = Q^\delta = Q = R^\tau$, whence $\delta^{-1}\tau\delta = \tau$. As $\tau \neq 1$, we obtain $\delta \in \mathfrak{Z}(S)$. Thus we have proved:

O3) Δ splits the set of points on $L(\Sigma)$ other than $P(\Sigma)$ and Q into two orbits of length $\frac{1}{2}(q - 1)$.

Using Maschke's theorem once more, we find a point A in $U(\Sigma)$ with $U(\Sigma) = L(\Sigma) + A$ and $A^\Delta = A$. Since Σ acts sharply transitively on the set of lines through $P(\Sigma)$ in $U(\Sigma)$ which are distinct from $L(\Sigma)$, we obtain similarly to O3)

O4) Δ splits the set of lines through $P(\Sigma)$ in $U(\Sigma)$ which are distinct from $A + P(\Sigma)$ and $L(\Sigma)$ into two orbits of length $\frac{1}{2}(q - 1)$.

Let B be a point on $U(\Sigma)$ with $B \not\subseteq L(\Sigma)$ and $B \notin A^H$. Assume $B^\eta = B$ for some $\eta \in H \backslash \{1\}$. Then $\eta \notin \Sigma$. Hence there exists $\mu \in H$ such that $\mu^{-1}\eta\mu \in \Delta$. We infer that B^μ is a fixed point of $\mu^{-1}\eta\mu$ which is distinct from A. Therefore $B^\mu + P(\Sigma)$ is a fixed line of $\mu^{-1}\eta\mu$. If $B^\mu + P(\Sigma) \neq A + P(\Sigma)$, then $\mu^{-1}\eta\mu \in \mathfrak{Z}(S)$ by O4). This yields $\eta \in \mathfrak{Z}(S)$. If $B^\mu + P(\Sigma) = A + P(\Sigma)$, then there exists $\tau \in \Sigma$ with $A^\tau = B^\mu$. This contradicts our assumption that B is not in the orbit of A under H. Thus we have proved $|B^H| = \frac{1}{2}q(q - 1)$. Hence we have:

O5) H splits the set of points of $U(\Sigma)$ into five orbits, namely $\{P(\Sigma)\}$, $L(\Sigma) \backslash \{P(\Sigma)\}$, A^H, and two orbits of length $\frac{1}{2}q(q - 1)$.

Let $X \in \pi$ contain $P(\Sigma)$. Then $X^\Sigma = X$ whence $X = L(\Sigma)$, as $L(\Sigma)$ is the only line fixed by Σ. This proves

O6) $L(\Sigma) \in \pi$ for all $\Sigma \in \mathrm{Syl}_p(S)$.

Next we prove

O7) $\mathfrak{P} = \{P(\Sigma) | \Sigma \in \mathrm{Syl}_p(S)\}$ consists of $q + 1$ points.

Assume $P(\Sigma) = P(T)$. Then $P(\Sigma)$ is fixed by the group generated by Σ and T. This yields $\Sigma = T$, since S does not fix a point by 2).

O8) $\mathfrak{L} = \{L(\Sigma) \mid \Sigma \in \mathrm{Syl}_p(S)\}$ *consists of* $q + 1$ *lines.*

Assume $L(\Sigma) = L(T)$. Then $L(\Sigma)$ is fixed by the group generated by Σ and T. Hence $\Sigma = T$, as S acts irreducibly.

Put $\pi_0 = \pi \backslash \mathfrak{L}$. Then $|\pi_0| = q(q-1)$.

O9) *No three points of* $\mathfrak{P}$ *are collinear.*

Assume that P, Q, R are three collinear points of $\mathfrak{P}$. Then the orbit of R under $S_{P,Q}$ contains exactly $\frac{1}{2}(q-1)$ points. Hence $P + Q$ carries at least $\frac{1}{2}(q-1) + 2 = \frac{1}{2}(q+3)$ points of $\mathfrak{P}$. Pick two distinct points X, Y in $\mathfrak{P}$ which are also distinct from these $\frac{1}{2}(q+3)$ points. Then $X + Y$ also carries $\frac{1}{2}(q+3)$ points of $\mathfrak{P}$, as S acts doubly transitively on $\mathfrak{P}$. Now $\frac{1}{2}(q+3) + \frac{1}{2}(q+3) = |\mathfrak{P}| + 2$, whence $P + Q$ and $X + Y$ have two points in common. Therefore $P + Q = X + Y$. This yields that all points of $\mathfrak{P}$ are collinear contradicting the fact that S acts irreducibly.

O10) *For* $\Sigma \in \mathrm{Syl}_p(S)$ *and* $X \in \pi_0$ *we have* $S_{P(\Sigma)} \cap S_X = \mathfrak{Z}(S)$.

Obviously $\mathfrak{Z}(S) \subseteq S_{P(\Sigma)} \cap S_X$. As $\mathfrak{L} \backslash \{L(\Sigma)\}$ is an orbit of $H = S_{P(\Sigma)}$, we have that $L(T) \cap U(\Sigma)$ is in an orbit of length q of H provided $T \neq \Sigma$. Hence $X \cap U(\Sigma)$ is in an orbit of length $\frac{1}{2}q(q-1)$ of H by O5). Let $\eta \in H \cap S_X$. Then η fixes $X \cap U(\Sigma)$. It then follows that $\eta \in \mathfrak{Z}(S)$.

O11) *No four points of* $\mathfrak{P}$ *are coplanar.*

Assume that there are five coplanar points $P_1, \ldots, P_5$ in $\mathfrak{P}$. Let E be the plane containing them. As S_{P_1,P_2} splits $\mathfrak{P} \backslash \{P_1, P_2\}$ into two orbits of length $\frac{1}{2}(q-1)$, there exists without loss of generality an $\eta \in S_{P_1,P_2}$ with $P_3^{\eta} = P_4$. It follows from O9) that $E^{\eta} = E$. By 48.3, there exists $X \in \pi$ with $X \subseteq E$. Hence $X^{\eta} = X$. By 1) and O10), we have $X \in \mathfrak{L}$, i.e., $X = L(\Sigma)$ for some $\Sigma \in \mathrm{Syl}_p(S)$. As η does not induce the identity on $\mathfrak{P}$, we have $P(\Sigma) \in \{P_1, P_2\}$. Starting with P_3 and P_4 instead of P_1 and P_2, we obtain the existence of a $T \in \mathrm{Syl}_p(S)$ with $L(T) \subseteq E$ and $P(T) \in \{P_3, P_4\}$. As E contains exactly one component of π, we finally reach the contradiction $P(\Sigma) = P(T)$.

Let P, Q, R be three points of $\mathfrak{P}$ and let E be the plane containing them. Again there exists $X \in \pi$ with $X \subseteq E$. Assume $X = L(\Sigma)$ for some $\Sigma \in \mathrm{Syl}_p(S)$. If $E = U(\Sigma)$, then all the points of $\mathfrak{P}$ are contained in E whence $E^S = E$, a contradiction. Hence $|E^{\Sigma}| = q$. Moreover two distinct planes of E^{Σ} have only $L(\Sigma)$ in common. Hence $sq = q$, where $s + 1$ is the number of points of $\mathfrak{P}$ contained in E. This yields the contradiction $s = 1$. Hence $X \in \pi_0$. In particular, X does not contain a point of $\mathfrak{P}$. Let n be the number of planes in the orbit of E under S_X. The proof of O10) together with O5) yields $|S_X| = q + 1$ or $|S_X| = 2(q + 1)$. Therefore S_X acts either transitively on $\mathfrak{P}$ or splits $\mathfrak{P}$ into two orbits of length $\frac{1}{2}(q+1)$. Hence $tn = q + 1$ or $tn = \frac{1}{2}(q + 1)$, if t is the number of points of $\mathfrak{P}$ in E.

Assume $t = 4$. Then $q \equiv 3 \bmod 4$. But then S_P acts transitively on the set of 2-subsets of $\mathfrak{P} \setminus \{P\}$. Hence $S_{E,P}$ still acts 2-homogeneously on the set of the three points of $\mathfrak{P}$ other than P which are in E. Hence 3 divides $|S_{E,P}| = |S_P \cap S_X|$ contradicting O10). This proves O11). Moreover $q \equiv -1 \bmod 3$. Therefore $\pi(V)$ is a Hering plane by 43.8 and 45.1.

Next we discuss the case $p = 2$. If σ is an element of order 2 in S, then $(x - 1)^2$ is the minimal polynomial of σ. Hence $\mathrm{rk}(A(\sigma)) = 2$ by 49.5. For $\Sigma \in \mathrm{Syl}_2(S)$ put $P(\Sigma) = \{v \mid v \in V, v^\sigma = v$ for all $\sigma \in \Sigma\}$. It follows from 49.1 that there exist $\sigma, \tau \in \Sigma \setminus \{1\}$ with $A(\sigma) \neq A(\tau)$. Thus $P(\Sigma)$ is a point. As S fixes no point by 2), we have therefore

E1) $\mathfrak{R} = \{P(\Sigma) \mid \Sigma \in \mathrm{Syl}_2(S)\}$ *is a set of* $q + 1$ *points.*

Put $U(\Sigma) = A(\sigma) + A(\tau)$ for $\sigma, \tau \in \Sigma \setminus \{1\}$. As Σ is abelian, $A(\sigma)^\tau = A(\sigma)$ and $A(\tau)^\sigma = A(\tau)$. Hence $U(\Sigma)$ is fixed by Σ and Σ fixes all the lines through $P(\Sigma)$ which are contained in $U(\Sigma)$ individually. If Σ fixes a line through $P(\Sigma)$ which is not contained in $U(\Sigma)$, then Σ fixes all the lines through $P(\Sigma)$, whence Σ consists only of transvections of V which is not the case. Hence $A(\rho) \subseteq U(\Sigma)$ for all $\rho \in \Sigma \setminus \{1\}$. Moreover $U(\Sigma)$ is fixed by $H = \mathfrak{R}_S(\Sigma)$. Let Z be a cyclic subgroup of order $q - 1$ of H. It follows from Maschke's theorem that $U(\Sigma)/P(\Sigma)$ is a completely reducible Z-module. Hence there are two lines $L_1(\Sigma)$ and $L_2(\Sigma)$ through $P(\Sigma)$ and contained in $U(\Sigma)$ which are fixed by Z.

E2) *Either* $L_1(\Sigma) \in \pi$ *or* $L_2(\Sigma) \in \pi$.

By 49.3, there exists exactly one $X \in \pi$ with $X \subseteq U(\Sigma)$. Hence $X^H = X$ whence in particular $P(\Sigma) \subseteq X$. If $X \neq L_1(\Sigma), L_2(\Sigma)$, then Z fixes all the lines through $P(\Sigma)$ which are in $U(\Sigma)$, as Z consists only of linear transformations. But there exists $\zeta \in Z$ with $\sigma^\zeta = \tau$ and hence $A(\sigma)^\zeta = A(\tau) \neq A(\sigma)$. This contradiction proves E2).

E3) Z *has exactly three fixed points in* $U(\Sigma)$ *one of them being* $P(\Sigma)$ *the other two lying on* $L_1(\Sigma)$ *respectively* $L_2(\Sigma)$.

It follows from the proof of E2) that all fixed points of Z are on $L_1(\Sigma)$ or on $L_2(\Sigma)$. Let Q and R be fixed points other than $P(\Sigma)$ on $L_i(\Sigma)$. As $A(\rho) \neq L_i(\Sigma)$ for all $\rho \in \Sigma$, there exists $\rho \in \Sigma$ with $Q^\rho = R$. Let $\zeta \in Z$. Then $Q^{\zeta^{-1}\rho\zeta} = Q^{\rho\zeta} = R^\zeta = R = Q^\rho$. Hence $\zeta^{-1}\rho\zeta = \rho$ for all $\zeta \in Z$. This yields $\zeta = 1$, i.e, $Q = R$. By Maschke's theorem, there exists a fixed point of Z on $L_i(\Sigma)$ other than $P(\Sigma)$. This proves E3).

$L_i(\Sigma)$ is fixed by Σ and by Z. Hence it is fixed by H. Thus we may label the $L_i(\mathrm{T})$ for $\mathrm{T} \in \mathrm{Syl}_2(S)$ in such a way that the $L_1(\mathrm{T})$'s as well as the $L_2(\mathrm{T})$'s form an orbit of S.

E4) Z *has exactly four fixed points.*

By Maschke's theorem, there exists a point F fixed by Z such that $V = U(\Sigma) \oplus F$. Pick $X \in \pi$ with $F \subseteq X$. Then $X^Z = X$. Hence $X \cap U(\Sigma)$ is a fixed point of Z on $U(\Sigma)$ which is distinct from $P(\Sigma)$ by E2). We may assume $L_1(\Sigma) \in \pi$. Then $X \cap U(\Sigma)$ is the second fixed point of Z on $L_2(\Sigma)$. In particular, all fixed points of Z which are not on $L_1(\Sigma)$ are on X. There exists $T \in \mathrm{Syl}_2(S) \backslash \{\Sigma\}$ with $Z \subseteq \mathfrak{N}_S(T)$. Assume $P(T) \subseteq L_1(\Sigma)$. Then $L_1(T) = L_1(\Sigma)$, whence $L_1(\Sigma)$ is fixed by $\langle \Sigma, T \rangle = S$, a contradiction. Hence $P(T) \not\subseteq L_1(\Sigma)$ which yields $P(T) \subseteq X$ by the remark made above. From this we infer $X = L_1(T)$.

If Z has more than four fixed points, then Z fixes $L_1(\Sigma)$ and $L_1(T)$ pointwise, as all the fixed points of Z are on $L_1(\Sigma)$ or on $L_1(T)$ and since there exists $\gamma \in \mathfrak{N}_S(Z)$ interchanging $L_1(\Sigma)$ and $L_1(T)$. The group Σ acts regularly on the set of points on $L_1(\Sigma)$ other than $P(\Sigma)$. Hence Σ is centralized by Z which is absurd. Thus Z has exactly four fixed points.

Put $\mathfrak{R}_i = \{L_i(\Sigma) \mid \Sigma \in \mathrm{Syl}_2(S)\}$.

E5) $\mathfrak{R}_1$ and $\mathfrak{R}_2$ are reguli opposite to each other.

If $L_i(\Sigma) = L_i(T)$ for $\Sigma \neq T$, then $L_i(\Sigma)$ is fixed by $\langle \Sigma, T \rangle = S$, a contradiction. Thus $|\mathfrak{R}_i| = q + 1$. We may assume $\mathfrak{R}_1 \subseteq \pi$. Then $\mathfrak{R}_1$ is a partial spread. Pick $L_2(\Sigma) \in \mathfrak{R}_2$. We have to show that $L_2(\Sigma)$ meets $L_1(T)$ for all $T \in \mathrm{Syl}_2(S)$. This is certainly true for $T = \Sigma$. Assume $T \neq \Sigma$. Then $Z = \mathfrak{N}_S(\Sigma) \cap \mathfrak{N}_S(T)$ is cyclic of order $q - 1$. By E4), the group Z has exactly four fixed points, three of them being points of $U(\Sigma)$, namely $P(\Sigma)$, a point F_1 on $L_1(\Sigma)$ and a point F_2 on $L_2(\Sigma)$. Moreover, Z has two fixed points Φ_1 and Φ_2 on $L_1(T)$. Hence one of them, say Φ_1, is equal to one of the points $P(\Sigma), F_1, F_2$. As $L_1(T) \cap L_1(\Sigma) = \{0\}$, we finally get $\Phi_1 = F_2$. This proves E5).

S acts triply transitively on $\mathfrak{P} = \{P(\Sigma) \mid \Sigma \in \mathrm{Syl}_2(S)\}$. Therefore no three points of $\mathfrak{P}$ are collinear, as S operates irreducibly on V. Let w be the maximal number of coplanar points of $\mathfrak{P}$ and let E be a plane containing w points of $\mathfrak{P}$. First of all $w \geqslant 3$. The group S_E acts triply transitively on the set of w points contained in E. It follows $S_E \cong \mathrm{SL}(2, 2^t)$ and $w = 2^t + 1$. Let $\mathfrak{Q}$ be the ruled quadric which carries $\mathfrak{R}_1$ and $\mathfrak{R}_2$. Then $E \cap \mathfrak{Q}$ is an oval of E which is fixed by S_E. Moreover, each orbit of S_E on $E \cap \mathfrak{Q}$ has length divisible by $2^t + 1$. Hence S_E fixes neither a secant nor a tangent of $E \cap \mathfrak{Q}$. Let X be the unique member of π contained in E. Then X is fixed by S_E. Hence X is an exterior line of $E \cap \mathfrak{Q}$. Denote by S^* the group of all projective collineations of E fixing $E \cap \mathfrak{Q}$. Then S^* is isomorphic to $\mathrm{SL}(2, q)$. Moreover $|S_X^*| = 2(q + 1)$ by 38.6. Hence $2^t(2^{2t} - 1)$ divides $2(q + 1)$. This yields $t = 1$; i.e., $w = 3$ and $q \equiv -1 \bmod 3$. Now it follows from 44.5 and 45.1 that $\pi(V)$ is a Schäffer plane.

It remains to consider the case where S acts reducibly but not completely reducibly on V. It follows from 2) that S fixes a line L of V. Furthermore, using 2) again, we see that S acts in its natural representation on L and on V/L. As all automorphisms of $\mathrm{SL}(2, q)$ are induced by semilinear mappings of the vector space $K^{(2)}$, we have that all elements of S may be represented by matrices of the form

$$\begin{bmatrix} a & b & 0 & 0 \\ c & d & 0 & 0 \\ e & f & a^\varphi & b^\varphi \\ g & h & c^\varphi & d^\varphi \end{bmatrix}$$

where φ is a fixed automorphism of K and $ad - bc = 1$. Moreover, using Maschke's theorem again, we may assume that $\rho_a = \text{diag}(a, a^{-1}, a^\varphi, a^{-\varphi})$ belongs to S.

Consider elements σ and τ of order p defined by

$$\begin{bmatrix} 1 & 0 & 0 & 0 \\ s & 1 & 0 & 0 \\ e_s & f_s & 1 & 0 \\ g_s & h_s & s^\varphi & 1 \end{bmatrix} \quad \text{resp.} \quad \begin{bmatrix} 1 & 0 & 0 & 0 \\ t & 1 & 0 & 0 \\ e_t & f_t & 1 & 0 \\ g_t & e_t & t^\varphi & 1 \end{bmatrix}.$$

Then $e_{s+t} = e_s + tf_s + e_t$, whence $tf_s = sf_t$, as $\sigma\tau = \tau\sigma$. Hence $f_s = sf_1 = sk$.

Assume $p = 2$. Then σ fixes a subspace of rank 2 vectorwise. Let $(x_1, x_2, x_3, x_4) \neq (0,0,0,x_4)$ be a fixed vector of σ. Then

$$sx_1 = 0, \qquad e_s x_1 + f_s x_2 = 0, \qquad g_s x_1 + h_2 x_2 + s^\varphi x_3 = 0.$$

This yields $x_1 = 0$. If $x_2 = 0$, then $x_3 = 0$, as $s^\varphi \neq 0$. But then $(x_1, x_2, x_3, x_4) = (0,0,0,x_4)$, a contradiction. Hence $f_s = 0$, i.e., $k = 0$ and $e_{s+t} = e_s + e_t$.

Conjugating σ by ρ_a yields $aa^{-\varphi}e_s = e_{a^2s}$. In particular $ae_1 = a^\varphi e_{a^2}$. Pick $a \in K^*$ with $a + 1 \neq 0$. This is possible, as $q \geq 4$. Then

$$a^\varphi e_{a^2} + e_1 = ae_1 + e_1 = (a+1)^\varphi e_{(a+1)^2} = (a^\varphi + 1)(e_{a^2} + e_1)$$
$$= a^\varphi e_{a^2} + a^\varphi e_1 + e_{a^2} + e_1.$$

Hence $0 = a^\varphi e_1 + e_{a^2} = (a^\varphi + aa^{-\varphi})e_1 = a^{-\varphi}(a^{2\varphi} + a)e_1$.

Assume $e_1 = 0$. Then $e_{a^2} = 0$ for all a and hence $e_s = 0$ for all s. But then $\sigma^2 = 1$ yields $h_s = 0$. Therefore the Sylow 2-subgroup of S consisting of the mappings

$$\begin{bmatrix} 1 & 0 & 0 & 0 \\ s & 1 & 0 & 0 \\ 0 & 0 & 1 & 0 \\ g_s & 0 & s^\varphi & 1 \end{bmatrix}$$

fixes the line $\{(0,x,0,y) \mid x,y \in K\}$ vectorwise. This contradicts the fact that S is not completely reducible (Theorem 49.1). Thus $a^{2\varphi} = a$ for all $a \in K$ and hence $a^\varphi = aa^{-\varphi}$ for all $a \neq 0$. Therefore $e_{a^2} = a^\varphi e_1 = a^{2\varphi^2}e_1$, whence $e_s = s^{\varphi^2}e_1$ for all $s \in K$.

Computing the coefficient in the lower left hand corner for $\sigma^2 = 1$ yields $sh_s = s^\varphi s^{\varphi^2}e_1$ for all $s \in K$. Hence $s^2 h_s^2 = s^{2\varphi}s^{2\varphi^2}e_1^2 = ss^\varphi e_1^2$. Therefore $h_s^2 = s^{-1}s^\varphi e_1^2$ and $h_s = s^{-\varphi}s^\varphi e_1$. From $h_{s+1} = h_s + h_1$ we deduce

$$(s+1)^{\varphi^2}e_1 = (s+1)^\varphi h_{s+1} = (s+1)^\varphi(h_s + h_1) = (s+1)^\varphi(s^{-\varphi}s^{\varphi^2} + 1)e_1.$$

Since $e_1 \neq 0$, we obtain $s^{\varphi^2} + 1 = s^{\varphi^2} + s^{\varphi} + s^{-\varphi}s^{\varphi^2} + 1$ and hence $s = s^{2\varphi} = s^{\varphi^2}$. This yields $s^{\varphi} = s^2$ for all $s \in K$ and hence $q = 4$. Moreover $h_s = s^{-2}se_1 = s^{-1}e_1 = s^2e_1$.

Let $\sigma \neq 1$ and let (x_1,x_2,x_3,x_4) be a fixed vector of σ. Then $x_1 = 0$ as we know already. Moreover $s^2e_1x_2 + s^2x_3 = 0$, whence $e_1x_2 + x_3 = 0$. This shows that there is a Sylow 2-subgroup of S fixing the line defined by the equations $x_1 = 0 = e_1x_2 + x_3$ vectorwise contradicting 49.1. Hence $p > 2$. But then 49.1 and 49.4 yield $p \geqslant 5$ and hence $q \geqslant 5$. Moreover $k = f_1 \neq 0$, as $(x - 1)^4$ is the minimal polynomial of all elements of order p of S. Conjugating σ by ρ_a^{-1} yields $ka^2s = f_{a^2s} = a^{-1}a^{-\varphi}f_s = ka^{-1}a^{-\varphi}s$. Hence $a^{\varphi} = a^{-3}$ for all $a \in K$.

Using the matrices

$$\begin{bmatrix} 1 & s & 0 & 0 \\ 0 & 1 & 0 & 0 \\ e'_s & f'_s & 1 & s^{\varphi} \\ g'_s & h'_s & 0 & 1 \end{bmatrix}$$

one obtains $a^{\varphi} = a$. Therefore $a = a^{-3}$; i.e., $a^4 = 1$ for all $a \in K^*$. Thus $q = 5$. It now follows from 46.12, 46.4, 46.10, 46.11 that $\pi(V)$ is isomorphic to one of the planes $\pi_1(V), \pi_{2,0}(V), \pi_{3,0}(V)$. □

50. The Collineation Groups of the Hering and Schäffer Planes

In this section we shall determine the collineation groups of the Hering and Schäffer planes as well as the number of isomorphism types of Schäffer planes of a given order.

50.1 Lemma. If $\pi(V)$ is a Hering or a Schäffer plane, then $\pi(V)$ does not possess a non-trivial shear.

PROOF. Let $S \cong SL(2,q)$ be the group used in the construction of $\pi(V)$. Then S splits π into an orbit $\mathfrak{R}$ of length $q + 1$ and two orbits of length $\frac{1}{2}q(q - 1)$, as we have seen in section 46.

Assume that σ is a non-trivial shear with axis X. We may assume $X \in \pi$. Assume $X \notin \mathfrak{R}$. The construction of $\pi(V)$ shows that there exists a homology of order 3 in S fixing X. As this homology does not commute with σ, there are at least 3 shears with axis X. Let G be the group generated by all shears of $\pi(V)$ whose axes pass through the origin. Then $G \cong SL(2,p^s)$, $G \cong S(2^{2s+1})$ or 4 does not divide the order of G by 35.10. (Remember that $q \equiv -1 \bmod 3$.) In the latter two cases, the characteristic of $\pi(V)$ is 2 whereas in the first case the characteristic is p. As there are at least 3 shears with axis X, we have that 4 divides $|G|$ if the characteristic is 2. Hence $G \cong SL(2,p^s)$ or $G \cong S(2^{2s+1})$. The number $n + 1$ of axes of non-trivial shears is $p^s + 1$ or $2^{2(2s+1)} + 1$ by 35.10. Therefore either $n + 1 = q^2 + 1$ or $n + 1 = q + 1 + \frac{1}{2}q(q-1) = \frac{1}{2}q(q-1) + 1$ or $n + 1 = \frac{1}{2}q(q+1) + \frac{1}{2}q(q-1) = q(q-1)$. But the latter two cases cannot occur, as $n = p^s$. Hence $n + 1 = q^2 + 1$ and $G \cong SL(2,q^2)$ or $G \cong S(q)$. In the first case, $\pi(V)$ is desarguesian by 38.12 which is obviously absurd. In the second case, $\pi(V)$ is a Lüneburg plane by 31.1. But 3 does not divide the order of the group of those collineations of the Lüneburg plane which are induced by linear mappings of V. Hence $X \in \mathfrak{R}$. Moreover $\mathfrak{R}$ is left invariant by the stabilizer of the origin in the full collineation group of $\pi(V)$ by what we have just seen. Pick $Y \in \mathfrak{R} \setminus \{S\}$. Then no element of $S_{X,Y} \setminus \mathfrak{Z}(S)$ fixes a third component belonging to $\mathfrak{R}$. Hence the only elements of $S_{X,Y}$ centralizing σ are the elements of $\mathfrak{Z}(S)$. Thus there are at least $\frac{1}{2}(q-1)$ non-trivial shears with axis X. This yields that the group of shears with axis X has order q. We deduce from this fact and 35.10 that the group G generated by all shears with axes through the origin is isomorphic to $SL(2,q)$. We infer from 49.5, 49.1, and 49.6 a) that $\pi(V)$ is desarguesian, a contradiction.   □

50.2 Lemma. Let $\pi(V)$ be a Hering or a Schäffer plane of order q^2 and let $S \cong SL(2,q)$ be the group used in the construction of $\pi(V)$. If $\mathfrak{R} \subseteq \pi$ is the orbit of length $q + 1$ of S, then $\mathfrak{R}$ is fixed by all the collineations of $\pi(V)$ which fix the origin.

PROOF. Assume to the contrary that there exists a collineation γ fixing the origin and an $X \in \pi \setminus \mathfrak{R}$ such that $X^\gamma \in \mathfrak{R}$. Then there exists a homology ρ of order 3 such that either X^γ is the axis or $X^\gamma \cap l_\infty$ is the centre of ρ. The group S_{X^γ} then moves either the centre or the axis of ρ whence there exists a non-trivial shear by 13.9 contradicting 50.1.   □

50.3 Theorem (Walker 1973, Schäffer 1975). *Let $\pi(V)$ be a Hering or a Schäffer plane of order q^2 with $q \neq 5$. Furthermore, let $S \cong SL(2,q)$ be the group used in the construction of $\pi(V)$. If G is the group of collineations of $\pi(V)$ fixing the origin, then $G = \mathfrak{R}_{\Gamma L(V)}(S)$.*

PROOF. We first show that $\mathfrak{R}_{\Gamma L(V)}(S) \subseteq G$. This is certainly true, if q is odd by 45.1. If q is even, then $\mathfrak{R}_{\Gamma L(V)}(S)$ permutes the two spreads left invariant by S. Hence it permutes the two reguli $\mathfrak{R}_1$ and $\mathfrak{R}_2$. Assume that there exists $\gamma \in \mathfrak{R}_{\Gamma L(V)}(S)$ with $\mathfrak{R}_1^\gamma = \mathfrak{R}_2$. Then $\mathfrak{R}_2^\gamma = \mathfrak{R}_1$. We may assume that γ fixes the points $(1,0,0,0)K$, $(0,0,0,1)K$ and $(1,1,1,1)K$ on the curve $\mathfrak{C}(\alpha)$. Put $e_1 = (1,0,0,0)$, $e_2 = (0,1,0,0)$, $e_3 = (0,0,1,0)$ and $e_4 = (0,0,0,1)$ and let β be the companion automorphism of γ. Then $(\sum_{i=1}^4 e_i x_i)^\gamma = \sum_{i=1}^4 e_i^\gamma x_i^\beta$. Moreover $e_1^\gamma = e_1 a$, $e_4^\gamma = e_4 b$ and $(\sum_{i=1}^4 e_i)^\gamma = (\sum_{i=1}^4 e_i)c$ whence $a = b = c$. We may assume $a = b = c = 1$.

The lines through $e_1 K$ belonging to $\mathfrak{R}_1$ resp. $\mathfrak{R}_2$ are the lines $\{(x,0,y,0) \mid x,y \in K\}$ and $\{(x,y,0,0) \mid x,y \in K\}$; hence $e_1 x^\beta + e_2^\gamma y^\beta = (e_1 x + e_2 y)^\gamma = e_1 x' + e_3 y'$ and $e_1 x^\beta + e_3^\gamma y^\beta = (e_1 x + e_3 y)^\gamma = e_1 x'' + e_2 y''$ for all $x,y \in K$. Putting $y = 0$ yields $e_1 x^\beta = e_1 x' + e_3 0'$ and $e_1 x^\beta = x_1 x'' + $

$e_2 0''$. Hence $0' = 0'' = 0$ and $x' = x'' = x$. Therefore, $e_2^\gamma y^\beta = e_3 y'$ and $e_3^\gamma = e_2 y''$. Now $\sum_{i=1}^4 e_i = (\sum_{i=1}^4 e_i)^\gamma = e_1 + e_3 1' + e_2 1'' + e_4$. Thus $1' = 1''$ $= 1$ and $e_2^\gamma = e_3$ and $e_3^\gamma = e_2$.

Now $(e_1 s s^\alpha + e_2 s^\alpha + e_3 s + e_4)^\gamma = e_1 s^\beta s^{\beta\alpha} + e_3 s^{\beta\alpha} + e_2 s^\beta + e_4$, as Aut$(K)$ is abelian. Hence there exists $t \in K$ such that $s^\beta s^{\beta\alpha} = tt^\alpha$, $s^\beta = t^\alpha$, $s^{\beta\alpha} = t$. This yields $s^\beta = s^{\beta\alpha^2}$ for all $s \in K$ whence $\alpha^2 = 1$, i.e., $q = 4 \not\equiv -1$ mod 3. This contradiction proves $\mathfrak{N}_{\Gamma L(V)}(S) \subseteq G$.

Assume there exists $\delta \in G \setminus \mathfrak{N}_{\Gamma L(V)}(S)$. Adjusting δ if necessary by an element of $\mathfrak{N}_{\Gamma L(V)}(S)$, we may assume that δ is linear. Then it follows from 43.3 and 44.1 that there exists $P \in \mathfrak{C}$ respectively $P \in \mathfrak{C}(\alpha)$ with $P^\delta \notin \mathfrak{C}$ resp. $\mathfrak{C}(\alpha)$. Let $\Pi \in \mathrm{Syl}_p(S)$ fix P and let $X \in \pi$ contain P. Then X^δ is fixed by Π^δ. Moreover there exists $Q \in \mathfrak{C}$ resp. $\mathfrak{C}(\alpha)$ with $Q \subseteq X$ by 50.2. Furthermore, $P^\delta \neq Q$. Let $\Pi^* \in \mathrm{Syl}_p(S)$ fix Q. Then $S^* = \langle \Pi^\delta, \Pi^* \rangle$ fixes X. As $P^\delta \neq Q$, the group S^* restricted to X is isomorphic to $SL(2,q)$. If S^* acts faithfully on X, then $q = 5$ by 49.6, since $\pi(V)$ is certainly not desarguesian nor a Hall plane. Thus S^* cannot act faithfully. The usual arguments now yield a non-trivial elation contradicting 50.1. Thus $G = \mathfrak{N}_{\Gamma L(V)}(S)$. $\qquad\square$

50.4 Corollary (Schäffer 1978). *Let $q = 2^r > 4$ and assume $q \equiv -1$ mod 3. Then there exist up to isomorphism exactly $\varphi(r)$ Schäffer planes.*

The proof is an easy exercise.

51. The Theorem of Cofman-Prohaska

In section 13, we studied the construction technique called derivation in order to obtain the Hall planes. The finiteness assumptions made there are not really necessary. If we want to generalize this construction to infinite planes, we have to make clear what we mean by a Baer subplane of an infinite projective plane: A proper subplane $\mathfrak{Q}$ of a projective plane $\mathfrak{P}$ is called a Baer subplane of $\mathfrak{P}$, if each point of $\mathfrak{P}$ is on at least one line of $\mathfrak{Q}$ and if each line of $\mathfrak{P}$ carries at least one point of $\mathfrak{Q}$. In the finite case, this definition is equivalent to the one given earlier, as is easily seen. Obviously, Baer subplanes are always maximal subplanes.

With this definition of Baer subplanes derivation carries over to arbitrary planes: Let $\mathfrak{A}$ be an affine plane, D a set of points on l_∞ and let $\mathfrak{B}$ be a set of Baer subplanes of $\mathfrak{A}$ such that:

1) If $\mathfrak{b} \in \mathfrak{B}$, then $\mathfrak{b} \cap l_\infty = D$.
2) If P, Q are distinct points of $\mathfrak{A}$ such that $PQ \cap l_\infty \in D$, then there exists $\mathfrak{b} \in \mathfrak{B}$ such that P and Q are points of $\mathfrak{b}$.

Let $\mathfrak{b}'$ be the plane generated by P, Q and D. Then $\mathfrak{b}' \subseteq \mathfrak{b}$. Let X be a point of $\mathfrak{b}$ which is not on PQ and does not belong to D. Then PX and QX

are two distinct lines meeting l_∞ in points of D. Hence PX and QX belong to $\mathfrak{b}'$ whence X belongs to $\mathfrak{b}'$. Therefore all points of $\mathfrak{b}$ not on PQ are points of $\mathfrak{b}'$ whence $\mathfrak{b} = \mathfrak{b}'$. This implies that there is exactly one $\mathfrak{b} \in \mathfrak{B}$ containing P and Q.

Now we define $\mathfrak{A}'$ in the same way we did in section 13. We want to show that $\mathfrak{A}'$ is an affine plane. By the remark just made, each pair of points of $\mathfrak{A}'$ is joined by exactly one line of $\mathfrak{A}'$. Let l be a line of $\mathfrak{A}'$ and Q a point not on l. Assume that l is also a line of $\mathfrak{A}$. Then $l \cap l_\infty \notin D$. As l intersects all the Baer subplanes in $\mathfrak{B}$ in a point of the projective closure of $\mathfrak{A}$ and since $l \cap l_\infty \notin D$, we see that l intersects all the Baer subplanes in $\mathfrak{B}$ in a point of $\mathfrak{A}'$. From this we infer that there exists exactly one line of $\mathfrak{A}'$ through Q which does not meet l in a point of $\mathfrak{A}'$, namely the line $Q(l \cap l_\infty)$ of $\mathfrak{A}$. Assume that $l \in \mathfrak{B}$. Then all the lines through Q which are also lines of $\mathfrak{A}$ meet l in a point of $\mathfrak{A}'$, as we have just seen.

The point Q is on exactly one line h of the Baer subplane l, as Q is not in l. Let g be a line of l parallel to but distinct from h. Furthermore, let j be a line of $\mathfrak{A}$ through Q with $h \neq j$ and $j \cap l_\infty \in D$. Put $X = j \cap g$. Then X does not belong to l, as otherwise j would be a line of l contradicting the uniqueness of h. There exists $m \in \mathfrak{B}$ with Q,X in m. Assume that m has a point Y in common with l which is not on l_∞. Pick a line i of l through Y which is not parallel to h. As h,g,i are also lines of m, it follows that $i \cap h$ and $i \cap g$ are distinct points common to m and l whence $m = l$, contradicting the fact that Q does not belong to l. Hence m and l have no point in common.

Let $m' \in \mathfrak{B}$ be such that Q is also a point of m' and l and m' have no point of $\mathfrak{A}'$ in common. Let Z be a point of m. Then Z is on a line z of l. As m and l have no point of $\mathfrak{A}'$ in common, z is parallel to h. Hence all the lines of l parallel to h are lines of m. Likewise we obtain that all these lines are also lines of m'. As all the lines of m through Q are also lines of m', we finally get that $m = m'$.

Finally, it is obvious that there is a triangle in $\mathfrak{A}'$. This proves that $\mathfrak{A}'$ is an affine plane.

The following theorem was first proved for finite planes by Prohaska 1972. The generalization to arbitrary planes and the proof given here are due to J. Cofman 1975.

51.1 Theorem. *Let $\mathfrak{A}$ be an affine plane, let D be a set of points on the line at infinity of $\mathfrak{A}$, and let $\mathfrak{B}$ be a set of Baer subplanes of $\mathfrak{A}$ such that the following is true*:

1) *If $\mathfrak{b} \in \mathfrak{B}$ then $\mathfrak{b} \cap l_\infty = D$.*
2) *If P,Q are distinct points of $\mathfrak{A}$ such that $PQ \cap l_\infty \in D$, then there exists $\mathfrak{b} \in \mathfrak{B}$ such that P and Q are points of $\mathfrak{b}$.*

Then all the planes belonging to $\mathfrak{B}$ are desarguesian.

PROOF. Let $\mathfrak{P}$ be the projective closure of $\mathfrak{A}$. Then D is a set of lines in the dual $\mathfrak{P}^d$ of $\mathfrak{P}$ all of them passing through the point l_∞ of $\mathfrak{P}^d$. Moreover $\mathfrak{B}$ is a set of Baer subplanes of $\mathfrak{P}^d$ with the following properties: If $\mathfrak{b} \in \mathfrak{B}$ then l_∞ is a point of $\mathfrak{b}$ and D consists of all the lines of $\mathfrak{b}$ which pass through l_∞. If l and m are lines of $\mathfrak{P}^d$ not passing through l_∞ such $l \cap m$ is on one of the lines of D, then there exists $\mathfrak{b} \in \mathfrak{B}$ such that l and m are lines of $\mathfrak{b}$.

Pick $a \in D$ and consider the following incidence structure $\mathfrak{a}$.

α) The points of $\mathfrak{a}$ are the points of $\mathfrak{P}^d$ other than l_∞ which are on lines of $D \setminus \{a\}$.

β) The lines of $\mathfrak{a}$ are those subsets of the set of points of $\mathfrak{a}$ which are intersections of lines of $\mathfrak{P}^d$ other than a with Baer subplanes belonging to $\mathfrak{B}$.

γ) The planes of $\mathfrak{a}$ are the point sets $l \setminus \{l_\infty\}$ with $l \in D \setminus \{a\}$ and $\mathfrak{b} \setminus \{a\}$ with $\mathfrak{b} \in \mathfrak{B}$. (Here we consider lines and planes as point sets.)

We shall prove that $\mathfrak{a}$ is a 3-dimensional affine space.

A) Let A, B, C be three distinct points of $\mathfrak{a}$ which are not collinear in $\mathfrak{P}^d$. Then there is exactly one plane of $\mathfrak{a}$ containing them.

At most one of the lines AB, BC, CA is in D. Hence there is at least one $\mathfrak{b} \in \mathfrak{B}$ containing A, B and C. By the remarks made before 51.1, the plane $\mathfrak{b}$ is uniquely determined.

B) If two planes b and c of $\mathfrak{B}$ have two distinct points A and B of $\mathfrak{a}$ in common, then $AB \cap b = AB \cap c$.

Case 1: $AB \notin D$. Then $AB \cap b = \{l \cap AB \mid l \in D \setminus \{a\}\}$ and $AB \cap c = \{l \cap AB \mid l \in D \setminus \{a\}\}$.

Case 2: $AB \in D$. Suppose $AB \cap b \neq AB \cap c$. Then we may assume without loss of generality that there is a point $C \in AB \cap b$ with $C \notin AB \cap c$. Pick a line h of the Baer subplane b of $\mathfrak{P}^d$ through C other than AB. Then h carries exactly one point H of c. As H is on a line $k \in D$, we see that $H = h \cap k$ is a point of b, too. Now $AH, BH \notin D$. Therefore, there is exactly one plane in $\mathfrak{B}$ such that AH and BH belong to this plane. Hence we reach the contradiction $c = b$.

(1) Any two distinct points of $\mathfrak{a}$ are joined by a unique line of $\mathfrak{a}$.

It follows from B) that there exists at most one line of $\mathfrak{a}$ joining the two distinct points A and B. As there is always a point C of $\mathfrak{a}$ such that A, B and C are not collinear in $\mathfrak{P}^d$, it follows from A) that there is at least one line of $\mathfrak{a}$ joining A and B.

(2) Each line of $\mathfrak{a}$ contains at least two points.

This is trivial.

(3) The line of $\mathfrak{a}$ joining two distinct points of a plane of $\mathfrak{a}$ is completely contained in this plane.

This follows from B).

(4) Any three distinct points of $\mathfrak{a}$ which are not collinear in $\mathfrak{a}$ are on a unique plane of $\mathfrak{a}$.

Let A,B,C be three points of $\mathfrak{a}$ which are not collinear in $\mathfrak{a}$. If A,B,C are not collinear in $\mathfrak{P}^d$, then (4) is true by A). Hence assume that A,B,C are collinear in $\mathfrak{P}^d$ and let l be the line carrying A,B,C. Then it follows that l belongs to D. From this, B) and the fact that A,B,C are not collinear in $\mathfrak{a}$, we infer that l is the only plane of $\mathfrak{a}$ containing A,B,C.

(5) Each plane of $\mathfrak{a}$ contains three points which are not on a line of $\mathfrak{a}$.

This is trivial.

(7) If two planes of $\mathfrak{a}$ have a point in common, then they have a line in common.

This is certainly true if at least one of the planes is induced by a line of D. We may therefore assume that the two planes are induced by elements $\alpha,\beta \in \mathfrak{B}$. By assumption α and β have a point P of $\mathfrak{a}$ in common. Suppose there is a line l of $\mathfrak{P}^d$ through P other than $l_\infty P$ which is a line of α and of β. Then $\{l \cap h \mid h \in D \setminus \{a\}\}$ is a line of $\mathfrak{a}$ common to the two planes of $\mathfrak{a}$ which are induced by α and β. Hence we may assume that there is no such line. Let Q be a point of $\mathfrak{a}$ on $l_\infty P$ belonging to β but not to α. Pick a line m of β through Q which is distinct from $l_\infty P$. Then m carries a point R of α which is distinct from P and l_∞. As R is on a line of D, it also belongs to β. Hence PR is a line of $\mathfrak{P}^d$ which belongs to both α and β in contradiction to our assumption. Hence $\beta \cap l_\infty P \subseteq \alpha \cap l_\infty P$. By symmetry $\alpha \cap l_\infty P \subseteq \beta \cap l_\infty P$. This proves (6).

C) The points and lines contained in a plane b of $\mathfrak{a}$ form an affine plane.

This is certainly true if b is induced by a $\beta \in \mathfrak{B}$. Hence we may assume that $b \cup \{l_\infty\} = l \in D \setminus \{a\}$. Pick $\gamma \in \mathfrak{B}$ and let X be a point on a which is not a point of γ. Denote by c the plane of $\mathfrak{a}$ induced by γ. For all $Y \in c$ define Y^γ by $Y^\gamma = XY \cap l$. Then γ is a bijection from c to b (as point sets). Using A) and B), it is easily seen that γ induces a mapping of the set of lines of $\mathfrak{a}$ contained in c onto the set of lines of $\mathfrak{a}$ contained in b. This proves C).

Two lines l and m of $\mathfrak{a}$ are said to be parallel if they are contained in a plane of $\mathfrak{a}$ and if either $l = m$ or $l \cap m = \emptyset$.

(8) Parallelism is an equivalence relation.

We have only to show that parallelism is transitive. Let l,m,n be three lines of $\mathfrak{a}$ such that $l \parallel m$ and $m \parallel n$. If l, m and n are all contained in the same plane of $\mathfrak{a}$, then $l \parallel n$ by C). Hence we may assume that they are not contained in one plane of $\mathfrak{a}$. This implies in particular that l, m and n are pairwise distinct. Let b be the plane containing l and m and let c be the plane containing m and n. Pick a point X on n. It follows from (4) and (3) that there exists a plane d of $\mathfrak{a}$ containing X and l. Let $p = d \cap c$. As l is not in c, we have by (6) that p is a line of $\mathfrak{a}$. Assume $p \neq n$. Then p meets m by C). As $p \cap l \subseteq c \cap b = m$, we have $p \cap l \subseteq m \cap p \cap l = \emptyset$. Hence p is parallel to l. Therefore m and p are parallels of l passing through the point $p \cap m$. As there is exactly one plane containing l and $p \cap m$, we reach the contradiction $p = m$. This contradiction proves $p = n$. If l and n are not parallel, then they meet by C). Therefore l and n are parallels of m having a point in common, whence $l = n$, again a contradiction.

(9) If l is a line of $\mathfrak{a}$, then there exists a line of $\mathfrak{a}$ parallel to and distinct from l.

This is trivial.

From (1) to (9) we conclude that $\mathfrak{a}$ is an affine space which is not an affine plane (see e.g. Lenz [1967, p. 136]). Moreover, the special form of (7) yields that the dimension of $\mathfrak{a}$ is 3.

Now, if $\beta \in \mathfrak{B}$ then the plane of $\mathfrak{a}$ induced by β is desarguesian, as all affine spaces are desarguesian. Hence β itself is desarguesian. $\square$

As an immediate consequence we have:

51.2 Corollary (Prohaska 1972). *If $\mathfrak{A}$ is a finite derivable plane, then the order of $\mathfrak{A}$ is a power of a prime.*

52. Prohaska's Characterization of the Hall Planes

We shall finish this book by proving the following theorem due to Prohaska 1972:

52.1 Theorem. *Let $\mathfrak{A}$ be a finite affine plane of order q^2 and let D be a set of $q + 1$ points on l_∞ such that $\mathfrak{A}$ is derivable with respect to D. If G is a rank-3-group of $\mathfrak{A}$ leaving invariant D and if $\mathfrak{A}'$ is the plane derived from $\mathfrak{A}$ with respect to D, then one of the planes $\mathfrak{A},\mathfrak{A}'$ is desarguesian and the other one is a Hall plane.*

Before we can prove this theorem, we have to establish a few lemmas.

52.2 Lemma. *Let $\mathfrak{A}$ be the desarguesian affine plane of order $q = p^r$ and let Π be a p-subgroup of the collineation group of $\mathfrak{A}$. If $|\Pi| \geqslant q$ and if Π fixes a point of $\mathfrak{A}$, then Π contains a shear of $\mathfrak{A}$.*

PROOF. Let O be the fixed point of Π and let Γ be the group of all collineations of $\mathfrak{A}$ fixing O. Then $\Gamma \cong \Gamma L(2,q)$. Hence

$$|\Gamma| = rq(q - 1)(q^2 - 1).$$

As the number of lines through O is $q + 1$, there is a line l through O fixed by Π. Let Σ be the group of all shears with axis l. Then $|\Sigma| = q$. Moreover Σ is normalized by Π, whence $\Pi\Sigma$ is a p-subgroup of Γ. Let p^s be the highest power of p dividing r. Then $|\Pi\Sigma|$ divides p^{s+r}. As $p^s \leqslant r < p^r = q$, we see that $|\Pi\Sigma| < q^2$. Therefore $|\Pi \cap \Sigma| |\Pi\Sigma| = |\Pi||\Sigma| \geqslant q^2$ yields $\Pi \cap \Sigma \neq \{1\}$.   □

52.3 Lemma. *Let $\mathfrak{A}$ be an affine plane of order $q^2 > 4$ and let D be a set of $q + 1$ points on l_∞ such that $\mathfrak{A}$ is derivable with respect to D. Denote by $\mathfrak{A}'$ the plane derived from $\mathfrak{A}$ with respect to D. If η is a collineation of $\mathfrak{A}$ fixing a point O of $\mathfrak{A}$ and each line through O which does not carry a point of D, then η is either a homology of $\mathfrak{A}$ or of $\mathfrak{A}'$.*

PROOF. First of all we note $D^\eta = D$. Hence η is a collineation of $\mathfrak{A}$ and of $\mathfrak{A}'$. If η fixes a point of $\mathfrak{A}$ other than O, then η fixes a quadrangle and hence a subplane of the projective closure of $\mathfrak{A}$ pointwise. As η has at least $q(q - 1)$ fixed points on l_∞, the order of this subplane is at least $q(q - 1) - 1$. As $q > 2$, it follows that the order of this subplane is greater than $q = \sqrt{q^2}$. Hence $\eta = 1$. Thus we may assume that η has no fixed point other than O.

Assume that η is not a homology of $\mathfrak{A}$. Then $\eta \neq 1$. Moreover, there exists a line l through O with $l^\eta \neq l$. Let $\mathfrak{b}$ be a Baer subplane of $\mathfrak{A}$ containing O and D. Then we have to show $\mathfrak{b}^\eta = \mathfrak{b}$. Assume $\mathfrak{b}^\eta \neq \mathfrak{b}$. Then $\mathfrak{b}$ and $\mathfrak{b}^\eta$ have only one affine point in common, namely O. Moreover, l is a line of $\mathfrak{b}$. Hence there exists a point P of $\mathfrak{b}$ other than O which is on l. As $l \neq l^\eta$, we have $P \neq P^\eta$. If the line PP^η carries a point of D, then PP^η is a line of $\mathfrak{b}$ as well as of $\mathfrak{b}^\eta$, since P is a point of $\mathfrak{b}$ and P^η is a point of $\mathfrak{b}^\eta$. But all affine lines common to both $\mathfrak{b}$ and $\mathfrak{b}^\eta$ pass through O. Hence $O \mathrel{I} PP^\eta$ whence $l = OP = OP^\eta$, i.e., $l = l^\eta$. This contradiction proves that PP^η does not carry a point of D. Let $F = l_\infty \cap PP^\eta$. Then $PP^\eta = FP = FP^\eta$ and hence $(PP^\eta)^\eta = PP^\eta$. This yields that η has a fixed point other than O in $\mathfrak{A}$ whence $\eta = 1$, a contradiction.   □

52.4 Lemma. *Let p be a prime and $q = p^r > p$. Furthermore, let $\mathfrak{B}$ be a set of length $q(q - 1)$ and S a permutation group of $\mathfrak{B}$ isomorphic to $PSL(2,q)$. If no element of order p of S fixes an element of $\mathfrak{B}$ and if all orbits of S have the same length, then there exists a partition $\mathfrak{p}$ of $\mathfrak{B}$ such that $|X| = 2$ for all $X \in \mathfrak{p}$ which is left invariant by S and such that S_X is a dihedral group of order $2|S|(q(q - 1))^{-1}$.*

This is proved easily with the help of 14.1.

PROOF OF 52.1. As G is a rank 3-group of $\mathfrak{A}$ as well as of $\mathfrak{A}'$, the planes $\mathfrak{A}$ and $\mathfrak{A}'$ are translation planes by 16.3. Moreover G contains the translation group T of $\mathfrak{A}$ and T is also the translation group of $\mathfrak{A}'$.

Let O be a point of $\mathfrak{A}$ and let $\mathfrak{L}$ be the set of lines through O which carry a point of D and denote by $\mathfrak{M}$ the remaining lines through O. Furthermore, let $\mathfrak{B}$ be the set of Baer subplanes containing O and D. Then G_O acts transitively on the sets $\mathfrak{L}, \mathfrak{M}$ and $\mathfrak{B}$ by 16.2. Moreover $|\mathfrak{L}| = q + 1$, $|\mathfrak{M}| = q(q - 1)$ and $|\mathfrak{B}| = q + 1$.

Put $\mathfrak{a} = \{P \mid O \neq P \, \mathrm{I} \, l$ and $l \in \mathfrak{L}\}$ and $\mathfrak{d} = \{P \mid O \neq P \, \mathrm{I} \, l$ and $l \in \mathfrak{M}\}$. Then $\mathfrak{a}$ and $\mathfrak{d}$ are orbits of $G_O = H$. Moreover $|\mathfrak{a}| = (q + 1)(q^2 - 1)$ and $|\mathfrak{d}| = q(q - 1)(q^2 - 1)$. Let $d = (2, q - 1)$. Then $d^{-1}q(q^2 - 1)^2$ is the least common multiple of $|\mathfrak{a}|$ and $|\mathfrak{d}|$. Hence $d^{-1}q(q^2 - 1)^2$ divides $|H|$.

Next we prove that H contains a collineation which is a shear of $\mathfrak{A}$ or of $\mathfrak{A}'$. Assume not. Pick a Sylow p-subgroup Π of H. Then q divides $|H|$. Moreover there exists $l \in \mathfrak{L}$ with $l^\Pi = l$, as $|\mathfrak{L}| = q + 1$. Also, there is $\mathfrak{b} \in \mathfrak{B}$ with $\mathfrak{b}^\Pi = \mathfrak{b}$. The group Π acts faithfully on $\mathfrak{b}$, since there exists no shear of $\mathfrak{A}'$ in H. Furthermore, $\mathfrak{b}$ is desarguesian by 51.1. Hence there exists $\pi \in \Pi$ which induces a non-trivial shear on $\mathfrak{b}$ by 52.2. As l is a line of $\mathfrak{b}$, it must be the axis of π considered as a shear of $\mathfrak{b}$, i.e., each point in $l \cap \mathfrak{b}$ is fixed by π.

Assume that H_l does not act faithfully on l. Then H_l contains a non-trivial homology with axis l of $\mathfrak{A}$. Let C be its centre. Then C is on the line at infinity of $\mathfrak{A}$. It follows from 13.9 that C is fixed by H_l. Hence $H_l \subseteq H_C$. Likewise $H_C \subseteq H_l$, whence $H_l = H_C$. This proves $C \in D$. Hence C is a point of the projective closure of $\mathfrak{b}$. But $\pi \in H_l$ yields $C^\pi = C$, whence π induces the identity on l, a contradiction.

As the assumptions on $\mathfrak{A}$ and $\mathfrak{A}'$ are symmetric, we also see that $H_\mathfrak{b}$ acts faithfully on $\mathfrak{b}$.

It follows from 16.2 that $H_\mathfrak{b}$ acts transitively on the set of points of $\mathfrak{b}$ other than O. From this we infer that $H_\mathfrak{b}$ acts transitively on D. Put $S = \langle \pi^\alpha \mid \pi \in \Pi, \alpha \in H_\mathfrak{b}, \pi$ induces a shear in $\mathfrak{b} \rangle$. Then either $S \cong \mathrm{SL}(2, q)$ or $p = 2$ and S is dihedral of order $2(q + 1)$, since $\mathfrak{b}$ is desarguesian and since $H_\mathfrak{b}$ acts faithfully on $\mathfrak{b}$.

As 52.1 is certainly true if $q = 2$ or 3, we may assume $q > 3$. Assume that S is dihedral of order $2(q + 1)$. Obviously, S is normal in $H_\mathfrak{b}$. Thus $|H_\mathfrak{b}|$ divides $2r(q + 1)$ where r is determined by $q = 2^r$. But $|H| = (q + 1) \cdot |H_\mathfrak{b}|$. Hence $q(q - 1)^2(q + 1)$ divides $|H_\mathfrak{b}|$, whence $q(q - 1)^2$ divides $2r$. This yields $q = 2$, a contradiction. Therefore $S \cong \mathrm{SL}(2, q)$.

Let $\Psi = \Pi \cap S$. Then Ψ is a Sylow p-subgroup of S, as S is normal in $H_\mathfrak{b}$. The line l is a Baer subplane of $\mathfrak{A}'$ and $\mathfrak{b}$ is a line of this plane. As Ψ acts faithfully on l, it induces a group of shears of order q on l. This yields that Ψ acts transitively on $\mathfrak{L} \backslash \{l\}$. Hence S acts transitively on $\mathfrak{L} \backslash \{l\}$. Therefore S contains a subgroup of index q. It then follows from 14.1 that $q = 3, 5, 7$, or 11. Hence $\mathrm{GF}(q)$ is in the kernel of $\mathfrak{A}$. Moreover the p-elements of S have minimal polynomial $(x - 1)^4$ by 49.5. Hence $\mathfrak{A}$ is one of the exceptional planes of order 25 by 49.6. But this yields the contradiction that $\mathfrak{b}$ is a line of $\mathfrak{A}$. Hence H contains a collineation which is a shear of $\mathfrak{A}$ or of $\mathfrak{A}'$.

We may assume that H contains a shear of $\mathfrak{A}$. Let S be the group generated by all these shears. It follows from $|\mathfrak{L}| = q + 1$ that the axes of these shears belong to $\mathfrak{L}$. Hence $S \cong \mathrm{SL}(2,q)$ or $p = 2$ and S is the dihedral group of order $2(q + 1)$.

Assume that S is the dihedral group of order $2(q + 1)$. Since S is normal in H, all orbits of S in $\mathfrak{M}$ have length t. Hence t divides $(2(q + 1), q(q - 1)) = 2$. (Remember that q is even.) Hence $t = 2$, as a shear does not fix a line in $\mathfrak{M}$. Let Z be the cyclic stem of S. Then $|Z| = q + 1$ and Z fixes all the lines of $\mathfrak{M}$. As Z acts transitively on $\mathfrak{L}$, we have by 52.3 that Z is a group of homologies of $\mathfrak{A}'$. This yields that $\mathrm{GF}(q^2)$ is the kernel of $\mathfrak{A}'$ whence $\mathfrak{A}'$ is desarguesian. Moreover $\mathfrak{A}$ is a Hall plane in this case.

Assume $S \cong \mathrm{SL}(2,q)$. If $q = p$, then 52.1 is a consequence of 49.6. Hence we may assume $q > p$. Then S induces a group S^* isomorphic to $\mathrm{PSL}(2,q)$ on $\mathfrak{M}$. Since H acts transitively on $\mathfrak{M}$ and since S is normal in H, we have the assumptions of 52.4. Let $\mathfrak{p}$ be the partition of $\mathfrak{M}$ whose existence is guaranteed by 52.4. Furthermore, let Z be the centralizer of S in H. Then Z fixes $\mathfrak{p}$ elementwise, as Z centralizes each dihedral subgroup of S^*. Moreover H/Z is isomorphic to a subgroup of $\mathrm{P}\Gamma\mathrm{L}(2,q)$, as this group is the automorphism group of S. Hence $|H|t = rq(q^2 - 1)|Z|$. On the other hand $|H| = sd^{-1}q(q^2 - 1)^2$, i.e., $std^{-1}q(q^2 - 1)^2 = rq(q^2 - 1)|Z|$. This yields $std^{-1}(q^2 - 1) = r|Z|$. From $dr < q - 1$ we infer $|Z| > q + 1$. Let Λ be the subgroup of Z fixing $\mathfrak{M}$ linewise. Then $|\Lambda| > \frac{1}{2}(q + 1)$. By 52.3, each $\lambda \in \Lambda$ is a homology of $\mathfrak{A}$ or of $\mathfrak{A}'$. As Λ is not the union of two proper subgroups, Λ consists entirely of homologies of $\mathfrak{A}$ or of $\mathfrak{A}'$. This implies that the kernel of $\mathfrak{A}$ or of $\mathfrak{A}'$ is $\mathrm{GF}(q^2)$, whence one of the planes is desarguesian and the other one is a Hall plane. □

Bibliography

André, J.

1954   Über nicht-Desarguessche Ebenen mit transitiver Translationsgruppe. *Math. Z.* **60**, 156–186.
1955   Projektive Ebenen über Fastkörpern. *Math. Z.* **62**, 137–160.

Baer, R.

1944   The fundamental theorems of elementary geometry. *Trans. Amer. Math. Soc.* **56**, 94–129.
1946a   Projectivities with fixed points on every line of the plane. *Bull. Amer. Math. Soc.* **52**, 273–286.
1946b   Polarities in finite projective planes. *Bull. Amer. Math. Soc.* **52**, 77–93.
1952   Linear Algebra and Projective Geometry. New York: Academic Press.
1963   Partitionen abelscher Gruppen. *Arch. Math.* **14**, 73–83.

Bruen, A. and J. C. Fisher.

1969   Spreads which are not dual spreads. *Can. Math. Bull.* **12**, 801–803.

Burmester, M. V. D. and D. R. Hughes.

1965   On the solvability of autotopism groups. *Arch. Math.* **16**, 178–183.

Burn, R. P.

1968   Bol quasifields and Pappus' theorem. *Math. Z.* **105**, 351–364.

Burnside, W.

1955   Theory of Groups of Finite Order. 2nd ed. New York: Dover Publications.

Cofman, J.

1975   Baer subplanes and Baer collineations of derivable projective planes. *Abh. Math. Sem. Hamburg.* **44**, 187–192.

Czerwinski, T.

1972   Finite translation planes with collineation groups doubly transitive on
          the points at infinity. *J. Algebra* **22**, 428–441.

Dembowski, P.

1966   Zur Geometrie der Suzukigruppen. *Math. Z.* **94**, 106–109.
1968   Finite Geometries. Berlin, Heidelberg, New York: Springer-Verlag.

Dickson, L. E.

1958   Linear Groups with an Exposition of the Galois Field Theory. New
          York: Dover Publications.

Ellers, E. and H. Karzel.

1964   Endliche Inzidenzgruppen. *Abh. Math. Sem. Hamburg* **27**, 250–264.

Foulser, D. A.

1964   Solvable flag transitive affine planes. *Math. Z.* **86**, 191–204.
1967a   A generalization of André's systems. *Math. Z.* **100**, 380–395.
1967b   A class of translation planes $\Pi(Q_g)$. *Math. Z.* **101**, 95–102.
1969   Collineation groups of generalized André planes. *Can. J. Math.* **21**,
          358–369.

Gleason, A. M.

1956   Finite Fano planes. *Amer. J. Math.* **78**, 797–807.

Gorenstein, D.

1968   Finite Groups. New York: Harper & Row.

Gorenstein, D. and J. H. Walter.

1965   The characterization of finite groups with dihedral Sylow 2-subgroups. I,
          II, III. *J. Algebra* **2**, 85–151, 218–270, 354–393.

Hall, M.

1943   Projective planes. *Trans. Amer. Math. Soc.* **54**, 229–277.
1949   Corrections to: Projective planes. *Trans. Amer. Math. Soc.* **65**, 473–474.

Hall, M., J. D. Swift and R. J. Walker.

1956   Uniqueness of the projective plane of order eight. *Math. Tables Aids
          Comp.* **10**, 186–194.

Hering, Ch.

1970   A new class of quasifields. *Math. Z.* **118**, 56–57.
1971   Über Translationsebenen, auf denen die Gruppe SL(2,q) operiert. Atti
          del Convegno di Geometria Combinatoria e sue Applicazione. Perugia:
          Ist. Mat. Univ. Perugia, 259–261.
1972a   On subgroups with trivial normalizer intersection. *J. Algebra* **20**,
          622–629.
1972b   On shears of translation planes. *Abh. Math. Sem. Hamburg* **37**, 258–268.

Higman, D. G.

1964   Finite permutation groups of rank 3. *Math. Z.* **86**, 145–156.

Hughes, D. R.

1962   Combinatorial analysis. *t*-designs and permutation groups. *Proc. Symp.
          Pure Math.* **6**, 39–41.

Hughes, D. R. and F. C. Piper.

1973   Projective Planes. New York-Heidelberg-Berlin: Springer Verlag.

Huppert, B.

1957   Zweifach transitive, auflösbare Permutationsgruppen. *Math. Z.* **68**, 126–150.
1967   Endliche Gruppen I. Berlin-Heidelberg-New York: Springer Verlag.

Johnson, N. L. and M. J. Kallaher.

1974   Transitive collineation groups on affine planes. *Math. Z.* **135**, 149–164.

Kallaher, M. J.

1969a   Projective planes over Bol quasifields. *Math. Z.* **109**, 53–65.
1969b   On finite affine planes of rank 3. *J. Algebra* **13**, 544–553.
1972   A note on finite Bol quasifields. *Arch. Math.* **23**, 164–166.
1974   A note on Z-planes. *J. Algebra* **28**, 311–318.

Kallaher, M. J. and T. G. Ostrom.

1970   Fixed point free linear groups, rank three planes, and Bol-quasifields. *J. Algebra* **18**, 159–178.

Lang, S.

1971   Algebra. 4th printing. Reading, Mass.: Addison Wesley Publ.

Lenz, H.

1967   Grundlagen der Elementarmathematik. 2. Aufl. Berlin: Deutscher Verlag der Wissenschaften.

Liebler, R. A.

1970   Finite affine planes of rank three are translation planes. *Math. Z.* **116** 89–93.
1972   A characterization of the Lüneburg planes. *Math. Z.* **126**, 82–90.

Lüneburg, H.

1964a   Charakterisierungen der endlichen Desarguesschen projektiven Ebenen. *Math. Z.* **85**: 419–450.
1964b   Finite Möbius planes admitting a Zassenhaus group as group of automorphisms. *Ill. J. Math.* **8**, 586–592.
1965a   Über projektive Ebenen, in denen jede Fahne von einer nichttrivialen Elation invariant gelassen wird. *Abh. Math. Sem. Hamburg* **29**, 37–76.
1965b   Die Suzukigruppen und ihre Geometrien. Berlin-Heidelberg-New York: Springer-Verlag.
1965c   Zur Frage der Existenz von endlichen projektiven Ebenen vom Lenz-Barlotti-Typ III-2. *J. reine angew. Math.* **220**, 63–67.
1969   Lectures on Projective Planes. Chicago: Dept. of Mathematics, University of Illinois.
1971   Über die Anzahl der Dickson'schen Fastkörper gegebener Ordnung. Atti del Convegno di Geometria Combinatoria e sue Applicazione. Perugia: Ist. Mat. Univ. Perugia.
1973   Affine Ebenen, in denen der Stabilisator jeder Geraden zweifach transitiv ist. *Arch. Math.* **24**, 663–668.
1974   Über einige merkwürdige Translationsebenen. *Geo. Ded.* **3**, 263–288.
1976a   On finite affine planes of rank 3. In: Foundations of Geometry, P. Scherk (ed.). Toronto: University of Toronto Press.

1976b   Über eine Klasse von endlichen Ebenen des Ranges 3. *Atti dei Convegni Lincei* **17**, 439–446.

Lüneburg, H. and T. G. Ostrom.

1975   Rang-3-Ebenen mit einer Bahn der Länge 2 auf der uneigentlichen Geraden. *Geo. Ded.* **4**, 249–252.

Ostrom, T. G.

1956   Double transitivity in finite projective planes. *Can. J. Math.* **8**, 563–567.
1964   Semi-translation planes. *Trans. Amer. Math. Soc.* **111**, 1–18.
1969   A characterization of generalized André planes. *Math. Z.* **110**, 1–9.
1970a   Linear transformations and collineations of translation planes. *J. Algebra* **14**, 405–416.
1970b   Finite translation planes. Berlin-Heidelberg-New York: Springer-Verlag.
1974   Elations in finite translation planes of characteristic 3. *Abh. Math. Sem. Hamburg* **41**, 179–184.

Ostrom, T. G. and A. Wagner.

1959   On projective and affine planes with transitive collineation groups. *Math. Z.* **71**, 186–199.

Ott, U.

1975   Eine neue Klasse von Translationsebenen. *Math. Z.* **143**, 181–185.

Passman, D. S.

1968   Permutation Groups. New York, Amsterdam: Benjamin.

Pickert, G.

1955   Projektive Ebenen. Berlin-Göttingen-Heidelberg: Springer-Verlag. A second edition appeared in 1975.
1959   Der Satz von Pappos mit Festelementen. *Arch. Math.* **10**, 56–61.

Piper, F. C.

1963   Elations of finite projective planes. *Math. Z.* **82**, 247–258.

Pollatsek, H.

1971   First cohomology groups of some linear groups over fields of characteristic two. *Ill. J. Math.* **15**, 393–417.

Prohaska, O.

1972   Endliche ableitbare affine Ebenen. *Geo. Ded.* **1**, 6–17.
1977   Konfigurationen einander meidender Kreise in miquelschen Möbius-ebenen ungerader Ordnung. *Arch. Math.* **28**, 550–556.

Qvist, B.

1952   Some remarks concerning curves of the second degree in a finite plane. *Ann. Acad. Sci. Fenn.*, **134**, 1–27.

Rao, M. L. N.

1973   A class of flag transitive planes. *Proc. Amer. Math. Soc.* **39**, 51–56.

Rao, M. L. N. and J. L. Zemmer.

1969   A question of Foulser on λ-systems of characteristic two. *Proc. Amer. Math. Soc.* **21**, 373–378.

Rink, R.

1977   Eine Klasse unendlicher verallgemeinerter André-Ebenen. *Geo. Ded.* **6**, 55–79.

Schaeffer, H. J.

1975   Translationsebenen, auf denen die Gruppe $SL(2,p^n)$ operiert. Diplom-arbeit Tübingen.

Schröder, E.

1968   Projektive Ebenen mit pappusschen Geradenpaaren. *Arch. Math.* **19**, 325–329.

Schulz, R.-H.

1971   Über Translationsebenen mit Kollineationsgruppen, die die Punkte der ausgezeichneten Geraden zweifach transitiv permutieren. *Math. Z.* **122**, 246–266.

Schur, I.

1907   Untersuchungen über die Darstellungen der endlichen Gruppen durch gebrochene lineare Substitutionen. *J. reine angew. Math.* **132**, 85–137.

Segre, B.

1959   On complete caps and ovaloids in three-dimensional Galois spaces of characteristic two. *Acta Arithm.* **5**, 315–332.

1961   Lectures on Modern Geometry. Roma: Cremonese.

Suzuki, M.

1962   On a class of doubly transitive groups. *Ann. Math.* **75**, 105–145.

Tits, J.

1962   Ovoides et groupes de Suzuki. *Arch. Math.* **13**, 187–198.

Wagner, A.

1964   A theorem on doubly transitive permutation groups. *Math. Z.* **85**, 451–453.

1965   On finite line transitive affine planes. *Math. Z.* **87**, 1–11.

Walker, M.

1973   On translation planes and their collineation groups. Ph. D. Thesis, University of London.

Wielandt, H.

1964   Finite Permutation Groups. New York, London: Academic Press.

Yaqub, J. C. D. S.

1966   On two theorems of Lüneburg. *Arch. Math.* **17**, 485–488.

Zassenhaus, H.

1936   Über endliche Fastkörper. *Abh. Math. Sem. Hamburg* **11**, 187–220.

Zsigmondy, K.

1892   Zur Theorie der Potenzreste. *Monatshefte Math. Phys.* **3**, 265–284.

Index of Special Symbols

Index